A B C'S
of
INTERPRETIVE LABORATORY DATA

SECOND EDITION — EIGHTH PRINTING

Seymour Bakerman, M.D., Ph.D.

TI0568602

Published by
Interpretive Laboratory Data, Inc.
Post Office Box 7066
Greenville, NC 27835-7066
919-756-6113

ISBN 0-945577-00-1

This book was started after Dr. Paul Bakerman made the observation that there was a need for a clear, concise, pocket-sized paperback textbook on interpretation of laboratory data for the users of the laboratory and laboratory personnel.

The format in the text is uniform throughout: Name of test, specimen, reference range, method and interpretation. Usually, specimen volume is not given because of the wide variation in requirements from one method to another even within the same laboratory. The reference ranges are those that are generally acceptable but not necessarily applicable to all methods. The methods are selective and not all inclusive. <u>The strength of this text is interpretation of the data</u>.

<u>First Edition</u>: I wish to acknowledge the contributions of the physicians, laboratory scientists, technologists, and medical students who made suggestions that improved the text of the first edition, 1983; these include: Dr. Paul R. Bakerman, Dr. Harold Bates, Dr. Bruce Berkowitz, Dr. Thomas Burkart, Dr. Robert Burress, Dr. Jose F. Caro, Dr. Roger Culpepper, Dr. David Day, Dr. Steven H. Grossman, Dr. Robert Hanrahan, Dr. Donald Hoffman, Brenda Humienny, Stanley "Stas" Humienny, Dr. Eurgia C. Land, Dr. Ginette Lapierre, Dr. Mike Messino, Dr. Susan Smith, Mario Turi, Dr. Murray Turner, Virginia Uy, Dr. Catherine Wood, the ECU medical students (Class of 1983), who took my review course, and the medical technologists in the Brody Medical Sciences Building and at Pitt County Memorial Hospital, especially the 3:00 to 11:00 P.M. shift. Ruth Carson dedicated herself to typing the text and was extremely helpful in drawing the Figures.

<u>Second Edition</u>: Since the first edition was published in July, 1983, I have had continued input by Dr. Paul Bakerman, who has made many important suggestions that have served to expand and improve this text.

As with the first edition, many of my colleagues have read this text and have made many valuable contributions; these include: Drs. C.V. Abraham, Paul Bakerman, Harold Bates, Robert Bolande, Thomas Burkart, Bruce Campbell, Jose F. Caro, Thomas Chaplinski, Jascha Danoff, Mark J. Ellison, Abbas Emami, Karen Filkins, Edward Flickinger, Roberta Gray, James Gutai, Jerome Haller, Robert Hanrahan, Donald Hoffman, C. Tate Holbrook, George S. Hughes, Arthur Kopelman, Eurgia C. Land, Gary Levine, Henry Marrow, Lynn Orr, Nicholas Patrone, Mark Phillips, John Rose, Joseph Russo, Robert Shaw, Dennis Sinar, Robert Sloss, Susan Smith, Paul Strausbauch, Roger Thomas, Edward Treadwell, Charles F. Willson.

Some of the best suggestions have come from medical students, particularly John Morrow ('85), Edward R. Setser ('85), Benson E.L. Timmons ('85), David Cook ('84) and the ECU Medical School Class of '84.

I wish to use this opportunity to thank the medical technologists who have contributed so much in the past, especially the 3:00-11:00 shift at Pitt County Memorial Hospital.

Dr. D.W. Skipp, Edmonton, Alberta was kind enough to send me the SI Manual in Health Care.

Phyllis Broughton has spent many arduous hours synthesizing this manual and putting up with me. She continued to do a wonderful job in the face of unrealistic impending deadlines. Jo Gillin has done remarkably well in proofreading this text.

This work could not have been done without the psychological, intellectual, and physical support of my wife, Winona, and my family, John, Paul and Beth.

Please feel free to consult on diagnostic laboratory problems, to critique and to make suggestions that would improve this text; my telephone number is 919-757-2801.

Seymour Bakerman, M.D., Ph.D.

S. Bakerman
TABLE OF CONTENTS

Subject	Page

PANELS

Subject	Page

BLOOD COLLECTION TUBES

Stopper Color	Anticoagulant	Comment
Red	No Anticoagulant	Serum or clotted whole blood; serum must be separated from cells within 45 minutes of venipuncture. <u>Most routine chemistries</u> are done on serum. <u>Most blood bank procedures</u>: ABO; Rh; antibody screen and identification; direct and indirect antiglobulin tests
Red (Barrier Vacuum Tube)	None	Fill tube; invert once to accelerate clotting; allow to clot for 30 minutes; centrifuge. If sample is to be mailed, pour serum into separate serum vial. Barrier tubes are <u>unacceptable</u> for <u>blood bank procedures</u>.
Lavender (Purple)	Ethylenediamine Tetraacetate (EDTA)	Most hematologic procedures such as <u>complete blood count</u>; RBC, WBC, platelet counts and platelet function tests, Hgb, Hct, red cell indices, differential; <u>erythrocyte sedimentation rate</u>(ESR); G-6-PD; <u>Hgb electrophoresis</u>; <u>reticulocyte count</u>; <u>sickle cell preparation</u>; <u>CEA</u>; <u>renin</u> (2 tubes on ice)
Blue	Citrate(must be full)	<u>Prothrombin time</u>(PT) and <u>partial thromboplastin time</u>(PTT); <u>thrombin time</u> (TT); <u>factor assays</u> (coagulation); <u>fibrinogen level</u>; <u>G-6-PD assay</u> (also 1 lavender top)
Green	Lt Heparin	<u>Blood gases</u>(pH, PCO_2, HCO_3, base excess; PO_2, % Sat.) collected in a heparinized syringe, cap and place on ice; transport to laboratory immediately. <u>Electrolytes</u>; <u>osmotic fragility</u>; certain <u>specific hematologic analyses</u>, e.g., <u>chromosomes</u>; <u>histocompatibility</u>; <u>ammonia</u> (on ice); <u>plasma hemoglobin</u>
Grey	Potassium Oxalate as Anticoagulant, Sodium Fluoride as Preservative	Blood glucose; fluoride exerts its action by inhibiting the enzyme system involved in glycolysis; <u>lactate</u> (on ice).
Yellow	ACD Solution	<u>Blood group and type</u>
Royal (Navy) Blue For Trace Metals	None	Serum or clotted whole blood; special tube for <u>trace metals</u>.

The <u>anticoagulants</u>, EDTA, citrate and oxalate, act to <u>prevent coagulation</u> by removing <u>calcium</u> ions from the blood. Heparin acts by <u>inactivating thrombin</u> and <u>thromboplastin</u>. Do <u>not</u> use anticoagulant containing sodium and potassium for electrolyte determinations; use lithium or ammonium salt; do <u>not</u> use anticoagulant containing ammonium for ammonia determination.

CRITICAL TEST LIMITS ("PANIC" VALUES)

Test	Low	High
Blood Hematocrit	<20%	>55%
Blood Hemoglobin	7g/dl	18g/dl
WBC	<2000/mm^3	>20,000/mm^3
Platelet Count	<30,000/mm^3	500,000/mm^3
Prothrombin Activity		>35 sec.
Partial Thromboplastin Time		>100 sec.
Fibrinogen	<100mg/dl	
Blood Gases-	All Results	
Serum Bilirubin, Total-(Newborn)		>5mg/dl
Serum Calcium	<7mg/dl	>12mg/dl
Serum Glucose	<50mg/dl	>300mg/dl
Serum Glucose-(Newborn)	<30mg/dl	>300mg/dl
Serum Magnesium	All Results on Obstetric Patients	
Serum Phosphorus	<2.0mEq/liter	
Serum Potassium	<3.0mEq/liter	>6.0mEq/liter
Serum Potassium-(Newborn)	<2.5mEq/liter	>8.0mEq/liter
Serum Sodium	<120mEq/liter	>160mEq/liter
Serum Bicarbonate	<10mEq/liter	>40mEq/liter
Positive Blood Culture	All	
Positive Cerebrospinal Fluid Gram Stain	All	
Positive Cerebrospinal Fluid Latex Agglutination or CIE	All	

METHOD: Commercial sources of kits for laboratory assays are given in the "Gold Book" (Laboratory Management 22, No. 2, Feb. 1984); these listings are updated every year.

ABO AND Rh TYPE

SPECIMEN: Red top tube, separate cells from serum; do not use serum separation tubes. Include diagnosis, history of recent and past transfusions, pregnancy and drug therapy.

REFERENCE RANGE: See Interpretation

METHOD: Red Blood Cell Typing (Forward): Reaction of patient's red blood cells with known antiserum (anti-A, anti-B, Anti-A,B). Anti-A reacts strongly with A_1 but weakly or not at all with A_2, A_3, A_0 or Am; anti-A,B reacts with Group A subgroups. The Rh test is performed with antiserum containing incomplete or blocking antibodies. (An incomplete blood group antibody does not react in saline). If the patient is Rh(D) positive, no further testing is done; if the patient is Rh negative, test for Rh variant and phenotyping. The Rh antigen is located on the red cell membrane only; the A and B antigens are located on the membrane of all cells.

Serum Typing (Reverse): ABO: Reaction of antibodies in patient's serum with red blood cells containing known surface antigens (known A_1, A_2, B and 0 cells).

The following conditions may interfere with typing: abnormal plasma proteins, cold autoagglutinins, positive direct Coombs test and some bacteria.

INTERPRETATION: ABO and Rh type is done on every ABO recipient before blood is issued for transfusion. The two antigens, A and B, on red blood cell membranes are responsible for the four blood groups. Cell type and serum antibody are:

Cell Antigen	Antibody in Serum
A	B
B	A
0	AB
AB	--

The majority of fatal transfusion reactions are due to ABO incompatibility.

Red Blood Cells: Reaction of patient's red blood cells with known antiserum (anti-A, anti-B, anti-A,B) is as follows:

Patient's Red Cells				ABO Blood			
	plus					American	
Anti-A	Anti-B	Anti A,B	Group of Patient	Whites	Blacks	Indians	Orientals
Negative	Negative	Negative	0	45	49	79	40
Positive	Negative	Positive	A	40	27	16	28
Negative	Positive	Positive	B	11	20	4	27
Positive	Positive	Positive	AB	4	4	<1	5

Some group A patients belong to subgroups of A; the approximate frequency of the major subgroups of A and B is given in the next Table (Sisson, J.A., Handbook of Clinical Pathology, J.B. Lippincott Company, Phila., 1976, pg. 345):

Major Subgroups of A and AB				
Group A Subgroup	Approx. Frequency	Reactions with Anti-A_1 Serum	Reactions with Anti-AB	Approx. % with Anti-A_1 in Serum
A_1	78	Strong	Positive	None
A_2	22	Neg. or Weak	Positive	1-2%
A_3	Rare	Neg. or Weak	Positive	About 50%
A_0	Very Rare	Weak or Neg.	Positive	Over 50%
Am	Ultra Rare	Negative	Negative	Anti-H

Very young infants do not have their own alloagglutinins, that is, A cells, anti-B; B cells, anti-A. Those that are present are primarily passively transferred maternal antibodies. Elderly patients may not have alloagglutinins.

Rh: In contrast to the ABO blood group system, there are no isoagglutins (no anti-Rh) normally present in the serum of patients. 85% of the population is Rh(D) positive and 15 percent react negatively. Routine Rh typing for blood donors and recipients involves only the antigen Rh_0(D).

S. Bakerman

ACETAMINOPHEN (TYLENOL, PARACETAMOL, DATRIL)

SPECIMEN: Red top tube, separate serum; or plasma (avoid heparin)
REFERENCE RANGE: Therapeutic range: 10-25mcg/ml (10-25mg/L); Time to Peak Plasma Level: 0.5-1.0 Hours; Half-Life: 2 to 4 hours; Time to Steady State: 10 to 20 hours; Toxic range: 100-250mcg/ml.
METHOD: Spectrophotometric technique (correct for salicylate). High pressure liquid chromatography. Ref: Sunshine, "Method Analytic Toxicology," C.R.C. Press, 1975, p. 14.
INTERPRETATION: Acetaminophen is the active ingredient of non-aspirin-containing analgesics. Acetaminophen is usually absorbed from the upper gastrointestinal tract; peak concentration occurs 1 hour after a therapeutic dose and 4 hours after an overdose. Unlike aspirin, acetaminophen has no significant effect on platelet aggregation.

This drug may cause hepatic failure. Hepatic toxicity may appear 3-5 days after ingestion of a toxic dose. Most of the acetaminophen is normally metabolized in the liver to sulfate and glucuronide conjugates; a small amount is metabolized by the P-450 system to a potentially toxic intermediate, acetimidoquinone, which is metabolized by reaction with glutathione. Acetimidoquinone arylates vital nucleophilic macromolecules within hepatocytes, resulting in hepatic necrosis. Serious toxicity is likely to occur if the ingested dose is more than 140mg/kg. Tablets contain 80mg to 500mg acetaminophen/tablet; fluid form may contain up to 1000mg acetaminophen/fluid ounce. Blood specimens are collected about 4 hours and 8 hours after ingestion. A nomogram of plasma or serum acetaminophen concentration versus time since acetaminophen ingestion is shown in the next Figure; (Rumack, B.H. and Peterson, R.G., Pediatr. Supp. 62, 898-903, 1978):

Nomogram of Plasma or Serum Acetaminophen Concentration versus Time Since Acetaminophen Ingestion

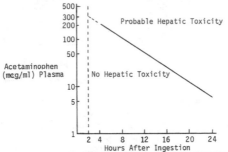

The half-life for acetaminophen elimination is 2 to 4 hours. The first sample is used to determine whether therapy with N-acetylcysteine (Mucomyst) should be started. Treatment should begin within 24 hours of ingestion. If the concentration of acetaminophen in the specimen collected 4 hours after ingestion is greater than 200mcg/ml, therapy should be begun using N-acetylcysteine in an effort to prevent the hepatic complications. N-acetylcysteine is believed to prevent hepatotoxicity because its sulfhydryl groups act as a glutathione substitute, binding to the metabolite. N-acetylcysteine (Mucomyst, Mead Johnson), available as a 20 percent solution, is best administered intravenously but may be given orally, diluted 1:3 with a soft drink or grapefruit juice to mask its taste. The most common side effect of N-acetylcysteine therapy is vomiting; if vomiting occurs within one hour of adminstration of a dose, repeat dose. Liver function tests should be done to follow the patient. Contact the Rocky Mountain Poison Center, Denver, Colorado (1-800-525-6115) for the most recent information on acetylcysteine therapy. (Prescott, L.F. and Critchley, J.A., Ann. Rev. Pharmacol. Toxicol. 23, 87-101, 1983).

Most of phenacetin (75-80%), a common constituent of formulations, is quickly converted to its active metabolite, acetaminophen.

ACETONE, SERUM

SPECIMEN: Red top tube, separate serum; perform test immediately or refrigerate specimen; specimen should be free from visible hemolysis.

REFERENCE RANGE: The clinical laboratory usually expresses acetone content in terms of dilutions; e.g., 1/2, 1/4, 1/8, 1/16, etc., in order of increasing concentration. Negative (normal, 0.3 to 2.0mg/dl). To convert traditional units in mg/dl to international units in micromol/liter, multiply traditional units by 172.2.

METHOD: Semi-quantitative: Nitroprusside (measures acetone and acetoacetate, not beta-hydroxybutyrate). Quantitative: Gas-liquid chromatography (specific for acetone).

INTERPRETATION: The assay of acetone is used as a measure of ketoacidosis. Acetone is formed from acetoacetate as shown in the next Figure:

Formation of Ketone Bodies

$$CH_3CCH_2COOH \xrightarrow{Spontaneously} CH_3CCH_3 + CO_2$$

Acetoacetate (20 percent)

Acetone (2 percent) / Carbon Dioxide

Beta-hydroxybutyrate Dehydrogenase +NADH + H+

$$CH_3CHCH_2COOH + NAD^+$$

Beta-Hydroxybutyrate (78 percent) / Nicotinamide Adenine Dinucleotide

Acetone is formed in the conditions listed in the next Table:
Causes of Acetonemia

Diabetes Mellitus, Uncontrolled
Children:
   Acute Febrile Illnesses
   Toxic States with Vomiting or Diarrhea
Alcoholism with Vomiting and Poor Food Intake
Starvation, Prolonged and Some Weight Reducing Diets
Isopropyl Alcohol (rubbing alcohol)
Secondary to acidosis in Von Gierke's disease
Stress

Ketosis occurs more frequently in the pregnant diabetic and develops at lower blood sugar levels (300 mg/dl-400 mg/dl range).

When ketone bodies accumulate, metabolic acidosis with anion gap develops. Beta-hydroxybutyrate is not assayed using nitroprusside. The extent of measured ketosis is dependent on the ratio of acetoacetate to beta- hydroxybutyrate. This ratio is low when a state of lactic acidosis coexists with ketoacidosis, because the reduced redox potential of lactic acidosis favors the production of beta-hydroxybutyrate. In this event, the level of ketosis as determined by the nitroprusside reaction appears to be inappropriately low for the degree of acidosis, and the additional diagnosis of lactic acidosis should be considered.

Furthermore, there is potential for confusion during the adequate therapy of diabetic ketoacidosis during which repeated measurements of serum ketones are performed. The failure of measured ketonemia to decline in the face of rising pH and falling blood sugar should not necessarily be cause for alarm. Beta-hydroxybutyrate levels fall rather rapidly, but this change is not detected by the nitroprusside reaction. In addition, beta-hydroxybutyrate can be metabolized into acetoacetate giving the false impression that the ketosis is worsening. (Turi, Mario, Personal Communication)

ACETONE, URINE

SPECIMEN: Random urine; specimen should be rejected if more than 3 hours old; false positives for semi-quantitative test are obtained with specimens containing bromosulphalein (BSP), L-dopa metabolites or phenylketones.

REFERENCE RANGE: Negative

METHOD: See ACETONE, SERUM

INTERPRETATION: Acetone is formed from acetoacetate as shown in the previous section, ACETONE, SERUM.

Acetone is formed in the conditions listed in the Table - See ACETONE, SERUM.

3

ACETYCHOLINE RECEPTOR ANTIBODY

SPECIMEN: Red top tube, separate serum
REFERENCE RANGE: <0.03 nanomoles/liter
METHOD: RIA
INTERPRETATION: Myasthenia gravis is manifested, clinically, by weakness of skeletal muscle.

Muscular contraction is mediated by release of acetylcholine from motor nerve terminals and its binding to receptor proteins on muscle; contraction is terminated by the enzyme acetylcholinesterase which catalyzes the hydrolysis of acetylcholine to choline plus acetate. Myasthenia gravis is an autoimmune disease in which antibodies to acetylcholine receptor proteins reduce the acetylcholine sensitivity of the receptor.

In 80 to 90 percent of patients with myasthenia gravis, there are antibodies to the acetylcholine receptor proteins. This assay may be used to monitor the response to therapy. (Lisak, R.P., Hospital Practice, pgs. 101-109, March, 1983.)

Over 75% of myasthenia gravis patients have an abnormal thymus; 15% of these contain thymomas and the remainder are hyperplastic glands. C.T. scan and linear tomography may help to distinguish tumor from hyperplasia (Janssen, R.S. et al., Neurology, 33, 534, 1983).

ACETYLSALICYLIC ACID, BLOOD (see SALICYLATE, BLOOD)

ACID MUCOPOLYSACCHARIDES (see MUCOPOLYSACCHARIDES)

ACID PHOSPHATASE, PROSTATIC, SERUM OR PLASMA

SPECIMEN: Red top tube, separate serum from cells within 30 minutes; or lavender (EDTA) top tube; separate plasma from cells within 30 minutes. Add 0.05ml, 20% acetic acid to 2.5ml serum or freeze within 45 minutes or add disodium citrate tablet.
REFERENCE RANGE: 0.1-0.8IU/L (variation with methodology).
METHOD: Thymophthalein monophosphate is relatively specific for prostatic acid phosphatase; this substrate is not useful to assay for acid phosphatase of Gaucher's disease. Also RIA.
INTERPRETATION: Serum acid phosphatase is used for the diagnosis of metastatic prostatic carcinoma and to monitor therapy with anti-neoplastic drugs in these patients. In about 3/4 of the patients, prostatic carcinoma arises in the posterior lobe of the prostate.

The causes of elevation of prostatic serum or plasma acid phosphatase are given in the next Table:

Causes of Elevation of Prostatic Acid Phosphatase
Carcinoma of the Prostate
Prostatic Conditions Other than Carcinoma
Prostatic Palpation; Hyperplasia of the Prostate;
Prostatic Infarction; Following Cystostomy;
Prostatic Surgery

Note that the substrate, thymophthalein monophosphate is insensitive to non-prostatic sources of acid phosphatase.
Carcinoma of the Prostate: Serum acid phosphatase is elevated in 10%-20% of patients without metastases if carcinoma is within the gland; in 20%-40% of patients with metastases but without bone involvement and in 70%-90% of patients with bone involvement. A high level of serum acid phosphatase practically always means that the tumor is no longer confined to the prostate. The undifferentiated carcinomas may not produce acid phosphatases.

The metastatic lesions to bone tend to be osteoblastic; serum alkaline phosphatase is elevated in approximately 85% of patients with carcinoma of the prostate with skeletal metastases (Schwartz, M.F. and Bodansky, O., N.Y. Acad. Sci., Vol. 166, 775-793, Oct. 14, 1969). Fifty percent of patients had alkaline phosphatase greater than two times the upper limit of normal. Serum glutamic oxalacetic transaminase (GOT) may be elevated in the serum of patients who have liver or other soft tissue metastases.

     Serial determinations of serum acid phosphatase in patients with carci-
noma of the prostate correlate with alterations of the clinical status with
spontaneous remissions, or remissions induced by hormonal therapy, i.e.,
testosterone, estrogens, or with castration effects. A decrease in the serum
level of serum acid phosphatase indicates relative effectiveness of therapy; an
increase indicates renewed activity of the metastases.
Prostatic Conditions Other than Carcinoma:
Prostatic Palpation: Pearson et al obtained blood samples for enzymatic and RIA
analysis for acid phosphatase before five minutes, one hour and 24 hours after
30 second prostatic massage in patients with prostatic carcinoma, prostatic
hyperplasia and controls. Acid phosphatase was elevated in 25 to 30 percent of
subjects by both methods. It was concluded that specimens for acid phosphatase
should be obtained before and at least 24 hours after prostatic examination.
(Pearson, J.C. et al., Urology 21, 37-41, 1983)
Hyperplasia of the Prostate: Serum prostatic acid phosphatase is slightly
elevated in 10-30 percent of patients with hyperplasia of the prostate.
Prostatic Surgery: Serum acid phosphatase usually increases markedly
immediately following surgery; the enzyme returns to normal in 72 hours.
(Pearson, J.C., et al., Urology 21, 37-41, 1983)

                                                                   Acid Phos.,
ACID PHOSPHATASE, TOTAL, PROSTATIC AND NON-PROSTATIC,   Total
SERUM
SPECIMEN: Red top tube, separate serum; or lavender (EDTA) top tube, separate
plasma. Separate from cells within 30 minutes. Add 0.05ml, 20% acetic acid to
2.5ml serum or freeze within 45 minutes or add disodium citrate tablet.
Specimen rejected if hemolyzed, lipemic or icteric.
REFERENCE RANGE: Depends on method; in the newborn, serum acid phosphatase is
almost two times the upper limit of normal for an adult; during the growth
period (up to age 13), serum acid phosphatase is approximately 1.5 times the
upper limit for an adult; the upper limit for adult males is slightly greater
than the adult level for females.
METHOD: P-nitrophenylphosphate; alpha-napthyl phosphate or other substrate
non-specific for acid phosphatase.
INTERPRETATION: Methods that are non-specific for acid phosphatase detect
enzyme originating from prostate, osteoclasts of bone, red and white blood
cells, platelets and the liver. This assay is most appropriately used in
patients with possible Gaucher's disease, Niemann-Pick disease and
reticulo-endotheliosis (hairy-cell leukemia).
     The conditions associated with elevation of serum or plasma acid
phosphatase are given in the next Table:

Conditions Associated with Prostatic and Non-Prostatic Acid Phosphatase
Carcinoma of the Prostate
Prostatic Conditions Other than Carcinoma: Prostatic Palpation;
Hyperplasia of the Prostate; Prostatic Infarction
Following Cystostomy; Prostatic Surgery
Marked Elevation of Serum Bone Alkaline Phosphatase Isoenzyme
Levels such as Occurs in Paget's Disease, Metastatic Carcinoma
and Osteoblastic Lesions of Bone, Hyperparathyroidism, etc.
Liver Disease
Hematogenic Conditions:
Hemolysis, Thrombocytopenia, Myeloproliferative Diseases
Gaucher's Disease and Niemann-Pick Disease

Gaucher's Disease: Gaucher's disease is a rare lipid storage disease caused by a
deficiency of the enzyme, glucocerebrosidase, which catalyzes the hydrolysis of
ceramide-glucose:

Ceramide-Glucose + H_2O $\xrightarrow{\text{Glucocerebrosidase}}$ Ceramide + Glucose

     In Gaucher's Disease, serum acid phosphatase is elevated. However, the
usual method of assay measures prostatic acid phosphatase and not the acid
phosphatase found in patients with Gaucher's disease. Therefore, it is
necessary to specifically indicate on the request form that assay of
non-prostatic acid phosphatase is needed.

ACID PHOSPHATASE, VAGINAL, EVIDENCE OF RECENT SEXUAL INTERCOURSE

SPECIMEN: An adequate amount of vaginal fluid is usually obtained by aspirating fluid from the vagina. A sample is collected by washing the vagina with a small amount, 2ml or less, of isotonic saline. Do not use cotton swabs because false positive results are obtained if the swab is saturated and false negative results are obtained if the swab does not contain sufficient specimen. After the saline wash is obtained, centrifuge the specimen. Draw off the supernatant for acid phosphatase determination; high concentrations of acid phosphatase in vaginal samples may be accepted as proof that semen is present. Use the sediment for examination of spermatozoa; spermatozoa may be detected on swabs for as long as 48 hours. When spermatozoa cannot be identified, serological tests using antisemen sera may give a definite result. Seminal blood group antigens may be detectable for at least 24 hours.
Reference: Editorial, Brit. Med. J. pg 154, July 15, 1978.

The specimens must be well identified and a clear chain of custody must be maintained. Specimens that are not assayed immediately for acid phosphatase may be stored at -20°C.

REFERENCE RANGE: 50U per sample or greater is considered "semen positive." Normal acid phosphatase activity in non-coital women is less than 10U/L (Lantz, R.K. and Eisenberg, R.B., Clin. Chem. 24, 486-488, 1978; Dahlke, M.B. et al., Am. J. Clin. Pathol. 68, 740-746, 1977). Acid phosphatase activity in the vagina remains relatively constant for about 14 hours; 40% of females are positive after 24 hours and 11% positive after 72 hours (Findley, T.P., Am. J. Clin. Pathol. 68, 238-242, 1977).

METHOD: Thymolphthalein monophosphate; this substrate is insensitive to non-prostatic sources of acid phosphatase.

INTERPRETATION: Assay of acid phosphatase is used to obtain evidence of recent sexual intercourse especially in those patients when it is no longer possible to detect spermatozoa. Detection of seminal fluid in the vagina means that recent sexual intercourse has taken place; detection of seminal fluid in other areas indicates sexual contact.

Acid phosphatase may be detected when spermatozoa are absent as following vasectomy.

ACQUIRED IMMUNODEFICIENCY SYNDROME (AIDS)

The laboratory abnormalities associated with AIDS relate to a damaged immune system and are given in the next Table:

Laboratory Abnormalities Associated With AIDS
Leukopenia and Lymphopenia
Immune Dysfunction:
Cutaneous Anergy
Decrease In-Vitro Lymphocyte Responses
Decreased Total T-Cells
Depletion of T-Helper(T4) Cells
Decreased T-Helper(T4)/T-Suppressor(T8)
Decreased Natural Killer Cell Numbers and Function
Increased Gamma Globulins, "Activated" B-Cells
Elevated Serum Beta-2 Microglobulin
Presence of Anti-HBc in 90% of AIDS Patients
Opportunistic Infections
HLA-DR5 Present in Approx. Two-Thirds of Cases of
Kaposi's Sarcoma

Weller, I., Brit. Med. J. 288, 136-137, Jan. 14, 1984.

<u>Suppressor T-Cells > Helper T-Cells</u>: There are two types of T-cells, helper
cells and suppressor cells. Helper cells help other immune cells eliminate
foreign organisms and suppressor cells inhibit this activity. <u>In normal adults,
helper cells occur at twice the level as suppressor cells</u>. In patients with
AIDS, the helper/suppressor T-cell ratio is reversed.
<u>Presence of Anti-HBc</u>: Anti-HBc has been found in approximately <u>90 percent</u> of
the blood of AIDS victims and is present in only <u>5 percent</u> of the general
population; it has been suggested that anti-HBc be used as a "surrogate test" as
a marker of AIDS infectivity in blood (Spira T., Science <u>219</u>, 271, 1983).
<u>Opportunistic Infections</u>: With breakdown in a person's natural immune system,
opportunistic infections occur. The presentation of patients is given in the
next Table; (Center for Disease Control, Acquired Immunodeficiency Syndrome
(AIDS) Update - United States MMWR <u>32</u>, 309-311, 1983):

Presentation of Patients with AIDS	
Presentation	Percentage
Pneumocystitis Carinii(PCP)	51
Kaposi Sarcoma with PCP	8
Kaposi Sarcoma without PCP	26
Other Opportunistic Infections	15

Other opportunistic infections that occur commonly in these patients
include <u>cytomegalovirus(CMV), herpes simplex virus(HSV), Chlamydia, viral
hepatitis, cryptococcosis, candidiasis, amoeba and N. avium-intracellular</u>
(Masur, H. et al., Ann. Int. Med. <u>97</u>, 533, 1982; Green, J.B. et al., ibid.
<u>97</u>, 539, 1982; McCue, J.D., Hosp. Pract. pgs. 179-186, Jan. 1983).
<u>Risk groups</u>: Risk groups are listed in the next Table; (CDC, MMWR <u>32</u>,
309-311, 1983):

Risk Groups	
Risk Groups	Percentage
Homosexual Men	71
Intravenous Drug Abusers	17
Immigrants from Haiti	5
Hemophiliacs	1
Other	6

(Mnemonic for Risk Groups: 4H's; H = Homosexual Men; H = Heroin Users; H =
Haitians; H = Hemophiliacs) Personal communication, Dr. Paul Bakerman.
    AIDS is transmitted to <u>hemophiliacs through factor VIII transfusion</u>;
factor VIII is prepared from the blood of thousands of blood donors.
    AIDS has been reported to occur in infants born of Haitian mothers (MMWR
<u>31</u>, 665-677, 1982; Jongas, J.H. et al., N. Engl. J. Med. <u>308</u>, 842, 1983).
    The cause of AIDS is not known. Hypotheses include a novel virus or
multiple factors (Sonnabend, J. et al., JAMA <u>249</u>, 2370-2374, 1983). Unusual
structures have been detected by electron microscopy of lymphoid cells of
homosexual men (Erving, E.P. et al., N. Engl. J. Med. <u>308</u>, 819-822, 1983;
Zucker-Franklin, D., N. Engl. J. Med. <u>308</u>, 837-838, 1983).
    Recent evidence from the National Cancer Institute suggests that the
causative agent in AIDS is a type of human T-cell lymphotropic virus(HTLV),
dubbed HTLV-III (Popovic, M. et al., Science <u>224</u>, 497-500, May 4, 1984; Gallo,
R.C. et al., ibid; pgs. 500-503; Schüpbach, J. et al., ibid, pgs. 503-505;
Sarngadharan, M.G., et al., ibid, pgs. 506, 508; Editorial Comment by Marx,
J.L., ibid, pgs. 475-477). A lymphadenopathy associated retrovirus(LAV) which
is similar to or identical to HTLV-III has been isolated in France from a
pre-AIDS patient (Vilmer, E. et al., The Lancet <u>1</u>, 753-757, April 7, 1984).
Review: Cohen, J., Brit. J. Hosp. Med. <u>31</u>, 250-259, April 1984.

**ACTIVATED PARTIAL THROMBOPLASTIN TIME (see PARTIAL
    THROMBOPLASTIN TIME)**

Activated
PTT

S. Bakerman

ACUTE ABDOMINAL PANEL

Laboratory tests for work-up of patients with an acute abdomen are listed in the next Table:

Acute Abdomen Panel
Complete Blood Count
Urinalysis
Serum Amylase and Urine Amylase
Serum Bilirubin
Urine Porphobilinogen for Acute Intermittent Porphyria
Serum Glucose
Gram Stain of Vaginal Secretions
Hemoglobin Electrophoresis if Sickle Status in Black Unknown

Causes of acute abdominal pain are listed in the next Table; (Thomas, R.E., Personal Communication):

Causes of Acute Abdominal Pain
Acute Appendicitis
Acute Diverticulitis
Acute Cholecystitis
Acute Pancreatitis
Perforation of Peptic Ulcer
Intestinal Obstruction, e.g., Intussusception
Ectopic Pregnancy
Acute Pelvic Inflammatory Disease(PID)
Strangulated Hernia
Acute Pyelonephritis
Renal Calculi
Right Lower Lobe Pneumonia (Referred Pain)
Liver Abscess
Hemolytic Crises in Hereditary Spherocytosis, Sickle Cell Disease
Diabetic Ketoacidosis
Ulcerative Colitis
Mesenteric Vascular Occlusion
Expanding or Leaking Abdominal Aneurysm
Perforated Intra-Abdominal Viscus
Peritonitis, Chemical or Bacterial
Torsion and Strangulation-Ovarian Cyst; Tumor; Omentum; Undescended Testes
Ruptured Esophagus
Acute Intermittent Porphyria
Henoch-Schönlein Purpura

ACUTE GLOMERULONEPHRITIS PANEL
(Madaio, M.P. and Harrington,J.T., N.Engl.J.Med. 309, 1299-1302,Nov. 24, 1983).
Proteinuria, Quantitative (Usually 500mg to 3g/day)
Hematuria; Red-Cell Casts

↓

Acute Glomerulonephritis
C3 Complement and Hemolytic Complement

Low Serum Complement Level:
  Systemic Diseases:
    Systemic Lupus Erythematosus
     (focal, approx.75%,
      diffuse, approx.90%)
    Subacute Bacterial Endocarditis
     (approx. 90%)
    "Shunt" Nephritis (approx.90%)
    Cryoglobulinemia (approx.85%)
  Renal Diseases:
    Acute Poststreptococcal
     Glomerulonephritis (approx.90%)
    Membranoproliferative
     Glomerulonephritis
      Type I (approx.50-80%)
      Type II (approx.80-90%)

Normal Serum Complement Level:
  Systemic Diseases:
    Polyarteritis Nodosa Group
    Hypersensitivity Vasculitis
    Wegener's Granulomatosis
    Henoch-Schönlein Purpura
    Goodpasture's Syndrome
    Visceral Abscess
  Renal Diseases:
    IgG-IgA Nephropathy (Berger's)
    Idiopathic Rapidly Progressive
     Glomerulonephritis
    Anti-Glomerular Basement
     Membrane Disease (Goodpasture's)
    Immune-Complex Disease(Membranous
     Glomerulonephritis)
    Negative Immunofluorescence Findings
     (Minimal Change Disease)

Comment: Determine serum creatinine and creatinine clearance.
    Nephrotic-range proteinuria plus low serum complement suggests lupus nephritis or membranoproliferative glomerulonephritis.
    Obtain a renal biopsy if the information will substantially alter the specific treatment or overall management.
(1) Acute Glomerulonephritis with Low Serum Complement Level: The proposed mechanisms for low serum complement levels in acute glomerulonephritis are as follows: Immune complex formation: consumption of complement in the kidney and in other organs in some systemic diseases exceeds the production of complement components and results in a depression of the serum complement level. In other patients, antibodies form against complement components e.g. membranoproliferative glomerulonephritis. Individuals with genetic deficiencies have a higher incidence of glomerulonephritis than patients with normal complement levels. A diagnostic approach to the differential diagnosis of acute glomerulonephritis with low serum complement is given in the next Figure:

Differential Diagnosis of Acute Glomerulonephritis with Low Serum Complement
Low Serum Complement
History; Physical Examination
Laboratory Data

Systemic Diseases

| Polyarthritis |
| Skin Rash |
| ANA Positive | → | Lupus |
| Anti-Native DNA | | Erythematosus |
| Positive |

Febrile		Bacterial
3 to 6 Blood Cultures		Endocarditis
Positive	→	-----or-----
Heart Murmur		Shunt
Emboli (Eyegrounds,etc.)		Nephritis

| Palpable Purpura |
| Polyarthralgias | → | Cryoglobulinemia |
| Cryoglob. (Mixed Type) |

Renal Diseases

| Antistreptolysin-O |
| Titer Increased |
| Transient (3 to 8 Week) |
| Decline of C3 |

↓

| Acute |
| Poststreptococcal |
| Glomerulonephritis |

| Continued Depression of |
| C3 |

| Membranoproliferative |
| Glomerulonephritis |

Comments on Renal Diseases; Antistreptolysin-O-Titer: The antistreptolysin-O titer is elevated in up to 70 percent of patients with <u>acute poststreptococcal glomerulonephritis</u> and in approximately 20 percent of patients with <u>membrano-proliferative glomerulonephritis</u>. Early use of penicillin prevents the antistreptolysin-O titer from rising. Many infectious agents, in addition to streptococcus, may cause postinfectious glomerulonephritis. <u>The titer does not rise after cutaneous streptococcal infection.</u>

Complement; Poststreptococcal Glomerulonephritis versus Membranoproliferative Glomerulonephritis:

	Complement			
Condition	C3	Alternate Pathway Activation	C3 Nephritic Factor	Classic Pathway Depressed C1,C4,C2
Poststreptococcal Glomerulonephritis	Transient Decline (3-6 Wks.)	Yes	No	Normal or Slight
Type I Membranoproliferative Glomerulonephritis	Declines	No	Yes (Approx. 33 Percent)	Yes
Type II Membranoproliferative Glomerulonephritis	Declines and Remains Depressed	Yes	Yes (Approx. 75 Percent)	Normal or Slight

(2) Differential Diagnosis of Acute Glomerulonephritis with Normal Serum Complement Level: See discussion in article by Madaio, M.P. and Harrington, J.T., N. Engl. J. Med. 309, 1299-1302, Nov. 24, 1983.
(3) <u>Specific Laboratory Tests</u>: Specific tests for glomerular disease are given in the next Table:

Tests for Glomerular Disease
Antistreptolysin-O Titer
Antinuclear Antibodies(ANA)
Anti-DNA Antibodies
Anti-Glomerular Basement Membrane(Anti-GBM)
Cryoglobulins
Bence-Jones Protein
Urine Protein Electrophoresis
Hepatitis-Associated Antigens
VDRL

ACUTE MYOCARDIAL INFARCTION PANEL
    (a)   Creatine Phosphokinase (CPK) and Isoenzymes (CPK-MB)
    (b)   Lactate Dehydrogenase (LDH) and Isoenzymes (LDH-1 and 2)
BLOOD SPECIMENS: Red Top Tube, separate serum. Obtain a minimum of three samples; at time of admission and at 12 and 24 hours after onset of symptoms of acute myocardial infarction. The changes in CPK, CPK isoenzymes, LDH and LDH isoenzymes are shown in the next Figure:

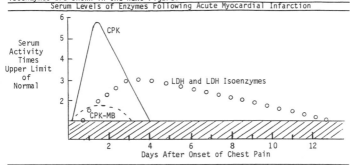

Serum Levels of Enzymes Following Acute Myocardial Infarction

The changes in the serum enzymes, as shown in this Figure, are summarized in the next Table:

Changes in Serum Enzymes Following Acute Myocardial Infarction

Enzyme	Beginning Increase (Hours)	Maximum (Hours)	Return to Normal (Days)
Creatine Phosphokinase (CPK)	2-12	24-36	3-5
Creatine Phosphokinase Isoenzymes (CPK-MB)	2-12	12-49	2-3
Lactate Dehydrogenase (LDH)	8-12	48-72	8-12
Lactate Dehydrogenase Isoenzymes (LDH-1 and 2)	8-12	48-72	10-15

CPK begins to rise in 2 to 12 hours following onset of myocardial infarction; it reaches a maximum in 24 to 36 hours and it remains elevated for three to five days.
CPK-MB begins to rise in 2 to 12 hours, reaches its peak in 12 to 40 hours and returns to normal in 2-3 days.
LDH begins to rise in 8 to 12 hours, reaches a maximum in 2 to 3 days, and returns to normal in 8 to 12 days.
The LDH isoenzymes (LDH-1 and 2) begin to rise 8 to 12 hours following onset of myocardial infarction, reach a maximum in two to three days, and return to normal in ten to fifteen days.
SGOT: Serum GOT begins to rise in 4 to 6 hours following onset of acute myocardial infarction; it reaches a maximum in one to two days, and it remains elevated for 4 to 6 days. SGOT has high sensitivity (about 95 percent) but low specificity for the diagnosis of myocardial infarction. It is elevated in many conditions other than myocardial infarction.

CPK and LDH isoenzyme patterns 24 to 36 hours after an acute myocardial infarction are shown in the next Figure:

CPK and LDH Isoenzyme Patterns Following Acute Myocardial Infarction

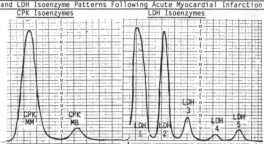

Direct current countershock, which is used to convert cardiac arrythmias to sinus rhythm, produces an increase in serum CPK in about 50 percent of patients; CPK-MM is elevated; occasionally, there is a mild elevation of CPK-MB. If CPK-MB is elevated, then the patient should be treated as if he has had a myocardial infarction.

CPK-MB is highly sensitive and specific as compared to other serum enzyme determinations for the diagnosis of acute myocardial infarction as shown in the next Table; (Wagner et al, Circulation 57, 263-269, 1973):

Parameter Sensitivity and Specificity for Diagnosis of Acute Myocardial Infarction (Wagner et al, 1973)				
Diagnostic Parameter	False-negative (%)	Sensitivity (%)	False-positive (%)	Specificity (%)
ECG	34	66	0	100
Total CPK	2	98	15	85
$LDH_1:LDH_2$	10	90	5	95
CPK-MB	0	100	1	99

Mortality following myocardial infarction has been correlated with the level of serum enzymes as shown in the following Table:

Fifty Percent (50%) Mortality and Serum Enzyme Levels Following Acute Myocardial Infarction		
Enzyme	Serum Enzyme Level Times Upper Limit of Normal	Serum Enzyme Levels in mIU
Glutamic Oxalacetic Transaminase (GOT)	7	(50)* 350
Lactate Dehydrogenase (LDH)	6	(200)* 1200
Creatine Phosphokinase (CPK)	15	(100)* 1500

*The numbers in parentheses are the upper limits of normal of the enzyme in mIU.

However, patients with a very small infarct may develop an arrhythmia and die and patients with large infarct may survive.

Protective Effects of Aspirin: One adult aspirin (325 mg., 5 grains) per day has a protective effect against acute myocardial infarction in men with unstable angina (Lewis, H.D. et al., N. Engl. J. Med. 309,396-403, 1983).

ACUTE PHASE REACTANTS (C-REACTIVE PROTEIN, ALPHA-1-ANTITRYPSIN, HAPTOGLOBIN, FIBRINOGEN, CERULOPLASMIN, C-3 COMPLEMENT)

SPECIMEN: Red top tube, separate serum; except for fibrinogen-use blue (citrate) top tube, separate plasma.
REFERENCE RANGE: See below plus individual tests.
METHOD: See individual tests.
INTERPRETATION: Cellular injury and inflammation evoke synthesis of a heterogeneous group of proteins, the so-called acute phase reactants (Gewurz, H., Hosp. Pract. pgs. 67-81, 1982); some of these are listed in the next Table:

Acute Phase Reactants
Concentration may increase a hundredfold to a thousand fold:
C-Reactive Protein: Normal plasma concentration: <0.5mg/dl; inflammation and host defense.
Concentration may increase twofold to threefold:
Alpha-1-Antitrypsin: Normal plasma concentration: 200-400mg/dl; protease inhibitor.
Haptoglobin: Normal plasma concentration: 40-180mg/dl; hemoglobin transport.
Fibrinogen: Normal plasma concentration: 200-450mg/dl; coagulation.
Concentration may increase by about 50%:
Ceruloplasmin: Normal plasma concentration: 15 to 60mg/dl; copper transport, free radical scavenger.
C-3 Complement: Normal plasma concentration: 80-170mg/dl; modification of inflammation, host defense.

S. Bakerman

ADRENAL CORTEX PANEL

Primary Adrenal Insufficiency: If you suspect that the patient has acute adrenal insufficiency, administer intravenous steroids immediately; then, <u>initiate</u> laboratory work-up.

Specific Tests for Primary Adrenal Insufficiency	
Test	Primary Adrenal Insufficiency
Plasma Cortisol	Decreased
Urinary Free Cortisol	Decreased
Plasma ACTH	Increased
Urinary 17-OH Corticosteroids 17-Keto- and 17-Ketogenic Steroids	Decreased
ACTH Infusion Test	Measure Blood Cortisol: No Change

In one version of the ACTH infusion test (Melby, J.C., N. Engl. J. Med. <u>285</u>, 735-739, 1971), a blood specimen is obtained for cortisol assay; then, inject intravenously 0.25mg of alpha 1-24 corticotropin (synthetic ACTH); two hours later, obtain blood specimen for cortisol assay. The normal individual responds with a 2-to 4-fold increase in plasma cortisol.

Other abnormal laboratory findings in primary adrenal insufficiency are shown in the next Table:

Other Abnormal Laboratory Findings in Primary Adrenal Insufficiency	
Test	Result
Serum Sodium (130mEq/liter or less)	Reduced
Potassium (6mEq/liter or more)	Elevated
BUN (>25mg/dl)	Increased
Glucose (<70mg/dl)	Reduced
Hematocrit (>45%)	Increased
Eosinophiles (normal 100 to 300/cumm)	Increased

Cushing's Syndrome: The tests that are used to answer the question, "Does the patient have Cushing's syndrome?" are listed in the next Table:

Does the Patient Have Cushing's Syndrome? - Tests	
Test	Result in Cushing's Syndrome
Plasma Cortisol	Increased
Urinary Free Cortisol [Best Test]	Increased
Urinary 17-OH Corticosteroids, 17-Keto-Steroids and 17-Ketogenic Steroids	Increased
Diurnal Variation of Plasma Cortisol	Loss of Diurnal Variation
Dexamethasone, 1mg at Midnight	8 A.M. Cortisol greater than 5 mcg/dl compatible with Cushing's
Dexamethasone, 0.5 mg every Six Hours for Two Days	Plasma Cortisol <5 mcg/dl at end of Second Day; 24 Hour Urinary 17-Hydroxycorticosteroid Less than 3 mg on the Second Day

At least three conditions, <u>depression</u>, <u>obesity</u> and <u>alcoholism</u> may produce results similar to that seen in Cushing's syndrome. Other abnormal laboratory findings in Cushing's syndrome are shown in the next Table:

Abnormal Laboratory Findings in Cushing's Syndrome
Serum Sodium Increased and Serum Potassium Decreased
Hyperglycemia-Abnormal Glucose Tolerance Test in about 80% of Patients
Lymphocytes and Eosinophils Decreased
Tendency Toward Metabolic Alkalosis (see Acid-Base Section)

What is the Type of Cushing's Syndrome? The usual problem is differentiation of <u>hyperplasia</u> (Cushing's disease: Bilateral adrenal hyperplasia secondary to a pituitary adenoma) from adenoma. The tests that may be done are given in the next Table:

Differentiation of Adrenal Hyperplasia from Adenoma		
Test	Hyperplasia	Adenoma
Plasma ACTH	Increased	Decreased
8mg/day of Dexamethasone	Suppression of Urinary Steroids	No Suppression of Urinary Steroids
Metopirone Test	Increase in Urinary Steroids	No Change in Urinary Steroids

Over 50 percent of patients with Cushing's disease have ACTH levels within normal range (Besser, G.M. and Edwards, C.R.W., J. Clin. Endocrinol. Metab. <u>1</u>, 451-490, 1972). The diagnosis and treatment of Cushing's disease at Duke University for the years 1977-1982 have been reviewed (Burch, W.M., North Carolina Medical Journal <u>44</u>, 293-296, No. 5, May, 1983).

14

ADRENOCORTICOTROPIC HORMONE (ACTH)

SPECIMEN: Fasting; collect blood in chilled plastic syringe containing heparin (1000 int. units/ml). Transfer to ice-cooled green top tubes (heparin); place tubes in ice; centrifuge in a refrigerated centrifuge; freeze plasma.

REFERENCE RANGE: 0800 hours: 20-140pg/ml; 2400 hours: approximately 50% of A.M. value. ACTH shows little variation with age or sex. To convert traditional units in pg/ml to international units in pmol/liter, multiply traditional units by 0.2202.

INTERPRETATION: ACTH blood levels are useful in differentiating the causes of Cushing's Syndrome; in differentiating the causes of adrenal insufficiency; and in assisting in the diagnosis of the adrenogenital syndrome.

The secretion of ACTH in various conditions is illustrated in the next Figure; (Catt. K.J., The Lancet, pg. 1275, June 13, 1970):

Secretion of ACTH in Adrenal Disorders (Catt, 1970)

F = Cortisol; S = Desoxycortisol

The change of ACTH and blood cortisol in adrenal disorders is summarized in the next Table:

Change in Plasma ACTH in Adrenal Disorders

Disorder	Plasma ACTH	Plasma Cortisol
Cushing's Syndrome:		
Adrenal Hyperplasia	Normal or Increased	Increased
Adrenal Adenoma or Carcinoma	Decreased	Increased
Nodular Hyperplasia	Decreased	Increased
Ectopic ACTH (such as oat cell carcinoma of the lung	Increased	Increased
Adrenal Insufficiency:		
Addison's Disease	Increased	Decreased
Pituitary Insufficiency	Decreased	Decreased
Adrenogenital Syndromes	Increased	Decreased

In Cushing's syndrome due to adrenal hyperplasia (Cushing's disease-pituitary-dependent adrenal hyperplasia) and ectopic ACTH, plasma cortisol is increased; however, over 50 percent of patients with Cushing's disease have ACTH levels within the normal range (Besser, G.M. et al., J. Clin. Endo. Met. 1, 451-490, 1972). When Cushing's syndrome is due to adrenal adenoma or carcinoma or nodular hyperplasia, plasma cortisol is increased but ACTH is low.

There are two tests that reflect secretion of ACTH; these are the dexamethasone suppression test and the use of metyrapone. Dexamethasone is a fluorinated steroid which has about 30 times the potency of cortisol and suppresses ACTH in normal individuals; metyrapone inhibits the 11-hydroxylating enzyme for the synthesis of cortisol from 11-deoxycortisol; these tests are discussed as DEXAMETHASONE SUPPRESSION TEST and METYRAPONE.

In adrenal insufficiency due to Addison's disease, plasma ACTH is increased and plasma cortisol is decreased; ACTH infusion may be used to measure the functional reserve of the adrenal. In pituitary insufficiency, plasma ACTH may be decreased and plasma cortisol is decreased; metopirone is used to measure pituitary reserve in patients with secondary adrenal insufficiency. Isolated ACTH deficiency is rare; there is only one report of isolated ACTH deficiency developing as a complication of post-partum hemorrhage (Stacpoole, P.W. et al., Am. J. Med. 74, 905-908, 1983).

In the adrenogenital syndromes, the plasma cortisol is decreased and the plasma ACTH is increased.

Normally, plasma ACTH undergoes a diurnal variation with levels in late P.M. being approximately 50% of the early A.M. levels. In Cushing's syndrome there is loss of diurnal variation of plasma ACTH.

15

ADRENOCORTICOTROPIC HORMONE(ACTH):
THE CORTICOTROPIN-RELEASING FACTOR(CRF) STIMULATION TEST

SPECIMEN: Inject corticotropin-releasing factor at a dose of 1 mcg per kg of body weight as in intravenous bolus injection. Collect blood at -15, 0, 5, 15, 30, 60, 90, 120, 150 and 180 minutes for measurement of ACTH and cortisol (see ACTH and cortisol assays for collection of blood specimens).

REFERENCE RANGE: See Interpretation

METHOD: RIA

INTERPRETATION: The corticotropin-releasing factor(CRF) stimulation test may be useful in differentiating pituitary from ectopic causes of Cushing's syndrome. Following I.V. CRF, patients with Cushing's disease (pituitary adenoma) develop a further increase in the already elevated levels of ACTH and cortisol. Patients with ectopic ACTH syndrome, who also have high basal plasma concentrations of ACTH and cortisol, have no ACTH or cortisol responses to CRF (Chrousos, G.P. et al., N. Engl. J. Med. 310, 622-626, 1984).

There are many potential uses of CRF: assessment of pituitary ACTH secretory capacity, differentiation between hypothalamic and pituitary causes of ACTH deficiency, evaluation of residual hypothalamic-pituitary functional abnormalities after various treatments for Cushing's syndrome, and possibly early detection of Cushing's disease or prediction of its recurrence (Orth, D.N., N. Engl. J. Med. 310, 649-651, 1984).

ADRENOCORTICOTROPIC HORMONE (ACTH) INFUSION TEST:
[PRIMARY (ADDISON'S) AND SECONDARY ADRENAL SUFFICIENCY]

PROCEDURE AND SPECIMENS (Screening Test): ACTH is given I.V. or I.M; plasma cortisol and plasma aldosterone are measured. Obtain a blood specimen (green top vacutainer) for plasma cortisol and aldosterone assays (time 0); give 0.25mg of alpha 1-24 corticotropin (Cortrosyn) I.V. or I.M.(in patients less than 2 years old, give 0.125mg). An occasional allergic reaction has been reported to synthetic corticotropin. Collect blood at 30 and 60 minutes.

INTERPRETATION: The results for normal subjects and for patients with primary Addison's disease) and secondary adrenal insufficiency are given in the next Figures:

Effect of ACTH on Plasma Cortisol and Aldosterone in Normal Subjects and Patients with Primary (Addison's Disease) and Secondary Adrenal Insufficiency

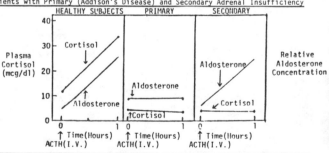

Normal: Following I.V. ACTH, a normal response is an increment of plasma cortisol greater than 10 mcg/dl and an increment of plasma aldosterone levels above control by at least 5 ng/dl.

Primary Adrenal Insufficiency: In primary adrenal insufficiency neither plasma cortisol nor plasma aldosterone increase following I.V. ACTH; the sensitivity is 100% and the specificity is 97% (Manu, P. and Howland, T., Clin. Chem. 29, 1450-1451, 1983).

Secondary Adrenal Insufficiency: In secondary adrenal insufficiency, <u>plasma cortisol shows no increase</u> following I.V. ACTH; however, <u>plasma aldosterone shows a normal increment.</u>

Procedure and Specimens (3 Day Test): 250 microgram of Cortrosyn is infused for 8 hours on each of 3 consecutive days; measure plasma cortisol (reference range: 25 to 50mcg/dl) and urinary 17-OHCS (2- to 5-fold rise: see 17-OHCS Test for urinary collection).

INTERPRETATION: Normally, plasma cortisol rises to 30-50 mcg/dl and urinary 17-OHCS excretion doubles or triples the normal baseline values. Patients with Addison's disease have all plasma values <u>less than 15 mcg/dl</u>; the urinary 17-OHCS value is less than 10 mg per 24 hours. Patients with adrenocorticotropin deficiency show a <u>sluggish</u> or <u>delayed response.</u>

Albumin

ALBUMIN
SPECIMEN: Red top tube, separate serum.
REFERENCE RANGE: Range: Adult: 3.5-5.0g/dl; newborn: 2.9-5.5g/dl; child: 3.8-5.4g/dl. To <u>convert</u> <u>traditional</u> units in g/dl to <u>international</u> units in g/liter, multiply traditional units by 10.0.
METHOD: Bromcresol green(BCG) or bromcresol purple(BCP) dye binding method.
INTERPRETATION: Albumin is decreased in conditions listed in the next Table:

Decrease in Serum Albumin
Subacute and Chronic Debilitating Diseases
Liver Disease
Malabsorption
Malnutrition
Loss:
Renal Disease; Nephrotic Syndrome
Gastrointestinal Loss
Third Degree Burns
Exfoliative Dermatitis
Dilution by I.V. Fluids
Genetic Variants such as Congenital Analbuminemia
and Bis-Albuminemia

Albumin is synthesized in the liver; its half-life is 15 to 19 days. The serum albumin level is considered a reliable index of severity and prognosis in patients with chronic hepatic disease; its main value lies in the follow-up therapy where improvement in the serum-albumin level is the best sign of successful medical treatment.

There may be loss of albumin in the gastrointestinal tract, loss of albumin in the urine by the damaged kidney and direct loss of albumin through the skin. More than 50% of patients with gluten enteropathy have depressed albumin.

The only cause of <u>increased</u> albumin is <u>dehydration</u>; there is no naturally occurring hyperalbuminemia.

(Alcohol see Ethanol)

ALCOHOL (see ETHANOL BLOOD)

ALCOHOLISM PANEL

The laboratory studies for patients with alcoholism are listed in the next Table; (Gambill, J.M., Personal Communication):

Laboratory Studies - Alcoholism

Hematology
    Complete Blood Count(CBC)
Blood Chemistry
    Enzymes:
        Serum Glutamic Oxalacetic Transaminase(SGOT)
        Serum Gamma Glutamyl Transpeptidase(GGT)
    Bilirubin
    Blood Urea Nitrogen(BUN)
    Blood, Glucose, Fasting
    Hospital Setting:
      Lactate
      Electrolytes, Na^+, K^+, Cl^-, CO_2 Content
      Arterial Blood Gases
      Chemistry Panel to include:
        Albumin
        Total Protein
        Inorg. Phosphorus
        Magnesium
        Creatine Phosphokinase(CPK)
Urinalysis
Microbiologic Studies:
    Gonorrhea Culture:
      Cervix, All Females
      Urethral Discharge, Males
    Pap Smear, Cervix, Females
Chest X-Ray

Complete Blood Count: Alcohol apparently has a direct toxic effect on bone marrow; low white cell counts and low platelets count may be observed in alcoholics. The changes in red blood cells that may be seen in alcoholics are given in the next Table:

Red Cell Changes in Alcoholics

Increased Mean Corpuscular Volume(MCV) without
    Folate Deficiency
Folate Deficiency
Sideroblastic Anemia
Iron Deficiency Anemia

Increased MCV without folate deficiency occurs in about 25 percent of alcoholics; these patients usually have a normoblastic bone marrow but occasionally have megaloblastic bone marrow. Alcohol has a direct toxic effect on the red cell.

The mechanism for low folate in chronic alcoholism is unknown but could be due to dietary deficiency, an effect of alcohol on folate absorption, a direct anti-folate effect of alcohol, enhanced utilization of folate as a co-factor in liver enzyme activity or a combination of these (Scott, J., The Lancet., pg. 1297, June 22, 1974-letter).

Sideroblastic anemia may be secondary to pyridoxine deficiency; pyridoxyl phosphate is involved in the synthesis of delta-aminolevulinic acid, a precursor in the synthesis of heme.

Anemia in alcoholics may be due to iron deficiency; iron deficiency may occur from chronic blood loss following repeated hemorrhages from ruptured esophageal varicies, gastritis or peptic ulcer.

Blood Chemistry: The sensitivity of laboratory tests in patients with alcoholism is given in the next Table; (Morse, R.M. and Hurt, R.D., JAMA 242, 2688-2690, 1979):

Sensitivity of Laboratory Tests in Patients with Alcoholism	
Tests	Sensitivity(%)
Gamma Glutamyl Transpeptidase(GGTP)	63
Serum Glutamic Oxalacetic Transaminase(SGOT)	48
Triglycerides	22
Serum Alkaline Phosphatase(Alk. Phos.)	16
Bilirubin	13
Uric Acid	10

GGTP is the most sensitive enzyme used to detect liver damage from excessive alcohol intake. GGTP is situated on the smooth endoplasmic reticulum; alcohol causes microsomal proliferation. (See Gamma Glutamyl Transpeptidase.)

Increased serum triglycerides may represent endogenous release of triglycerides and increased hepatic synthesis. The lipoprotein electrophoretic pattern is that of secondary Type IV hyperlipidemia.

Hyperuricemia is secondary to increased lactate; lactate acid interferes with the renal excretion of uric acid. The hyperuricemia may precipitate attacks of gout.

Hypoglycemia may develop in those alcoholics who are drinking and not eating; liver glycogen stores may be depleted. Liver glycogen is depleted by a 72-hour fast.

Low BUN, (< 5 mg/dl) is found in advanced cirrhosis.

Arterial Blood Gases: Patients who have acute alcoholic intoxication, delirium tremens or chronic liver disease may develop disturbances as listed in the next Table:

Acid-Base Imbalance in Acute Alcoholic Intoxication
Respiratory Alkalosis
Metabolic Acidosis
Ketoacidosis
Lactic Acidosis
Renal Tubular Acidosis

The most common disturbance is respiratory alkalosis.

Laboratory findings in alcoholic ketoacidosis are given in the next Table; (Jatlow, P., Am. J. Clin. Pathol. 74, 721-724, 1980):

Laboratory Findings in Alcoholic Ketoacidosis
Acidosis:
Anion Gap Increased
Plasma Beta-Hydroxybutyrate Increased
Plasma Lactate Increased
Negative or Weak Acetest (Nitroprusside) in Serum and Urine
Plasma Glucose Low Normal or Slightly Increased
Increased Plasma Free Fatty Acids

Lactate: Lactic acidosis may occur in acute alcohol intoxication; lactic acidosis is usually not severe and is usually of short duration. The arterial pH is usually not less than 7.2 and the blood lactate level is usually not greater than five times the upper limit of normal. Spontaneous recovery usually occurs as alcohol is oxidized and the lactate ions are converted by the liver or kidney to glucose or are oxidized. The increased NADH/NAD ratio, secondary to the metabolism of alcohol, favors the conversion of pyruvate to lactate, as follows:

$$\text{Pyruvate} + \text{NADH} + \text{H}^+ \xrightarrow[\substack{\text{Dehydrogenase} \\ \text{(LDH)}}]{\text{Lactate}} \text{Lactate} + \text{NAD}^+$$

Electrolytes: Serum potassium levels in patients during acute alcohol withdrawal are given in the next Table; (Vetter, W.R. et al., Arch. Intern. Med. 120, 536-541, 1967):

Serum Potassium Levels in Acute Alcohol Withdrawal	
Range of Serum Potassium Concentrations	Percent of Patients Within Each Range
1.5 - 2.5	18
2.6 - 3.4	46
3.5 - 3.9	20
4.0 - 4.5	16

Eighteen percent of patients were severely depleted of potassium; 46 percent of patients were moderately depleted and 20 percent were low normal. There was a high incidence (45%) of cardiac arrhythmias in the moderately and severely potassium depleted groups and a low incidence (11%) in the normokalemic group.

Possible causes of potassium depletion include the following: low potassium content of inexpensive fortified wines and distilled liquors, poor dietary intake, and increased gastrointestinal loss of potassium through vomiting and diarrhea.

Other Chemistry Tests: The chemistry tests may include those tests listed in the next Table:

Chemistry Tests
Albumin
Total Protein
Inorg. Phosphorus
Magnesium
Creatine Phosphopkinase(CPK)

Albumin: Albumin may be decreased in the alcoholic secondary to poor nutrition and/or liver disease. Albumin has a <u>relatively slow turnover</u> (half-life, 15 to 19 days) and does <u>not</u> <u>respond rapidly</u> to changes in diet or protein requirements.

Total Protein: Liver disease acts as a stimulus for the production of immunoglobulins; alcoholic liver disease tends to be associated with elevated IgA levels while other types of liver disease are not.

Inorganic Phosphorus: The hypophosphatemia in acute alcoholics is most likely caused by <u>administration of glucose</u> with a <u>shift</u> in <u>phosphate</u> from the extracellular space to the intracellular space.

Serum Magnesium: <u>Hypomagnesemia</u> is a <u>common</u> finding in chronic alcoholics, especially in delirium tremens. However, there is <u>no clear clinical correlation</u> between hypomagnesemia and the complications of alcoholism, e.g., delirium tremens or cirrhosis in the sense that alcoholics with normal serum magnesium levels may or may not develop these complications. The incidence of low serum magnesium in alcoholics is given in the next Table; (Sullivan, J.F. et al., N.Y. Acad. Sci. <u>162</u>, 947-962, 1969):

Incidence of Low Serum Magnesium	
Diagnosis	Percentage with Low Magnesium
Alcoholism	30
Delirium Tremens	86
Post-Alcoholic Cirrhosis	37
Non-Alcoholic Patients	2

Creatine Phosphokinase(CPK): Alcohol has a direct toxic effect on skeletal muscle; increased serum CPK may be observed during hypophosphatemia during administration of glucose.

ALDOLASE, SERUM OR PLASMA
SPECIMEN: Red or green (heparin) top tube; separate serum or plasma
immediately. Do not use hemolyzed specimen. Perform assay within 5 hours;
otherwise, specimen can be stored for 5 days in the refrigerator or frozen for 2
weeks.
REFERENCE RANGE: 0-3 days: 4-24 IU/liter; 4 days-11 years: 2-12 IU/liter; 12
years-adult: 1-6 IU/liter.
METHOD: Aldolase is an enzyme occurring in the glycolytic pathway, glucose to
pyruvate; it catalyses the conversion of fructose-1, 6-diphosphate into
dihydroxyacetone phosphate and glyceraldehyde-3-phosphate. The products are
assayed using auxiliary and indicator reactions.
INTERPRETATION: Aldolase is increased in the serum in conditions listed in
the next Table:

Conditions Associated with Increased Serum Aldolase
Skeletal Muscle Conditions:
Progressive (Duchenne) Muscular Dystrophy
Dermatomyositis
Polymyositis
Limb-Girdle Dystrophy
Myotonia Dystrophy
Rhabdomyolysis
Liver Disease:
Hepatic Necrosis from any Cause, e.g., Viral Hepatitis
Hepatotoxic Drugs
Carcinoma Metastatic to the Liver
Other Conditions:
Acute Myocardial Infarction
Acute Pancreatitis
Prostatic Tumors
Other Neoplasms
Delirium Tremens
Injections of Drugs, e.g., Cortisone, Desoxycorticosterne
and ACTH

Serum aldolase is not increased in neurogenic atrophies.
The assay of CPK has replaced that of aldolase for the evaluation of
patients with skeletal muscle disorders. In acute viral hepatitis, the serum
aldolase level tends to parallel the change in SGPT.

ALDOSTERONE, SERUM

SPECIMEN: Fasting, green (heparin) top tube, separate plasma; or red top tube, separate serum; place specimen on ice and deliver to laboratory immediately; spin specimen and place 3ml of the plasma or serum in plastic vial and freeze (Renin often obtained at same time - see RENIN ACTIVITY, PLASMA).

REFERENCE RANGE: (On ad lib sodium intake) Supine: 3-10ng/dl; obtain specimen from recumbent patient in early A.M. prior to time that patient arises. Upright: (2-5 times supine) 5-30ng/dl; obtain specimen at 9:00 A.M. after patient is upright for two hours. On low sodium diet, serum aldosterone is 2-5 fold increase over ad lib sodium diet. Aldosterone is decreased in patients on heparin therapy. To convert traditional units in ng/dl to international units in pmol/liter, multiply traditional units by 27.74.

METHOD: RIA

INTERPRETATION: The usual reason for determining serum aldosterone is in the work-up of patients with hypertension for possible primary aldosteronism. The relative levels in serum and urine of aldosterone and renin in patients with primary and secondary hyperaldosteronism are shown in the next Table:

Aldosterone and Renin Levels in Primary and Secondary Aldosteronism		
Aldosteronism	Aldosterone	Renin
Primary	↑(Start Here)	↓
(60% Adenoma)		
Secondary	↑	↑(Start Here)

The triad of polyuria, hypokalemia, hypertension tends to occur in primary aldosteronism; patients tend to have hypokalemic alkalosis. The combination of elevated aldosterone secretory rate and low plasma renin activity is practically pathognomonic of primary aldosteronism.

Potential screening tests to detect primary aldosteronism with aldosterone producing adenoma are listed in the next Table (reviewed by Hiramastu, K. et al, Arch. Intern. Med. 141, 1589-1593, 1981):

Potential Screening Tests to Detect Primary Aldosteronism		
Screening Test	Expected Value in Primary Aldosteronism	Remarks
Serum Sodium	Increased	Valueless
Serum Potassium	Decreased	False positives and false negatives
Measurement of Plasma Renin Activity (PRA) after diuretic (furosemide) and upright posture	Low	Twenty-five percent of patients with essential hypertension, have low renin values
Measurement of Plasma Aldosterone concentration after I.V. of 2 L of normal saline over four hours	Increased	False positive rate = 48%
Measurement of Ratio of Plasma Aldosterone to Plasma Renin Activity (A-PRA)	Ratio Above 400	Outpatient screening test; convenient, economical reliable outpatient screening test for aldosterone producing adenoma; sensitivity - 100%

Ratio of Plasma Aldosterone to Plasma Renin Activity (A-PRA Ratio); (Hiramatsu, K., et al., Arch. Intern. Med. 141, 1589-1593, Nov. 1981; Carey, R.M., Arch. Intern. Med. 141, 1594, Nov. 1981): The A-PRA ratio is a convenient, economical and reliable outpatient screening test for primary aldosteronism. Unlike PRA and aldosterone concentration, the A-PRA ratio is not influenced by variations of sodium intake, diuretics, total-body deficit of potassium or diurnal variation. The test is done as follows: Collect blood for the assay of plasma renin activity in EDTA; place immediately in plastic vial. The normal range of plasma renin activity is 1.9 to 15ng/ml/hr. Collect blood for assay of aldosterone in red stoppered vacutainer; separate serum and freeze serum. The normal range of aldosterone is 4 to 20ng/dl (40 to 200pg/ml). Aldosterone-(in pg/ml-PRA (in ng/ml/hr) ratio greater than 400 suggests aldosterone producing adenoma. The ratio of plasma aldosterone to plasma renin activity in patients with hypertension is shown in the next Figure:

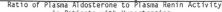

Ratio of Plasma Aldosterone to Plasma Renin Activity
in Patients with Hypertension

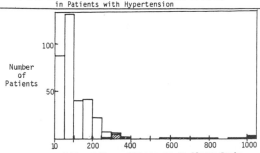

Ratio of Plasma Aldosterone to Plasma Renin

▪ Black area indicates patients with aldosterone-producing adenoma
▨ Shaded area indicates patients patients with moderately high aldosterone
to plasma renin activity but no adenoma

Elevation of the concentration of aldosterone in the serum is paralleled by those of urine.

In Bartter's syndrome, there is hyperplasia of the juxta-glomerular apparatus, increased renin, increased aldosterone, hypokalemia but without hypertension.

Four subgroups of patients with primary aldosteronism have been characterized; the causes and percent incidence are shown in the next Table; (Carey, R.B., Arch. Intern. Med. 141, 1594, 1981):

Subgroups of Patients with Primary Aldosteronism	
Subgroup	Percent
Aldosterone Producing Adenoma (APA)	60
Idiopathic Hyperaldosteronism (IHA)	40
Glucocorticoid Remediable Hyperaldosteronism	Rare
Indeterminant Hyperaldosteronism	Rare

Only patients with APA predictably respond to surgical therapy.

ALDOSTERONE, URINE

Aldosterone,
Urine

SPECIMEN: 24 hour urine; place 10 grams of boric acid as preservative into container prior to collection. Instruct the patient to void at 8:00 A.M. and discard the specimen. Then, collect all urine including the 8:00 A.M. specimen at the end of the 24 hour collection period; refrigerate urine during collection. Measure 24 hour volume and record on test request form. Mix urine and take the pH. Adjust the pH to 4.0 with glacial acetic acid. Obtain a 50ml aliquot and freeze.

REFERENCE RANGE: Concentration in urine is dependent on sodium intake; low sodium (10mEq): 20-80 micrograms/24 hours; normal sodium diet (100-200mEq): 3-19 microgram/24 hours; high sodium diet (more than 200mEq): 2-12 microgram/24 hours; Prepubertal Children: normal sodium intake: 1-8 microgram/24 hours or 4-22mcg/g creatinine. To convert traditional units in micrograms/24 hours to international units in nmol/day, multiply traditional units by 2.774.

METHOD: RIA

INTERPRETATION: The usual reason for determining urine aldosterone is in the work-up of patients with hypertension for possible primary aldosteronism. Elevations of the concentration of aldosterone in the urine are parallel to those of serum. For interpretation see ALDOSTERONE, SERUM.

S. Bakerman

ALKALINE PHOSPHATASE (ALK. PHOS.)

<u>SPECIMEN:</u> Red top tube, separate serum; or green (heparin) top tube, separate plasma; do not freeze. Alkaline phosphatase activity increases on standing.

<u>REFERENCE RANGE:</u> Adult values: 30-125mU/ml;

The alkaline phosphatases vary with <u>age</u> and <u>pregnancy</u>. With <u>age</u>, the alkaline phosphatases are <u>elevated in growing children</u>, decrease to an adult level, and then increase slightly in older people. The change of alkaline phosphatase with age is shown in the next Figure:

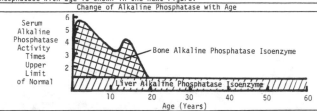

It rises rapidly during the first few weeks of life and reaches <u>values five to six times normal</u> in about one month. Then, it decreases <u>slowly</u> until <u>puberty</u> and then there is <u>another increase</u> at puberty. This is followed by a decrease to adult level at 16-20 years of age. It remains constant until about age 40 to 50 and then it begins to rise to about ten percent above the adult level in older individuals.

The 95th percentile values for serum alkaline phosphatase for males and females are given in the next Table; (Meites, S., Editor, Pediatric Clinical Chemistry, page 72, 2nd. ed., Am. Assoc. Clin. Chem., 1981):

95th Percentile Limits for Alkaline Phosphatase		
Age Group (Years)	Males	Females
Up to 1	477	442
1-2	333	415
2-5	291	341
5-7	291	341
7-10	316	354
10-13	362	393
13-15	-	322
15-17	365	-
15-18	-	122
17-19	176	-
18 and over	-	86
19 and over	126	-

During the <u>third trimester of pregnancy</u> the alkaline phosphatases elevate to two times normal level, rise to three times normal level during labor and return to normal in three to four weeks. This <u>increase</u> in alkaline phosphatase is derived from the placenta.

METHOD:

p-Nitrophenylphosphate + H_2O $\xrightarrow{\text{Alkaline Phosphatase}}$ p-Nitrophenylate + Phosphate (410nm)

INTERPRETATION: The alkaline phosphatases are elevated in diseases involving the liver and diseases associated with stimulation of osteoblasts of bone. Conditions associated with elevated serum alkaline phosphatase are listed in the next Table:

Conditions Affecting Liver and Bone Associated with Elevated Serum Alkaline Phosphatase	
Liver	(Osteoblastic) Bone Conditions
Cholestasis; Intrahepatic and Extrahepatic	Paget's Disease
	Malignancy (Osteoblastic)
Liver Disease	Renal Disease (Secondary Hyperparathyroidism)
Congestive Heart Failure	
Infectious Mononucleosis	Osteomalacia (Secondary Hyperparathyroidism)
Malignancy with Liver Metastasis	
Acute Pancreatitis (occ.)	Malabsorption (Secondary Hyperparathyroidism)
Cytomegalovirus Infections	
	Rickets
	Primary Hyperparathyroidism (Late)
	Healing Fractures
	Hyperthyroidism

- -

Other:
  Sarcoidosis
  Amyloidosis
  Ulcerative Colitis
  Perforation of Bowel
  Pulmonary and Myocardial Infarction (Late)
  Sepsis

In liver disease, alkaline phosphatases are elevated in both obstructive and hepatocellular processes. It is moderately to markedly elevated in obstructive jaundice, biliary cirrhosis, cholangiolitic hepatitis, occlusion of bile duct, and space occupying lesions; it is slightly to moderately elevated in viral hepatitis, infectious mononucleosis and cirrhosis, both portal and post-necrotic.

Serum alkaline phosphatase is elevated in 78 percent of patients with cancer and with metastases to the liver; gamma glutamyltranspeptidase (GGTP) is elevated in 97.3 percent of patients with metastases to the liver (Kim, N.K. et al., Clin. Chem. 23, 2034, 1977).

Marked elevation of serum alkaline phosphatase (4 to 10 times normal) and gamma glutamyl transpeptidase (GGTP) (5 to 13 times normal), little elevation of bilirubin (normal to 4 times normal) and minimal elevation of serum transaminases were found in patients with major systemic infections; the mechanism responsible for the marked elevation of alkaline phosphatase and GGTP is unknown (Fang, M.H. et al., Gastroenterology 78, 592-597, 1980).

The elevation of alkaline phosphatase in bone conditions reflects increased osteoblastic activity of bone. In bone disease, alkaline phosphatases are markedly elevated in osteitis deformans; moderately elevated in rickets, osteomalacia, hyperparathyroidism and metastatic bone disease; slightly elevated in healing fractures.

Patients with chronic renal failure develop osteomalacia and elevated serum alkaline phosphatase; during therapy with calcitriol, serum alkaline phosphatase levels usually decrease, paralleling symptomatic response. Normal values are often obtained after six to 12 months of treatment (Voigts, A.I. et al., Arch. Intern. Med. 143, 1205-1211, June 1983).

Conditions associated with decrease in serum alkaline phosphatase are malnutrition and hypophosphatasia.

ALKALINE PHOSPHATASE ISOENZYMES

SPECIMEN: Red top tube, separate serum.
REFERENCE RANGE: There are four alkaline phosphatase isoenzymes; these are liver isoenzyme, bone isoenzyme, placental or Reagan isoenzyme and intestinal isoenzyme. Alkaline phosphatase isoenzymes in different conditions are shown in the next Table:

Alkaline Phosphatase Isoenzymes		
Children	Adult	Pregnant Female
Bone	Liver Only	Liver
Liver		Placental

Liver Alkaline Phosphatase Isoenzyme: The activity of this isoenzyme remains relatively constant throughout our lifespan; it is the only isoenzyme that is present in the serum of all - young and old.
Bone Alkaline Phosphatase Isoenzyme: This isoenzyme is derived from osteoblasts of bone; it is found during the growth periods when bone osteoblastic activity is greatest. The increase of bone alkaline phosphatase during the growth period is given in the previous section on ALKALINE PHOSPHATASE.
Placental Alkaline Phosphatase Isoenzyme: Alkaline phosphatase is produced by the placenta; it appears in maternal blood. The change in alkaline phosphatase during pregnancy in the post-partum period is shown in the next Figure:

Change in Total Serum Alkaline Phosphatase
in Pregnancy and in the Post-Partum Period

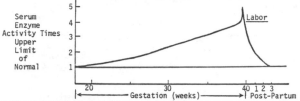

Change of Alkaline Phophatase During Pregnancy: The total serum alkaline phosphatase begins to increase above normal level around 15 to 20 weeks of gestation. It increases progressively until the end of gestation; at the time of labor, it increases suddenly and then begins to decrease immediately following delivery; three to five weeks following delivery, the serum alkaline phosphatase returns to normal. The increase in serum alkaline phosphatase is due to increase in placental alkaline phosphatase.
    Measurements of maternal serum placental alkaline phosphatase are of no value in predicting fetal distress, the apgar score at birth, or dysmaturity.
"Reagan" Alkaline Phosphatase Isoenzyme: The Reagan alkaline phosphatase isoenzyme is an example of ectopic production of enzymes by cancers; it may be similar to ectopic production of hormones. The chemical, physical, and immunological properties of the Reagan alkaline phosphastase isoenzyme are identical to that of the placental alkaline phosphatase isoenzyme.
Intestinal Alkaline Phosphatase Isoenzyme: The activity of this enzyme in the serum depends on three factors; these are blood group, secretory status and influence of fatty meal. (Secretory Status-Soluble ABO antigens are demonstrable in the secretions, e.g., saliva of about 80 percent of the population.)
    The serum enzyme activity is blood group dependent; the relationship of serum intestinal alkaline phosphatase isoenzyme activity and blood group is as follows: O> B> AB> A, that is, subjects with blood group O are more likely to have serum intestinal alkaline phosphastase than those of blood group B et cetera. Serum enzyme is also dependent on secretor status; the serum activity of alkaline phosphatase is higher in patients who are secretors. The activity is increased especially two to three hours following a fatty meal.
    The normal level of serum intestinal alkaline phosphatase is up to 25 percent of the total alkaline phosphatase; that is, if normal serum alkaline phosphatase is 60 mIU/ml, then intestinal alkaline phosphatase may add 15 additional mIU/ml for a total of 75 mIU/ml.

The normal level of serum intestinal alkaline phosphatase is up to 25 percent of the total alkaline phosphatase; that is, if normal serum alkaline phosphatase is 60 mIU/ml, then intestinal alkaline phosphatase may add 15 additional mIU/ml for a total of 75 mIU/ml.

METHOD: Polyacrylamide gel electrophoresis (California Immuno Diagnostics, San Marcos, CA, Tel. #800-821-0395)

INTERPRETATION: Alkaline phosphatase isoenzymes are determined in order to differentiate the source of an elevated alkaline phosphatase, whether due to bone or liver disease. Assay of alkaline phosphatase isoenzymes should never be done for children or during adolescence because both liver and bone isoenzymes are present. Isoenzyme studies should not be done for adults when only the liver isoenzyme is normally present. The conditions for which serum alkaline phosphatase might be done in order to assign the cause of an elevated serum alkaline phosphatase are shown in the next Table:

Indications for Alkaline Phosphatase Isoenzymes
Paget's Disease
Malignancy (Osteoblastic)
Secondary Hyperparathyroidism
Renal Disease
Osteomalacia
Malabsorption
Rickets
Primary Hyperparathyroidism
Healing Fractures
Hyperthyroidism

ALPHA-1-ANTITRYPSIN

SPECIMEN: Red top tube; separate serum.

REFERENCE RANGE: 85-215mg/dl; spuriously elevated in heterozygous deficient patients during concurrent infection, pregnancy, estrogen therapy, steroid therapy and with exercise. Elevated levels may be seen in rheumatoid arthritis, bacterial infections, vasculitis and carcinomatosis. To convert traditional units in mg/dl to international units in g/liter, multiply traditional units by 0.01.

METHOD: Radial immunodiffusion

INTERPRETATION: Alpha-1-antitrypsin deficiency is associated with chronic obstructive lung disease (emphysema) and less frequently with hepatic cirrhosis in infants and respiratory distress of the newborn. Alpha-1- antitrypsin is a protease inhibitor; it inhibits the action of the naturally occurring proteolytic enzymes, trypsin, elastase, chymotrypsin, collagenase, leukocytic proteases, plasmin and thrombin, which may be released during inflammatory reactions in the lung. In the absence of alpha-1-antitrypsin, inhibition of these enzymes does not take place, and these enzymes may digest pulmonary parenchyma.

A protocol for laboratory investigation of alpha-1-antitrypsin deficiency is given in the next Table:

Protocol for Laboratory Investigation of Alpha-1-Antitrypsin Deficiency
Serum Protein Electrophoresis; Carefully Examine Alpha-1-Band
Assay Alpha-1-Antitrypsin when Alpha-1-Band is Decreased or Absent (Usual Technique is Radial Immunodiffusion)
Perform Phenotyping if Indicated

Alpha-1-antitrypsin deficiency is inherited as an autosomal trait. The frequency of the heterozygote is about one in twenty and the frequency of the homozygote is about one in 1600.

Acquired decrease in alpha-1-antitrypsin occurs in the following conditions: nephrosis, malnutrition and cachexia.

Increase in alpha-1-antitrypsin occurs as an acute phase response to tissue necrosis and inflammation. (See ACUTE PHASE REACTANTS).

ALPHA-1-ANTITRYPSIN PHENOTYPING

SPECIMEN: Red top tube, separate serum

REFERENCE RANGE: Absence of homozygous ZZ; alpha-1-antitrypsin is usually associated with the Z state but other genotypes such as SS and SZ are also affected.

METHOD: Isoelectric focusing in polyacrylamide gels.

INTERPRETATION: The Z allele is associated with neonatal cirrhosis or pulmonary emphysema. The frequency of protease inhibitor (Pc) (alpha-1-antitrypsin) types in the population of England and Wales is given in the next Table; (Cook, P.J.L., Postgrad. Med. J. 50, 362-364, 1974):

Frequency of Protease Inhibitor (Pi) Types in England and Wales	
Type	Frequency (%)
MM	86
MS	9
MZ	3 (Partial Enzyme Deficiency)
SS	0.25
SZ	0.2
ZZ	0.029 (2.9 out of 10,000)(Severe Enzyme Deficiency)

86% of the population is homozygote for MM and is normal. Genetic variants with at least one protease inhibitor Z (PiZ) allele have been associated with liver disease (Hodges, J.R., et al., N. Engl. J. Med. 304, 557-560, 1981). The ZZ allele is associated with severe alpha-1-antitrypsin deficiency; the MZ allele is associated with partial enzyme deficiency, the S allele may also be associated with deficiency.

ALPHA-FETOPROTEIN(AFP)

SPECIMEN: Red top tube, separate serum.
REFERENCE RANGE: <25ng/ml. To convert traditional units in ng/ml to
international units in mcg/liter, multiply traditional units by 1.00.
METHOD: RIA
INTERPRETATION: Serum alpha-fetoprotein level is useful in following the
response to therapy of patients with hepatocellular carcinoma, non-seminiferous
germ cell testicular tumors, certain ovarian tumors and in differential
diagnosis of neonatal hepatitis versus biliary atresia in newborns and in
maternal blood and amniotic fluid in certain fetal abnormalities (Davenport, D.
and Macri, J., Am. J. Obstet. and Gynecol. 146, 657-661, July 15, 1983).
Hepatocellular Carcinoma: The blood level of AFP in 70% of patients with
hepatocellular carcinoma is higher than 500ng/ml (upper limit <25ng/ml).
Testicular Tumors: Alpha-fetoprotein and human chorionic gonadotropin (HCG) in
testicular tumors are given in the next Table:

Histopathologic Classification of Testicular Tumors, Alpha-Fetoprotein and Human Chorionic Gonadotropin (HCG)				
Testicular Tumor	Alpha-FP	HCG	Frequency Of Markers	Other Hormones
Germinal Cell Origin:				
Embryonal Carcinoma	+	-		
Embryonal Carcinoma with STGC	+	+		
Adult Type	+	+	88%	
Infantile Type (Yolk-Sac Tumor or Endodermal Sinus Tumor)	+	-		
Choriocarcinoma	-	+	100%	
Teratoma	-	-		
Compound Tumors				
Embryonal Carcinoma with Teratoma	+	-		
Any Combination of the Above	+	+	(if STGC present)	
Seminoma	-	-		
Seminoma with STGC	-	+	7.7%	FSH
Nongerminal Cell Origin:	-	-		
Interstitial (Leydig) Cell Tumor; Sertoli Cell Tumor; Gonadal-Stroma Tumor; Compound Tumors				

STGC=Syncytiotrophoblastic Giant Cell
     The tumors above the first dashed line are associated with an elevated serum
human chorionic gonadotropin (HCG) or alpha-fetoprotein (alpha-FP) or both;
these are the non-seminomatous germ cell tumors. Only 7.7% of the patients with
seminomas showed elevated tumor marker in the serum and these were associated
with STGC. Tumors of nongerminal cell origin were not associated with tumor
markers. Anderson et al (Ann. Int. Med. 90, 373-385, 1979) reported that 85% to
90% of patients with nonseminomatous germ cell tumors had circulating tumor
markers, e.g., alpha-FP and/or HCG.
Ovarian Tumors: Alpha-fetoprotein is found in the serum of patients with germ
cell tumors of the ovary; these are listed in the next Table:

Ovarian Tumors Secreting Alpha-Fetoprotein
Yolk Sac Tumor (Endodermal Sinus Tumor); Embryonal Carcinoma; Mixed Germ Cell Tumor

Neonatal Hepatitis versus Biliary Atresia in Newborns: Serum alpha-fetoprotein
is elevated in neonatal hepatitis but is usually normal in biliary atresia.
Alpha-Fetoprotein and Fetal Abnormalities: Alpha-fetoprotein is used to detect
fetal central nervous system disorders. The fetal abnormalities associated with
an increase in alpha-fetoprotein (AFP) in amniotic fluid are shown in the next
Table:

Fetal Abnormalities Associated with An Increase in Alpha-Fetoprotein (AFP) in Amniotic Fluid
Spina Bifida (a Vertebral Gap)
Anencephaly
Conditions Causing Increased AFP Early in Pregnancy: Exomphalos; Congenital Nephrosis; Sarcococcygeal Teratoma; Duodenal Atresia; Intrauterine Death
Conditions Causing Increased AFP in the Third Trimester: Esophageal Atresia; Fallot's Tetralogy; Hydrocephaly; Severe Rh Isoimmunization

     The incidence of neural tube defects is about five per thousand births in
England and one per thousand births in the U.S.; between 3,000 and 6,000

affected infants are born each year. Approximately one-half survive the first
24 hours; of these, 75 percent will have severe physical handicaps and 20
percent moderate or severe mental retardation (Milunsky, A. and Alpert, E., N.
Engl. J. Med. 298, 738-739, 1978; Brock, D.J.H., The Lancet, pgs. 1281-1282,
June 16, 1979).

Maternal Serum Alpha-Fetoprotein Screening: Fetal anencephaly and open spina
bifida are associated with elevated maternal serum alpha-fetoprotein. A
screening program has been undertaken in the United Kingdom. Refer to the
article by Milunsky and Alpert, 1978, already cited, for current thoughts on
mass screening of obstetrical patients in the United States.

    Screening for serum alpha-fetoprotein is done as follows: Maternal serum is
obtained between 16 and 18 weeks of gestation; the distributions of maternal
serum alpha-fetoprotein levels during the second trimester are given in the next
Table; (Haddow, J.E., Advanced Clin. Chem. No. ACC-32 1979):

Distributions of Maternal Serum Alpha-Fetoprotein Levels During the Second Trimester			
Week Post-Last Menstrual Period	Median (ng/ml)	2 x Median (ng/ml)	2.5 x Median (ng/ml)
15	30	60	75
16	35	70	88
17	37	74	93
18	41	82	103
19	45	90	113
20	54	108	135

    About five percent of patients lie above 2 x median; about three percent of
patients lie at 2.5 x median. If the alpha-fetoprotein is elevated, a second
serum specimen is obtained in 7 to 10 days following the first; if the result of
the second test is elevated, ultrasound examination is done. Ultrasound is used
to determine the gestation more exactly, and, as a function of that value, the
serum alpha-fetoprotein is reevaluated. If the serum alpha-fetoprotein is
elevated, amniocentesis is done. The concentration of alpha-fetoprotein in
amniotic fluid is 200 times greater than the serum level.

    When amniotic fluid alpha-fetoprotein is elevated, assay of
acetylcholinesterase is helpful in confirming the diagnosis of an open neural
tube defect.

    The American College of Obstetricians and Gynecologists and the American
Academy of Pediatrics recommend that AFP blood testing be reserved for women
with family histories of neural tube defects (Editorial, Medical World News 24,
59, July 15, 1983). However, such limitation will miss over 90% of neural tube
defects.

ALPHA-HYDROXYBUTYRATE DEHYDROGENASE (ALPHA-HBD)

SPECIMEN: Red top tube, separate serum; reject if specimen hemolyzed.
REFERENCE RANGE: Up to 170 mU/ml; method dependent.
METHOD:

$$\text{Alpha-Oxobutyrate} + NADH + H^+ \xrightleftharpoons{HBD} \text{Alpha-Hydroxybutyrate} + NAD^+$$

INTERPRETATION: Alpha-hydroxybutyrate dehydrogenase(alpha-HBD) activity
represents the activity of isoenyzmes of LDH, primarily LDH-1. Conditions
associated with an increase in alpha-HBD are given in the next Table:

Conditions Associated with Increase in Alpha-HBD
Acute Myocardial Infarction
Hemolytic Anemia
Megaloblastic Anemia
Progressive Muscular Dystrophy (Duchenne Type)
Cancer

    The elevation of alpha-HBD in disease reflects its tissue concentration,
e.g., heart, erythrocytes and kidney; liver and skeletal muscle show little
alpha-HBD.

    Following acute myocardial infarction, alpha-HBD begins to increase in
10-12 hours, reaches a peak in 2-3 days and returns to normal in about two
weeks.

AMIKACIN

SPECIMEN: Red top tube, separate serum and freeze. Obtain serum specimens as follows:
(1) 24-48 hours after starting therapy if loading dose is not given.
(2) 5 to 30 minutes before I.V. amikacin (trough).
(3) 30 minutes following completion of a 30 minute I.V. infusion of amikacin (peak).

Do not use heparinized collection tubes.

REFERENCE RANGE: Given I.V. or I.M., Recommended Dose: 10-15mg/kg/day, I.M. or I.V.; Therapeutic Range: 15-25mcg/ml; Toxic Level, Peak: >35mcg/ml; Toxic Level, Trough: >5mcg/ml; Half-Life: 2-3 hours; Time to Steady State: Adults, 5-35 hours; Children, 3.5-15 hours; varies with dosage; may be significantly prolonged in patients with renal dysfunction.

Draw Samples as Follows: Peak: 60 minutes post I.M. injection (normal renal function). Peak: 30 minutes after end of 30 minute I.V. infusion; directly after 60 minute I.V. infusion (normal renal function). Trough: Immediately prior to next dose. Sampling is often done prior to attainment of steady state to help prevent toxicity and to ensure therapeutic efficacy.

High concentrations of beta-lactam antibiotics (penicillins, cephalosporins) inactivate aminoglycosides (gentamicin, streptomycin, amikacin, tobramycin and kanamycin); to reduce this interaction, specimens containing both classes of antibiotics should either be assayed immediately using a rapid method or stored frozen.

METHOD: RIA; EMIT; SLFIA (Ames); Fluorescence Polarization (Abbott).

INTERPRETATION: Amikacin is an aminoglycoside antibiotic (gentamicin, streptomycin, amikacin, tobramycin and kanamycin-parenteral therapy) which is used frequently in hospitals to treat patients who have serious gram-negtive bacterial infections, especially septicemia and staphylococcal infections, untreatable with penicillins. Amikacin is the aminoglycoside of choice when gentamicin resistance is prevalent (Edson, R.S. and Keys, T.F., Mayo Clin. Proc. 58, 99-102, 1983).

The first serum level of amikacin is obtained when amikacin has reached steady-state serum concentrations; steady-state is reached in 5 to 13 hours (adults) and 3.5 to 15 hours (children) after starting therapy if the patient has not received a loading dose (steady state = 5-7 drug half-lives; half-live of amikacin is 2 to 3 hours). The following times should be recorded on the laboratory requisition form and on the patient's chart:

Trough Specimen Drawn_____(Time) (5 to 30 minutes before Amikacin)
Amikacin Started _____(Time)
Amikacin Completed _____(Time) (30 minutes I.V. infusion)
Peak Specimen Drawn _____(Time) (30 minutes after I.V. Amikacin)

The three main toxic side effects of amikacin are ototoxicity, nephrotoxicity and neuromuscular blockage; it is important to control the dose given by monitoring peak and trough levels of the drug, particularly in patients with any degree of renal failure. As renal function declines, drug half-life increases.

To minimize risk of toxicity, it has been recommended that peak levels not exceed 35mcg/ml and that trough levels should fall below 5mcg/ml.

Amikacin is eliminated exclusively by renal excretion; excessive serum concentrations may occur and lead to further renal impairment. The renal damage is to the renal proximal tubules and is usually reversible if discovered early.

Ototoxicity is usually due to vestibular damage and is often not reversible.

Aminoglycosides

AMINOGLYCOSIDES (SEE AMIKACIN, TOBRAMYCIN)

S. Bakerman

AMINO ACID ANALYSIS

SPECIMEN: Urine: 24 hour collection preferred; the specimen should be kept refrigerated during collection; add 10ml of 6N HCl/liter to pH 2-3. A 50ml aliquot should be forwarded for analysis; note total 24 hour volume. Otherwise, random urine. Serum or Plasma: Red top tube, separate serum and freeze immediately; or green (heparin) top tube, separate plasma and freeze immediately
REFERENCE RANGE: MetPath: Serum or Plasma: Ref. Levels(max.) in micromoles/dl

Amino Acid	Children	Adults	Amino Acid	Children	Adults
Taurine	11.5	14.0	Cystine	8.0	14.0
Aspartic Acid	2.5	5.0	Methionine	2.0	4.0
Hydroxyproline	--	--	Isoleucine	3.0	10.0
Threonine	9.5	25.0	Leucine	18.0	16.0
Serine	11.0	19.0	Tyrosine	7.0	9.0
Asparagine	2.0	5.0	Phenylalanine	6.0	12.0
Glutamic Acid	25.0	12.0	Beta-aminoisobutyric Acid	--	--
Glutamine	40.0	80.0	Tryptophan	--	--
Proline	45.0	44.0	Ornithine	9.0	13.0
Glycine	22.0	50.0	Lysine	15.0	26.0
Alanine	30.0	50.0	1-methyl-histidine	--	--
Citruline	3.0	5.0	Histidine	9.0	12.0
Alpha-aminobutyric Acid	4.0	4.0	3-methyl-histidine	--	--
Valine	28.0	33.0	Arginine	9.0	15.0

METHOD: Quantitative: Gas-liquid chromatography; ion-exchange chromatography; Qualitative: Thin-layer chromatography
INTERPRETATION: Conditions associated with increased amino acids in plasma and urine are given in the next Table; (BioScience):

Amino Acids	PKU	Maple Syrup Urine Disease (MSUD)	Cystinuria	Homocystinuria	Hartnup	Arginino-Succinicaciduria	Histidinemia	Hyperprolinemia Type A	Citrullinuria
Leucine, Isoleucine		P;U			U				
Phenylalanine	P;U				U				
Valine, Methionine		P;U		P					P
Tryptophan, Beta-Amino Isobutyrate					U				
Tyrosine					U				
Proline	P							P;U	
Alanine, Ethanolamine					U		U		U
Threonine, Glutamate					U				U
Homocitrulline, Glycine, Serine, Hydroxyproline, Aspartic Acid, Glutamine, Citrulline	P				U	U		U	P;U
Homocystine, Asparagine				U					
Argininosuccinic Acid, Histidine, Arginine, Lysine, Ornithine, Cystathionine, Cystine, Cysteine, Hydroxylysine			U		U	U	P;U		U

P = Plasma
U = Urine

Generalized aminoaciduria is also found in Wilson's disease, Lowe's syndrome, galactosemia, cirrhosis of the liver and renal tubular abnormalities.

AMINOPHYLLINE (see THEOPHYLLINE)

AMITRIPTYLINE (see TRICYCLIC ANTIDEPRESSANTS)

AMMONIA, BLOOD

SPECIMEN: Green (heparin) top tube or heparinized syringe; red top tube for serum may be used; the tube must be filled completely; kept tightly stoppered at all times. Specimen must be placed on ice immediately. Blood ammonia increases rapidly at room temperature. Test should be performed within 60 minutes of the venipuncture. Ammonia is stable in frozen plasma for several days at -20°C. Hemolyzed specimen should be rejected.

REFERENCE RANGE: Ammonia varies with age. (Units in micromol/liter): Newborn, 65-105; 0-2 Weeks, 55-90; >1 Month, 20-50; Adult, 10-32. To convert international units in micromol/liter to traditional units in microgram/dl, multiply international units by 1.7.

METHOD:

$$NH_4^+ + \text{Alpha-ketoglutarate} + NADPH \xrightarrow[\text{Dehydrogenase}]{\text{Glutamate (340nm)}} \text{L-glutamate} + NADP^+ + H_2O$$

INTERPRETATION: The most common cause of elevated blood ammonia is severe liver disease; blood ammonia is also elevated in Reye's syndrome.

Normally, ammonia is produced in the intestine by bacterial action on protein. Ammonia is then transported through the portal venous blood to the liver. The hepatocytes metabolize ammonia to urea via the Krebs-Henseleit cycle: e.g.

$$\text{Ammonia} \xrightarrow{\text{Krebs-Henseleit cycle}} \text{Urea}$$

The formation of urea from protein is illustrated in the next Figure:

Formation of Urea from Protein

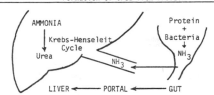

Other sources of ammonia are the kidney and muscle.

In liver disease, blood ammonia may be increased because of increase in collateral circulation which by-passes the liver and by failure of the diseased hepatocytes to metabolize ammonia to urea.

In Reye's syndrome, it has been suggested that peak plasma ammonia levels are predictive of disease severity and patient survival. When peak ammonia levels are less than 5 times normal, the survival rate was 100%, peak ammonia levels in excess of five times normal are associated with significant mortality. Peak ammonia levels occurred within four hours of admission in 88% of patients. (Fitzgerald, J.F. et al., "The Prognostic Significance of Peak Ammonia Levels in Reye's Syndrome," Pediatrics 70, 997-999, Dec. 1982). Other laboratory findings in Reye's Syndrome are two-fold elevation of SGOT; prolongation of prothrombin time (PT); serum bilirubin <3mg/dl; normal CSF; hypoglycemia is more commonly found in infants less than two years of age.

AMNIOCENTESIS PANEL

Tests that are done on amniotic fluid are given in the next Table:

Tests on Amniotic Fluid
Bilirubin Amniotic Fluid: Evaluation of Hemolytic Disease of the Newborn
Lecithin/Sphingomyelin (L/S) Ratio and Phosphatidylglycerol (PG): Fetal Lung Maturity
Indications for <u>Genetic</u> Amniocentesis (Sullivan, M.M., Rawnsley, B.E. and Hatch, D., Check Sample, Cytopathology <u>10</u>, Number 8, 1982)
Maternal Age > 35 Years
Previous Child with Chromosome Anomaly
Previous Child with Neural Tube Defect
Elevated Serum Alpha-Fetoprotein Screen
Previous Child or Known Carrier - Inborn Error of Metabolism
Family History of Sex-Linked Recessive Disease

<u>Maternal Age > 35 Years</u>: Children born to mothers over 35 years of age are at high risk for <u>Down's syndrome</u>. Down's syndrome occurs in 1 in 600 births and four-fifths of all affected children are born to mothers over 35 years of age.

More than 90 percent of patients with Down's syndrome have trisomy 21; the appearance of chromosomes in trisomy 21 is shown in the next Figure:

Trisomy 21, Down's Syndrome

<u>Down's Syndrome:</u>
(Extra Chromosome, Group G, Trisomy 21)

The characteristics of Down's syndrome are as follows: Patients are mentally retarded, have short stature, are <u>susceptable to bacterial infections</u> and increased incidence of acute leukemia, an <u>increased incidence of congenital lesions of the heart</u>, most commonly ventricular septal defects; hypotonic as newborns, hypermobility of joints. <u>Head</u>: flattening of occiput, fine sparce hair, epicanthic folds (a prolongation of a fold of skin of the upper eyelid over the angles of the eye), low nasal bridge, open mouth, large tongue, low-set ears, short neck. <u>Hands and Feet</u>: Hands have <u>short metacarpals and phalanges</u>; <u>a typical pattern of palmer creases with one crease in fifth finger</u>; feet have a wide gap between first and second toes.

<u>Previous Child with Chromosome Anomaly</u>: There is an increased incidence of chromosome anomalies in children whose mothers gave birth to a previous child with a chromosome abnormality.

<u>Previous Child with Neural Tube Defect</u>: If a mother has given birth to a child with a neural tube defect, the <u>rates markedly increase</u> in subsequent siblings.

<u>Elevated Serum Alpha-Fetoprotein Screen</u>: See ALPHA-FETOPROTEIN, SERUM.

Autosomal Recessive Inheritance

Cystic Fibrosis (1:2,000 in Whites)
Sickle Cell Anemia (1:500 in Blacks)
   Substitution of Valine for Glutamic Acid in the Beta Chain
Phenylketonuria (PKU) (1:10,000 in Whites) (Metabolic Pathway--Part D)
   Deficiency in the Enzyme Phenylalanine Hydroxylase
Tay-Sachs Disease; a lipid storage disease (1:3,600 in Jews)
   Deficiency in the Enzyme, Hexosaminidase A
Cretinism, Familial
Beta Thalassemia Major (Deficiency of Beta Chains)
Albinism (Deficiency of Tyrosinase) (1:10,000)
Alpha-1-Antitrypsin Deficiency
Familial Hypercholesterolemia (Carrier 1:500)
Hepatolenticular Degeneration (Wilson's Disease)

Rare Conditions:
   Alkaptonuria (Deficiency of Homogentisic Oxidase)
   Tyrosinosis (Deficiency of Parahydroxyphenyl Pyruvic Oxidase)
   Lipid Storage Diseases:
      Gaucher's Disease (Deficiency of Glucocerebrosidase)
      Niemann-Pick Disease (Deficiency of Sphingomyelinase)
   Adrenogenital Syndrome (Deficiency of 21, 11 and 17 Hydroxylase)
   Ataxia-Telangiectasia (1:40,000)
   Homocystinuria
   Lawrence-Moon-Biedl Syndrome
   Galactosemia
   Reily-Day Syndrome (Familial Dysautonomia)
   Werdnig-Hoffman Disease (Progressive Spinal Muscular Atrophy)
   Agammaglobulinemia, Swiss Type

(2) Autosomal Dominant Inheritance: Conditions associated with autosomal
dominant inheritance are given in the next Table:

Autosomal Dominant Inheritance

Gilbert's Disease (Hyperbilirubinemia)
Polycystic Kidney Disease
Hereditary Spherocytosis
Ehlers-Danlos Syndrome
Huntington's Disease (Killed Woody Guthrie)
Marfan's Syndrome (Arachnodactyly)
Medullary Carcinoma of Thyroid
Neurofibromatosis
Retinoblastoma
Multiple Intestinal Polyposis
Renal Cell Carcinoma
Acute Intermittent Porphyria
Myotonic Dystrophy
Hereditary Hemorrhagic Telangiectasia (Osler's Telangiectasia)
von Willebrand's Disease (Defective Platelet Adhesiveness and low
   Factor VIII)

Rare Conditions:
   Osteogenesis Imperfecta ("Brittle Bones")
   Achondroplasia
   Milroy's Disease (Congenital Lymphedema)
   Alport's Syndrome (Hereditary Nephropathy with Deafness)
   Aniridia
   Tuberous Sclerosis

Family History of Sex-Linked Recessive Disease: Sex-linked conditions are
listed in the next Table:

Sex-Linked Conditions
Color Blindness
Hemophilia A (Factor VIII Deficiency)
Hemophilia B (Factor IX Deficiency)
Muscular Dystrophy, Duchenne Type (1:36,000)
Glucose-6-Phosphate Dehydrogenase Deficiency
Rare Conditions:
Fabry's Disease
Lesch-Nyhan Syndrome (Deficiency of enzyme hypoxanthine-guanine phosphoribosyl transferase)
Diabetes Insipidus
Ectodermal Dysplasia, Anhidrotic Type
Ocular Albinism
X-linked Ichthyosis
Agammaglobulinemia (Bruton Disease)
Wiskott-Aldrich Syndrome (an Immunodeficiency State)

AMNIOTIC FLUID BILIRUBIN (see BILIRUBIN, AMNIOTIC FLUID)

**AMPHETAMINES, DEXTROAMPHETAMINE AND METHAMPHETAMINE
("SPEED" OR "CRYSTAL", "UPPERS" OR "PEP PILLS")**
SPECIMEN: Red top tube, separate serum; and/or 50ml random urine.
REFERENCE RANGE: Negative
METHOD: Thin layer chromatography of urine; EMIT.
INTERPRETATION: Amphetamines are stimulants; stimulants increase alertness,
reduce hunger and induce a feeling of well-being. It has been used to suppress
appetite and to reduce fatigue.

AMYLASE, CLEARANCE

SPECIMEN: Red top tube, separate serum; and random urine specimen; timed urine specimen is not necessary.

REFERENCE RANGE: 5%; some laboratories have a value less than 5%.

METHOD: Serum and urine assays

INTERPRETATION: Amylase clearance is increased in patients with acute pancreatitis but is normal in other causes of hyperamylasemia. Amylase clearance has been useful in differentiating the causes of increased serum amylase, and in confirming or excluding the diagnosis of acute pancreatitis. Amylase clearance is the renal clearance of amylase, expressed as a percentage of creatinine clearance, and is given by the following equation:

$$\frac{\text{Amylase Clearance}}{\text{Creatinine Clearance}} = \frac{\dfrac{\text{Urine Amylase}}{\text{Serum Amylase}} \times \text{Urine Volume per Unit Time}}{\dfrac{\text{Urine Creatinine}}{\text{Serum Creatinine}} \times \text{Urine Volume per Unit Time}} \times 100$$

This equation simplifies to the following form:

$$\frac{\text{Amylase Clearance (\%)}}{\text{Creatinine Clearance}} = \frac{[\text{Urine Amylase}]}{[\text{Serum Amylase}]} \times \frac{[\text{Serum Creatinine}]}{[\text{Urine Creatinine}]} \times 100$$

As seen by inspection of the above equation, the clearance ratio, expressed as percentage, is calculated simply from the concentrations of amylase and creatinine in serum and urine samples obtained simultaneously. No timed collections are necessary.

In normal subjects, the value of the ratio amylase clearance/creatinine clearance is usually less than 4%. Increased amylase clearance occurs in patients with acute pancreatitis. The mean value in acute pancreatitis is about 3 times that of the normal value. An elevated ratio reflects defective proximal tubular reabsorption of amylase which occurs in virtually all patients with acute pancreatitis. Other conditions that are associated with acute defective tubular function are burns and diabetic acidosis; these conditions may also cause an elevation of the ratio.

The amylase clearance decreases during acute pancreatitis; the mean value for different time intervals post-onset of acute pancreatitis are as follows: Days 0-4, 6.8; days 5-8, 4.5; days 9-15, 4.

The current methods for measuring amylase are imprecise and precision must improve before the amylase-creatinine clearance can be applied with confidence. Macroamylasemia: In macroamylasemia, the amylase clearance is decreased to about 0.5%.

AMYLASE, SERUM

SPECIMEN: Red top tube, separate serum; or green top tube (heparin), separate plasma.

REFERENCE RANGE: Adult: 15-200U/dl; practically no amylase activity is present in neonates; measurable enzyme activity is detected at approximately two months of age and increases slowly to adult values by the age of 1 year.

METHOD: Nephelometry; enzymatic

INTERPRETATION: Causes of increased serum amylase are given in the next Table:

Causes of Increased Serum Amylase
Acute Pancreatitis
Pseudocyst of Pancreas
Non-Pancreatic Conditions: Acute Small Bowel Obstruction; Macroamylasemia; Common Duct Stone; Acute Cholelithiasis Perforated Duodenal Ulcer into the Pancreas; Ectopic Pregnancy; Renal Failure; Mumps of Other Inflammation of Salivary Glands; Morphine Administration; Cerebral Trauma; Type I Hyperlipoproteinemia

The usual cause for increased serum amylase in a child is mumps or other inflammation of salivary glands.

S. Bakerman

The change in serum amylase, serum lipase and urine amylase <u>in acute</u>
<u>pancreatitis</u> is illustrated in the next Figure:

<u>Serum Amylase, Serum Lipase and Urine Amylase Following Acute Pancreatitis</u>

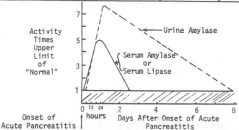

Following onset of acute pancreatitis, <u>serum amylase begins to rise in 2</u>
<u>to 6 hours</u>, reaches a <u>maximum</u> in <u>12 to 30 hours</u> and <u>remains elevated</u> for <u>2 to 4</u>
<u>days</u>. Urine amylase begins to rise 4 to 8 hours following onset of acute
pancreatitis; this is several hours after the initial increase of serum amylase.
<u>Urine amylase remains elevated</u> for <u>7 to 10 days</u>, about five days after the serum
amylase returns to normal. <u>Serum lipase</u> changes in a manner similar to that of
serum amylase following onset of <u>acute</u> pancreatitis; the changes are summarized
in the next Table:

Enzyme Changes in Acute Pancreatitis			
Enzyme	Beginning of Increase (hrs.)	Maximum (hrs.)	Return to Normal (Days)
Serum Amylase	2-6	12-30	2-4
Urine Amylase	4-8	18-36	7-10
Serum Lipase	2-6	12-30	2-4

It is important to note that <u>urine amylase is elevated when serum amylase</u>
<u>is normal</u>. This occurs because renal glomerular filtration for amylase is
increased in acute pancreatitis; thus, amylase appears in the urine several days
after the serum amylase returns to normal.
<u>Pancreatic Pseudocyst</u>: Pancreatic pseudocyst is the most common complication of
pancreatitis, occurring in <u>2% to 10%</u> of patients. Pseudocysts form as a result
of <u>alcoholic pancreatitis</u>(60%), <u>choledocholithiasis</u>(20%), <u>trauma</u>(10%); <u>blunt</u>
<u>abdominal trauma is the most common cause of pseudocysts in children</u>. Urine
amylase is elevated in 60% of patients and serum amylase is elevated in 50% of
patients. About one-third of patients have abnormal liver function tests,
anemia, increased serum glucose, and elevated white blood cells.
<u>Definitive diagnosis</u> of pseudocyst is made by <u>radiology</u>: <u>Plain film</u> of the
<u>abdomen</u> for <u>calcification</u> or mass lesion(40%) <u>ultrasonography</u> and <u>contrast</u>
<u>examination</u> of the upper gastrointestinal tract (90%); <u>CT scan</u>; IVP and contrast
studies of the biliary tract (Van Landingham, S.B. and Roberts, J.W., Hospital
Medicine, pgs. 71-88, Jan. 1984).

ANAEROBIC CULTURE
SPECIMEN: The usual sources of anaerobic specimens are as follows: abscesses;
body fluids: cerebrospinal fluid, pleural fluid, peritoneal fluid, pericardial
fluid, synovial fluid, culdocentesis, amniotic fluid; tissues: surgical and
autopsy specimens, placenta; wounds; sinus tracts; transtracheal aspirates; bone
marrow; duodenal aspirates; middle ear aspirates; mastoid aspirates; eye;
bronchial brushings.
   The relative incidence of anaerobic bacteria in various infections is given
in the next Table; (Allen, S.D. and Siders, J.A. in Manual of Clinical
Microbiology, 3rd ed., Am. Soc. Microbiology, Wash., D.C., 1980, pg. 398):

| Relative Incidence of Anaerobic Bacteria in Various Infections | |
Type of Infection	Incidence (%)
Bacteremia	10-20
Central Nervous System	
Brain Abscess	89
Subdural Empyema	50
Meningitis	low
Head and Neck	
Chronic Sinusitis	approx. 50
Peridontal Abscess	100
Pleuropulmonary	
Aspiration Pneumonia	85-90
Lung Abscess	93
Necrotizing Pneumonia	85
Empyema	76
Intraabdominal	
Peritonitis and Abscess	90-95
Liver Abscess	>50
Female Genital Tract	
Salpingitis; Pelvic Peritonitis	56 (or higher)
Tuboovarian Abscess	92
Vulvovaginal Abscess	74
Septic Abortion and Endometritis	73
Soft Tissue	
Gas Gangrene (Myonecrosis)	100
Creptitant Cellulitis	High
Necrotizing Fasciitis	High
Urinary Tract	
Cystitis	1
Urethritis	<1

   Materials which should not be routinely cultured for anaerobic bacteria
(because anaerobes occur as normal flora) are as follows (Allen, S.D. and
Siders, J.A., already cited): throat or nasopharyngeal swabs; gingival swabs;
expectorated sputum; bronchoscopic specimens not collected by a protective
double lumen catheter; gastric contents, small bowel contents, feces, rectal
swabs, colocutaneous fistulae, and colostomy stomata; surface material from
decubitus ulcers, swab samples of other surfaces, sinus tracts and eschars;
material adjacent to skin or mucous membranes other than the above which have
not been properly decontaminated; voided urine; vaginal or cervical swabs.
   Some anaerobes are killed by contact with oxygen for only a few seconds.
Great care should be taken to reduce contamination from adjacent surfaces.
Transport of Specimens: Porta-A-Cul or B-D anaerobic specimen collection tubes
may be used for transport of specimens to the laboratory. A syringe may be used
to collect and transport the specimens; expel all air, cap needle and transport
to laboratory immediately.
Gram-Stain: Gram-stain is helpful for quality control in that preliminary
information may be obtained. Characteristics of some anaerobic bacteria with
Gram-stain are given in the next Table:

Anaerobes and Gram-Stain	
Characteristic	Anaerobe (Possible)
Gram Positive, Large, Broad Rods with Blunted Ends in Suspected Gas Gangrene	C. Perfringens
Gram Negative, Irregular Staining with Bipolar Staining; Specimen from Abscess	Bacteroides or Fusobacterium Species
Gram Negative, Filamentous Slim Rods with Tapered Ends	F. Nucleatum
Gram Positive, Cocci, Clusters and Chains, within Neutrophilic Exudate from Postoperative Intraabdominal Wound	Peptococcus Peptostreptococcus
Sulfur Granules from a Cervicofacial Lesion	Actinomycetes

Culture Systems: Most low-volume laboratories rely on observed growth in the depths of thioglycollate broth tubes as the initial clue that an anaerobic bacterium may be present.

The CDC has developed a formulation for anaerobe culture.

There are different methods to generate an anaerobic atmosphere. A H_2-CO_2 generator (Gas Pak, BBL Microbiology Systems) may be used.

Identification of anaerobes can be done using gas-liquid chromatography.

INTERPRETATION: The percentage distribution of anaerobic bacteria recovered from human clinical infections is given in the next Table; (Allen, S.D. and Siders, J.A. in Manual of Clinical Microbiology, 34 ed., Am. Soc. Microbiology, Wash. D.C., 1980, pg. 405):

Percentage Distribution of Anaerobic Bacteria	
Group	Percent
Gram-Negative Nonsporeforming Bacilli	43
Bacteroides Fragilis Group	23
Gram-Positive Nonsporeforming Bacilli	23
Gram-Positive Cocci	21
Gram-Positive Sporeforming Bacilli	11
Clostridium Perfringens	5
C. Ramosum	2
C. Difficile	1
Gram-Negative Cocci	2

A recent survey of cases of bacteremia in England and Wales showed that 6.5 percent of all cases were due to anaerobic organisms, mainly Bacteroides (B. fragilis) species.

ANDROGEN PANEL [TESTOSTERONE, DEHYDROEPIANDROSTERONE-SULFATE (DHEA-S)]

SPECIMEN: Red top tube, separate and freeze serum.
REFERENCE RANGE: Testosterone: Normal reproductive females: 10-60ng/dl; Prepubertal children: <10ng/dl; Adult male: 350-800ng/dl.
Testosterone, Free: Normal reproductive females: 0.3-1.9ng/dl; Adult male: 9-30ng/dl.
Dehydroepiandrosterone sulfate (DHEA-S): Normal reproductive females: 5-35ng/dl; Normal adult male: 15-55ng/dl.
METHOD: RIA
INTERPRETATION: Testosterone, free testosterone and DHEA-S aid in the evaluation of androgen deficiency or excess (hirsutism and/or virilism). The physiologically active testosterone is the free testosterone. The free testosterone gives a more accurate measure of the testosterone available to target cells. The total testosterone is influenced by the level of the testosterone binding protein; the testosterone binding protein is changed by the factors in a manner similar to those that change the concentration of thyroxine-binding proteins. Two conditions, hyperthyroidism and syndromes of androgen resistance are associated with increased total testosterone; however, free testosterone is normal. The clinical causes of hirsutism are given in the next Table:

Causes of Hirsutism

Common
  Polycystic ovary syndrome (Stein-Leventhal Syndrome)
  Idiopathic
Rare
  Adrenal Origin
    Cushing's Syndrome
    Congenital adrenal hyperplasia
    Androgen-secreting tumors
  Ovarian Origin: Hilus Cell Tumor; Androblastoma; Teratoma
  Hypothyroidism
  Acromegaly
Drugs
  Phenytoin; Diazoxide; Minoxidil;
    Menopausal mixtures containing
      androgens, corticosteroids (rarely)

Hirsutism in the female can be of adrenal or ovarian origin and may be caused by excess androgens. DHEA-S is a good indicator of adrenal function; testosterone is a good indicator of ovarian function. By measuring testosterone and DHEA-S, one can localize the lesion to the ovaries or the adrenal. Plasma testosterone over 200ng/dl is usually indicative of an ovarian abnormality.
Stein-Leventhal Syndrome (Polycystic Ovaries): The Stein-Leventhal Syndrome is the most common hormonal cause of hirsutism. Evidence of menstrual irregularity may point towards polycystic ovarian disease. Borderline elevations of the androgens, testosterone, and 17-ketosteroids are present; raised LH levels and increased LH/FSH ratios are consistent with this disease.

ANDROSTENEDIONE   Androstenedione
SPECIMEN: Red top tube, separate serum and store frozen.
REFERENCE RANGE: Adult female: 85-275 ng/dl; Post-menopausal: 30-120 ng/dl. Adult male: 75-205 ng/dl. Prepubertal children: 10-50 ng/dl. To convert conventional units in ng/dl to international units in nmol/liter, multiply conventional units by 0.035.
METHOD: RIA
INTERPRETATION: Androstenedione is produced primarily by the ovaries and to a lesser extent by the adrenals. Assay of androstenedione is done to evaluate the cause of hirsutism and/or virilization; the causes of hirsutism are given in the Androgen Panel.

Serum androsternedione is elevated in polycystic ovarian disease but normal in hyperthecosis. Polycystic ovarian disease may be differentiated from hyperthecosis as follows: Measure androsternedione before and after chorionic gonadotropin stimulation; androstenedione is markedly increased in patients with polycystic ovarian disease but remains unchanged in patients with hyperthecosis (Abraham, G.E. and Buster, J.E., Obstet. Gynecol. 47, 581, 1976).

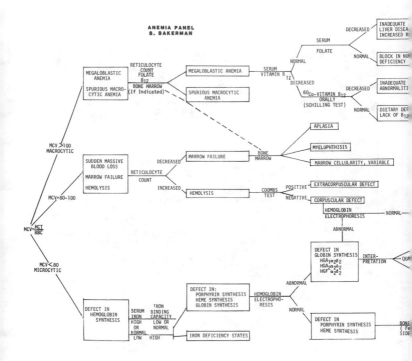

ANEMIA PANEL
S. BAKERMAN

42

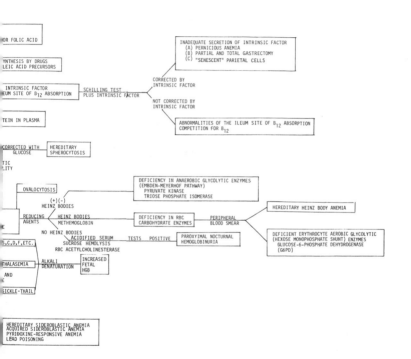

OR FOLIC ACID

YNTHESIS BY DRUGS
LEIC ACID PRECURSORS

INTRINSIC FACTOR
EUM SITE OF B$_{12}$ ABSORPTION — SCHILLING TEST PLUS INTRINSIC FACTOR

INADEQUATE SECRETION OF INTRINSIC FACTOR
(A) PERNICIOUS ANEMIA
(B) PARTIAL AND TOTAL GASTRECTOMY
(C) "SENESCENT" PARIETAL CELLS

CORRECTED BY
INTRINSIC FACTOR

NOT CORRECTED BY
INTRINSIC FACTOR

TEIN IN PLASMA

ABNORMALITIES OF THE ILEUM SITE OF B$_{12}$ ABSORPTION
COMPETITION FOR B$_{12}$

CORRECTED WITH
GLUCOSE — HEREDITARY SPHEROCYTOSIS

TIC
LITY

OVALOCYTOSIS

DEFICIENCY IN ANAEROBIC GLYCOLYTIC ENZYMES
(EMBDEN-MEYERHOF PATHWAY)
PYRUVATE KINASE
TRIOSE PHOSPHATE ISOMERASE

(+)(-)
HEINZ BODIES

REDUCING
AGENTS — HEINZ BODIES
METHEMOGLOBIN

DEFICIENCY IN RBC
CARBOHYDRATE ENZYMES — PERIPHERAL BLOOD SMEAR

HEREDITARY HEINZ BODY ANEMIA

S,C,D,F,ETC.

NO HEINZ BODIES
ACIDIFIED SERUM   TESTS   POSITIVE
SUCROSE HEMOLYSIS
RBC ACETYLCHOLINESTERASE

PAROXYIMAL NOCTURNAL
HEMOGLOBINURIA

DEFICIENT ERYTHROCYTE AEROBIC GLYCOLYTIC
(HEXOSE MONOPHOSPHATE SHUNT) ENZYMES
GLUCOSE-6-PHOSPHATE DEHYDROGENASE
(G6PD)

THALASEMIA
AND
E — ALKALI DENATURATION — INCREASED FETAL HGB

SICKLE-THAIL

HEREDITARY SIDEROBLASTIC ANEMIA
ACQUIRED SIDEROBLASTIC ANEMIA
PYRIDOXINE-RESPONSIVE ANEMIA
LEAD POISONING

ANEMIA PANEL

The _initial_ laboratory tests in differentiating causes of anemia are given in the next Table; (Wallerstein, R.O., Consultant, pgs, 65-70, August 1980):

Initial Laboratory Tests - Anemia
Hemoglobin, Hematocrit, Red Blood Count
Red Cell Indices(MCHC, MCV, MCH)
Peripheral Blood Film
Reticulocyte Count
Platelet Count
White Cell Count

BLOOD FILM: Morphologic alterations, characteristic of certain red blood cell conditions, are given in the next Table:

Morphologic Alterations of Red Blood Cells	
Alteration	Differential Diagnosis
Spherocytes	Hereditary Spherocytosis
	Autoimmune Hemolytic Anemia
	Acute Alcoholism
	Hemoglobin C Disease
	Following Severe Burns
	Hemolytic Transfusion Reactions
	Severe Hypophosphatemia
	Acute Oxidant Injury in Hexose Monophosphate Shunt Defects
	Clostridium Welchii Septicemia
Oval Cells(Elliptocytes)	Megaloblastic Anemia
	Myelofibrosis
	Refractory Normoblastic Anemia
	Hereditary Elliptocytosis
	Thalassemia Syndromes
Fragmented Red Cells (Schistocytes)	I. Microangiopathic Hemolytic Anemias
	Thrombotic Thrombocytopenia Purpura(PTT)
	Hemolytic Uremic Syndrome(HUS)
	Disseminated Intravascular Coagulopathy (DIC)
	Secondary to Immune Mechanisms
	Giant Hemangioma
	Metastatic Carcinoma
	Malignant Hypertension
	Eclampsia
	II. Macroangiopathic Hemolytic Anemia
	Prosthetic Valve Replacement, etc.
Spur Cells (Acanthocytes)	Abetalipoproteinemia
	Cirrhosis/Hepatic Necrosis
	Pyruvate Kinase(PK) Deficiency
	Uremia
	Infantile Pyknocytosis
Stippling	Lead and Arsenic Poisoning
	Thalassemia Syndromes
	Sideroblastic Anemia
	Other Severe Anemias
	Unstable Hemoglobins
	Pyrimidine 5'-Nucleotidase Deficiency
Target Cells	Thalassemias
	Hemoglobin C, S
	Liver Disease

RED CELL INDICES: Diagnoses, derived from indexes, are shown in the next Table:

Indexes and Anemia	
Index	Diagnosis
MCV>100	Megaloblastic Anemia: Folate or Vitamin B_{12} Deficiency
	Hemolytic Anemia (Elevated Reticulocyte Count)
	Liver Disease
MCV<80 and MCHC<30	Iron Deficiency Anemia or Thalassemia Minor
MCHC>36	Hereditary Spherocytosis

CLASSIFICATION OF ANEMIA: A classification of anemia is given in the next Table;
(Cohn, L.S., Diagnosis, pgs. 125-137, Oct. 1983):

Classification of Anemia	
Classification	Index
Microcytic Hypochromic Anemia	MCV<80; MCHC<30
Normocytic Normochromic Anemia	MCV and MCHC, Normal
Macrocytic Anemia	MCV>100

Anemia is detected by decreased hemoglobin, hematocrit or erythrocyte
count. Anemia is classified as follows: microcytic hypochromic anemia;
normocytic normochromic anemia, e.g., hemolytic anemia; and macrocytic anemia.

MICROCYTIC HYPOCHROMIC ANEMIA: Microcytic hypochromic anemia is characterized by
MCV<80 & MCHC<30. Causes of microcytic hypochromic anemia are given in the next
Table:

Causes of Microcytic Hypochromic Anemia
Iron Deficiency Anemia
Thalassemia Syndromes
Defects in Porphyrin Synthesis
Hereditary Sideroblastic Anemia
Acquired Sideroblastic Anemia
Pyridoxine-Responsive Anemia
Lead Poisoning
Severe Protein Deficiency
Chronic Infection

Iron Deficiency Anemia: Causes of iron deficiency anemia are given in the next
Table:

Causes of Iron Deficiency Anemia
Nutritional Iron Deficiency
Chronic Blood Loss, Usually G.I., Uterine, (e.g. Menstrual Blood Loss-Most Common Cause in Women), Hookworm
Achlorhydria and Gastrectomy
Defective Absorption e.g., Sprue or Steatorrhea; Billroth II Procedure for Peptic Ulcer Disease
Increased Demand, e.g., Pregnancy

Tests for iron deficiency anemia are given in the next Table:

Tests for Iron Deficiency Anemia		
Test	Finding	Specimen
Peripheral Smear	Microcytic, Hypochromic	Fresh Capillary or Venous Blood; Fresh EDTA-Anticoagulated Blood May also be Utilized
Red Cell Indices	Decreased	Lavender Top Tube
Serum Ferritin	Decreased	Red Top Tube
Zinc Protoporphyrin(ZPP)	Increased	Lavender Top Tube
Serum Iron	Decreased	Red Top Tube
Serum Iron Binding Capacity	Increased	Red Top Tube
Bone Marrow	Absence of Storage Iron	

Progessive Stages of Iron Deficiency: (1) Depletion of Iron Reserves: Bone
marrow iron, dec.; serum ferritin, dec.; (2) Impaired Erythropoiesis: Serum
iron, dec.; transferrin (total iron binding capacity) inc.; transferrin
saturation, dec.; zinc protoporphyrin(ZPP), inc.; MCH, normal or dec.; MCV,
normal or dec.; (3) Anemia: Hemoglobin, dec.; MCH, dec.; MCV, dec. (Finch, C.A.
JAMA 251, 2004, April 20, 1984).

Iron Deficiency Anemia Versus Thalassemia Minor: Thalassemia minor and iron
deficiency are associated with a marked decreased in the MCV. Patients with
iron deficiency and beta-thalassemia trait may be differentiated using serum
ferritin and MCV (Hershko, C., Acta Haematol. 62, 236-239, 1979) and zinc
erythrocyte protoporphyrin(ZPP). The characteristics of iron deficiency anemia
and thalassemia minor are given in the next Table:

Iron Deficiency Anemia and Thalassemia Minor		
Parameter	Iron Deficiency Anemia	Thalassemia Minor
MCHC(gm/dl)	<32.5	Slightly Reduced
MCV(cubic microns)	<80	Usually significantly reduced below 80
MCV/RBC	>13	<13
Serum Ferritin	Decreased	Normal
Zinc Erythrocyte Protoporphyrin(ZPP)	Increased	Normal

The anemia of thalassemia minor is always mild; hemoglobin is usually between 10 and 13 gm/dl. The red cell count may be higher than normal. The diagnosis is made by finding an elevated HgA$_2$ or HgF on hemoglobin electrophoresis.

In iron deficiency anemia, the serum iron is low, serum iron binding capacity is elevated and the serum ferritin is low. There is reduced or no iron in the bone marrow.

The MCV/RBC is useful in differentiating iron deficiency anemia and thalassemia minor; MCV/RBC is >13 in iron deficiency anemia but <13 in thalassemia minor. In addition, the zinc protoporphyrin test(ZPP) is normal in thalassemia minor but elevated in iron deficiency anemia; serum ferritin is normal in thalassemia minor but decreased in iron deficiency anemia.

Anemia of Chronic Disease(ACD): See IRON, SERUM for differential diagnosis of iron deficiency anemia, ACD and chronic renal disease.

NORMOCYTIC NORMOCHROMIC ANEMIA: Normocytic, normochromic anemia is characterized by normal MCV (76-99) and normal MCHC (>32). The tests to be considered in normocytic, normochromic anemia are given in the next Table:

Tests in Normocytic, Normochromic (Hemolytic) Anemia
Complete Blood Count with Morphology
Reticulocyte Count
Direct Antiglobulin (Coombs) Test
Hemoglobin Electrophoresis
Bone Marrow - If Reticulocyte Count Low
Red Cell Survival
Osmotic Fragility
Serum Bilirubin, Total Direct and Indirect
Serum LDH and Isoenzymes
Serum Haptoglobin
Urinalysis
Renal Function Testing
Blood Type and Cross-Match

The causes of normocytic, normochromic anemia are given in the next Table:

Causes of Normocytic, Normochromic Anemia
Sudden Massive Blood Loss (Elevated Reticulocyte Count)
Hemolytic Anemia (Elevated Reticulocyte Count):
Acquired: (Morphologic Abnormalities)
Immune Mediated
Coombs Positive (Autoantibody)
Paroxysmal Nocturnal Hemoglobinuria(PNH)
Microangiopathic
Disseminated Intravascular Coagulopathy(DIC)
Hemolytic-Uremic Syndrome
Congenital:
Defect in Hemoglobin: (Morphologic Abnormalities)
Hemoglobinopathies
Unstable Hemoglobins
Defect of Red Blood Cell Membrane: (Morphologic abnormalities)
Spherocytosis
Elliptocytosis
Stomatocytosis
Deficiencies of Red Blood Cell Enzymes: (Normal Morphology)
Embden-Myerhof Pathway, Most Common is
Pyruvate Kinase(PK) Deficiency
Pentose Pathway, Most Common is Glucose-6-
Phosphate Dehydrogenase Deficiency
Underproduction (Normal or Low Reticulocyte Count)
Pure Red Cell Aplasia
Drugs
Leukemia
Aplastic Anemia
Myelophthisis
Refractory Anemia
Chronic Disease, e.g., Infection
Renal Failure
Liver Disease (also associated with Macrocytosis)
Hypothyroidism (also associated with Macrocytosis)

Normocytic, normochromic red cells are seen with <u>acute blood loss,</u>
<u>hemolysis or bone marrow failure.</u>
<u>Reticulocyte Count:</u> The <u>reticulocyte count</u> is useful in differentiating <u>bone</u>
<u>marrow failure</u> from <u>other causes of normocytic, normochromic anemia</u> as
illustrated in the next Figure:

Reticulocyte Count

Reticulocyte Count
(Normal 25,000-75,000
micro/liter;
Adult, 0.5-2.0%
Newborn, up to 5%)

Low(0.5%) → Bone Marrow Failure
Elevated   200,000 to 500,000/microliter:
(>2.0%)   Hemolytic Anemia
                   Acute Blood Loss
                   Therapy for Iron and Vitamin
                     B₁₂ Deficiency
                   >1,000,000/microliter:
                     Autoimmune Hemolytic Anemia
                     (rarely, pyruvate kinase deficiency)

An absolute reticulocyte count is obtained by multiplying the percentage of
reticulocytes by the red cell count (Normal reticulocyte count is 25,000 to
75,000 per microliter).
<u>Hemolytic Anemia:</u> The usual findings in hemolytic anemia are: reticulocyte
count, elevated; haptoglobin, decreased; serum indirect bilirubin, elevated;
serum lactic dehydrogenase, elevated. Specific tests for hemolytic anemias are
given in the next Table:

Specific Tests for Hemolytic Anemias	
Condition	Test
Autoimmune Hemolytic Anemia.......Direct Coombs Test	
Hemoglobinopathy..................Hemoglobin Electrophoresis(Hgb S, C, etc.)	
Hereditary Spherocytosis..........Osmotic Fragility, Autohemolysis	
Hereditary Nonspherocytic.........Glucose--6-Phosphate Dehydrogenase(G-6PD)	
Hemolytic Anemia Pyruvate Kinase(PK)	
Paroxysmal Nocturnal..............(see Paroxysmal Nocturnal Hemoglobinuria	
Hemoglobinuria (PNH) Screen on Ham Acid Hemolysis	
Test or Sugar or Sugar Water Hemolysis	
Test)	

<u>Underproduction:</u> Aplastic anemia, myelophthisis and malignancy are associated
with bone marrow failure; <u>bone marrow failure</u> is secondary to replacement of
normal myeloid tissue as listed in the next Table:

Bone Marrow Failure	
Condition	Replacement
Aplastic Anemia	Fat
Myelofibrosis	Fibrotic Tissue
Malignancy	Multiple Myeloma, Leukemia or Lymphoma

The anemia of <u>chronic disease, infection, inflammation or malignant disease</u>
is characterized by low serum iron, low TIBC, normal or increased serum ferritin
and normal or increased bone marrow iron stores. Definitive diagnosis of bone
marrow failure is made by bone marrow biopsy.
<u>Coombs-Positive Hemolytic Anemia:</u> Coombs-positive hemolytic anemia is <u>detected</u>
<u>by the direct Coombs test;</u> this test detects <u>antibody</u> or <u>complement</u> on the
surface of the red cells. Coombs-positive hemolytic anemias are divided into
<u>warm-antibody</u> and <u>cold-agglutinin disease.</u>
<u>Warm-antibody disease</u> is generally due to <u>IgG antibody,</u> occasionally
<u>complement</u> and sometimes <u>both</u> are present. Some of the causes of warm-antibody
disease are given in the <u>next</u> Table:

Causes of Warm-Antibody Hemolytic Anemia
Drug-Induced, e.g., Alpha Methyl Dopa, Quinidine
and Penicillin
Collagen Vascular Disease
Lymphoproliferative Disorders
Viral Infection
Autoimmune

<u>Cold-agglutinin disease</u> is generally due to IgM antibody, but IgM is
generally not detected; however, a positive Coombs test is due to <u>complement</u>
that remains on the cell surface after IgM has been eluted. This disease is
identified by <u>elevated cold agglutinin titer</u> plus a positive Coombs test.

MACROCYTIC ANEMIA: Macrocytic anemia is characterized by MCV>100. The causes of macrocytosis are given in the next Table:

Causes of Macrocytosis
Megaloblastic Anemia
Folate Deficiency
Vitamin B_{12} Deficiency
Spurious Macrocytic Anemia: Aplastic; Myelofibrosis
Macrocytosis
Reticulocytosis Associated with some Hemolytic
Anemia and Blood Loss
Liver Disease
Hypothyroidism
Normal Newborn

Folate deficiency is much more common than vitamin B_{12} deficiency. The laboratory tests to be considered in macrocytic anemia are given in the next Table:

Laboratory Aids in Macrocytic Anemia	
Test	Result
Screening Tests:	
Hematologic Tests:	
Red Cell Indices:	
MCV	Increased
MCHC	Normal
Blood Counts:	
Red Cell Count(RBC)	Decreased
White Cell Count(WBC)	Decreased
Platelets	Decreased
Hemoglobin and Hematocrit	Decreased
Peripheral Blood Smear	Hypersegmentation of the Nuclei of the Polymorphonuclear Leukocytes:Normal Segmentation is as Follows:
	Lobes Percent of Polymorphs
	2 20 to 40
	3 40 to 50
	4 15 to 25
	There is an average of 3.42 segmentations per 100 cells; in megaloblastic anemia segmentation is increased. The differential diagnosis of hypersegmentation: Megaloblastic anemia; Congenital hypersegmentation; Patients with chronic renal disease.
Chemical Tests:	
Lactic Dehydrogenase (LDH)	Increase
Lactic Dehydrogenase (LDH) Isoenzymes	Increase in LDH-1 and LDH-2
Indirect Bilirubin	Increased (1.0 to 1.5mg/dl)
Serum Iron	Increased
Plasma Volume	Increased
Specific Tests:	
Folate,Serum	Decreased in Folic Acid Deficiency
B_{12}	Decreased in Vitamin B_{12} Deficiency (Approximately 10% False +)
Other Tests in B_{12} Def.:	
Gastrin	Increased
Antibody to Intrinsic Factor	Present in About 2/3 of Patients
Achlorhydria	Almost all Patients with Pernicious Anemia

ANGIOTENSIN-1-CONVERTING ENZYME (ACE)
SPECIMEN: Red top tube, separate serum and freeze.
REFERENCE RANGE: Varies with methodology. This test should not be done on subjects less than age 20; these subjects have a markedly increased ACE.
METHOD: Radiochemical assay
INTERPRETATION: SACE is elevated in some patients with active sarcoidosis; it is usually not elevated in patients with other lung diseases such as fungal or tuberculous disease or beryllium disease. The serum level of SACE in granulomatous diseases of unknown cause is given in the next Figure; (Katz, P. et al., Ann. Int. Med. 94, 359-360, 1981):

Serum Angiotensin-Converting Enzyme Activity in Granulomatous Disease of Unknown Cause

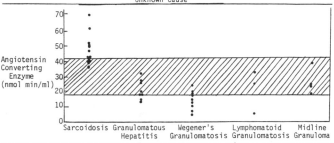

These data indicate that serum angiotensin-converting enzyme is useful in distinguishing sarcoidosis from these other granulomatous diseases.

The activity of SACE reflects the stage of sarcoidosis. Sixty-seven percent of patients with Stage 1 disease (bilateral hilar adenopathy) had elevated SACE activity; 88 percent of patients with Stage 2 disease (bilateral hilar adenopathy plus parenchymal infiltrates) had elevated SACE activity; 95 percent of patients with Stage 3 disease (parenchymal infiltrates) had elevated SACE activity (Rohrbach, M.S. and DeRemee, R.A., Clinical Laboratory Annual, 1, 1981).

Serial SACE determinations are valuable for following the clinical course of sarcoidosis (Lawrence, E.C. et al., Am. J. Med. 74, 747-756, 1983; Lufkin, E.G. et al., Mayo Clinic. Proc. 58, 447-451, 1983). However, SACE levels are not reliable for following sarcoid disease activity during a change in steroid therapy (Baughman, R.P. et al., Am. Rev. Respir. Dis. 128, 631-633, 1983).

SACE is produced by endothelial cells of blood vessels. In sarcoidosis, ACE values are increased in lymph nodes and it is suggested that the sarcoid granuloma particularly epitheliod cells and giant cells may be actively synthesizing ACE.

In addition to sarcoidosis, angiotensin-converting enzyme activity is increased in patients with Gaucher's disease and in patients with leprosy and in some patients with amyloidosis, myeloma and primary biliary cirrhosis.

Sarcoidosis has an incidence of about 30 to 40 per 100,000 of population in the United States; about 80,000 Americans have sarcoidosis. It affects about twice as many women as men and affects blacks about 10 times as commonly as whites. It has a mortality rate of 5 percent.

ANION GAP (Na$^+$, Cl$^-$, CO$_2$ CONTENT)

SPECIMEN: Red top tube, separate serum.

REFERENCE RANGE: Upper limit of normal: 15mmol/liter; borderline: 15-20mmol/liter; increased: >20mmol/liter; lower limit of normal: <5mmol/liter.

METHOD: Na$^+$ and K$^+$ by ion-specific electrodes or flame emission photometry; CO$_2$ content by electrode, colorimetrically or enzymatically.

INTERPRETATION: The causes of increased anion gap are given in the next Table:

Causes of Increased Anion Gap
Renal failure
Diabetic ketoacidosis
Starvation ketosis
Toxic agents
Salicylates poisoning
Ethylene glycol
Methyl alcohol
Paraldehyde (rarely)
Propyl Alcohol
Lactic acidosis

A mnemonic for anion gap acidosis is "A MUDPIE"; A=aspirin, M=methyl alcohol, U=Uremia, D=diabetic ketoacidosis, P=paraldehyde, I=idiopathic lactic acidosis, E=ethylene glycol.

Only four clinical conditions are associated with high anion gap metabolic acidosis; these are renal failure, ketoacidosis, drugs or toxins and lactic acidosis. In the absence of renal failure or intoxication with drugs and toxins, an increase in anion gap is assumed to be due to ketoacids or lactate accumulation.

The anion gap is calculated by the equation:

$$\text{Anion Gap} = (Na^+) - [(Cl^-) + CO_2 \text{ Content}]$$

The normal anion gap is obtained by substituting the "normal" values for Na$^+$, Cl$^-$ and CO$_2$ content into the above equation.

$$\text{Normal Anion Gap} = (140) - (100 + 25)$$
$$\text{Anion Gap} = 15 \text{ mmol/liter}$$

Conditions associated with low anion gap are listed in the next Table:

Conditions Associated with Low Anion Gap
Multiple Myeloma
Hyponatremia
Bromide Ingestion

ANTI-ACETYLCHOLINE RECEPTOR (see ACETYLCHOLINE RECEPTOR ANTIBODY)

ANTIBODY TO EYE MUSCLE ANTIGEN IN GRAVE'S OPHTHALMOPATHY
SPECIMEN: Red top tube, separate serum.
REFERENCE RANGE: Negative
METHOD: ELISA (Enzyme-linked immunosorbent assay) (Kodama, K. et al., The
Lancet, II, 1353-1356, Dec. 18, 1982). Commercial kits are not available; this
test is not available in reference laboratories.
INTERPRETATION: Grave's ophthalmopathy is thought to be an autoimmune
disorder; the eye changes range from mild lid lag and lid retraction to
involvement of the entire orbital content and loss of sight.

Grave's ophthalmopathy occurs before, during or after thyroid dysfunction
in 90 percent of cases; thus, the disorder occurs in the absence of any thyroid
abnormality in 10 percent of cases.

Circulating autoantibody against a soluble eye-muscle antigen has been
detected in almost 75 percent of the patients with Grave's ophthalmopathy,
7 percent of patients with Hashimoto's thyroiditis and 20 percent of patients
with subacute thyroiditis, in no patients with Grave's hyperthyroidism without
eye disease, and in none with multinodular goiter (Kodama, K. et al., The Lancet
II, 1353-1356, Dec. 18, 1982). Immunosuppression therapy with azothioprine to
prevent Grave's ophthalmopathy has been used by some physicians before the
appearance of signs and symptoms; treatment has been ineffective in established
lesions. It may be that detection of antibodies may help to predict the
evolution of severe exophthalamus (Editorial, The Lancet II, 1378-1379, Dec. 18,
1982).

ANTIBODY SCREEN (see COOMBS, INDIRECT)

Antibody
Screen

Anticentromere
Antibodies

ANTICENTROMERE ANTIBODIES
SPECIMEN: Red top tube, separate serum.
REFERENCE RANGE: Negative
METHOD: Indirect immunofluorescent staining of metaphase cells (Fritzler,
M.J. and Kinsella, T.D., Am. J. Med. 69, 520-526, 1980).
INTERPRETATION: Antibodies to centromeric chromatin are found in 80 to 90
percent of patients with the CREST syndrome: CREST is a mnemonic for the
characteristics of this syndrome: calcinosis cutis, Raynaud's phenomenon,
esophageal motility abnormalities, sclerodactyly and telangiectasia.

CREST is a variant of scleroderma or progressive systemic sclerosis (PSS).
CREST patients differ from patients with PSS in that they do not have widespread
involvement and the skin changes are confined to the hands and face.

Conditions associated with anticentromere antibodies are listed in the next
Table; (Fritzler, M.J. and Kinsella, T.D., Am. J. Med. 69, 520-526, 1980):

Anticentromere Antibodies	
Condition	Percent
CREST	80-90
Systemic Sclerosis	12
Raynaud's Disease	29
Systemic Lupus Erythematosus (SLE)	2
Mixed Connective Tissue Disease (MCTD)	7

The titers of the anticentromere antibodies are significantly less in
patients with other diseases as compared to those with CREST.
References: Tan, E.M., Hospital Practice, pgs. 79-84, Jan. 1983; Snaith, M.L.,
Brit. Med. J. 287, 377-378, Aug. 6, 1983.

ANTI-DESOXYRIBONUCLEASE-B (ANTi-DNAse-B)
SPECIMEN: Red top tube, separate and refrigerate serum
REFERENCE RANGE: Preschool, <60 units; School age, <170 units; Adult, <85 units.
METHOD: The antigen, desoxyribonuclease, is incubated with patient serum dilutions, and a specific substrate (DNA methyl green) to the antigen is added to the mixture. The results are read as that dilution of patient's serum that inhibits the reaction of substrate and antigen. The end-point is a color change from green to decolorization.
INTERPRETATION: Elevation of DNAse-B antibodies is especially evident in streptococcal pyodermal infections and acute glomerulonephritis, whereas the ASO response is weak. The most consistent antigenically is desoxyribonuclease-B. Streptococcal infections will result in a positive test for anti-DNAse-B in 80 to 85 percent of patients.

ANTI-GLIADIN ANTIBODIES (see GLIADIN ANTIBODIES)

ANTI-GLOMERULAR BASEMENT MEMBRANE (GBM) ANTIBODY
SPECIMEN: Red top vacutainer, separate serum and freeze; send 1ml of frozen serum shipped in dry ice (Reference Laboratory: Scripps-BioScience)
REFERENCE RANGE: None detected
METHOD: RIA; method uses radiolabeled human GBM antigens.
INTERPRETATION: Anti-glomerular basement membrane antibody is present in the serum of patients with Goodpasture's syndrome (glomerulonephritis and pulmonary hemorrhage) or anti-basement membrane antibody-induced glomerulonephritis alone. These antibodies tend to be present early in the course of the disease and last weeks to several years. Measurement of antibodies is useful in following the response to therapy. Some patients with lupus erythematosus have anti-glomerular basement membrane antibodies.

Immunofluorescent studies on renal biopsies of patients with Goodpasture's syndrome demonstrate a linear fluorescent pattern.

ANTI-HUMAN GLOBULIN TEST (see COOMBS, DIRECT)

ANTI-MICROBIAL SUSCEPTIBILITY, SERUM BACTERICIDAL TITER (SBT)(see SCHLICHTER TEST)

ANTI-MITOCHONDRIAL ANTIBODIES (see MITOCHONDRIAL ANTIBODIES)

ANTI-NATIVE DNA
SPECIMEN: Red top tube, separate serum
REFERENCE RANGE: Depends on method used by reference laboratory
METHOD: Crithidia immunofluorescent test or Farr Assay. The specificity and
sensitivity of the immunofluorescent test was 100 percent and 67 percent
respectively while the specificity and sensitivity of the Farr assay was 91
percent and 90 percent respectively. (Burdash, N.M. et al., Ann. Clin. Lab.
Science 13, No. 1, 1983).
INTERPRETATION: High levels of antibody to native double-stranded
desoxyribonucleic acid (DNA) are found in patients with active systemic lupus
erythematosus (SLE) but uncommonly in other diseases. Anti-DNA was elevated in
about 70 percent of SLE patients when first seen and in 90 percent at some time
during the clinical course (Weinstein, A. et al., Am. J. Med. 74, 206-216,
1983). There is good correlation between active SLE (particularly lupus
nephritis) and elevated levels of anti-DNA antibodies; exacerbations of SLE are
associated with a homogenous or rim pattern on immunofluorescence. Inactive SLE
usually is accompanied by low or absent serum levels of anti-DNA antibodies.
     Low levels of antibody to double stranded DNA are found in a number of
other connective tissue diseases.
     In individual patients, reappearance of anti-DNA antibodies in serum, or a
rising titer of these antibodies, or persistence of elevated levels of anti-DNA
antibodies, may correlate closely with subsequent exacerbation of disease;
disappearance of the antibodies or a fall in titer may indicate remission.
     In drug-induced lupus (hydralazine, procainamide), antibodies to
double-stranded DNA are absent.

                                                    Anti-Nuclear
                                                    Antibody(ANA)

ANTI-NUCLEAR ANTIBODY (ANA, FANA)
SPECIMEN: Red top tube, separate serum. The serum should be separated from
cells as soon as possible and stored at 2°C to 4°C if the specimen is to be
analyzed within 24 hours; otherwise, the serum specimen should be frozen. Do
not use grossly hemolyzed serum. If the specimen is sent by mail, the serum
should be frozen.
REFERENCE RANGE: Negative at 1:20 dilution
METHOD: In the FANA test, the patient's serum is tested for the presence of
antibody by detecting the affinity of the antibody for a nuclear preparation.
     The peroxidase method may be used to detect anti-nuclear antibodies, but
patterns are not obtained.
INTERPRETATION: This is a screening test for connective tissue diseases. A
characteristic finding in the serum of patients with systemic connective tissue
diseases is antibodies to nuclear antigens; these are referred to as antinuclear antibodies (ANA). The connective tissue diseases associated with the
presence of anti-nuclear antibodies in the serum are given in the next Table:

Connective Tissue Diseases with Anti-Nuclear Antibodies
Systemic Lupus Erythematosus
Drug-Induced Lupus-Like
Procainamide
Hydralazine
Mixed Connective Tissue Disease (MCTD)
Sjögren Syndrome
Scleroderma
Polymyositis-Dermatomyositis
Rheumatoid Arthritis

     There are four fluorescent patterns by immunofluorescence; these are
homogenous, peripheral, speckled and nucleolar patterns. These patterns reflect
the presence of many types of anti-nuclear antibodies. The correlations of
pattern, antibodies and connective tissue diseases are given in the next Table:

S. Bakerman

Correlations with Fluorescent Antinuclear Antibody Patterns	
Pattern	Correlation
Homogenous	Systemic Lupus Erythematosus and Other C.T. Diseases
Peripheral	Systemic Lupus Erythematosus
Speckled	Systemic Lupus Erythematosus Mixed C.T. Disease (MCTD) Sjögren Syndrome Polymyositis-Dermatomyositis Scleroderma
Nucleolar	Scleroderma Sjögren Syndrome

The peripheral pattern is relatively specific for lupus erythematosus; the nucleolar pattern is found in scleroderma and Sjögren syndrome.

The speckled pattern is associated with a variety of nuclear antigens including antibodies to a specific nuclear antigen-the centromere of chromosome spreads (Burnham, T.K. and Kleinsmith, D'AM., Sem. Arthr. Rheum. 13, 155-159, 1983) and RNP and non-histone proteins.

Drugs that are associated with false positive ANA tests are as follows: p-aminosalicylic acid(PAS), carbamazepine(Tegretol), chlorpromazine(Thorazine), ethosuximide(Zarontin), griseofulvin, hydralazine(Apresoline), isoniazid(INH), mephenytoin(Mesantoin), methyldopa(Aldomet), penicillin, phenylbutazone (Butazolidin), phenytoin(Dilantin),-hydantoin group, primidone(Mysoline), procainamide, propylthiouracil, trimethadione(Tridone). Also some drugs in the following categories: heavy metals, iodides, oral contraceptives, tetracyclines, thiazide diuretics and thiourea derivatives including sulfonamides.

ANTI-RIBONUCLEAR PROTEIN (ANTI-RNP)
SPECIMEN: Red top tube, separate serum as soon as possible and store at 2°C to 4°C if the specimen is to be analyzed within 24 hours; otherwise, the serum specimen should be frozen. Do not use grossly hemolyzed serum. If the specimen is sent by mail, the serum should be frozen.
REFERENCE RANGE: Negative
METHOD: Antibodies may be identified by immunodiffusion; titers are determined by hemagglutination. The Sm antigen is not destroyed by ribonuclease while RNP is destroyed by ribonuclease; this is used to distinguish between antibodies to Sm and RNP in the hemagglutination test.

Sm and RNP antigens are obtained from rabbit thymus; these antigens are used to coat tanned sheep red blood cells. An aliquot of the coated cells is treated with ribonuclease. Then, hemagglutination test is done using the patient's serum.
INTERPRETATION: Very high titers of antibody to nuclear RNP are found in mixed connective tissue disease(MCTD) occurring in 95% to 100% at titers > T:10:000 (Tan, F.M., Hospital Practice, pgs. 79-84, Jan. 1983), see CONNECTIVE TISSUE DISEASE PANEL.

Antibody to nuclear RNP may also be present in systemic lupus erythematosus and progressive systemic sclerosis (scleroderma).

ANTI-SMITH(Sm) ANTIBODIES
SPECIMEN: Red top tube, separate serum.
REFERENCE RANGE: Negative
METHOD: Antibodies may be identified by immunodiffusion; titers are determined by hemagglutination. The Sm antigen is not destroyed by ribonuclease while RNP is destroyed by ribonuclease; this is used to distinguish between antibodies to Sm and RNP in the hemagglutination test.

Sm and RNP antigens are obtained from rabbit thymus; these antigens are used to coat tanned sheep red blood cells. An aliquot of the coated cells is treated with ribonuclease. Then, hemagglutination test is done using the patient's serum.
INTERPRETATION: Antibodies to Sm antigens are found almost exclusively in systemic lupus erythematosus (SLE), and are found in 20 to 30 percent of patients with SLE. Clinical data indicate that there is an association of the presence of anti-Sm antibody with a higher incidence of vasculitis, resulting in peculiar visceral manifestations, responding poorly to therapy (Beaufils, M., et al., Amer. J. Med. 74, 201-205, 1983).

ANTI-SMOOTH MUSCLE(ASM) ANTIBODY

SPECIMEN: Red top tube, separate serum and refrigerate
REFERENCE RANGE: Negative; positive at serum dilution of 1:10
METHOD: Indirect immunofluorescence technique. The patient's serum is
diluted, usually 1:10, and added to fresh (frozen) tissue from mouse stomach or
rat uterus or other appropriate tissue containing smooth muscle. The serum and
tissue are incubated, and fluorescein-conjugated antiglobulin is added. The
sections are examined by fluorescence microscopy.
INTERPRETATION: This test is useful in differentiating chronic active
hepatitis(CAH) (lupoid hepatitis) from extrahepatic biliary obstruction,
drug-induced liver disease, viral hepatitis and other conditions involving the
liver. Conditions associated with increased anti-smooth muscle antibody are
given in the next Table:

Increased Anti-Smooth Muscle Antibody		
Condition	Percent Postive at 1:10 Dilution	Titer Suggestive of Diagnosis
Chronic Active Hepatitis(CAH) (Lupoid Hepatitis Type)	50-80	Generally, Titer Between 1:80 to 1:320 and Persists
Viral Hepatitis,(Acute) Viral, Infectious Mononucleosis)	1-2	Titers Below 1:80 and Transient
Biliary Cirrhosis	0-50	Titers Between 1:10 to 1:40
Intrinsic Asthma	20	Low Titers

In chronic active hepatitis, lupoid hepatitis type, the titers of ASM
antibody are generally high and persistent.

ANTI-STREPTOCOCCAL EXOZYMES (see STREPTOZYME)

ANTI-STREPTOCOCCAL-O (ASO)   ASO

SPECIMEN: Red top tube, separate serum; a fasting specimen is preferred.
Interferences: hemolysis, lipemia, contamination.
REFERENCE RANGE: Negative: <200I.U.
The reference ranges for children and adults for ASO and other antibodies
associated with streptococcal antigens are given in the next Table:

Serum Antibodies Associated with Streptococcal Antigens			
Streptococcal Antibodies	Upper Limit for Normal Population 5-12 Years	Young Adults	Lower Limits for Rheumatic Fever
Anti-Streptococcal-O(ASO)	333	200	250
Anti-DNAse		80	300
Anti-Hyaluronidase	110	80	300

METHOD: Latex Agglutination test
INTERPRETATION: A rise in titer of serum antibodies to streptolysin O (an
oxygen labile hemolysin derived from group A streptococci) is a sensitive test
and may be the single best test to document antecedent streptococcal infections.
ASO titer does not rise after cutaneous streptococcal infection.
The antistreptolysin-O titer is elevated in up to 70% of patients with
acute poststreptococcal glomerulonephritis and in approximately 20% of patients
with membranoproliferative glomerulonephritis. Early use of penicillin prevents
the antistreptolysin-O titer from rising (Madaio, M.P. and Harrington, J.T.,
N. Engl. J. Med. 309, 1299-1302, Nov. 24, 1983).
ASO titers are clinically useful only if serum is obtained at 2 to 3 week
intervals. A marked rise in titer or a persistently elevated titer indicates
that a focus of streptococcus infection or poststreptococcal sequalae is
present. A rise in titer begins about one week after infection and peaks two to
four weeks later. Evidence suggestive of a recent group A streptococcal
infection is a four fold or greater rise in titer between acute and
convalescent phase sera. In the absence of complications or reinfection, the
ASO titer will usually fall to preinfection levels within 6 to 12 months.
Over 80% of patients with acute rheumatic fever and 95% of patients with
acute glomerulonephritis have elevated titers of ASO.

ANTI-THROMBIN III IMMUNOLOGIC AND FUNCTIONAL ASSAY (HEPARIN CO-FACTOR)

<u>SPECIMEN</u>: Blue (citrate) top tube; separate plasma and freeze in a plastic vial immediately; serum activity is approximately 1/3 less than that of plasma. Patient must be off heparin for at least 6 hours before drawing specimen.

<u>REFERENCE RANGE</u>: Immunologic: 17-30mg/dl; functional: 80 to 120%; newborn babies have about half the normal adult activity and the adult anti-thrombin III concentration is reached by six months of age (McDonald, M.M., et al., Thrombosis Haemostas 47, 56-58, 1982).

<u>METHOD</u>: <u>Immunologic</u>: radial immunodiffusion (RID); <u>Functional</u>: synthetic substrate using thiobenzl substrate.

<u>INTERPRETATION</u>: Anti-thrombin III levels between 50 and 75% indicate a moderate risk for thrombosis and levels less than 50% indicate a significant risk of thrombosis.

Conditions in which there is a <u>decrease</u> in the level of antithrombin III are listed in the next Table:

Decrease in Antithrombin III
Chronic Liver Disease
Nephrotic Syndrome
Disseminated Intravascular Coagulation (DIC)
Fibrinolytic Disorders
I.V. Heparin for greater than 3 days
Carcinoma
Acute Leukemia
Post-Surgical Trauma (Major)
Deep Venous Thrombosis
Thrombophlebitis
Gram Negative Septicemia
Women on Contraceptive Pills
Pregnancy
Familial Antithrombin III Deficiency

Antithrombin III (AT-III) <u>reacts</u> with the negatively charged mucopolysaccharide anticoagulant, <u>heparin</u>. Heparin alone has minimal anticoagulant effects but when combined with antithrombin III, the <u>inhibitory action</u> of <u>antithrombin III</u> on coagulation enzymes results in the <u>inhibition of thrombus propagation</u>, e.g., <u>venous thrombosis</u> and prevention of pulmonary <u>embolism</u>. The most sensitive <u>coagulation enzymes</u> to the <u>inhibitory</u> effects of the heparin-AT III complex are <u>factors Xa and thrombin</u>.

<u>Familial Antithrombin III Deficiency</u>: Familial anti-thrombin III deficiency is rare (prevalence between 1 in 2000 and 1 in 5000). High-risk patients are as follows: Patients under the age of 35 who have had massive thrombosis after surgery or trauma; or any thrombosis in the course of a minor illness; all patients with mesenteric thrombosis; individuals with a strong family history of thrombosis; women in whom thrombosis develops while they are taking oral contraceptives or early pregnancy; infants born to AT-III deficient mothers; relatives of AT-III deficient patients; and patients difficult to heparinize. (Editorial, The Lancet, pgs. 1021-1022, May 7, 1983).

Antithrombin III concentrations in plasma are <u>elevated</u> in normal survivors of myocardial infarction and in patients taking <u>warfarin</u> (Coumadin; Vitamin K antagonist) and <u>depressed</u> in those receiving <u>heparin</u>. It is <u>significant</u> if the antithrombin III level is low in a patient taking <u>warfarin</u> indicating that Warfarin is not working effectively.

Ref.: Fareed, J. et al., Seminars in Thrombosis and Hemostasis, 84, 288, 1982.

ANTI-THYROGLOBULIN ANTIBODY (see THYROID ANTI-THYROGLOBULIN ANTIBODY)

ANTI-THYROID MICROSOMAL ANTIBODY (see THYROID ANTI-MICROSOMAL ANTIBODY)

APOLIPOPROTEIN A-I
SPECIMEN: Lavender(EDTA) top tube, separate plasma.
REFERENCE RANGE: Greater than 140mg/dl
METHOD: RIA (Maciejko, J.J. and Mao, S.J.T., Clin. Chem. 28, 199-204, 1982).
INTERPRETATION: Apolipoprotein A-I is the major protein component of high density lipoprotein (HDL); the composition of HDL is given in the next Table:

Composition of High Density Lipoprotein (HDL)

Composition	Percent
Lipid	50
Phospholipids	40-50
Cholesterol	32
Triglycerides	10
Protein	50
Apolipoprotein A-I	65
Apolipoprotein A-II	30

HDL contains 50 percent protein and 50 percent lipid. Apolipoprotein A-I (65 percent) and apolipoprotein A-II (30 percent) are the major protein components of the high-density-lipoprotein (HDL).

The plasma levels of apolipoprotein A-I, serum level of HDL-cholesterol, total cholesterol and serum triglycerides are given in the next Table;(Maciejko, J.J. et al., N. Engl. J. Med. 309, 385-389, 1983):

Serum Levels (Mean Values) of Apolipoprotein A-I, HDL-Cholesterol, Cholesterol and Triglycerides in Patients with Single-, Double-, or Triple- Coronary Artery Disease

| | | Coronary Artery Disease | | |
| | | Single- | Double- | Triple- |
Constituent (mg/dl)	Control	Vessel	Vessel	Vessel
Apolipoprotein A-I	147	104	101	92
HDL Cholesterol	46	30	33	22
Total Cholesterol	218	217	227	217
Triglycerides	159	184	169	170

Apolipoprotein A-I by itself is more useful than HDL-cholesterol for identifying patients with coronary-artery disease (Maciejko, J.J. et al., N. Engl. J. Med. 309, 385-389, 1983). It is not known whether the apolipoprotein A-I level is useful as a screening test for predicting the future occurrence of coronary-artery disease in the general population (Blackburn, H., N. Engl. J. Med. 309, 426-428, 1983).

Appendicitis
Acute, Panel

APPENDICITIS, ACUTE, PANEL
The test panel for acute appendicitis is given in the next Table:

Test Panel for Acute Appendicitis

White Cell Count(WBC)
White Cell Differential for Percent and Band Neutrophils
C-Reactive Protein(CRP)

Automated methods for white cell differential, e.g., (cytochemical reaction, Technicon) may not separate band neutrophils from mature polymorphonuclear leukocytes.

The total white blood count (>10,500), and the percent neutrophils (>75%) have the highest sensitivities (81-84%) for the diagnosis of acute appendicitis. The sensitivity (positivity in disease) and specificity (negativity in subjects without the disease) of combinations of laboratory tests in the diagnosis of acute appendicitis are given in the next Table;(Marchand, A. et al., Am. J. Clin. Pathol. 80,369-374, 1983):

Sensitivity and Specificity of Laboratory Tests in Diagnosis of Acute Appendicitis

| | | Test Combinations | | | |
WBC (Cells/cu mm)	Cytochem. Neut.	Manual bands	C-Reactive Protein (mg/dl)	Sensitivity Percent	Specificity Percent
>10,500	>75%	---	>1.2	97	42
>10,500	---	>11%	>1.2	100	47
>10,500	7,880/cu mm	---	>1.2	97	53
>10,500	---	1,150/cu mm	>1.2	100	47

Appendicitis   S. Bakerman

These data indicate that when the results of three tests (WBC, neutrophils, CRP) are within reference internal, the patient is unlikely to have acute appendicitis. However, increase in total white cell count is a late finding. The incidence of perforation in infants with acute appendicitis is very high, probably reflecting the difficulty in interpreting clinical signs and symptoms. The clinical history to establish the exact hour of onset is paramount (Thomas, R.E. personal communication). Keep in mind that the morbidity and mortality of missing a case of acute appendicitis with subsequent abcess formation and rupture outweigh the morbidity and mortality associated with removal of a normal appendix.

APT TEST FOR FETAL HEMOGLOBIN
SPECIMEN: Bloody rectal discharge from infant.
REFERENCE RANGE: Negative for fetal blood
METHOD: Mix specimen with an equal quantity of tap water; centrifuge or filter. Supernatant must have pink color to proceed. Add 1 part of 0.25N NaOH (1g NaOH, add water to 100ml) to 5 parts of supernatant.
Fetal Hemoglobin: Pink color persists over two minutes.
Adult Hemoglobin: Pink color becomes yellow in two minutes or less.
INTERPRETATION: This test is used to distinguish ingested maternal blood from gastrointestinal lesions of infants as a cause of bloody rectal discharge.
The theoretical basis for this test is the fact that fetal hemoglobin is relatively resistant to alkali denaturation as compared to adult hemoglobin. (Apt, L. and Downey, W.S., J. Pediatr. 47, 6, 1955)

ARSENIC, HAIR OR TOENAILS
SPECIMEN: Hair: Hair samples must have the root end and distal ends oriented and labeled for acute and subacute poisoning. Collect 0.5g of hair. Nails: Clippings at the end of toe nails represent deposition of arsenic six months prior. Collect 0.5g of nail from all ten toe nails.
REFERENCE RANGE: Hair: >1mcg/g(ppm); Nail: >2mcg/g(ppm). To convert traditional units in micrograms/g to international units in nmol/g, multiply traditional units by 13.35.
METHOD: Atomic absorption; anodic stripping voltammetry
INTERPRETATION: Arsenic has a high affinity for the keratin in hair or nails. Hair: The earliest excess arsenic detectable in emerging hair (hair next to the root) appears two weeks after a dose of arsenic and may persist for months or years. Hair grows 1/2 inch per month (Weisaman, W., "Laboratory Aids in Toxicological Problems", BioScience Laboratories).
Toe nails: Toenails require from 6 to 9 months to grow so clippings obtained from the end represent deposition of arsenic six months prior.
Following exposure to arsenic, transverse white strips (Mees' lines) about 1mm wide and extending across the entire base of the nails appear in about two months; this band contains a very high concentration of arsenic.

ARSENIC, SERUM
SPECIMEN: Use Sarstedt syringes; draw blood into the syringe; cap the syringe; allow blood to clot and then centrifuge the specimen in the Sarstedt syringe; after centrifugation, pour the serum into a second Sarstedt syringe. Cap the syringe.
METHOD: Anodic stripping voltammetry; atomic absorption
REFERENCE RANGE: <0.07mcg/ml
INTERPRETATION Arsenic combines with proteins, especially sulfhydryl groups; this affinity for intracellular proteins is responsible for the rapid removal of arsenic from the blood. Arsenic clears the blood within four days after a significant dose. Blood is usually not the specimen of choice. See ARSENIC, URINE.

ARSENIC, URINE
SPECIMEN: Collect 24 hour urine in acid washed-containers. Use plastic containers (borosilicate, polyethylene or polypropylene); add 10% HCl solution to the container and allow to "soak" for 10 minutes; rinse with five volumes of tap water and then five volumes of deionized or distilled water. No preservative is needed. The patient should urinate at 8:00 A.M. and the urine is discarded. Then, urine is collected for 24 hours including the next day 8:00 A.M. specimen. Indicate 24 hour volume. A 50ml aliquot is used for analysis.
REFERENCE RANGE: Up to 20mcg/liter (seafood) diet: up to 100mcg/liter; industrial exposure: up to 2000mcg/liter.
METHOD: Qualitative: Reinsch test; Quantitative: Atomic absorption; Anodic stripping voltammetry.
INTERPRETATION: Arsenic is the most common acute heavy metal poisoning and is second only to lead as a chronic poison.
     Urine is the preferred specimen from the living patient because high levels of arsenic persists in the urine for about a week after acute poisoning and up to about a month after chronic poisoning (McBay, A.J., Clin. Chem. 19, 361-365, 1973). Urinary excretion of arsenic fluctuates and several negative urines are required to exclude arsenic poisoning. Arsenic combines with the proteins, especially sulfhydryl groups; this affinity for intracellular proteins is responsible for the rapid removal of arsenic from the blood. Blood is usually not the specimen of choice.
Acute: The acute symptoms relate to the central nervous, gastrointestinal and renal systems. There is headache, giddiness, dizziness, convulsions and coma; nausea, vomiting and profuse watery diarrhea ("rice-water" stools); shock and anuria.
Chronic: The chronic symptoms are multisystem: symmetrical hyperkeratosis of the hands and feet, symmetrical pigmentation, conjunctivitis, tracheitis, polyneuritis with sensory and motor involvement.
     Elemental arsenic is not toxic; inorganic trivalent arsenic trioxide is especially toxic; pentavalent arsenic is partially toxic. Inorganic arsenic is found in insecticides, herbicides, rodenticides, fruit sprays, moonshine, paint and smelters. Seafoods, particularly shellfish (oysters and mussels) are rich in arsenic and cause moderately elevated urinary arsenic levels; fortunately, the shellfish concentrate organic forms of arsenic which are apparently non-toxic.
     2,3 Dimercaptopropanol (BAL, British anti-Lewisite) is used to treat arsenic poisoning; BAL is a reducing agent that converts arsenic to a less toxic form. The maximum excretion of arsenic in the urine occurs between three and four hours after intramuscular injection of BAL.

ARTHRITIC (JOINT) PANEL

The diseases that involve joints are listed in the next Table:

Diseases Involving Joints
Osteoarthritis (Degenerative Joint Disease)
Traumatic Arthritis
Pseudogout
Gout
Rheumatic Fever
Acute Bacterial Arthritis
Tuberculous Arthritis
Rheumatoid Arthritis
Systemic Lupus Erythematosus

The laboratory tests for these conditions are as follows:

Laboratory Tests for Arthritis
Blood Tests:
Complete Blood Count
Erythrocyte Sedimentation Rate (ESR)
Serum Uric Acid
Immunologic Joint Disease:
Serum Protein Electrophoresis
Quantitation of Immunoglobulins
Serum Rheumatoid Factor (RF)
Serum Antinuclear Antibodies (ANA)
Serum Complement
Streptozyme and ASO Tests
Synovial Fluid Analysis:
Appearance
Mucin Clot Test
Glucose (Simultaneous Serum Glucose)
Microscopic: Cells; Crystals (Polarized Light)
Gram Stain
Culture

The diseases and useful serum tests (also skin test for T.B.) are given in the next Table:

Tests, Arthritis	
Disease	Serum Tests
Osteoarthritis	None
Traumatic Arthritis	None
Pseudogout	None
Gout	Serum Uric Acid Usually Elevated
Rheumatic Fever	Immunological Response to Group A, beta hemolytic streptococci; tests: Streptozyme and Antistreptolysin O(ASO) tests.
Bacterial Arthritis	Blood Culture
Tuberculous Arthritis	T.B. Skin Test
Rheumatoid Arthritis	RF; ANA Positive, see Connective Disease Panel
Lupus Erythematosus	ANA; Anti-Double-Stranded DNA; Anti-Smith; Serum Complement

Acute Bacterial Arthritis: Only a few bacteria are responsible for acute bacterial arthritis which is most often the result of bacteremia; the causes of septic arthritis, with age of patient, are given in the next Table (Parker, R.H. in Infectious Diseases, Hoeprich, P.D., ed., Harper and Row, Publishers, N.Y., First Ed., 1972, pg. 1204):

Bacterial Causes of Bacterial Arthritis and Age				
	Age (years)			
Organism	<2 (%)	2-15 (%)	16-50 (%)	>50 (%)
Staphlococcus Aureus	40	50	15	75
Streptococcus Pyogenes	15	25	5	5
Streptococcus Pneumoniae (Pneumococci)	10	10		5
Hemophilus Influenzae	30	2		
Niesseria Gonorrhoeae		5	75	
Enterobacteriaccae and Pseudomonas	3	5	5	10
Other	2	3	-	5

In <u>children</u>, the most common causes of septic arthritis are <u>Staphlococcus
aureus</u>, <u>Streptococcus pyogenes</u>, <u>Streptococcus pneumoniae</u> and <u>H. influenzae</u>. In
<u>adults</u>, the most common cause of septic arthritis is <u>Neisseria gonorrhoeae</u>. In
<u>older adults</u>, the most common cause is <u>Staphlococcus aureus</u>; in the <u>aged</u>, septic
arthritis is usually imposed on some kind of chronic bone disease.
<u>Tuberculous Arthritis</u>: Tuberculous arthritis is usually associated with
osteomylitis.

<u>Lactic Acid</u>: The level of lactic acid correlates with the presence of septic
arthritis when levels exceed 250mg/dl (Borenstein, D.G. et al., Arthritis and
Rheumatism <u>25</u>, 947-953, 1982).

<u>Synovial Fluid Analysis</u>: Diseases and synovial fluid analysis are given in the
next Table:

Diseases and Synovial Fluid Analysis					
Disease	Appearance	Mucin Clot	Glucose	Cells/cumm	Crystals
Normal	Straw, Clear	Good	70-110 mg/dl	10-600 25% polys	
Osteoarthritis	Yellow, Clear	Good to Fair	Decreased (-10)	1000 50% Lymphs	
Traumatic Arthritis	Cloudy, Bloody	Good to Fair	Decreased (-10)	20,000 Many RBCs	
Pseudogout	Slightly Yellow Cloudy	Fair to Poor	Decreased (-20)	6000 75% polys	Polarized:* Rods, Rectangles or Rhomboids
Gout	Yellow-White, Milky	Fair to Poor	Decreased (-20)	10,000 75% polys	Polarized:** Rods or Needles
Rheumatic Fever	Yellow,Slightly Cloudy	Good to Fair	Decreased (-10)	10,000 50% polys	
Bacterial Arthritis	Grey-Red, Turbid	Poor	Decreased (-60)	80,000 90% polys	
Tuberculous Arthritis	Yellow, Cloudy	Poor	Decreased (-60)	25,000 50% polys	
Rheumatoid Arthritis	Yellow-Green Cloudy	Fair to Poor	Decreased (-30)	8,000- 40,000 70% polys	
Lupus Erythematosus		Good to Fair	Decreased (-10to20)	3,000 10% polys	

*Weakly positive birefringent; **Strongly negative birefringent
Specimens: Divide specimen into 4 aliquots as follows:
(1) Red top tube for gross appearance and mucin clotting
(2) Heparin tube for microscopic examination
(3) Sterile tube for culture
(4) Heparin tube for chemical analysis

<u>Appearance</u>: Clear fluid indicates noninflammatory arthritis.

<u>Mucin Clot (Ropes) Test</u>: The mucin clot test reflects polymerization of
hyaluronate. In the mucin clot test, a few drops of synovial fluid are added to
20ml of 5% acetic acid. Normally, a mucin clot forms within 1 minute; a poor
clot indicates inflammation (see Table).

<u>Constituents of Normal Synovial Fluid</u>: The constituents in normal synovial
fluid are shown in the next Table:

Constituents of Normal Synovial Fluid	
Constituent	Synovial Fluid
Protein	1-3g/dl
Albumin	55-70%
Hyaluronate	0.3-0.4g/dl
Glucose	70-110mg/dl
Uric Acid	
Males	2-8mg/dl
Females	2-6mg/dl
Lactate	10-20mg/dl (1-2mmol/liter)

Normally about 0.1 to 2.0 ml of synovial fluid is present in the knee joint. Synovial fluid protein is less than that of plasma. Hyaluronate is a polymer of repeating disaccharide units and is not found in the plasma; it gives synovial fluid its high viscosity. Glucose and uric acid concentrations are like that of plasma; lactate concentration is like that of venous blood. Synovial fluid does not normally clot.

Glucose: A difference of more than 30mg/dl between serum and synovial fluid glucose concentrations suggests an inflammatory cause.

Crystals: Pseudogout: Using polarized light, the calcium pyrophosphate dihydrate crystals appear as rods, rectangles or rhomboids and weakly positive birefringent.
Gout: Crystals are seen in about 90% of patients during attacks of acute gouty arthritis and in about 75% between attacks. Using polarized light, urate crystals appear as birefringent rods or needles, strongly negative birefringent; the crystals appear yellow when parallel to the axis of the red compensator and blue when perpendicular to the axis. (Mnemonic: "U Pay Peb"; Urate crystals = Parallel, Yellow; Perpendicular, Blue. Turi, M., personal communication). The pyrophosphate crystals of pseudogout have the opposite orientation. Note whether crystals are intracellular or extracellular. Intracellular crystals suggest urate as cause of acute arthritis. Incubate with uricase to confirm impression (McCarty, D.J. and Hollander, J.L., Ann. Intern. Med. 54, 452, 1961).

ASCITIC FLUID ANALYSIS

SPECIMEN: A catheter is introduced through a trocar. Collect 3 tubes; purple (EDTA) for cell count; special Bactec-vials for microbiological studies, blue (aerobic), yellow (anaerobic); red or green (heparin) top tube for chemistries. Refrigerate; record volume and color. Culture, Gram stain and Ziehl-Neelsen staining should always be done on a centrifuged specimen.

If no fluid is aspirated, then one liter of normal saline or Ringer's lactate (10 to 20ml/kg body weight) is infused over 15 to 20 minutes. After body movement to disperse fluid, the <u>lavage fluid</u> is siphoned back from the peritoneal cavity into the original container and examined.

REFERENCE RANGE: See below; normal volume, <50ml.

METHOD: The tests that should be done are given in the next Table:

Tests of Ascitic Fluid
Specific Gravity
Total Protein (Serum and Ascitic Fluid)
Lactate Dehydrogenase (LDH) of Serum and Ascitic Fluid
White Blood Cell Count and Differential (Vij, D. et al., JAMA 249, 636-638, 1983)
Red Cell Count
Culture, Gram Stain, Ziehl-Neelsen Stain for Mycobacteria
Cytologic Examination
CEA for Tumor
Alpha-Fetoprotein (if Hepatoma or Endodermal Sinus Tumor of Ovary Suspected)
Amylase and Lipase (if Pancreatitis Suspected)
Glucose (Suspected Rheumatoid Disease or Infection)

INTERPRETATION: The causes of ascites are given in the next Table (Krieg, A.F. in Henry, J.B., Clinical Diagnosis and Management, W.B. Saunders, Phila., 1979, Vol. I, pg. 668):

Causes of Ascites	
Transudates	Exudates
Hepatic Cirrhosis	Malignancy: Metastatic Carcinoma;
Congestive Heart Failure	Lymphoma; Hepatoma; Mesothelioma;
Congestive Pericarditis	Pancreatitis
Hypoproteinemia	Infections
(e.g., Nephrotic Syndrome)	Tuberculosis
Myxedema	Primary Bacterial Peritonitis (May be
	Superimposed on Transudate)
	Secondary Bacterial Peritonitis (e.g.,
	Appendicitis, Intestinal Infarct)
	Trauma
	Bile Peritonitis
	Chylous Effusion
	Damage or Obstruction to Thoracic Duct,
	e.g., Trauma, Lymphoma, Carcinoma,
	Tuberculosis, Parasitic Infestation

The most common cause of ascites is liver disease. Malignancy, congestive heart failure, tuberculosis and chronic pancreatitis are other important possibilities. Differentiation of transudate from exudate is given in the next Table; (Boyer, T.D. et al., Arch. Intern. Med. 138, 1103-1105, 1978; Editorial, Brit. Med. J., 282, 1499, May 9, 1981):

Differentiation of Transudate from Exudate		
Characteristic	Transudate	Exudate
Specific Gravity	<1.015	>1.018
Protein	<2 to 3gms/dl	>3gms/dl
Ascitic Fluid Protein / Serum Protein =	<0.5	>0.5
Ascitic Fluid LDH / Serum LDH =	<0.6	>0.6
WBC Count	<500/cu mm <25% Polymorphonuclear Nuclear Leukocytes	>500/cu mm (suggests Bacterial and Tuberculous Peritonitis)

Analysis of Lavage Fluid (Kreig, A.F. in Henry, J.B., Clinical Diagnosis and Management, W.B. Saunders, Phila., 1979, Vol. I, pgs. 667-674): Blood in lavage fluid may reflect intraperitoneal hemorrhage or traumatic tap. Traumatic tap is characterized by clearing on continued aspiration. The quantity of intraperitoneal hemorrhage is reflected by gross and microscopic examination; on gross examination of lavage fluid in bottle: gross blood, >100ml/liter; bright red, 5-100ml/liter; pink, 2ml/liter; pale pink, 8 drops/liter. On microscopic examination, 100,000 erythrocytes/cu mm is considered "positive" and consistent with over 20ml whole blood in peritoneal cavity.

Leukocyte count of over 500/cu mm in lavage fluid is considered abnormal but not diagnostic for peritonitis.

ASCORBIC ACID (see VITAMIN C)

ASPERGILLUS ANTIBODY SERUM
SPECIMEN: Red top tube, separate and then refrigerate serum. If titer is obtained, paired sera advisable to detect rising titers.

REFERENCE RANGE: Negative

METHOD: The immunodiffusion(ID; double diffusion) test using A. niger, A. fumigatus and A. flavus antigens is most often utilized. The formation of one to four lines of identity between patient's serum and reference serum versus A. antigen indicates a positive reaction. The I.D. test will detect precipitins in over 90 percent of patients with pulmonary aspergilloma and in about 70 percent of patients with allergic bronchopulmonary aspergillosis.

Serologic tests, other than immunodiffusion(ID) that are useful in confirming a diagnosis of noninvasive aspergillosis are as follows: complement fixation(CF), counterimmunodiffusion(CIE), enzyme-linked immunosorbent assay (ELISA) and passive hemagglutination assay(PHA).

IgE and IgG antibodies against A. fumigatus are measurable by radioimmunoassay (Wang, J.L.F. et al., Am. Rev. Respir. Dis. 117, 917-927, 1978).

INTERPRETATION: Aspergilli are fungi, saprophytic (obtaining food by absorbing dissolved organic material) molds; the main species of clinical significance are A. fumigatus and A. flavus.

The predisposing condition to infection by aspergilli is the immunocompromised patient; this may occur in patients on immunosuppressive therapy and in patients with debilitating diseases. Up to 5 percent of patients with leukemia may become infected with aspergilli.

Allergic Bronchopulmonary Aspergillosis(ABPA); (Ricketti, A.J. et al., Arch. Intern. Med. 143, 1553-1557, 1983): The eight diagnostic criteria of ABPA are given in the next Table:

Diagnostic Criteria of Allergic Bronchopulmonary Aspergillosis (ABPA)
Laboratory Findings:
(1) Precipitating Antibodies Against A.Fumigatus Antigen
(2) Elevated Serum IgE and IgG Antibodies to A.Fumigatus
(3) Blood Eosinophilia (>1000/cumm)
(4) Elevated Serum IgE Concentration
Clinical Findings:
(1) Episodic Bronchial Obstruction (Asthma)
(2) Immediate Skin Reaction to A.Fumigatus Antigen
(3) History of Pulmonary Infiltrates (Transient or Fixed)
(4) Central Bronchiectasis

Other findings in ABPA may include positive sputum culture of A.fumigatus, a history of expectoration of brown plugs and late (Arthus type) skin reactivity to intracutaneous testing with A.fumigatus antigen.

The stages in the natural history in patients with ABPA are as follows: acute remission, recurrent exacerbation, corticosteroid-dependent asthma and fibrosis.

Pulmonary Aspergilloma: The ID test will detect precipitins in over 90 percent of patients with pulmonary aspergilloma. This is a chronic infection generally arising in an old tubercular cavity or bronchial dilatation. The combination of caseated material and tightly branched radiating hyphae is called a "fungus" ball.

ASPIRIN, BLOOD (see SALICYLATE BLOOD)

ASTHMA PANEL

The diagnostic procedures in asthma are given in the next Table:

Diagnostic Procedures in Asthma
1. History
2. Physical Examination
3. Basic Laboratory Evaluation
a. Chest X-Ray
b. Complete Blood Count
c. Spirometry
4. Optional as Indicated
a. Allergen Skin Tests
b. Alpha$_1$-Antitrypsin
c. Blood Gases
d. Complete Pulmonary Function Tests
e. Electrocardiogram(ECG)
f. Exercise Challenge
g. Quantitative Immunoglobulins
h. Radioallergosorbent Test
i. Sinus Roentgenograms
j. Sputum Exam and Curve
k. Sweat Test
l. Trial of Allergen Avoidance
m. Tuberculin Skin Test
n. Stool for Ova and Parasites
5. Special Procedures
a. Inhalation Challenge
b. Occupational Challenge
c. Oral Challenge

The physiologic alterations in bronchial asthma are given in the next Table:

Physiologic Alterations in Bronchial Asthma
1. Increased airway resistance, decreased expiratory flow, (increased bronchial smooth muscle tone, mucosal edema, mucus secretion).
2. Airway closure at higher than normal distending pressure (decreased vital capacity).
3. Increased lung volume (RV and FRC always)(TLC often).
4. Increased pulmonary artery pressure relative to pleural pressure(ECG evidence of right ventricular strain or P pulmonale).
5. Increased negative pleural pressure resulting in marked respiratory variations in arterial pressure (pulsus paradoxus).
6. Increased ventilation-perfusion imbalance (decreased PO_2, increased deadspace, increased pCO_2 in severe cases).

The spirographic tracing of forced expiration is given in the next Figure:

Spirographic Tracing of Forced Expiration, Normal and Asthma

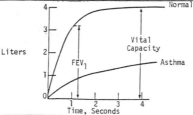

Volume in the spirometer is plotted against time. The vital capacity is represented by the total volume expired. One-second forced expiratory volume(FEV_1) is the volume expired during the first second.

AUTOHEMOLYSIS

SPECIMEN: Two green (heparin) top tubes
REFERENCE RANGE: No Sugar: 24 hours, 0.05-0.5%; 48 hours, 0.4-4.5%;
With Sugar: 24 hours, 0-0.4%, 48 hours, 0.3-0.6%.
METHOD: When blood is incubated at 37°C, a small amount of lysis of the red
blood cells occurs; glucose partially inhibits lysis.
INTERPRETATION: This test is most often used for diagnosis of hereditary
spherocytosis; the results of autohemolysis are shown in the next Table:

Autohemolysis Test		
Condition	No Glucose	Glucose
Hereditary Spherocytosis	Markedly Increased	Corrected
G-6-P-D Deficiency	Moderately Increased	Corrected
Pyruvate Kinase Deficiency	Markedly Increased	Not Corrected

B_{12} (see VITAMIN B_{12})

BACTERIAL MENINGITIS ANTIGENS

SPECIMEN: One ml CSF (Other specimens are as follows: Red top tube, separate
serum; 10ml urine; one ml pleural fluid; one ml of joint fluid); maintain
sterility; store and forward to reference laboratory refrigerated or frozen.
REFERENCE RANGE: Negative
METHOD: Counterimmunoelectrophoresis(CIE), latex agglutination(LA) and ELISA.
The sensitivity of ELISA methodology is greater than that of CIE.
CIE: The arrangement of the electrodes and the direction of flow of antibody
and antigen are shown in the next Figure:

Counterimmunoelectrophoresis

(1) Place serum antibody into appropriate wells:

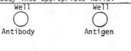

Antigens Assayed
Neisseria Meningitidis
Hemophilus Influenzae
Streptococcus Pneumoniae
Listeria Monocytogenes
Group B Streptococcus

(2) Electrophorese:

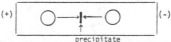

LA: Latex particles coated with appropriate antigens are available; the
specimen (CSF, urine, etc.) is placed in contact with particles. Commercial
sources of kits are listed in the next Table:

Commercial Source of Kits for Latex Agglutination	
Antigen Tested	Source
Group B Streptococci (Wellogen Strep B)	Wellcome Diagnostics, 800-334-8570; in N.C. call 919-541-9090, ext. 4617
N. Meningitidis Groups A,B,C,Y and W135	Pharmacia Diagnostics 800-526-3617 in N.J. 201-457-8000
H. Influenzae, Type b	
S. Pneumoniae, all 83 Capsular Types	
H. Influenzae, Types a,c,d,e,f	

The Wellcome Strep B latex assay rapidly identified all cases of
culture-positive sepsis and meningitis for group B Streptococcus disease; the
quantitation of antigen concentration combined with the peripheral WBC count
proved helpful in predicting outcome (Friedman, C.A. et al., Pediatrics
73,27-30, 1984).
INTERPRETATION: Rapid and specific diagnosis of bacterial meningitis is done
by latex agglutination(LA) tests and counterimmunoelectrophoresis(CIE) for assay
of antigens of microorganisms; the tests, available as of May, 1983, are listed
in the next Table:

Latex Agglutination(LA) and Counterimmunoelectrophoresis(CIE) for Bacterial Antigens		
Antigen	Latex Agglutination	Counterimmunoelectrophoresis
Neisseria Meningitidis	Yes	Yes
Hemophilus Influenzae, Polyvalent (a-d)	Yes	Yes
Streptococcus Pneumoniae	Yes	Yes
Listeria Monocytogenes		Yes
Group B Streptococcus	Yes	Yes

Bacteria Causing Meningitis: The organisms most commonly responsible for bacterial meningitis by age-group are given in the next Table:

Usual Organisms by Age in Bacterial Meningitis	
Age	Organisms
Newborns	Gram Negative Bacilli, Group B Streptococci, Listeria Monocytogenes
Children, >2 months	Haemophilus Influenzae, Type b
Healthy Adults	Neisseria Meningitidis (Meningococci)
Elderly	Streptococcus Pneumoniae (Pneumococci)

Barbituates

BARBITUATES

SPECIMEN: Red top tube, separate serum or 50ml random urine.
REFERENCE RANGE: The therapeutic ranges and toxic values for barbituates are given in the next Table:

Therapeutic Ranges and Toxic Values for Barbituates				
Barbituates	Therapeutic Range(mcg/ml)	Toxic (mcg/ml)	Half-Life	Time to Peak Plasma Levels	Time to Steady State
Short-Acting (3-6 hours)	1-5	over 5			
Pentobarbital (Nembutal)			15-48 hrs.	--	--
Secobarbital (Seconal)			19-34 hrs.		
Intermediate Acting (4-8 hours)	5-14	over 30	8-42 hrs.		
Amobarbital (Amytal)				--	--
Long-Acting (8-16 hours) Phenobarbital; Barbital	15-40	over 40	Adults: 50-120 hr. Children: 40-70 hr.	6-18 hours	Adults: 11-25 days Children: 8-15 days

METHOD: EMIT, gas-liquid chromatography, high-pressure liquid chromatography, RIA, thin-layer chromatography.

INTERPRETATION: Barbituates are one of the leading causes of accidental poisoning death in children and adults and are a leading mode of suicide attempt. The majority of barbituate overdose cases have involved the use of phenobarbital, pentobarbital and secobarbital.

The symptoms of acute barbituate intoxication affect primarily the central nervous system and the cardiovascular system. The fatal plasma concentration is greater in the long-acting barbituates than in the short-acting barbituates.

The depth and severity of coma in barbituate poisoning are graded; the grading system is given in the next Table:

Grading of Coma in Barbituate Poisoning			
Grade	Characteristics		
	Deep Tendon Reflexes	Painful Stimuli	Vital Signs
I	Intact	Withdraws	
II	Intact	No Response	
III	Reduced or Absent	No Response	
IV	Absent	No Response	Respiratory Depression and/or Circulatory Instability

A flat EEG and abnormal reflexes may occur during acute intoxication; these findings do not necessarily indicate structural damage to the CNS. Electrolytes and serum levels of calcium should be followed; it may be advisable to monitor pulmonary wedge pressure.

Barbital, which is used primarily in laboratory buffers and is not available as a prescription or over-the-counter drug in the United States, has been abused and detected in drug overdose.

Barbituates are hypnotics and sedatives that have the following disadvantages: dependence, unnatural sleep, hangover and hepatic enzyme induction; barbituates have been largely replaced by benzodiazepines.

BENCE-JONES PROTEIN, URINE (LIGHT CHAIN EXCRETION)

Determine concentration of protein in urine; if <100mg/24 hrs., then do not perform electrophoresis except in those patients with known monoclonal gammopathy. Note: A urine protein electrophoresis is done prior to IEP.

SPECIMEN: 10ml of first morning specimen; refrigerate.

REFERENCE RANGE: None detected

METHOD: The urine is concentrated; then, electrophoresis is done followed by immunodiffusion.

INTERPRETATION: This is a test for qualitative identification of the light chains, kappa or lambda. Increased light chains (kappa or lambda chains) occur in malignant conditions such as myelomatosis, Waldenström's disease, lymphatic leukemias and lymphomas and in non-malignant conditions, primary amyloidosis (Solomon, A., N. Engl. J. Med. 294, 17, 91, 1976) and very rarely in patients with benign monoclonal gammopathies (Dammacco, F. and Waldenström, J., Acta. Med. Scand. 184, 403, 1968). Light chains are excreted in the urine when cells acquire an imbalance in heavy and light chain synthesis and produce too many light chains. Bence-Jones proteins are found in 60 percent of patients with paraproteins. In about 15% of myelomatosis, only light chains are synthesized and released.

Light chains may occur in association with any one of the myelomatosis. Light chains can pass into the glomerular filtrate because they have a relatively low molecular weight (25,000 daltons if monomers and 50,000 daltons if dimers); they do not usually accumulate in the serum except in renal failure.

Only small amounts of Bence-Jones proteins are found in amyloidosis; about 60% are lambda light chains.

Light chain disease is characterized as follows: (1) On serum protein electrophoresis, there is absence of M-spike but hypogammaglobulinemia is found as in multiple myeloma. (2) The peripheral smear fails to show rouleaux formation because of the absence of hypergammaglobulinemia. (3) Renal disease with elevated BUN and creatinine may be the initial laboratory findings. (4) 2.5% of cases are associated with amyloidosis. (5) Thrombocytopenia, anemia and leukopenia are absent.

BENZODIAZEPINES (VALIUM, LIBRIUM AND SERAX)

SPECIMEN: Random urine
REFERENCE RANGE: Negative
METHOD: EMIT
INTERPRETATION: The benzodiazepines are used primarily as therapy for anxiety and insomnia and are the most commonly prescribed drugs for this purpose. The drugs in this class are listed in the next Table; (Mack, R.B., N.C.M.J., pgs. 505-506 July, 1982):

Common Benzodiazepines			
Drug	Common Use	Half-Life (hours)	Long-Acting (L); Short-Acting (S)
Diazepam (Valium)	Anti-anxiety, Anti-convulsant	20-50	L
Chlordiazepoxide (Librium) Chlordiazachel, A-poxide, SK-Lygen	Anti-anxiety	5-30	L
Flurazepam (Dalmane)	Hypnotic	24-100	L
Chlorazepate (Tranxene)	Anti-anxiety	36-200	L
Chonazepan (Clonipin)	Anti-convulsant	24-48	L
Lorazepam (Ativan)	Anti-anxiety	10-20	S
Oxazepam (Serax)	Anti-anxiety	5-10	S
Prazepam (Centrax)	Anti-anxiety	36-200	L
Temazepam (Restoril)	Anti-anxiety	9-12	S
Halazepam (Paxipam)	Anti-anxiety	7	S
Alphrazolam (Xanax)	Anti-anxiety	12-19	S
Triaxolam (Halcion)	Hypnotic	2.3	S

The most commonly prescribed benzodiazepines are Valium, Librium and Dalmane; Valium is one of the most commonly prescribed drugs in the United States. In the blood, the drugs are strongly bound to protein (85%-95%) and are metabolized in the liver. Most of the benzodiazepine derivatives, except oxazepam and lorazepam, have relatively long half-lives, 30-200 hours.

Death is rare in overdose.

Diazepam: Diazepam acts rapidly when taken as a single dose; it is absorbed rapidly from the gastrointestinal tract, and since it is soluble in lipids, it quickly reaches therapeutic concentrations in the central nervous system. Diazepam is metabolized by oxidation and has active metabolites; it is "long-acting" with a half-life greater than 24 hours. It may be more hazardous for the elderly and for patients with liver disease.

Current Status of Benzodiazepines: The current status of benzodiazepines has been reviewed (Greenblatt, D.J. et al, N. Engl. J. Med. 309, 354-358; 410-416, 1983).

Beta-1c/Beta-1a Globulin

BETA-1c/BETA-1a GLOBULIN (see COMPLEMENT C3)

Beta-2-Microglobulin

BETA-2-MICROGLOBULIN

SPECIMEN: Serum, CSF, and Urine; for serum specimen, use red top tube, separate serum and freeze; 1 ml CSF and freeze; 1 ml of random urine, freeze.
REFERENCE RANGE: Serum: up to 4.45 mg/liter; Urine: up to 475 mcg/liter.
METHOD: RIA and EIA available from Pharmacia Diagnostics.
INTERPRETATION: Beta-2-microglobulin is used to diagnose leukemia or lymphoma of the central nervous system (Mavligit, G.M. et al, N. Engl. J. Med. 303, 718-722, 1980). The simultaneous measurement of CSF and serum level of beta-2-microglobulin is useful in detecting central nervous system involvement with leukemia or lymphoma. When the CSF level of beta-2-microglobulin was significantly higher than the serum level in patients with acute leukemia and lymphoma, the CNS was involved.

Beta-2-microglobulin is a cell-membrane component closely associated with HLA antigens; elevated serum levels of beta-2-microglobulin is believed to reflect increased turnover of the lymphopoietic cells.

Elevated serum levels of beta-2-microglobulin has been detected in a patient with acquired immunodeficiency syndrome (AIDS) (Francioli, P., and Clement, F., N. Engl. J. Med. 307, 1402-1403, 1982).

S. Bakerman

Beta-2-microglobulin has a molecular weight of 11,800 and is partially filtered at the glomerulus; after glomerular filtration, it is normally 99.9% reabsorbed. In tubular injury, urinary excretion of beta-2-microglobulin is increased and the urinary excretion of beta-2-microglobulin has been used as a marker of renal tubular cell injury. This protein is also elevated in the urine in patients with renal parenchymal disease (Sherman, R.L. et al., Arch. Intern. Med. 143, 1183-1185, June 1983).

BETA-GLUCOSIDASE, (GLUCOCEREBROSIDASE)
(FOR GAUCHER'S DISEASE)

SPECIMEN: The preferred specimen is the fibroblast. Obtain a skin biopsy (4mm punch). Use culture media for transport; maintain specimen at 4°C.
    White blood cells are used as an alternate specimen.
REFERENCE RANGE: Normal enzyme activity
METHOD: Culture; measure enzyme; report results in terms of units/gram cellular protein; the enzyme assay reaction is as follows:

4-Methylumbelliferyl- + H_2O $\xrightarrow{\text{Beta-Glucosidase}}$ D-Glucose + 4-Methyl-
    Beta-D-Glucoside   umbelliferone
                                                        (Fluorescent)

The activity of beta-glucosidase is measured by following the release of the fluorescent compound 4-methylumbelliferone.
INTERPRETATION: Gaucher's disease is the most common lipid storage disease; it is transmitted by an autosomal recessive gene. The carrier rate is 1/25; 1/625 couples are at risk; their risk is 1/4; thus, the incidence is 1/2,500. The disease affects Jews of eastern European origin; the incidence in other groups is not well documented. There is a deficiency of the enzyme, glucocerebrosidase, which catalyzes the reaction:

Beta-D-Glucosyl Ceramide + H_2O $\xrightarrow{\text{Glucocerebrosidase}}$ D-Glucose + Ceramide

The glucolipid, ceramide-glucose (glucocerebroside) accumulates in reticuloendothelial cells of the liver, spleen and bone marrow causing enlargement of the liver and spleen and erosion of the cortices of the long bones and pelvis.
    There are three forms of the disease, Type I: Adult; Type II, Infantile; and Type III: Juvenile.
Type I; Adult: Most patients with adult-type Gaucher's disease are asymptomatic. The most common symptoms are progressive splenic and hepatic enlargment, leukopenia, thrombocytopenia, bleeding tendency, bone pain, fractures and anemia. Neurologic involvement does not occur. There is a moderate reduction in the enzyme beta-glucosidase(10-40%) of mean normal values. Increased serum acid phosphatase activity (non-prostatic) is a frequent finding in Type I.
    The Gaucher cell may be seen on bone marrow; the macrophages have fibrillated cytoplasm, characteristic of Gaucher cell.
Type II, Infantile: The usual onset occurs acutely at 2-3 months; neuropathic abnormalities are severe and death usually occurs within two years.
Type III; Juvenile: The onset is subacute; neuropathic involvement occurs.
    In Type II and Type III, there is a marked reduction of the enzyme, beta-glucosidase, in tissues.

BETKE-KLEIHAUER STAIN (see FETAL HEMOGLOBIN STAIN)

BILIRUBIN, AMNIOTIC FLUID (AMNIOTIC FLUID BILIRUBIN)

SPECIMEN: 2ml amniotic fluid. The specimen is centrifuged for 10 minutes and the supernatant is removed; the specimen is kept in the dark prior to scan since bilirubin is unstable to light.

REFERENCE RANGE: See below

METHOD: Spectrophotometric scan of amniotic fluid between 300nm and 600nm. Bilirubin absorbs at 450nm. In order to obtain the absorbance at 450nm, draw a straight line between the absorbance at 550nm and 365nm. Draw a vertical line at 450nm from the top of the absorbance curve to intersect the straight line. The height from the curve to the straight line at 450nm is the absorbance and is called ⌴450nm. A scan of amniotic fluid from 600nm to 300nm and clinical significance of   absorbance is shown in the next Figure:

Spectrophotometric Scan of Amniotic Fluid and Clinical Significance of Absorbance Values

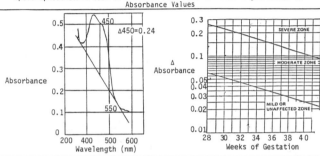

INTERPRETATION: The   450nm of 0.24 after 32 weeks of gestation indicates severe fetal distress.

The presence of fetal erythroblastosis fetalis in-utero is assessed by assay of amniotic fluid for bilirubin pigment; the amount of bilirubin pigment reflects the degree of hemolysis in the fetus. Erythroblastosis fetalis is due to Rh(-) mother and Rh(+) fetus, and occurs as follows: The mother, on exposure to Rh(+) fetal red blood cells, develops antibodies to the Rh(+) antigens; the antibodies pass from the maternal blood into the fetus. In the fetus, the antibodies combine with the Rh antigens of the fetal red blood cells; complement combines with antibody. Macrophages or lymphocytes attach to the Fc region of the immunoglobulin of the red blood cells. Sphering and then lysis of the red cell occur; antibody-coated red cells may also be destroyed by polymorphonuclear leukocytes or by phagocytes. The pathogenesis of erythroblastosis fetalis is illustrated in the next Figure:

Lysis of Red Blood Cells in Erythroblastosis Fetalis

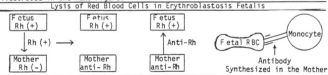

Diagnosis of erythroblastosis fetalis in the newborn is made as follows: mother, Rh(-); infant, Rh(+); Coombs test (+), then jaundice is due to Rh imcompatibility.

BILIRUBIN, DIRECT (CONJUGATED, SOLUBLE)

SPECIMEN: Red top tube

REFERENCE RANGE: 0-0.3mg/dl. To convert from conventional units in mg/dl to international units in micromol/liter, multiply conventional units by 17.10.

METHOD: Reaction of bilirubin with diazonium salt

INTERPRETATION: Direct bilirubin is elevated in the conditions listed in the next Table:

Causes of Elevated Serum Direct (Conjugated) Bilirubin
Liver Disease:
Hepatocellular Disease
Biliary Tract Obstruction, Extrahepatic or Intrahepatic
Dubin-Johnson Syndrome
Rotor's Syndrome
Hepatic Storage Syndrome

Conjugated bilirubin is elevated in hepatocellular disease, biliary tract obstruction and some conditions involved in bilirubin metabolism.

Direct bilirubin is bilirubin conjugated with glucuronic acid at the smooth endoplasmic reticulum of the hepatic cell to form bilirubin-diglucuronide. This form of bilirubin is soluble in aqueous media. Healthy subjects have no detectable conjugated bilirubin; however, technical problems with available assays explain why laboratories detect some conjugated bilirubin. Direct conjugated bilirubin is the most sensitive test in the diagnosis of liver disease.

BILIRUBIN, TOTAL

SPECIMEN: Red top tube, separate serum

REFERENCE RANGE: 0.1 to 1.4mg/dl. To convert from conventional units in mg/dl to international units in micromol/liter, multiply conventional units by 17.10.

METHOD: Bilirubin + Diazotized Sulfonilic Acid $\longrightarrow$ AzoDye (600nm)

INTERPRETATION: This test is useful in detecting and following liver disease and hemolytic disorders and is elevated in the conditions listed in the next Table:

Causes of Increased Serum Bilirubin
Cholestasis; Intrahepatic and Extrahepatic
Liver Disease
Congenital Hyperbilirubinemia:
Crigler-Najjar, Dubin-Johnson, Gilbert's Disease
Hemolytic Anemia
Reabsorbing Extravasated Blood
Malnutrition
Starvation - 24+ hours
Infection - Subacute and Chronic
Hyperthyroidism
NOTE: Increased with Exercise, Estrogens, Oral Contraceptives,
Menstruation, Hemolysis

The major source of bilirubin is hemoglobin catabolism resulting either from the destruction of adult circulating red blood cells or as a result of ineffective erythropoiesis within the bone marrow. In normal man, the sources of bilirubin are listed in the next Table; (Berk, P.D. et al. J. Lab. Clin. Med. 87, 967, 1976):

Sources of Bilirubin	
Source	Percent
Death of Senescent, Circulating Red Blood Cells	70
Ineffective Erythropoiesis	10
Other Principally Hepatic	20

BILIRUBIN, TOTAL (NEONATE)

SPECIMEN: 2 capillary tubes

REFERENCE RANGE: Newborn: Less than 12mg/dl decreasing to adult levels by end of first month. To convert from <u>conventional</u> units in mg/dl to <u>international</u> units in micromol/liter, multiply conventional units by 17.10.

METHOD: Neonatal Serum + PO_4^{-3} (buffer) $\longrightarrow$ Absorption 452nm.
Absorbance at 452nm is due to bilirubin and, if present, hemoglobin. At 540nm, bilirubin does <u>not</u> absorb while <u>hemoglobin</u> exhibits the <u>same absorbance</u> as it does at 452nm. The use of 540nm as the blanking wavelength, eliminates any hemoglobin contribution from the total absorbance at 452nm.

INTERPRETATION: The <u>causes</u> and characteristics of the conditions associated with jaundice in the <u>newborn</u> are given in the next Table:

Causes and Characteristics of the Conditions Associated with
Jaundice in The Newborn

Causes	Onset of Jaundice	Comments
Physiological Jaundice	After 24 hours of birth	
Red Cell Incompatibility	Within 24 hours of birth	Diagnosis: mother, Rh(-); infant, Rh(+), Coombs test positive; measure bilirubin every 5 to 8 hours and plot values.
Infection (Septicemia):	Usually after 4th day of birth	Commonly infective causes are septicemia and urinary tract infection.
Low-Intake of Milk by Breast-Fed Infants		Resolves at end of one week
Other Causes: Viral Hepatitis Atresia of the Bile Ducts Hypothyroidism Galactosemia		

Hepatic Immaturity: Hepatic immaturity or physiological jaundice is due to a temporary deficiency of the enzyme glucuronyl transferase; there is increased unconjugated bilirubin. In <u>full-term infants</u>, the jaundice <u>always appears after the first 24 hours of life</u> and reaches a peak on the 4th or 5th day; in <u>preterm</u> infants, it usually begins 48 hours after birth, peaks at day 5, and may last up to 2 weeks.

Red Cell Incompatibility: The main causes of red cell incompatibility jaundice are <u>Rh incompatibility</u> e.g., Rh(-) mother, Rh(+) fetus; and <u>ABO incompatibility</u>, the mother's blood is usually group 0 and the infant's group A or group B.

Infection: The common causes of <u>jaundice</u> and <u>infections</u> in the newborn are <u>septicemia</u> and <u>urinary tract infection</u>; jaundice usually begins after the 4th day of life; in urinary tract infections, the jaundice is of hepatic origin.

Low Intake of Milk by Breast-Fed Infants: This type of jaundice usually resolves at the end of the first week, but fatalities have been reported in this condition; it is <u>caused by a deficiency of vitamin K</u> and is due to deficiency of vitamin K dependent clotting factors. This condition is characterized by spontaneous hemorrhage in an otherwise well baby. This disorder is almost completely confined to breast fed babies. The clotting defect is rapidly corrected by administration of vitamin K; formula milk contains substantially more vitamin K than human breast milk. In some places, vitamin K is given routinely to every newborn baby (McNinch, A.W., The Lancet, pgs. 1089-1090, May 14, 1983; Letters to the Editor, The Lancet, pgs. 1439-1440, June 25, 1983).

Rare cases of jaundice are as follows: <u>hypothyroidism</u>, <u>galactosemia</u>, <u>viral</u> <u>hepatitis</u>, and <u>atresia of the bile ducts</u>; <u>jaundice lasts more than 10 days in these conditions</u>. Glucose-6-phosphate dehydrogenase (G-6-P-D) deficiency produces a clinical picture similar to blood group incompatibility. In galactosemia, the urine gives a positive result on testing with Clinitest (Ames: oxidation-reduction reaction-test for reducing substances) but the Clinistix test may be negative

(Ames: glucose + O_2 + H_2O + $\xrightarrow{\text{Glucose}}$ H_2O_2 + gluconic acid;
                                  Oxidase

H_2O_2 + reduced orthotolidine $\xrightarrow{\text{Peroxidase}}$ oxidized orthotolidine+H_2O).

G-6-P-D deficiency occurs most often in infants of African, Mediterranean or Chinese descent.

Treatment: If the serum bilirubin concentration is greater than 15mg/dl, exchange transfusion or phototherapy may be needed. Phototherapy converts bilirubin into a colorless compound which does not have long-term effects on the infant.

Bilirubin Encephalopathy; (Karp, W.B., Pediatrics 64, No. 3, 361-368, 1979; Gitzelmann-Cumarasamy, N., and Kuenzle, C.C., Pediatrics 64, No. 3, 375-377, 1979): Theoretically, the level of free bilirubin in blood should be the most sensitive indicator of kernicterus, since only free bilirubin can directly affect the brain cells by diffusion across cell membranes; free bilirubin is soluble in the fat of cell membranes. It is thought that with increasing bilirubin, the bilirubin is increasingly bound to albumin until the three binding sites (one high affinity and two low affinity binding sites) are occupied; excess bilirubin appears free in the plasma and is then deposited in the brain. The relationship of free bilirubin is shown in the next Figure:

Free Bilirubin

Red Blood Cells

Hemoglobin
↓
R.E. System
↓
Albumin:High Affinity Binding Sites
Albumin:Low Affinity Binding Sites
Albumin:Non-Specific Binding Sites
↓ Excess Bilirubin
Free Plasma Unconjugated
Bilirubin
Brain   Liver

Predictive Value for Neonatal Bilirubin Encephalopathy: The possible parameters that may yield information regarding predictive value for neonatal bilirubin encephalopathy are given in the next Table:

Possible Parameters for Neonatal Bilirubin Encephalopathy
Total Bilirubin
Bilirubin Loosely Bound to Albumin
Free Bilirubin
Binding Capacity of Albumin for Bilirubin
Bilirubin Bound to Albumin

Total serum bilirubin is the easiest parameter to measure. Kernicterus is frequent when total serum bilirubin exceeds 30 mg/dl but rare when it is less than 20 mg/dl. However, bilirubin encephalopathy may occur at values lower than 20 mg/dl, especially in sick, low birth weight premature infants.

Free bilirubin, the value of which might be most useful to predict kernicterus, cannot be reliably estimated. A large number of tests have been investigated, and references to these may be found in the articles by Karp, W.B. (Pediatrics 64, 361-368, 1979 and Gitzelmann-Cumarasamy, N. and Kuenyle, C.C., Pediatrics, 64, No.3, 375-377, 1979).

BIOLOGICAL MARKERS FOR DEPRESSION

Biological markers for depression are listed in the next Table; (Keffer, J.H., Am. Soc. Clin. Pathologists, Check Sample, Clin. Chem. No. CC83-8, 23, No. 8, 1983; Gold, M.S. et al., JAMA 245, 1562-1564, 1981):

Biological Markers for Depression
Dexamethasone Suppression Test
Thyrotropin-Releasing Hormone(TRH) Stimulation Test
Prolactin Response to Thyrotropin-Releasing Hormone(TRH)
3-Methoxy-4-Hydroxyphenylglycol(MHPG) Urine
Platelet Monamineoxidase(MAO) Activity
Phenylacetic Acid(PAA) Urine
Conditions Associated with Psychiatric Manifestations:
Hyperthyroidism and Hypothyroidism (Gold, M.S. et al., JAMA 245, 1919-1922, 1981)
Primary Hyperparathyroidism (Tibblin, S., et al., Ann. Surg. 197, 135-138, 1983)
Cushing's Syndrome
Acute Intermittent Porphyria(AIP)
Vitamin B_{12} Deficiency (Spivak, J.L., Arch. Intern. Med. 142, 2111-2114, 1982; Evans, D.L., et al., Am. J. Psychiatry 140, 218-221, 1983)

Dexamethasone Suppression Test: The dexamethasone suppression test has been used to distinguish endogenous from nonendogenous depression (see DEXAMETHASONE SUPPRESSION TEST AS DIAGNOSTIC AID IN DEPRESSION)

Thyroid Releasing Hormone(TRH) Stimulation Test: The TRH stimulation test has been used to distinguish unipolar from bipolar depression. In this test, TRH is injected intravenously; serum specimens are obtained at 15 to 20 minute intervals for 1 hour (see THYROID STIMULATING HORMONE(TSH). Agitated patients typically demonstrate a blunted TRH response, a reaction also characteristic for 77% of unipolar depressives; 17% of patients with bipolar depression demonstrate a blunted TRH response (Loosen, P.T. and Prange, A.J., Jr., Am. J. Psychiatry, 139, 405-416, 1982; Kirstein, L., et al., J. Clin. Psychiatry 43, 191-193, 1982; Sternbach, H.A. et al., JAMA 249, 1618-1620, 1983).

Prolactin Response to Thyrotropin-Releasing Hormone(TRH): (see PROLACTIN; Judd, L.L. et al., Arch. Gen. Psychiatry 39, 1413-1416, 1982).

3-Methoxy-4 Hydroxyphenylglycol(MHPG) in Urine: The urinary level of MHPG reflects the central nervous system metabolism of norepinephrine and has been correlated with types of depression (Schildkraut, J.J. et al., Psychopharma. Bull. 17, 90-91, 1981; Rosenbaum, A.H. et al., Am. J. Psych. 140, 314-317, 1983).

Platelet Monamineoxidase(MAO) Activity: see Weinshilboum, R.M., Mayo Clin. Proc. 58, 319-330, 1983.

BLASTOMYCES ANTIBODY, SERUM

SPECIMEN: Red top tube, separate and then refrigerate serum. If titer is obtained, paired sera advisable to detect rising titer.

REFERENCE RANGE: Negative

METHOD: The immunodiffusion(ID) and complement fixation(CF) tests are used to detect antibody. Yeast- and mycelial-phase antigens are used; yeast-phase antigens are probably more reliable (Penn, R.L. et al., Arch. Intern. Med. 143, 1215-1220, 1983).

Complement Fixation(CF) Test: A titer of 1:8 is suggestive of infection; a titer of 1:32 or greater is indicative of active disease. The sensitivity of this test is less than 50 percent; false positive results have been obtained in patients with histoplasmosis and coccidiomycosis immitis.

Immunodiffusion(ID): The sensitivity of this test has been improved and approaches 80 percent; false-positive results have not been detected (Williams, J.E., Murphy, R., Standard, P.G., Phair, J.P., Serologic Response in Blastomycosis; Diagnostic Value of Double Immunodiffusion Assay, Am. Rev. Respir. Dis. 123, 209-212, 1981).

INTERPRETATION: Blastomyces dermatitidis is a fungus which causes pulmonary blastomycosis when inhaled. The fungus may form a solitary focus of consolidation and extend to regional lymph nodes; the fungi may then disseminate throughout the lung. Systemic dissemination may lead to infection in the skin, brain, bones and other tissues in the body; the skin is involved in 25 percent of cases.

BLEEDING TIME

SPECIMEN: This test is done at the bedside. The test is not performed if the patient has taken aspirin or an aspirin containing preparation within 5 days of the test. It is not done if the platelet count is less than 50,000/ul.

REFERENCE RANGE: 2-6 minutes; the reference interval for children is 2-8 minutes.

METHOD: (1) Place a blood pressure cuff on the arm above the elbow and inflate to 40mmHg. (2) Clean an area which is free of visible veins on the forearm with alcohol. (3) Puncture the skin with a sterile blade (Bard-Parker No.11) using adhesive tape fixed to the blade to allow a wound 2mm deep. A mechanical device, the Simplate from General Diagnostics, permits a standard incision, 5 mm long and 1 mm deep. (4) With a round piece of filter paper blot off the blood every 30 seconds. Only the tip of the drop should touch the filter paper.

INTERPRETATION: The bleeding time is an in-vivo test of platelet function; the bleeding time is proportional to the platelet count below 100,000/ul; it is prolonged in the conditions listed in the next Table:

Prolongation of Bleeding Time
Low Platelet Count
Abnormal Platelet Function
Glanzmann's Thrombasthenia
Uremia and Myeloproliferative Diseases
Aspirin Ingestion Within 5 Days;
other drugs, i.e., Phenylbutazone,
Indomethacin (Indocin),
and Moxalactam (Cephalosporin),
high doses of Penicillin, Carbenicillin
and other Beta-Lactam Antibiotics
von Willebrand's Disease or Syndrome

When platelet counts are low, the expected bleeding time is calculated from the formula; (Harker, L.A. and Slichter, S.J., N. Engl. J. Med. 287, 155, 1972).

$$\text{Bleeding Time} = 30.5 - \frac{\text{Platelet Count/Cubic mm}}{3850}$$

A bleeding time longer than that calculated from platelet numbers alone suggests defective function in addition to reduced number. A bleeding time shorter than that calculated suggests the presence of active young platelets.

The bleeding time is prolonged in patients following aspirin ingestion and von Willebrand's disease or syndrome.

In Glanzmann's thrombasthenia, there is a normal platelet count but abnormal clot retraction and prolongation of bleeding time.

Moxalactam can prolong bleeding time and may result in clinical bleeding in certain patients (Weitekamp, M.R. and Aber, R.C., JAMA 249, 69, Jan. 7, 1983).

BLOOD COMPONENT THERAPY

The blood components available for transfusion and the indications for the use of these components are given in the next Table:

Blood Component	Indications
Red Cells	Chronic Anemia; Slow Blood Loss; Acute Blood Loss up to 30% Blood Volume; Exchange Transfusions.
Red Cells, Washed, Leukocyte and Plasma Poor	Same as for Red Cells; Leukocyte Poor: Prevent Febrile Reaction Due to Leukocyte Antibodies; Plasma Poor: Prevent IgA Sensitization.
Red Cells, Frozen-Thawed	Same as above plus Autotransfusions and Rare Blood Types.
Whole Blood	Massive Blood Loss
Plasma, Fresh Frozen	Treatment of Bleeding Patients with Multiple Factor Deficiencies; Hypovolemia
Albumin, 25% (25g/dl)	Burns; Exchange Transfusion of Infancy; Prevention and Treatment of Cerebral Edema
Albumin, 5% (5g/dl)	Burns, Shock
Platelet Concentrates	Bleeding due to Thrombocytopenia or Platelet Function Abnormality
Granulocyte (Leukocyte) Concentrate	Granulocytopenia with Sepsis
Cryoprecipitate	Factor Deficiency: Hemophilia A, von Willebrand's Disease and Factor XIII Deficiency; Hypofibrinogenemic States
Factor VIII Concentrate	

Red cell transfusions have largely superceded the use of whole blood. All blood products must be administered through a filter.

Preoperative Crossmatch Guidelines: For patients without preoperative complications undergoing elective surgery, the following guidelines are recommended: Blood is collected from the patient and testing is done for ABO, Rh(D) and antibody screen; cross-match is not recommended. If the patient should require transfusion, blood of the same ABO group and Rh(D) type is selected; on initial spin, crossmatch is performed and the blood released in approximately 10-15 minutes. The standard crossmatch is completed while the patient is being transfused. This approach allows for a reduction in blood inventory and patient costs without sacrificing patient safety.

The ^{51}Cr-labeled red blood cell survival test has been used to evaluate in-vivo survival of a small sample of donor red blood cells in patients with blood compatibility problems (Pineda, A.A. et al., Mayo Clin. Proc. 59, 25-30, 1984).

RED CELLS: Red blood cells are prepared by centrifugation or sedimentation of a unit of whole blood from a hematocrit of 40% to a hematocrit of 70% to 80%. Red blood cells transfusions are indicated for chronic anemias, slow blood loss, acute blood loss up to 30% blood volume and for exchange transfusions.

Red blood cell transfusions are superior to whole blood transfusions for patients with cardiac disease, chronic anemia and for patients with liver or kidney disease who require restricted sodium or citrate intake.

A unit (hematocrit, 70 to 80%) which contains 200ml of red blood cells, should increase recipients hematocrit about 3% in a 70kg adult. Cells must be administered through a filter; do not exceed 4 hours/unit.

The side effects and hazards of red blood cell transfusion are like that of the transfusion of whole blood.

The number of units of red blood cells required for transfusion to achieve a desired hematocrit in a nonbleeding patient is calculated from the formula:

$$\frac{(\text{Blood vol. x Hct}) + (\text{No. of Units Transfused x 200ml}) \times 100}{\text{Blood Vol. + No. of Units Transfused X 300ml}} = \text{New Hct.}$$

In this formula, the volume of red blood cells is 300ml and the red cell mass is 200ml.

ADSOL(AS-1) Red Blood Cells: AS-1 red blood cells is a new preparation of red blood cells; ADSOL refers to ADditive SOlution System; ADSOL has the following advantages:

(1) This preparation has an improved red blood cell survival with extension of red blood cell shelf life to 49 days.

(2) The infusion rate is similar to whole blood; it is <u>not</u> necessary to add
    saline at the time of transfusion.
(3) When the product is prepared from the donor, the yield of platelets and
    plasma factors is increased.

The system is generically called Additive Solution System because it uses a
<u>second</u> preservative solution in addition to the anticoagulant solution, citrate
phosphate dextrose(CPD) needed for whole blood collection. The additive
consists of denine in 100ml. of normal saline, with a small amount of citrate
anticoagulant remaining. The plasma is separated from the red blood cells prior
to the addition of the additive solution. Patients with documented mannitol
hypersensitivity should not receive this product.

Because of the small amount of mannitol (4% of the usual therapeutic dose)
and the slightly increased volume per unit (60ml.), rare patients with <u>cardiac
or renal failure</u> may require volume reduction just prior to transfusion. In
these patients, the order must be written as "<u>volume reduced red blood cells.</u>"
The AS-1 unit of red blood cells will be centrifuged in the Blood Bank just
before transfusion and most of the liquid removed, resulting in a product with a
hematocrit of about 80% which will not flow nearly as well as the AS-1 units.
An order for packed cells without specifying "volume reduced" will be filled
with the usual AS-1 RBC's.

A comparison of AS-1 RBC's with CPDA-1 RBC's and CPDA-1 whole blood is shown
in the next Table:

Comparison of AS-1 RBC's with CPDA-1 RBC's and CPDA-1 Whole Blood			
Parameters Measured After 35 Days Storage	AS-1 RBC's	CPDA-1 RBC's	CPDA-1 Whole Blood
Maximum Storage Period, Days (FDA)	35	35	35
Volume per Unit	310	250	510
Hematocrit, Percent	60	75	40
Viscosity, Relative to Whole Blood	1.2	2.6	1.0
Post-Transfusion Survival (Percent Recovery)	86	78	76
Supernatant Hgb (mg/dl)	112	900	43
Percent of RBC's Lysed During Storage	0.2	0.9	0.2
Supernatant Potassium (mEq/liter)	45	86	27
pH	6.70	6.58	6.72
ATP, Percent of Day Zero	74	52	62
Glucose, Percent of Day Zero	65	15	2

CP2D(AS-1 Nutricel): This new preparation has an improved red cell survival
of 35 days; there is increase in dextrose concentration. The additive consists
of saline, dextrose and adenine.

RED CELLS, WASHED (LEUKOCYTE AND PLASMA-POOR):
Leukocyte Poor: Washed red cells help to prevent <u>febrile reactions</u>. Recipients
who have had multiple <u>transfusions</u> or <u>pregnancies</u> may develop <u>antibodies</u> to <u>HLA</u>
or other <u>antigens</u> on <u>leukocytes</u>. Subsequent transfusions with cells carrying
these antigens may cause a <u>febrile</u> (non-hemolytic) <u>reaction</u> in 0.5% to 1% of
<u>transfusions</u>. After washing, less than 20% of the leukocytes remain and the
unit contains more than 80% of the original red blood cells. Leukocyte poor
preparations should be given to patients that <u>require frequent transfusions</u>,
such as <u>leukemia</u> and <u>aplastic anemia</u>.
Plasma Poor: Three conditions in recipients are prevented by removing plasma;
these are listed in the next Table:

Conditions Prevented by Using Plasma-Poor Red Cells	
Component(s) Removed	Condition Prevented
IgA	Anaphylactic Reaction in IgA Deficient Patients: IgA deficient patients develop antibodies to IgA following exposure to IgA in a previous transfusion.
Complement (C-3)	Hemolytic Episodes in PNH: Patients with paroxysmal nocturnal hemoglobinuria (PNH) have an acquired intrinsic defect in that their red blood cells bind more C-3 than do normal cells.
Electrolytes and Metabolic Products	Electrolyte imbalance

Each unit contains 185ml packed red blood cells in a 200-250ml volume. There are two disadvantages to the use of plasma poor red cells; one is the cost and the other is the shelf-life which is only 24 hours.

DEGLYCEROLIZED RED CELLS (FROZEN-THAWED RED BLOOD CELLS): Red blood cells may be frozen for up to 3 years with glycerol. These preparations are particularly useful for autotransfusions and rare blood types.

WHOLE BLOOD: (Volume 520 ± 45ml, Hct. 40%; smaller volumes for pediatric patients; administer through a filter) (Donor and recipient must be ABO identical crossmatch): Whole blood is given to patients who have had massive blood loss or who are bleeding profusely. Whole blood must be through a filter; transfusion should not be slower than 4 hours/unit.

FRESH FROZEN PLASMA (SINGLE DONOR): Fresh frozen plasma, which may be stored up to one year, is used in the treatment of bleeding patients with multiple clotting factor deficiencies; fresh frozen plasma contains XI, IX, VIII, X, VII, V, II and XIII. Stored whole blood also contains all of these factors but factors V and VIII are unstable in whole blood and are markedly decreased in activity by the second or third day of storage. Infusion of a unit of fresh frozen plasma causes all coagulation factors to rise about 8 percent; fibrinogen level rises about 13mg/dl. Fresh frozen plasma is stored at -20°C or below and requires 30 minutes to thaw in a 37°C water bath.

Fresh frozen plasma is used in the treatment of patients with conditions listed in the next Table:

Conditions Treated with Fresh Frozen Plasma
Disseminated Intravascular Coagulopathy (DIC)
Other Defibrination Syndromes
Liver Disease

The disadvantages of the use of fresh frozen plasma are as follows: excessive volume causing circulatory overload, possibility of hepatitis, allergic reactions with chills and fever and blood group antibodies; thus, fresh frozen plasma must be ABO-group specific with the recipients red blood cells.

Deficiency of factor VIII and fibrinogen (factor I) alone should be treated with cryoprecipitated factor VIII concentrates or purified factor VIII concentrates.

Single donor plasma is used instead of pooled plasma because of the high incidence of hepatitis in pooled plasma. The plasma is administered through a filter.

ALBUMIN (25% and 5%): Albumin (25%, 25g/dl) is used in the treatment of burns and other causes of fluid loss to counteract fluid and sodium loss. The dosage is given so as to maintain the circulating plasma level of albumin at 2.5 ± 0.5g/dl (Tullis, J.L., JAMA 237, 460, 1977).

Albumin, 25%, may be given in the exchange transfusion of neonates with hyperbilirubinemia to help protect against bilirubin toxicity (kernicterus). It may also be used in the prevention and treatment of cerebral edema.

Albumin (5%, 5g/dl) is used in the treatment of burns and other causes of fluid loss. It may be used in the early treatment of shock due to hemorrhage until blood is available. The usual minimal effective adult dose is 250-500ml. (Borucki, D.T., Blood Component Therapy, Am. Ass. Blood Banks, 3rd Ed., 1981).

PLATELET CONCENTRATES: Platelet concentrates are given to patients who are bleeding due to thrombocytopenia (<80,000) or due to abnormal platelet function (thrombocytopathia); or if platelet counts are below 20,000. One platelet

concentrate increases the platelet counts of a patient with a <u>5000ml blood</u>
volume by <u>5,000 per cu mm</u>; the usual dose is six to eight units of platelet
concentrates.

ABO and Rh compatibility are desirable because of red blood cell
contamination.

Platelet concentrates are <u>not</u> given to patients with thrombocytopenia
caused by accelerated platelet destruction, e.g., <u>thrombocytopenic purpura</u> or
<u>disseminated intravascular coagulation (DIC)</u>.

<u>Transfusion risks are hepatitis, immunization</u> to HLA and red blood cell
antigens, febrile reaction, <u>allergic reactions</u> with urticaria.

Advanced notification when possible is recommended due to short life span
of platelets. Life span is 3-5 days depending on type of bag used for storage.

Platelet concentrates may be obtained by use of <u>plateletpheresis</u>;
plateletpheresis is the centrifugal separation of platelets from whole blood
with either continuous or intermittent return of platelet-poor red blood cells
and plasma to the donor. <u>Plateletpheresis involves some risk to the donor.</u>

<u>GRANULOCYTE (LEUKOCYTE) CONCENTRATE:</u> Granulocyte transfusions are given to
patients who are <u>granulocytopenic</u> (<500 neutrophils/cu mm) and have <u>infections</u>,
e.g., septicemia, that are not responsive to antibiotics or other forms of
therapy. Leukocytes are administered daily, usually for 5 days in succession,
or until WBC rises above 500 granulocytes.

Leukocyte concentrates must be ABO-group compatible and are administered
through a blood filter. Transfusion risks include chills, fever, allergic
reactions, viral hepatitis, cytomegalic virus infection and in immunodeficient
or immunosuppressed patients, graft versus host reaction. The presence of red
blood cells can result in a hemolytic reaction.

Granulocyte concentrates may be obtained by use of <u>leukapheresis</u>;
leukapheresis is the centrifugal separation of leukocytes from whole blood with
either continuous or intermittent return of leukocyte-poor cells and plasma to
the donor. <u>Leukapheresis involves some risk to the donor.</u>

<u>CRYOPRECIPITATE:</u> Cryoprecipitate is a source of Factors VIII, XIII and
<u>fibrinogen</u> and is used to treat the conditions listed in the next Table:

Conditions Treated with Cryoprecipitate
Factor Deficiency
Hemophilia A (Factor VIII Deficiency)
von Willebrand's Disease
Factor XIII Deficiency
Hypofibrinogenemic States, e.g.,
Disseminated Intravascular Coagulopathy (obstetrical complications
e.g., abruptio placenta and amniotic fluid embolism; endotoxin in
bacterial sepsis; severe intravascular hemolysis; carcinomatosis)

The <u>primary use of cryoprecipitate is in the treatment of hemophilia A</u>. It
is more advantageous to use cryoprecipitate rather than fresh frozen plasma
because cryoprecipitate is given in a much smaller volume (about 10ml) than
fresh frozen plasma. Cryoprecipitate contains 250mg of fibrinogen, 80 to 100
units of Factor VIII, some Factor XIII and von Willebrand's factor; 1ml of
normal plasma contains 1 unit of Factor VIII activity.

A patient with hemophilia A who is bleeding usually has a Factor VIII level
of about 2 to 10 percent; the desired therapeutic level is 10 to 100 percent.
Cryoprecipitate should be given every 12 hours to a severely bleeding
hemophiliac or to a hemophiliac who is about to undergo surgery.

<u>Side effects</u> of cryoprecipitate include viral hepatitis, fibrile and
allergic reactions.

Cryoprecipitate may be stored for 1 year at -20°C or below and requires 15
minutes to thaw in a 37°C water bath.

<u>FACTOR VIII CONCENTRATE:</u> Factor VIII is indicated for the treatment of
hemophilia A patients for the prevention and control of hemorrhagic episodes.
In hemophilia, inherited as a sex-linked recessive, there is a deficiency of a
part of the Factor VIII molecule, called Factor VIII C. Factor VIII concentrate
is obtained from a large number of donors and recipients have developed <u>acquired</u>
<u>immunodeficiency syndrome (AIDS)</u>. There is also a very high incidence of non-A,
non-B hepatitis in patients receiving factor VIII concentrate (Andes, W.A., JAMA
<u>249</u>, 2331, May 6, 1983).

TRANSFUSION REACTIONS: Transfusion reaction rates are given in the next Table;
(Goldfinger, D. and Lowe, C., Transfusion 21, 277, 1981):

Transfusion Reaction Rates	
Reaction	Rate (per thousand Units)
Febrile, Nonhemolytic	3.0
Urticarial	1.6
Hemolytic, Delayed	0.7
Hemolytic, Immediate	0.06
	5.36

FEBRILE REACTIONS: Febrile reactions usually occur 1 to 2 hours after the
transfusion has been started and lasts for several hours to 24 hours. A febrile
reaction is defined as a 2 degree increase in body temperature or a one degree
increase in temperature accompanied by shaking chills. The differential
diagnosis of febrile reactions to blood transfusion products is given in the
next Table:

Differential Diagnosis of Febrile Reactions to Blood Products
Reactions to:
Leukocyte Antigens
Platelet Antigens
Plasma Proteins
Hemolytic Reactions
Bacterial Contamination

In the most common cause of a febrile reaction, antibodies in the patient
react with antigens present on donor leukocytes.

Signs and symptoms of febrile reactions are as follows: febrile reactions,
mild to severe; chills, nausea, vomiting, hypotension, cyanosis, tachycardia,
transient leukopenia, chest pain and dyspnea.

Nurses should proceed as follows:
(1) Stop transfusion. Keep I.V. open with slow saline drip.
(2) Recheck patient identification on tie tag and patient arm band, and
recheck blood unit numbers.
(3) Notify physician and blood bank.
(4) Send clotted sample and blood container (without removing recipient
set) to the blood bank and collect a post-transfusion urine specimen
to send to the laboratory marked "Transfusion Reaction Specimen".

Physicians should proceed as follows:
(1) Evaluate patient for possible signs of hemolytic transfusion reaction.
(2) Ordinarily, order the transfusion discontinued.
(3) Administer antihistamine, antipyretics and possibly steroids as
indicated.

Prevention: Use washed (leukocyte poor) red cells for subsequent transfusion;
use pretransfusion antihistamine in patients who have had previous febrile
reactions.

ALLERGIC-ANAPHYLACTOID REACTIONS: Usually, these reactions are mild. Clinical
signs and symptoms include pruritis, urticaria, occasional facial and
periorbital edema; rarely bronchospasm and anaphylactoid shock. Urticaria is
not seen with hemolytic reactions; therefore, a work-up for hemolysis is not
necessary.

Usually, the transfusion is discontinued and a antihistamine is
administered I.V. (diphenyldramine). If laryngeal or bronchospasm occur, give
epinephrine. If anaphylactoid shock occurs, establish an airway, give I.V.
fluid, aminophylline, corticosteroids, and cardio-supportive drugs.

If the reaction is very mild, urticaria only, then slow transfusion rate
and administer antihistamines I.V. If there is no clinical improvement, stop
transfusion.

Prevention: Patients with a history of allergic transfusion reactions should
receive oral Benadryl 30 minutes prior to transfusion. Transfuse packed red
blood cells or washed red blood cells instead of whole blood.

Causes: Most of the allergic reactions are probably caused by many different
proteins.

Anaphylactoid transfusion reactions are rare; they most often occur in
patients with IgA deficiency. About 1 in 850 individuals are deficient in IgA
and about 25 percent have IgA antibodies. The diagnosis is made by finding
anti-IgA in the recipient plasma and IgA in the donor plasma.

81

HEMOLYTIC, DELAYED: These reactions usually occur <u>several days</u> after administration of red cells and are <u>most commonly</u> caused by <u>antibody in the recipient</u> reacting with <u>donor red cell antigens</u>. These reactions occur most often in patients who have had <u>multiple transfusions</u> or <u>women</u> previously sensitized to red cell antigens <u>from pregnancy</u>.

The <u>signs and symptoms</u> of delayed hemolytic reaction are: <u>fever</u>, <u>jaundice</u> and occasionally <u>hemoglobinuria</u>; <u>lack</u> of appropriate <u>increase</u> in <u>hematocrit</u> or <u>hemoglobin</u>.

Obtain blood specimens as follows:

Blood Specimens for Delayed Hemolytic Reaction	
Specimen	Assay
Purple (EDTA) or Blue (citrate) Top Tube	Direct Coombs Test
Red Top Tube	Antibody Screening Test
Red Top Tube	Antibody Titer
Purple (EDTA) Top Tube	Complete Blood Count (Hgb, Hct, RBC, Indices), Smear
Red Top Tube	Bilirubin, Haptoglobin

The management of delayed hemolytic reaction is <u>supportive</u>; treatment of hypotension or renal failure is similar to acute hemolytic reaction. The usual responsibility of the blood bank is to search for antibody in the recipient.

HEMOLYTIC, IMMEDIATE: The most common cause of <u>fatal</u> hemolytic reactions is <u>clerical</u> errors. The most severe hemolytic reactions result from <u>antibody</u> in the <u>recipient</u> reacting with donor cells.

The features are as follows: fever, chills, <u>hypotension</u>, hemoglobinuria, oozing from incisional sites, nausea, anxiety, back pain, tachycardia, <u>oliguria</u>, shock, <u>acute renal failure</u>; death can occur.

Guidelines for evaluation of a suspected immediate hemolytic reaction are as follows:
(1) <u>Stop transfusion immediately</u>; notify and send blood to blood bank; keep I.V. line open with saline.
(2) Recipient: <u>Record and monitor</u> (a) <u>vital signs</u>: temperature, pulse, blood pressure, and respirations; (b) <u>urine output</u> and <u>fluid</u> intake.
(3) Verify that recipient has received proper unit by checking donor numbers against recipient identification bracelet and transfusion request slip.
(4) Return complete I.V. set intact and unused portion of donor unit to blood bank:
    (a) Repeat ABO-Rh typing, antibody screen, compatibility testing
    (b) Bacterial culture
(5) Obtain blood specimens from recipient at a location away from the infusion site:
    (a) <u>Blood Bank</u>: (1 blue (citrate) and red top tube) Repeat ABO-Rh typing, antibody screen, direct Coombs test and donor compatibility testing.
    (b) <u>Hematocrit</u>: Lavender (EDTA) top tube
    (c) <u>Blood Culture</u>: Sterile yellow stoppered tube
(6) Urinalysis for blood assay.
<u>Treatment</u>: <u>Treat shock</u> with vasopressors, appropriate IV fluids and corticosteroids. Start fluid balance sheet, charting intake and output. Maintain fluid balance. Maintain renal blood flow with IV furosemide (20-80mg) and/or mannitol. Prevent propagation of DIC, if present, with 10,000 U heparin, IV. Administer appropriate compatible blood components as necessary, i.e., red blood cells, platelets, cryoprecipitate, fresh frozen plasma. Follow status of patient for possible renal failure requiring renal dialysis.

SIDE EFFECTS AND HAZARDS: Side effects associated with transfusion of whole
blood are given in the next Table:

Side Effects Associated with Blood Transfusion	
Side Effect	Comment
Hemolytic Transfusion Reactions	Incompatibility between donor red blood cells and recipient plasma.
Transmission of Infectious Disease	Viral hepatitis: Eight percent; mostly non-A Non-B hepatitis. Other diseases that may be transmitted are malaria, brucellosis, trypanosomiasis, Epstein-Barr virus, cytomegalovirus, toxoplasmosis, syphilis in the serological negative phase, Colorado tick fever
Immunization of the Recipient	Recipients who have had transfusions or pregnancies may develop antibodies to HLA or other antigens on leukocytes, or platelets or red blood cells. Subsequent transfusions with cells carrying these antigens may cause a febrile reaction in 0.5% to 1.0% of transfusions; or an anamnestic reaction and a delayed hemolytic reaction (autoimmune hemolytic anemia-like) with a positive direct antiglobulin test.
Allergic Reactions	Urticaria occasionally accompanied by chills and fever; fever; 3% of recipients; may be prevented in patients with known history by premedication with antihistamines. Anaphylactoid Reaction: About 1 in 500 individuals lack IgA and develop anti-IgA antibodies; on subsequent transfusion, an anaphylactoid reaction may occur; treat with adrenalin and corticosteroids.

Other Reactions: Circulatory Overload; Iron Overload; Metabolic Complications
Trisodium citrate is an anticoagulant in blood; one unit of whole blood contains
17mEq of citrate; one unit of packed cells contains 5mEq of citrate. Following
transfusions, citrate is metabolized to $[HCO_3^-]$ in equimolar amounts; that
is 17mEq of citrate is converted to 17mEq of $[HCO_3^-]$ and 5mEq of citrate is
converted to 5mEq of $[HCO_3^-]$. Normally, the $[HCO_3^-]$ load is excreted by
the kidney. However, there are certain conditions in which patients receive
many units of blood and in which the excretion of $[HCO_3^-]$ by the kidney is
compromised; these are given in the next Table:

Blood Transfusions and Compromised Excretion of $[HCO_3^-]$ by the Kidney
Conditions in which many Blood Transfusions may be given:
Shock
Trauma
Sepsis
Open heart Surgery
Compromised Excretion of $[HCO_3^-]$ by the Kidney:
(Enhanced Renal HCO_3^- Reabsorption)
ECF Volume Contraction
Potassium Depletion
Hypercapnia
Acute or Chronic Renal Disease
Oliguria

An example of a consultation request for suspected transfusion reaction is
given in the next Figure; (McCord, R.G. and Myhre, B.A., Laboratory Medicine, 9,
39-46, March 1978):

Blood Component
Therapy   S. Bakerman
_____
              Consultation Request for Suspected Transfusion Reaction
           AM
  / /   PM
_____ ml
Date   Time   No. of Units   Vol. Transfused

Clinical Diagnosis Prior to Transfusion_____
_____
Donor No. of Last Unit Given_____   _____
_____

          PRE TRANSFUSION   POST TRANSFUSION
     Temp._____   Temp._____
     B/P_____   B/P_____
     Pulse_____   Pulse_____
_____
                      CHECK THOSE WHICH APPLY
  ___ Pulse   ___ Chest Pain   ___ Back Pain   ___ Urine Output
  ___ B/P   ___ Headache   ___ Jaundice   ___ Heat at I.V. Site
  ___ Chills   ___ Dyspnea   ___ Rash   ___ Pain at I.V. Site
  ___ Fever   ___ Hemoglobinuria   ___ Flushing   ___ Delirium
  ___ Coma   ___ Nausea   ___ Pruritus   ___ Muscle Tenderness
  ___ Syncope   ___ Vomiting   ___ Urticaria   ___ Petechiae
  ___ Other (Specify)_____

_____
     LIST DONOR NO. FROM ALL SUSPECTED UNITS IN THIS TRANSFUSION SERIES

| Donor No. | Date | Time | | Amount | Whole Blood, Packed | Reaction |
		Started	Stopped	Given	cells, Platelets, etc.	

Complete this portion of the form and return to the TRANSFUSION LABORATORY with
post-reaction samples of the patient's urine and blood (clotted). Return the
blood bags from all units in this transfusion series.

     A standard report, used by residents in the Department of Pathology at UCLA
School of Medicine, for febrile transfusion reactions and allergic transfusion
reactions is as follows:
          Standard Consultation Report for Febrile Transfusion Reactions
According to the above information and the chart entry of ___/___/___, this age
year old man-woman-child experienced chills and a fever (degrees F)
during-shortly after the transfusion of a unit of packed cells-whole blood-
platelets. There was no complaint of back pain or signs of dyspnea, shock or
hemoglobinuria as is often the case in a hemolytic reaction. Furthermore,
inspection of pre-transfusion and post-transfusion sera revealed that they were
straw-colored, rather than pink or red as one sees with acute intravascular
hemolysis. Repeat crossmatch and compatibility testing revealed that compatible
blood was administered; vis., ABO, RH type blood was given to a(n) ABO, Rh type
recipient. In addition, antibody screening failed to demonstrate the presence
of any unexpected antibodies. On the basis of these negative findings, we
suspect that the patient experienced a FEBRILE TRANSFUSION REACTION due to
pyrogenic material in the unit of blood he-she received. This material usually
originates in the granulocytes contained in the blood. With regard to future
transfusion of blood to this patient, please read the "Pathology Resident's Note
on Febrile Transfusion Reactions" which has been taped into the progress notes.
     The appropriate choice of underlined words is selected.

          Standard Consultation Report for Allergic Transfusion Reactions
According to the above information and the chart entry of ___/___/___, this age
year old man-woman-child experienced flushing and pruritis and urticaria
during-shortly after the transfusion of a unit of packed cells-whole blood-
plasma-cryoprecipitate. There was no complaint of back pain or signs of
dyspnea, shock or hemoglobinuria as is often the case in a hemolytic reaction.
Furthermore, inspection of pre-transfusion and post-transfusion sera revealed

that they were straw-colored, rather than pink or red as one sees with acute
intravascular hemolysis. Repeat crossmatch and compatibility testing revealed
that compatible blood was administered; vis., ABO, Rh type blood was given to
a(n) ABO, Rh type recipient. In addition, antibody screening failed to
demonstrate the presence of any unexpected antibodies. On the basis of these
negative findings, we suspect that the patient experienced an ALLERGIC
TRANSFUSION REACTION due to allergenic material in the unit of blood he-she
received. With regard to future transfusion of blood to this patient, please
read the "Pathology Resident's Note on Allergic Transfusion Reactions" which has
been taped into the progress notes.

Standard Note on the Causes and Prevention of Febrile Transfusion Reactions
Pathology Resident's Note on Febrile Transfusion Reactions

(If you have not already done so, please see the "Consultation Request for
Suspected Transfusion Reaction", which has been inserted into the CONSULTATION
section of this patient's chart.)

Granulocytic pyrogenic material can be released in two ways. First, the
recipient may have antigranulocytic antibodies (usually leukoagglutinins) in his
blood. These antibodies can attach to receptor sites on the granulocyte
membrane and, with the aid of the complement system, cause lysis of the
granulocyte in vivo and concomitant release of pyrogenic substances. Secondly,
stored refrigerated granulocytes normally undergo autolysis in vitro within 24
hours releasing substances which are pyrogenic. Hence, pyrogenic material can
be present in stored blood in the absence of immune lysis.

The problem of febrile reactions due to granulocytic pyrogenic material*
can be minimized by giving the patient washed packed erythrocytes. However,
washing a fresh unit of packed red cells only removes about 80% of the
granulocytes. The remaining cells (if the unit is used immediately) may be
subject to immune lysis as described above. If the washed packed cells are
stored, the remaining granulocytes will undergo autolysis and again release
pyrogenic material.

Febrile reactions due to granulocytic pyrogenic material can be eliminated
by ordering packed red cells which have been refrigerated or frozen. When
ordering refrigerated packed red cells, be certain to request that the units be
washed. The washing will remove the pyrogenic substance that has been released
by granulocyte lysis. In addition, washing will remove any potassium or
phosphate that has leaked out of the stored red cells, as well as any lactic
acid generated by red cell metabolism. (Washing units of blood introduces the
possibility of bacterial contamination, but we feel that the risk is minimal
since Federal regulations require that blood be used within 24 hours after
thawing and this should preclude significant bacterial growth). If frozen
packed red cells are ordered, they will be routinely washed since washing is
required to remove the cytoprotective agent, glycerol.

When the administration of whole blood is not vital, the BLOOD BANK
recommends the use of frozen packed red cells for the following reasons:

1. These units are extensively washed to remove the glycerol. The
   pyrogenic material from the lysed granulocytes will be removed during
   the washing procedure.
2. The process of freezing and thawing destroys many granulocytes which
   would survive a normal washing procedure.
3. Glycerol is bacteriostatic, therefore greatly decreasing the chance of
   bacterial growth in the unit of blood.
4. The freezing process is believed to attenuate or even destroy hepatitis
   virus.

*Administration of platelets may also cause a febrile reaction, but this
discussion is limited to correction of anemia only.

Standard Note of the Causes and Prevention of Allergic Transfusion Reactions
Pathology Resident's Note on Allergic Transfusion Reactions

(If you have not already done so, please see the "Consultation Request for
Suspected Transfusion Reaction", which has been inserted into the CONSULTATION
section of this patient's chart.)

An allergic transfusion reaction is an example of an "immediate-type"
hypersensitivity reaction. This kind of reaction is antibody-mediated and
occurs within minutes of exposure to an antigen. (This is in contradistinction
to the "delayed-type" hypersensitivity reaction which is cell-mediated and

occurs within hours or days.) Allergic transfusion reactions (flushing, pruritus, urticaria) are most commonly seen in people who are atopic or who have a congenital lack of IgA antibody; however, they may occur in anyone. Approximately 10% of the population are atopic and are sensitive to a variety of environmental allergens. They experience allergic reactions which are mediated by IgE antibodies. The other type of severe allergic transfusion reaction is the interaction between transfused IgA and anti-IgA antibodies in the recipient's plasma. In this case, the recipient has a congenital lack of IgA and has actually formed IgG antibody against the IgA in the transfused blood. About 1 in 500 recipients lack IgA and could theoretically be at risk.

Prevention of allergic transfusion reactions involves elimination of the offending allergen. When correction of anemia is the only hematologic problem facing the clinician, the BLOOD BANK recommends use of <u>washed packed red</u> cells or frozen reconstituted packed cells. The washing process should <u>remove</u> sufficient allergens (vis., environmental allergens or IgA antibody) to prevent an allergic reaction. In addition, washing removes any potassium or phosphate that has leaked out of the stored red cells, as well as any lactic acid generated by red cell metabolism. If frozen packed red cells are ordered, they will routinely be washed since washing is required to remove the cryoprotective agent, glycerol. (Washing units of blood introduces the possibility of bacterial contamination, but we feel that the risk is minimal since Federal regulations require that blood be used within 24 hours after thawing and this should preclude significant bacterial growth).

When it is absolutely necessary to administer whole blood, plasma or cryoprecipitate to a person who is known to have experienced mild allergic reactions, prophylactic administration of antihistamine should be considered. Furthermore, this patient has a potential risk of development of laryngeal edema. Consequently, it is recommended that a physician be nearby for approximately one half hour post-transfusion; a nurse should continue surveillance for at least 12 hours post-transfusion.

When the administration of whole blood is not vital, the BLOOD BANK recommends the use of frozen packed red cells for the following reasons:

1. <u>These units are extensively washed</u> to remove the glycerol. The washing will remove any allergenic material in the unit. (Note: If a patient has ever experienced an anaphylactic reaction due to sensitivity to transfused IgA antibody, then transfusion of frozen red cells is absolutely essential, since routine washing does not remove enough plasma to eliminate the reaction.)

2. Polymorphonuclear leukocyte breakdown produts will also be washed out, thus removing the pyrogenic material responsible for febrile reactions.

3. Glycerol is bacteriostatic, therefore greatly decreasing the chance of bacterial growth in the unit of blood.

4. The freezing process is believed to attenuate or even destroy hepatitis virus.

BLOOD GASES, ARTERIAL

Test Includes: pH, PCO_2, PO_2, Total CO_2, Bicarbonate, Base Excess, Oxygen
Saturation

SPECIMEN FOR BLOOD GASES, ARTERIAL: Blood specimens are obtained from the
radial artery at the wrist, or from the brachial artery accessible at the
junction of the arm and forearm; an artery of the nondominant arm is
recommended. Occasionally, the specimen is obtained from the femoral artery.
Infrequently, local anesthesia with one percent plain lignocaine, 2ml-3ml,
infiltrated around the artery may precede puncture. When blood is obtained from
the radial artery, the patency of the ulnar artery should also be determined.

   Use a 22 to 25 gauge, one to 1.5 inch disposable needle. Blood may be
drawn into plastic disposable syringes which have better seals than glass
syringes. Vacutainers may also be used. A tourniquet is not required. Blood
is collected using heparin as anticoagulant. Adequate volume for oxygen
saturation, PO_2, PCO_2 and pH is obtained by using a 1 ml tuberculin syringe
lubricated with heparin (100,000 units/ml). When the artery is penetrated,
pulsations are noted when using a syringe. Multiple air bubbles and froth
render the sample useless. After collection of the blood, the tip of the
syringe should be sealed with an appropriate cap. (Fleming, W.H. and Bowen,
J.C. "Complications of Arterial Puncture," Military Medicine, pgs. 307-308,
April, 1974.) After arterial puncture, pressure must be maintained on the
arterial puncture site for at least five minutes.

   Blood obtained by deep puncture from a vasodilated capillary bed (ear lobe
or finger) is practically equivalent to arterial blood when capillary blood is
not stagnant (Gambino et al, Am. J. Clin. Path. 46, 376-381, 1966).

   Ice water is used to preserve the samples if whole blood cannot be measured
within 15 minutes; storage on ice will protect the specimen for 1 to 2 hours. Do
not centrifuge cold preparation. If the specimen is not maintained cold for a
sufficient length of time, glycolysis will occur and lactic acid will be
produced; a fall in pH and a rise in PCO_2 will occur.

ARTERIAL pH
REFERENCE RANGE: 7.36 to 7.44
METHOD: pH electrode
INTERPRETATION: The relationship of pH and [H^+] is given in the next Table:

Relationship of pH and [H^+]	
pH	[H^+] Concentration (nmol/liter)
6.8	160 (Four Times Normal)
7.1	80 (Twice Normal)
7.4	40 (Normal)
7.7	20 (Half Normal)

The pH is decreased in metabolic acidosis and respiratory acidosis; the pH is
increased in metabolic alkalosis and respiratory alkalosis.

METABOLIC ACIDOSIS: In metabolic acidosis, HCO_3^- is decreased. Rule: A
decrease in pH of 0.15 is the result of a decrease in bicarbonate of
10mEq/liter. If compensation occurs, PCO_2 decreases. A diagnostic approach to
the differential diagnosis of metabolic acidosis based on anion gap and serum
potassium, is given in the next Figure; (Narins, R.B., Diagnostic Dialog, 3,
14-15, Jan. 1, 1981):

Diagnostic Approach to Differential Diagnosis of Metabolic Acidosis

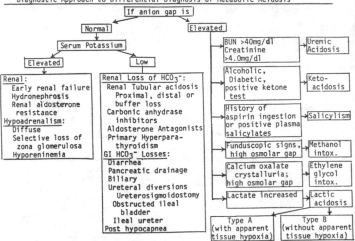

RESPIRATORY ACIDOSIS: In respiratory acidosis, PCO_2 is increased. <u>Rule:</u> An increase of PCO_2 of 10mmHg is associated with a <u>decrease</u> of pH of 0.08 units. If compensation occurs HCO_3^- is increased. Causes of <u>acute</u> respiratory acidosis are given in the next Table:

Causes of Acute Respiratory Acidosis	
Condition Affecting Organ or System	Examples
Brain Depression	Sedative, Opiate, Other Drug and Anaesthetic Overdose Comatose States, e.g., Cerebrovascular Accidents Neurosurgery and Head Injury
Spinal Cord	Injury
Neuromuscular System Disorders	Respiratory Paralysis, Acute Neuromuscular Disease, Myasthenia Gravis, Poliomyelitis, Guillain-Barre Syndrome Spinal Cord Trauma, Cervical Vertebral Fracture
Chest Wall Limitation	Rib Fracture with Flail Chest
Upper Airways	Airway Obstruction, e.g., Aspiration of Foreign Body Laryngospasm
Lower Airways and Lungs	Acute Severe Pulmonary Edema Severe Pulmonary Infections, e.g., Severe Pneumonia Open Chest Wounds Prolonged Open Chest and Open Heart Operations Pneumothorax, Hemothorax, Atelectasis
Other	Mechanical Underventilation Abdominal Distention from Ascites and Peritonitis

Acute respiratory acidosis may occur when a patient with <u>chronic pulmonary disease</u> is <u>given</u> an <u>excessive</u> amount of <u>oxygen</u>. The respiratory center is normally very sensitive to changes in arterial PCO_2. When the PCO_2 rises above 65mmHg, the respiratory center becomes insensitive to PCO_2. Hypoxia then becomes the main stimulus to respiration (carotid and aortic bodies); <u>these bodies</u> respond to <u>lowered arterial oxygen saturation</u>.

Causes of chronic respiratory acidosis are given in the next Table:

Causes of Chronic Respiratory Acidosis

Severe Lung Disease
  Emphysema
  Chronic Obstructive Lung Disease, e.g., Chronic Bronchitis
  Chronic Asthma
  Bronchiectasis
  Pulmonary Fibrosis
Hypoventilation
  Extreme Obesity, i.e, Pickwickian Syndrome
  Chest Deformity: Kyphoscoliosis and Injury to Thoracic Cage

These patients <u>also</u> have low PO_2 and <u>cyanosis may be present</u>.

<u>METABOLIC ALKALOSIS</u>: In metabolic alkalosis, HCO_3^- is increased. <u>Rule</u>: An increase in pH of 0.15 is the result of an <u>increase</u> in bicarbonate of 10mEq/liter. If compensation occurs, PCO_2 is increased. Metabolic alkalosis is <u>associated</u> with conditions that cause a <u>decrease in potassium</u> and by conditions that cause a <u>loss of hydrogen ions</u>, by <u>excess bicarbonate</u> and by <u>chloride losing diarrhea</u>; these conditions are given in the next Table:

Causes of Metabolic Alkalosis

(1) Decrease of (K^+):

Mechanisms in Hypokalemia

Urine Loss	G.I. Loss	Movement into the Intracellular Space
Diuretics; Thiazides;	Vomiting	Diabetic Kketoacidosis
Lasix (Furosemide); Edecrine	Nasogastric Suction	(treated)
(Ethacrynic Acid)	Pyloric Obstruction	Familial Periodic
Antibiotics		Hypokalemic Paralysis
Carbenicillin		
Amphotericin B		Other
Magnesium Depletion		
Increased Mineralocorticoid		Decreased K^+ Intake
Licorice Abuse		Acute Myeloid Leukemia
Bartter's Syndrome		

(2) Loss of (H^+):
  (a) <u>Gastrointestinal Tract</u>: Vomiting or gastric aspiration
  (b) <u>Kidney</u>: <u>Hypercalcemia</u> from causes other than Primary Hyper-parathyroidism
(3) Excess of (HCO_3^-): Patient taking excessive $NaHCO_3$ or other alkaline salts
(4) Chloride Losing Diarrhea

Differential Diagnosis of Metabolic Alkalosis; (Harrington, J.T. and Cohen, J.J., N. Engl. J. Med. 293, 1241-1248, 1975). The common causes of metabolic alkalosis are two-fold: (1) those associated with <u>loss of hydrogen ion and extracellular fluid volume</u> (chloride responsive type) and (2) those associated with pathologically <u>excessive circulating levels of mineralocorticoid hormone</u> (chloride resistant type).

Differential Diagnosis of Metabolic Alkalosis

Chloride Responsive (Urine Chloride <10mEq/liter)	Chloride Resistant (Urine Chloride >20mEq/liter)
Diuretic Therapy	Adrenal Disorders
Gastrointestinal Causes	Hyperaldosteronism
Vomiting	Cushing's Syndrome
Nasogastric Suction	Pituitary, Adrenal, Ectopic
	ACTH
Chloride-Wasting Diarrhea	Exogenous Steroid
Villous Adenoma-Colon	Gluco- or Mineralcorticoid
Rapid Correction of Chronic	Licorice Ingestion (Glycyrrhizic
Hypercapnea	acid has mineralocorticoid-like
Carbenicillin	activity)
Reduced Dietary Intake of	Carbenoxalone
Chloride	Bartter's Syndrome
	Alkali Ingestion

The measurement of urinary chloride is valuable in differentiation of causes of persistent metabolic alkalosis. The most common causes of persistent metabolic alkalosis, e.g., loss of gastric juice and diuretic induced, are due to chloride depletion and are responsive to the administration of chloride adequate to replace body stores.

RESPIRATORY ALKALOSIS: In respiratory alkalosis, PCO_2 is decreased. Rule: A decrease of PCO_2 of 10mmHg is associated with an increase of pH of 0.08 units. If compensation occurs, HCO_3^- is decreased. Disorders which cause excessive pulmonary elimination of carbon dioxide result in respiratory alkalosis. Promptly, after the onset of hyperventilation, hydrogen ions from intracellular sources enter the extracellular fluid and bicarbonate moves into the red blood cells in exchange for chloride, thus minimizing extracellular alkalosis.

The causes of respiratory alkalosis are given under four headings:

> Respiratory Center Stimulation
> Hypermetabolic States
> Artificial Ventilation
> Mechanisms Unknown

In these patients, hyperventilation occurs with an increased tidal volume and increased alveolar ventilation.

The specific conditions causing respiratory alkalosis are given in the next Table:

Causes of Respiratory Alkalosis
(1) Respiratory Center Stimulation

(1) Respiratory Center Stimulation
   Central nervous system disease; Encephalitis; Brain stem lesions
      (cerebral-vascular accidents); Intracranial surgery
   Drugs: Salicylate intoxication (early stage); Infants of Heroin
      addicted mothers; Adults on withdrawal from Heroin; Paraldehyde
      intoxication; 2,4-Dinitrophenol intoxication
   Anxiety and hysteria (hyperventilation syn.)
   Hypoxia: Anemia from any causes; High altitude residents;
      Congestive heart failure;
      Pulmonary Disease (Pulmonary Fibrosis with Alveolar-
            capillary Block)
         Pulmonary Emboli; Boeck's sarcoid; Beryllium granulomata;
         Asbestosis; Pulmonary scleroderma; Alveolar cell carcinoma;
         Diffuse metastatic carcinoma; Pulmonary fibrosis
      Right to left shunt: Congenital heart disease; Pulmonary
         atelectasis
   Early septic shock
   Reflex hyperventilation: Pneumothorax; Pulmonary hypertension
   Pregnancy
(2) Hypermetabolic States: Thyrotoxicosis; Febrile States; Exercise
(3) Artificial Ventilation: Excessive ventilation; Mechanical overventilation
   using low PCO_2
(4) Others: Cirrhosis of the liver; Extracorporeal circulation; Beriberi;
   Transient phase following diabetic ketoacidosis; Alcohol intoxication--
   delirium tremens

Common causes of respiratory alkalosis are underlined.

ACID-BASE (MIXED STATES)
(Narins, R.G. and Emmett, M., Medicine 59, 161-187, 1980):
(a) Metabolic Alkalosis and Respiratory Alkalosis: Increased $[HCO_3^-]$,
decreased PCO_2, Increased pH.

(1) <u>Critically Ill Surgical Patients</u>: The combination of metabolic alkalosis and respiratory alkalosis may occur in critically ill surgical patients. The causes of combined metabolic alkalosis and respiratory alkalosis are given in the next Table:

Causes of Combined Metabolic and Respiratory Alkalosis	
Metabolic Alkalosis	Respiratory Alkalosis
Vomiting or Nasogastric Suction	Excessive Mechanical Ventilation
Massive Blood Transfusions	Hypoxemia
Lactated Ringer's Solution and	Sepsis
and High Dose Antacid	Hypotension
	Neurologic Damage
	Liver Disease
	Pain
	Drugs

<u>Blood Transfusions</u>: Trisodium citrate is an anticoagulant in blood; one unit of whole blood contains 17mEq of citrate; one unit of packed cells contains 5mEq of citrate. Following transfusions, citrate is metabolized to $[HCO_3^-]$ in equimolar amounts; that is 17mEq of citrate is converted to 17mEq of $[HCO_3^-]$ and 5mEq of citrate is converted to 5mEq of $[HCO_3^-]$. Normally, the $[HCO_3^-]$ load is excreted by the kidney. However, there are certain conditions in which patients receive many units of blood and in which the excretion of $[HCO_3^-]$ by the kidney is compromised; these are given in the next Table:

Blood Transfusions and Compromised Excretion of $[HCO_3^-]$ by the Kidney
Conditions in which many Blood Transfusions may be given:
Shock
Trauma
Sepsis
Open Heart Surgery
Compromised Excretion of $[HCO_3^-]$ by the Kidney:
(Enhanced Renal HCO_3^- Reabsorption)
ECF Volume Contraction
Potassium Depletion
Hypercapnia
Acute or Chronic Renal Disease
Oliguria

(b) <u>Respiratory Acidosis and Metabolic Alkalosis</u>: Increased $[HCO_3^-]$, increased PCO_2, possibly normal pH. This combination is the <u>most frequently encountered mixed acid-base disturbance</u>. It occurs in patients with chronic hypercapnia (such as chronic obstructive pulmonary disease, COPD) who have received diuretics (such as the thiazides) or who have had their $PaCO_2$ abruptly lowered.

(c) <u>Mixed Metabolic Acidosis and Respiratory Alkalosis</u>: See Narins, R.G. and Emmett, M. Medicine <u>59</u>, 161-187, 1980: Gary, N.E., Resident and Staff Physician, pgs. 32-38s, June, 1981). This combination usually occurs in patients with uremia, lactic or ketoacidosis and liver or heart failure; bacteremic shock and salicylate intoxication may cause this combination.

(d) <u>Mixed Metabolic Alkalosis and Acidosis</u>: This combination may occur when there is a loss of bicarbonate due to diarrhea resulting in metabolic acidosis accompanied by vomiting resulting in metabolic alkalosis. Another cause of mixed metabolic alkalosis and acidosis is overzealous sodium bicarbonate treatment of acidosis.

PCO$_2$:

REFERENCE RANGE: Sea level, Men: 34 to 45mmHg (4.52kPa to 5.99kPa); Women: 31 to 42mmHg (4.12kPa to 5.59kPa); pregnancy: PCO$_2$ decreased; the normal adult respiratory volume of 5 liters per minute is increased in pregnancy to 10 liters per minute owing to the effects of progesterone which causes a chronic hyperventilation; during labor, the ventilatory rate may increase to 35 liters per minute with a further decrease of PCO$_2$ (Huch, R. and Huch, A., "Fetal and Maternal Pt$_c$O$_2$ Monitoring", Crit. Care Med. 9, 694-697, 1981).
One Mile Altitude: 34 to 38mmHg.

METHOD: The PCO$_2$ electrode consists of two basic components. There is a membrane permeable only to CO$_2$ gas between the solution to be measured and the pH electrode.

INTERPRETATION: PCO$_2$ is altered in the conditions listed in the next Table:

Alterations of PCO$_2$
Increased PCO$_2$:
Respiratory Acidosis
Compensation in Metabolic Alkalosis
Decreased PCO$_2$:
Respiratory Alkalosis
Compensation in Metabolic Acidosis

The causes of each of these conditions are given in the previous section.

BICARBONATE (HCO$_3^-$):

REFERENCE RANGE: 20 to 26mEq/liter

METHOD: [HCO$_3^-$] is obtained by calculation by inserting the values of the pH and PCO$_2$ into the equation:

$$[H^+] = 24 \frac{PCO_2}{[HCO_3^-]}$$

INTERPRETATION: Bicarbonate (HCO$_2^-$) is altered in the conditions listed in the next Table:

Alteration of Bicarbonate (HCO$_3^-$)
Increased HCO$_3^-$:
Metabolic Alkalosis
Compensation in Respiratory Acidosis
Decreased HCO$_3^-$:
Metabolic Acidosis
Compensation in Respiratory Alkalosis

The causes of each of these conditions are given in the previous section.

BASE EXCESS:

REFERENCE RANGE: -3 to +3mMol/liter

METHOD: Base excess is obtained directly from the blood gas analyzer from the PCO$_2$, pH and HCO$_3^-$ values.

INTERPRETATION: Base excess suggests the presence of metabolic alkalosis; base deficit suggests metabolic acidosis. Base excess gives, in mmol/liter , the surplus of base or acid present in the blood or expressed in another way, base excess or deficit shows directly how many mmol/liter of acid or base excess must be added to "normalize" the pH of the blood.

P_aO_2, ARTERIAL:

SPECIMEN: 1ml arterial blood in heparinized syringe, capped and iced.

REFERENCE RANGE: Normal Adult: 80 to 100mmHg at sea level; 65 to 75mmHg at one mile altitude. The arterial blood gas values in mmHg in the fetus and first week of life are as follows; (Weisberg, H.F., Ann. Clin. Lab. Sci. 12, 245-253, 1982): Fetus: 16-24; Birth, umbilical artery, 8-24; 5-10 minutes, 33-75; 20 minutes, 31-85; 60 minutes, 38-83; after 1 hour, 55-80; one day, 54-95; one week, 57-94.

Pregnancy: (at term): 85-115; during labor, the ventilatory rate can increase to as much as 35 liters per minute, from the pre-pregnancy rate of 5 liters per minute, resulting in an increased P_aO_2.

METHOD: PO₂ electrode

INTERPRETATION: Conditions that are associated with decreased P_aO_2 and that require acute oxygen therapy are listed in the next Table:

Conditions Associated with Decreased P_aO_2 and That May Require Acute Oxygen Therapy
Carbon Monoxide Poisoning (give 100 percent Oxygen)
All Acute Pulmonary Disorders, e.g., Adult Respiratory Distress Syndrome Pulmonary Edema; Severe Bronchial Asthma; Pulmonary Thromboembolism; Severe Pneumonia
Myocardial Infarction (Cardiogenic Shock)
Congestive Heart Failure
Drug Overdose, e.g., Barbituate Poisoning
Head and Musculoskeletal Trauma, e.g., Crushed Chest Injury
Hepatic Failure; Acute Pancreatitis; Shock; Sepsis; Hemorrhage; Pneumonia

Acute oxygen therapy, if prolonged, causes changes in alveolar structures (Davis, W.B. et al., N. Engl. J. Med. 309, 878-883, 1983).

In acute conditions, a low P_aO_2 is usually associated with a low or normal arterial carbon dioxide pressure.

Those conditions associated with hypoventilation and increased PCO₂ may be associated with hypoxia; these conditions were listed under the heading respiratory acidosis and are repeated here for convenience.

Causes of Acute Respiratory Acidosis	
Condition Affecting Organ or System	Examples
Brain Depression	Sedative, Opiate, Other Drug and Anaesthetic Overdose Comatose States, e.g., Cerebrovascular Accidents Neurosurgery and Head Injury
Spinal Cord	Injury
Neuromuscular System Disorders	Respiratory Paralysis; Acute Neuromuscular Disease; Myasthenia Gravis; Poliomyelitis; Guillain-Barre Syndrome; Spinal Cord Trauma; Cervical Vertebral Fracture
Chest Wall Limitation	Rib Fracture with Flail Chest
Upper Airways	Airway Obstruction, e.g., Aspiration of Foreign Body Laryngospasm
Lower Airways and Lungs	Acute Severe Pulmonary Edema Severe Pulmonary Infections, e.g., Severe Pneumonia Open Chest Wounds Prolonged Open Chest and Open Heart Operations Pneumothorax; Hemothorax; Atelectasis
Other	Mechanical Underventilation Abdominal Distention from Ascites and Peritonitis

The causes of chronic respiratory acidosis, which are associated with low PO₂ are given in the next Table:

Causes of Chronic Respiratory Acidosis Plus Low PO₂
Severe Lung Disease
Emphysema
Chronic Obstructive Lung Disease, e.g., Chronic Bronchitis
Chronic Asthma
Bronchiectasis
Pulmonary Fibrosis
Hypoventilation
Extreme Obesity, i.e, Pickwickian Syndrome
Chest Deformity: Kyphoscoliosis and Injury to Thoracic Cage

S. Bakerman

BLOOD GASES, VENOUS

Venous blood should be collected as follows: light tourniquet pressure for a short time; no movement of extremity; release tourniquet before sampling is complete.

Normal values and ranges for venous and arterial blood gases and acid-base values are given in the next Table:

Gases and Acid-Base Values		
Parameter	Venous	Arterial
pH	7.36(7.31-7.41)	7.40(7.36-7.44)
PCO₂ (mm Hg)	46(40-52)	40(34 to 45)
HCO₃⁻ (mEq/liter)	25(22-28)	24(22-26)
PO₂ (mm Hg)	40(30-50)	90(80-100)
O₂ Saturation (%)	75(60-85)	95 or Greater

The values for males are not significantly different than that of females.

Venous pH is slightly less than that of arterial pH while that of venous PCO₂ is mildly greater than that of arterial PCO₂. There is no significant difference between venous and arterial HCO₃⁻. Venous blood may be used in acid-base studies. However, in certain common clinical conditions, venous blood does not reflect whole body acid-base status; these conditions are congestive heart failure, shock, poor tissue perfusion, and in the newborn period. Venous blood taken from an extremity reflects the acid-base status of that extremity and not that of the whole body; this is especially true if there is poor circulation to that extremity. With stasis, PCO₂ increases 20 mm to 40 mm in one minute.

The oxygen content of venous blood is significantly less than that of arterial blood.

BONE PANEL

The tests that are done to reflect bone pathology are shown in the next Table:

Tests to Reflect Bone Pathology
Bone Alkaline Phosphatase
Hydroxyproline
Serum Calcium
Serum Phosphorus

Bone alkaline phosphatase is derived from osteoblasts of bone and is elevated in osteoblastic lesions. Conditions associated with osteoblastic lesions of bone are given in the next Table:

Conditions Associated with Osteoblastic Lesions of Bone
Paget's Disease
Malignancy (Osteoblastic)
Secondary Hyperparathyroidism
Renal Disease
Osteomalacia
Malabsorption
Rickets
Primary Hyperparathyroidism
Healing Fractures
Hyperthyroidism

Hydroxyproline is a measure of resorption of bone collagen; it is increased in the serum and urine in conditions listed below:

Causes of Increased Urinary Hydroxyproline
Growth Period
Bone Diseases:
Primary and Secondary Hyperparathyroidism
Paget's Disease
Myeloma
Osteoporosis
Marfan's Syndrome

Mnemonic for Tumors Metastatic to Bone: "P.T. Barnum Loves Kids"; P=Prostate; T=Thyroid; B=Breast; L=Lung; K=Kidney.

BOWEL INFARCTION PANEL

The diagnostic clues to bowel infarction are hemoconcentration, hyperamylasemia, hyperphosphatemia and metabolic acidosis (Jamieson, W.G. et al., Surg. Gynecol. Obstet., 148, 334-338, 1979). However, laboratory tests were found not to be helpful in the diagnosis of bowel infarction on an acute medical service; laboratory findings in bowel infarction and in disease mimicking bowel infarction are given in the next Table; (Cooke, M. and Sande, M.A., Am. J. Med. 75, 984-992, Dec. 1983):

Laboratory Findings (Mean Values) in Bowel Infarction and in Disease Mimicking Bowel Infarction		
Test	Bowel Infarction	No Bowel Infarction
Hematocrit (percent)	39.3	36.3
White blood cell count (per mm^3)	16.2	9.6
Blood urea nitrogen (mg/dl)	41.0	62.0
Amylase (IU/liter)	56.0	62.0
Lactate dehydrogenase (IU/liter)	382.0	768.0
Alkaline phosphate (IU/liter)	153.0	192.0
Aspartate aminotransferase (IU/liter)	121.0	432.0
Serum phosphate (mg/dl)	5.7	5.0
pH	7.33	7.33
Serum bicarbonate (meq/liter)	16.5	13.9

Almost one-half of the patients had metabolic acidosis and almost one-half had respiratory alkalosis; a few patients had normal arterial pH values. The usual admitting medical diagnoses in bowel infarction were as follows: acute myocardial, upper gastrointestinal bleeding and sepsis. The usual symptom was acute abdominal pain followed by collapse.

BUN (see UREA NITROGEN)   BUN

CALCITONIN (THYROCALCITONIN)   Calcitonin
SPECIMEN: Fasting; green (heparin) top tube; plasma should be obtained in a refrigerated centrifuge, and then freeze.
REFERENCE RANGE: Male, equal to or less than 0.155ng/ml; female, equal to or less than 0.105ng/ml (Mayo Clinic).
METHOD: RIA
INTERPRETATION: Serum calcitonin may be elevated in the conditions listed in the next Table:

Causes of Elevated Serum Calcitonin
Medullary Carcinoma of the Thyroid (10% of cases familial, usually associated with Multiple Endocrine Neoplasia, Type II Sipple Syndrome)
Zollinger-Ellison Syndrome; Pernicious Anemia; Pregnant Women at Term; Newborns
Associated with Some Cancers: Lung, Breast, Pancreas
Chronic Renal Failure

Plasma calcitonin assays are useful to detect the presence of medullary carcinoma of the thyroid and in revealing the recurrence of treated medullary thyroid cancer. Very high serum calcitonin levels (>2000 pg/ml) are virtually diagnostic of medullary carcinoma of the thyroid. The familial form of medullary thyroid carcinoma is inherited as an autosomal dominant trait often associated with pheochromocytoma and parathyroid adenoma or hyperplasia (multiple endocrine adenomatosis, Type II). To detect familial medullary thyroid carcinoma, tests provocative of calcitonin secretion (infusion of calcium or pentagastrin or both) may be done. Graze et al studied the natural history of familial medullary thyroid carcinoma using provocative tests (N. Engl. J. Med. 299, 980-985, 1978; Hillyard, C. and Stevenson, J., The Ligand Review 2, 31-33, 1980).
Calcium Infusion Test: At time 0, obtain a blood specimen for calcium and calcitonin levels. Inject calcium (15mg calcium/kg) and obtain blood samples at 3 and 4 hours for calcium and calcitonin levels. Reference Range: Maximum value has not exceeded, Male: 0.265ng/ml; Female: 0.120ng/ml.
Pentagastrin Injection: (Do not give pentagastrin to patients who have known, diagnostically, high basal calcitonin levels). At time 0, obtain a blood specimen for calcium and calcitonin levels. Inject pentagastrin (Peptavlon, Ayerst, 0.5microgram/kg, I.V.). Obtain blood samples at 1.5, 2 and 5 minutes for calcitonin levels. Reference Range: Maximum value has not exceeded, Male: 0.210ng/ml; Female: 0.105ng/ml.

S. Bakerman

CALCIUM EXCRETION TEST

SPECIMEN: A serum specimen and two urine specimens. Serum: Red top tube, separate serum for analysis of serum calcium, creatinine, uric acid and phosphorus. Urine: Determination of 2 hours urine volume, calcium, creatinine, uric acid and magnesium on each of the two urine specimens.

Instructions: The patient should avoid, for one week, foods that contain a high calcium content such as dairy products (<400mg calcium/day for seven days).

1. The patient must not take any food or drink after midnight. The test is not performed if the patient has had any food, solid or liquid.
2. The patient must arrive at the laboratory at 7:30 A.M. Blood is collected for serum calcium, creatinine, uric acid and phosphorus.
3. The patient should void and discard all urine upon arrival.
4. The patient takes 600ml of water (measured) and drinks the total volume.
5. All urine excreted is collected over the next two hours; this specimen is labeled "fasting 2 hr. urine". The following determinations are done:

Determination	Calculation of Total Excreted
Total Urine Volume (ml)	
Calcium (mg/dl)	x Volume = _____ (mg/2 hours)
Creatinine (mg/dl)	x Volume = _____ (mg/2 hours)
Uric Acid (mg/dl)	x Volume = _____ (mg/2 hours)
Magnesium (mg/dl)	x Volume = _____ (mg/2 hours)

Calculate: $\dfrac{\text{Calcium (mg/dl)}}{\text{Creatinine (mg/dl)}}$

6. After the two hour fasting urine is collected, then the patient ingests by mouth; the following:
   a. 3 ounces of "Calcitest"
   b. 300ml. of water with 40ml. of "Neo-Calglucon"
   Note: Calcitest and neo-calglucon are obtained from the Pharmacy.
7. Collect all urine excreted over the next two hours and this specimen is labeled "2 hr. Urine 1 gm Calcium Load". The following determinations are done:

Determination	Calculation of Total Excreted
Total Urine Volume (ml)	
Calcium (mg/dl)	x Volume = _____ (mg/2 hours)
Creatinine (mg/dl)	x Volume = _____ (mg/2 hours)
Uric Acid (mg/dl)	x Volume = _____ (mg/2 hours)
Magnesium (mg/dl)	x Volume = _____ (mg/2 hours)

Calculate: $\dfrac{\text{Calcium (mg/dl)}}{\text{Creatinine (mg/dl)}}$

REFERENCE RANGE:

CONDITION	FASTING Serum Ca	FASTING Ca/Cr ratio	POST 1 GRAM Ca^{++} Load Serum Ca	POST 1 GRAM Ca^{++} Load Ca/Cr Ratio
Primary Hyperparathyroidism	↑	>0.11	↑	>0.20
Absorptive Hypercalciuria	N	<0.11	N	>0.20
Renal tubular Hypercalciuria	N	>0.11	N	>0.20
Normocalciuric Nephrolithiasis	N	<0.11	N	>0.13
Normal	N	<0.11	N	<0.13

METHODS: See individual methodology

INTERPRETATION: The data that are necessary for the work-up of subjects for hypercalciuria and renal stones are given in protocol in the next Table; data collected from a patient who was being worked-up for renal stone formation are written in script in order to give a practical application of the use of this protocol:

Protocol for Renal Stone Formers

	Fasting 2 Hour Urine	2 Hour Urine with 1 gm Calcium Load
Urine Collection Started (time)	0810	1010
Completed (time)	1010	1210
Urine Volume	475 ml/2 hr	400 ml/2 hr
Calcium Concentration	1.4 mg/dl	3.5 mg/dl
Total Calcium	6.6 mg/2 hr	14.0 mg/2 hr
Creatinine Concentration	15.6 mg/dl	18.2 mg/dl
Total Creatinine	74.1 mg/2 hr	72.8 mg/2 hr
Calcium/Creatinine Ratio	0.09	0.19
	Normal = <0.11	Normal = <0.20
Uric Acid Concentration	11.1 mg/dl	10.4 mg/dl
Total Uric Acid	52.7 mg/2 hr	41.6 mg/2 hr
	Normal = <97 mg/2 hr	Normal =<86 mg/2 hr
Magnesium Concentration	1.3 mg/dl	2.2 mg/dl
Total Magnesium	6.2 mg/2 hr	8.8 mg/2 hr
Magnesium/Calcium Ratio	93.9	62.9
Mg x 100/Ca		Normal = >40

The causes of hypercalciuria are given in the next Table:

Causes of Hypercalciuria

Absorptive Hypercalciuria
Resorptive Hypercalciuria
    Primary Hyperparathyroidism, Malignancy,
    Hyperthyroidism, Renal Tubular Acidosis
Renal Tubular Hypercalciuria
Normocalciuric Nephrolithiasis

Absorptive Hypercalciuria: In absorptive hypercalciuria, there is normocalcemia, increased intestinal absorption of calcium and increased calcium in the urine. The primary abnormality is an increased intestinal absorption of calcium; the possible mechanism is illustrated in the next Figure:

Possible Mechanism of Absorptive Hypercalciuria

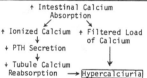

As shown, increased intestinal absorption of calcium increases the total body calcium and inhibits PTH leading to reduced reabsorption of filtered calcium by the renal tubule and leading to increased urinary excretion. This pathogenic sequence is associated with an idiopathic form, vitamin D intoxication and sarcoidosis (probably represents an increased sensitivity to vitamin D). Sodium cellulose phosphate (SCP) has been approved by the FDA for the treatment of absorptive hypercalciuria, type I (FDA Drug Bulletin, 13, No.1, April 1983). Absorptive hypercalciuria, type I, is associated with increased intestinal absorption of calcium and persists despite a low-calcium (200mg/day) diet.

Renal Tubular Hypercalciuria (Renal Leak Hypercalciuria): Calcium is filtered at the glomerulus; it is largely reabsorbed in the proximal tubule. In renal tubular hypercalciuria, a defect in calcium reabsorption of unknown cause occurs; a possible mechanism of renal tubular hypercalciuria is given in the next Figure:

Possible Mechanism of Renal Tubular Hypercalciuria

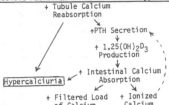

Thus far, the pathogenesis of renal leak hypercalciuria (renal calcium leak) in all patients is idiopathic.

References: Pak, C.Y.C. et al., N. Engl. J. Med. 292, 497-500, 1975; Broadus, A.E. et al., J.Clin. Endocrinol. Metab. 47, 751, 1978; Sakhall, K. et al., Urology, 14, 251-255, 1979.

CALCIUM IONIZED (FREE)

SPECIMEN: Red top tube; separate serum; keep specimen in refrigerator and forward on ice (Do not freeze).

REFERENCE RANGE: At pH 7.4, free calcium = 3.9 to 4.5mg/dl. To convert conventional units in mg/dl to international units in mmol/liter, multiply conventional units by 0.2495.

METHOD: Reaction with cresolphthalein complexone.

INTERPRETATION: Ionized (free) calcium is elevated in conditions listed in the next Table:

Causes of Elevated Ionized Calcium
Primary Hyperparathyroidism
Ectopic PTH-Producing Tumors
Excess Intake of Vitamin D
Various Malignancies

Calcium circulates in the serum in three forms: free (46%), protein bound (40%), and diffusible calcium complexes (14%). About 80 percent of the calcium binding protein in serum is albumin; the remaining 20 percent is globulin. Calcium forms complexes with bicarbonate, citrate, phosphate and sulfate. About 20 percent of the amino acids of albumin are glutamic and aspartic acids; binding of calcium to albumin probably occurs through these acidic amino acids. The free fraction is most important physiologically.

The serum forms of calcium are illustrated in the next Figure:

Serum Forms of Calcium

$$\begin{array}{ccc}
\begin{matrix}
-\overset{O}{\underset{}{C}}-O^-Ca^{++} \\
-\overset{O}{\underset{}{C}}-O^-Ca^{++} \\
-\overset{O}{\underset{}{C}}-O^-Ca^{++} \\
-\overset{O}{\underset{}{C}}-O^-Ca^{++}
\end{matrix}
&
\begin{matrix}
Ca^{++} \\
Ca^{++} \\
Ca^{++} \\
Ca^{++}
\end{matrix}
&
\begin{matrix}
HCO_3^- \\
Citrate \\
PO_3^{-3} \\
SO_4^{-2}
\end{matrix}
\end{array}$$

Albumin-Ca^{++}	Free-Ca^{++}	Complex-Ca^{++}
(40%)	(46%)	(14%)

Effects of pH on Free Calcium: Acidosis causes an increase in free calcium while alkalosis causes a decrease in free calcium; the reactions are given in the next chemical equations:

$$\text{Alkalosis:} \quad Ca^{++} + 2OH^- \longrightarrow Ca(OH)_2$$
$$\text{Acidosis:} \quad H_3O^+ + OH^- \longrightarrow 2\,H_2O$$

The change in percent of free calcium with pH is shown in the next Figure:

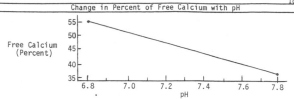

Change in Percent of Free Calcium with pH

At a serum pH of 7.4, 46 percent of the total serum calcium is free calcium. As the pH of the serum increases, the percent of free calcium decreases. Acute respiratory alkalosis may precipitate tetany because of sudden decrease in free calcium.

Systemic acidosis has been associated with osteomalacia and rickets because low pH interferes with mineralization of newly formed bone and acidosis may interfere with the production of 1,25di(OH) cholecalciferol.

CALCIUM, TOTAL, SERUM

Calcium,
Total

SPECIMEN: Red top tube.
REFERENCE RANGE: 8.5 to 10.5mg/dl in adults; premature: 7-10.0mg/dl, full term: 7.0-12.0mg/dl; children: 8.0-11.0mg/dl. Panic Values: less than 7mg/dl or greater than 12mg/dl. To convert conventional units in mg/dl to international units in mmol/liter, multiply conventional units by 0.2495.
METHOD: Reaction with cresolphthalein complexone
INTERPRETATION: The causes of hypercalcemia, arranged in decreasing order of incidence, are given in the next Table:

Causes of Hypercalcemia
(1) Primary Hyperparathyroidism
(2) Malignancy (Especially Metastatic Breast and Lung Tumors; Multiple Myeloma)

(3) Drug Induced Hypercalcemia
    Thiazides; Calcium Containing Antacids
    Calcium Containing Antacids plus Milk (Milk-Alkali Syndrome);
    Vitamin D

(4) Familial Benign Hypercalcemia (Familial Hypocalciuric Hypercalcemia,FHH)
(5) Thyrotoxicosis (occasionally)
(6) Sarcoidosis (occasionally)
(7) Tuberculosis (Kitrow et al., Ann. Int. Med. 96, 255-256, 1982)
(8) Idiopathic Hypercalcemia of Infancy
(9) Immobilization (rarely)
(10) Acute Adrenal Insufficiency (rarely)

Mnemonic for Hypercalcemia: "CHIMPS"; C=Cancer; H=Hyperthyroidism; I=Iatrogen (Drugs); M=Multiple Myeloma; P=Primary Hyperparathyroidism; S=Sarcoid. (Personal communication, Dr. Paul Bakerman).

Primary hyperparathyroidism, malignancy and drug-induced hypercalcemia are, by far, the most common causes of hypercalcemia. In the differential diagnosis of hypercalcemia, the rare causes, e.g., immobilization, Paget's disease and acute adrenal insufficiency, occur infrequently. Coma may be seen with levels above 13mg/dl; acidosis may intensify the effects of hypercalcemia.

The most useful tests in the differential diagnosis of hypercalcemia are PTH, chloride, phosphorus, and alkaline phosphatase.
Primary Hyperparathyroidism: The diagnosis of primary hyperparathyroidism may be suggested by the following findings: increased serum calcium, decreased serum phosphorus, increased serum chloride, increased PTH, normal hematocrit; these findings will correctly classify primary hyperparathyroidism in 98% of the patients with hypercalcemias (Lafferty, F.W., Arch. Int. Med. 141, 1761-1766, 1981).
Malignancy: The malignancies associated with hypercalcemia are shown in the next Table:

Tumors Associated with Hypercalcemia

Breast
Multiple Myeloma
Carcinoma of Squamous Lined Surfaces:
     Lung (Squamous Cell)
     Head and Neck
     Esophagus
     Cervix
Lung (Squamous Cell, Large Cell Anaplastic, Adenocarcinoma)
Prostate; Kidney; Lower Genitourinary Tract
Thyroid; Leukemia; Malignant Lymphoma; Melanoma; Chondrosarcoma

Drug-Induced Hypercalcemia: The administration of thiazide diuretics to patients may result in increased serum calcium. The hypercalcemia is due to an increase in both total and ionized calcium. The serum calcium may begin to rise after 1 day of thiazides and increases gradually; the increase in serum calcium is usually moderate.

Familial Benign Hypercalcemia (Familial Hypocalciuric Hypercalcemia(FHH):
Familial benign hypercalcemia is frequently confused with primary hyperparathyroidism. Familial benign hypercalcemia is characterized by autosomal-dominant inheritance, asymptomatic life-long hypercalcemia, unexpectedly low urinary calcium excretion with a lower ratio of calcium clearance to creatinine clearance than in hyperparathyroidism and inappropriately "normal" levels of immunoreactive parathyroid hormone. All patients suspected of having primary hyperparathyroidism should routinely have 24-hour urinary calcium and renal calcium to creatinine clearance ratio determination to rule out familial benign hypercalcemia; a value below 0.01 for this ratio suggests familial benign hypercalcemia (Marx, S.J. et al., Medicine 60, 397-412, 1981). High-resolution parathyroid ultrasonography frequently demonstrates enlarged parathyroid glands in patients with primary hyperparathyoidism; smaller glands are found in patients with familial benign hypercalcemia (Law, W.M. et al., Mayo Clin. Proc. 59, 153-155, 1984).

The causes of hypocalcemia are listed in the next Table:

Causes of Hypocalcemia

(1) Renal Failure
(2) Hypoparathyroidism
(3) Pseudohypoparathyroidism (Compare to pseudopseudo-)
(4) Magnesium Deficiency
(5) Drug-Induced Hypocalcemia: Anti-convulsives
(6) Hypocalcemia in Susceptible Neonates
(7) Rickets: Hypophosphatemic Vitamin D Resistant
                 (Refractory) Rickets
           Vitamin D Dependent Rickets
           Gluten Enteropathy
(8) Acute Pancreatitis
(9) After Multiple ACD Blood Transfusions
(10) Wilson's Disease (Carpenter, T.O. et al., N. Engl. J. Med.
        309, 873-877, 1983)

The first three conditions in the previous Table, renal failure, hypoparathyroidism, and pseudohypoparathyroidism, are associated with hyperphosphatemia as given in the next Table:

Conditions of Hyperphosphatemia and Hypocalcemia

Renal Failure
Hypoparathyroidism
Pseudohypoparathyroidism

Calcium in the "free" state is the important physiologically active form; calcium circulates in the serum in three forms: free (46%) protein-bound (40%), and diffusible calcium complexes (14%). Serum calcium level cannot be properly interpreted without serum albumin level, an equation for correction of serum calcium in patients with depressed albumin is:

Adjusted Calcium = Serum Calcium - Serum Albumin + 4.0

This equation corrects serum calcium so that serum calcium is increased by 1.0mg/dl for every 1.0g/dl that the albumin concentration is below 4.0g/dl.

pH effects free calcium; acidosis (metabolic or respiratory) causes an increase in free calcium while alkalosis (metabolic or respiratory) causes a decrease in free calcium.

CALCIUM, URINE
SPECIMEN: 24 hour urine collection. Instruct the patient on 24 hour urine collection as follows: Void at 8:00 A.M. and discard specimen. Collect all urine during the next 24 hours including the 8:00 A.M. specimen the next morning. Add 15ml of 6N HCl to pH 1 to 2. Refrigerate urine at 4°C.
REFERENCE RANGE: Varies with diet; low calcium, <50mg/24hr. Average diet, 50-150mg/24hr; high calcium, 100-300mg/24hrs. Urinary excretion of calcium varies greatly with its intake.
METHOD: Reaction with cresolphthalein complexone.
INTERPRETATION: Urine calcium is increased in the conditions listed in the next Table:

Increase in Urine Calcium
Primary Hyperparathyroidism (Increased in 2/3 of patients)
Vitamin D Toxicity
Hyperthyroidism
Renal Tubular Acidosis (Type I, Distal)
Osteolytic Conditions
Multiple Myeloma
Tumor Metastases
Paget's Disease
Idiopathic

A cause of decreased urinary excretion of calcium is the thiazide diuretic.

CALCULATIONS AND FORMULAS
SENSITIVITY: Sensitivity = Positivity in Disease. Sensitivity indicates the probability of a positive test result when the disease is present and is given by the relationship:

$$\text{Sensitivity(\%)} = \frac{\text{True(+)}}{\text{Total Tested with Disease}} \times 100$$

$$= \frac{\text{True Positives(TP)}}{\text{True positives(TP) + False Negatives(FN)}}$$

TP = True positives; the number of patients with the disease correctly classified by the test.
FN = False negatives; the number of patients with the disease misclassified by the test or normal test results in diseased patients.
The larger the number of false negatives, the less the sensitivity.

SPECIFICITY: Specificity = Negativity in Health. Specificity indicates the probability that the test result will be negative when the disease is not present; specificity is given by the relationship:

$$\text{Specificity (\%)} = \frac{\text{True(-)}}{\text{Total Tested Without Disease}}$$

$$= \frac{\text{True Negatives(TN)}}{\text{True Negatives(TN) + False Positives(FP)}} \times 100$$

TP = True positives; the number of patients with the disease correctly classified by the test.
FN = False negatives; the number of patients with the disease misclassified by the test or normal test results in diseased patients.
The larger the number of false negatives, the less the sensitivity.

PREDICTIVE VALUE OF A POSITIVE TEST: Predictive value of a positive test indicates the probability that the disease is present when the test or procedure is positive. In other words, it reflects the probability that a positive test indicates disease. It is simply the number of true positive results expressed as a fraction of all positive test results, both true and false. It may be expressed as follows:

Predictive Value(%) = $\dfrac{\text{with a Positive Test Result}}{\text{Number of All Subjects with a Positive Test Result}}$ x 100

$= \dfrac{\text{True Positives(TP)}}{\text{True Positives(TP) + False Positives(FP)}}$

The larger the number of false positives, the less the predictive value for a positive test.

Prevalence: Prevalence = $\dfrac{\text{Total with Disease}}{\text{Total Tests}}$

References: Griner, P.F. et al, Ann. Intern. Med. 94, 553-600, 1981; Gottfried, E.L. and Wagar, E.A., Disease-A-Month, August, 1983.

PROBLEM: Calculate sensitivity, specificity and predictive value of a positive test from the following information:

Enzyme Test Results in Coronary Intensive Care Unit			
Enzyme Test	Patients with M.I.	Patients without M.I.	Total
Positive Test	450 (TP)	20 (FN)	470
Negative Test	50 (FN)	480 (TN)	530
	500	500	

Answer:

Sensitivity = $\dfrac{TP}{TP + FN}$ x 100 = $\dfrac{450}{450 + 50}$ x 100 = 90%

Specificity = $\dfrac{TN}{TN + FP}$ x 100 = $\dfrac{480}{480 + 20}$ x 100 = 96%

$\begin{matrix}\text{Predictive}\\\text{Value of}\\\text{Positive Test}\end{matrix}$ = $\dfrac{TP}{TP + FP}$ x 100 = $\dfrac{450}{450 + 20}$ x 100 = 96%

CORRELATION COEFFICIENT: The coefficient of correlation (r) ranges from +1 (perfect positive correlation) to 0 (no correlation) to -1 (perfect negative correlation).

Problem: Which one of the following correlation coefficients shows the strongest relationship between two variables? +0.85, +0.50, +1.25, -0.95, 0.00.

Answer: -0.95.

CREATININE CLEARANCE:

Creatinine clearance reflects glomerular filtration rate. The equation for the calculation of creatinine clearance is as follows:

Clearance $\dfrac{ml}{min}$ = $\dfrac{\text{Urine Volume (ml/min) x Urine Creatinine}}{\text{Serum Creatinine}}$ x $\dfrac{1.73}{A}$

where A=Body Surface Area.

EXCRETION FRACTION OF FILTERED SODIUM:

This is a sensitive and specific test for acute tubular necrosis (Espinel, C.H. and Gregory, A.W., Clin. Nephrol. 13, 73-77, 1980). The measurements that are done are as follows: Serum and urine sodium, serum and urine creatinine; no timed specimen is necessary. The excretion fraction of filtered sodium is given by the equation:

Excretion Fraction of Filtered Sodium = $\dfrac{\text{Urine Na}^+ \text{ x Plasma Creatinine}}{\text{Plasma Na}^+ \text{ x Urine Creatinine}}$ x 100

AMYLASE CLEARANCE/CREATININE CLEARANCE:

Amylase clearance is <u>increased</u> in patients with <u>acute pancreatitis</u> but is normal in other causes of hyperamylasemia. Amylase clearance is useful in <u>differentiating the causes of increased serum amylase</u>; it is <u>especially useful in confirming or excluding the diagnosis of acute pancreatitis</u>. Amylase clearance is the renal clearance of amylase, expressed as a percentage of creatinine clearance, and is given by the equation:

$$\frac{\text{Amylase Clearance}}{\text{Creatinine Clearance}} = \frac{\dfrac{\text{Urine Amylase}}{\text{Serum Amylase}} \times \text{Urine Volume per Unit Time}}{\dfrac{\text{Urine Creatinine}}{\text{Serum Creatinine}} \times \text{Urine Volume per Unit Time}} \times 100$$

This equation simplifies to the form:

$$\frac{\text{Amylase Clearance (\%)}}{\text{Creatinine Clearance}} = \frac{[\text{Urine Amylase}]}{[\text{Serum Amylase}]} \times \frac{[\text{Serum Creatinine}]}{[\text{Urine Creatinine}]} \times 100$$

As seen by inspection of the above equation. the clearance ratio, expressed as percentage, is calculated simply from the concentrations of amylase and creatinine in serum and urine samples obtained simultaneously. <u>No timed collections are necessary.</u>

CORRECTION OF SERUM CALCIUM FOR DEPRESSED ALBUMIN:

Serum calcium level cannot be properly interpreted without serum albumin level; an equation for correction of serum calcium in patients with depressed albumin is:

Adjusted Calcium = Serum Calcium - Serum Albumin + 4.0

This equation corrects serum calcium so that serum calcium is increased by 1.0mg/dl for every 1.0g/dl that the albumin concentration is below 4.0g/dl.

CONVERSION FORMULAS:
To convert <u>drug</u> concentrations to or from international units, determine the conversion factor(CF) using the following formula:

$$\text{Conversion Factor(CF)} = \frac{1000}{\text{Mol. Wt.}}$$

Traditional Units to International Units:
(microgram/ml) x C.F. = I.V.(micromol/liter)
International Units to Traditional Units:
I.V.(micromol/liter)/CF = (microgram/ml)

CANCER MARKERS PANEL

Tumor markers are used to detect tumor recurrences and to assess therapy. There are three basic classes of tumor markers; these are tumor-specific or tumor-associated antigens, enzymes and hormones. The tumor markers that are most often used are listed in the next Table:

Tumor Markers

Tumor-Specific or Tumor-Associatecd Antigens:
Carcinoembryonic Antigen(CEA)
Alpha-Fetoprotein(Alpha-FP)
Enzymes: Acid Phosphatase
Hormones: Human Chorionic Gonadotropin(HCG)

Conditions that have been associated with an increase in serum concentration and the approximate sensitivities of CEA in these conditions are given in the next Table:

Conditions Associated with Increase in Serum Concentration of Carcinoembryonic Antigen(CEA)

Carcinomas:	Sensitivities(%)	Non-Malignant Conditions	Sensitivities(%)
Colon and Rectum	70-80	Pulmonary Emphysema	20-50
Pancreas	60-90	Active Ulcerative Colitis	10-25
Lung	65-75	Alcoholic Cirrhosis	25-70
Stomach	30-60	Cholecystitis	6-20
Breast	50-65	Rectal Polyps	4-20
Ovary	25-50	Benign Breast Disease	4-15
Thyroid	20		
Non-Carcinoma Malignancy			

The concentration of alpha-fetoprotein in the serum in various conditions is shown in the next Table:

Frequency of Alpha-Fetoprotein Concentrations in the Serum in Different Conditions

Condition	<20ng/ml	>20ng/ml	>1000ng/ml
Normal Individuals	100		
Hepatoma	10	90	65
Nonseminomatous Germ Cell Testicular Tumors	25	75	62
YoTk "Sac" Tumors of Ovary		High Values	
Pancreatic Carcinoma	77	23	
Gastric Carcinoma	82	18	
Colonic Carcinoma	95	5	
Breast Carcinoma	100		
Viral Hepatitis	69	31	<1
Chronic Active Hepatitis	67	33	2
Ataxia Telangiectasia	0	100	10
Tyrosinemia	0	100	

The elevation of serum acid phosphatase in patients with carcinoma of the prostate is shown in the next Table:

Serum Acid Phosphatase in Patients with Carcinoma of the Prostate

Condition	Percent of Patients
Carcinoma but no metastases	10-20
Metastases without bone involvement	20-40
Metastases with bone involvement	70-90

HCG is not found in the sera of healthy non-pregnant persons and is not subject to any feedback-loop control. Monitoring levels of HCG are useful in following patients with tumors as listed in the next Table:

Tumors Associated with an Increase in Human Chorionic Gonadotropin(HCG)

Gestational Trophoblastic Neoplasia
Hydatidiform Mole; Invasive Mole;
Choriocarcinoma
Testicular Tumors: Choriocarcinoma;
Embryonal Carcinomas with STGC;
Seminoma with STGC
Ovarian Tumors: Embryonal Carcinoma;
Some Mixed Germ Cell Tumors
Ectopic Production of HCG
Pancreatic Islet-Cell Carcinomas

STGC = Syncytiotrophoblastic Cell

Tumors other than those listed in the previous Table may be associated with increased serum HCG; these are cancers of the liver(21 percent), stomach(22 percent), pancreas(33 percent).

Other Tumor Markers: Tumor markers, other than CEA, AFP, acid phosphatase and HCG have been used and are being developed. This has been done in an attempt to obtain greater sensitivity and specificity.

Tumor-specific and tumor-associated antigens are listed in the next Table:

Tumor-Specific or Tumor-Associated Antigens:
CA 125 (Ovarian Carcinomas)
Breast-Cystic Fluid Protein(BCFP)
Colon Mucoprotein Antigen(CMA)
Colon-Specific Antigen(CSA)
Zinc Glycinate Marker(ZGM)(Colon)
Carbohydrate Antigen 19-9 (Pancreatic Carcinoma, Hepatoma Gastric Carcinoma)
Pancreatic Oncofetal Antigen(POA)
Prostate-Specific Antigen(PSA)
S-100 Protein(Malignant Melanoma)
Sialoglycoprotein (Wide Variety of Cancers)
Tissue Polypeptide Antigen(TPA)(Wide Variety of Cancers)
B-Protein (Wide Variety of Antigens)
"Tennessee" Antigen Glycoprotein (Wide Variety of Cancers)

This list is not all inclusive. Rising and falling levels of CA 125 correlated with progression or regression of epithelial ovarian cancer in 93 percent of patients (Bast, R.C., et al. N. Engl. J. Med. 309, 883-887, 1983; Lewis, J.L. N. Engl. J. Med. 309, 919-921. 1983). Centocor, Inc. (244 Great Valley Parkway, Malvern, PA 19355) has introduced a kit for assay of CA 125; currently, this kit is available for research purposes only. Other tumor markers in this Table show varying degrees of specificity for different tumors.

Tumor-specific enzymes are, in general, the least specific of the different types of tumor markers; a list of enzymes that have been used to monitor cancers is given in the next Table:

Enzymes As Tumor Markers
Alkaline Phosphatase, Lactate Dehydrogenase(LDH) and Gamma Glutamyl Transpeptidase
Muramidase
Creatine Phosphokinase-"Brain" Isoenzyme(CPK-BB)
Galactosyl Transferase Isoenzyme II(GT-II)
Beta-Glucuronidase
Sialytransferase
Terminal Deoxynucleotidly Transferase
Ribonuclease
Histaminase (Medullary Carcinoma of the Thyroid)
Amylase
Cystine Aminopeptidase

Finally, hormones that can be associated with non-endocrine tumors are listed in the next Table:

Hormones Associated with Non-Endocrine Tumors	
Tumor Marker	Common Tumor Association
ACTH	Lung (Usually Oat Cell)
PTH	Lung (Usually Epidermoid Type)
Insulin	Lung
Glucagon	Pancreas
Gastrin	Stomach and Other Carcinomas
Calcitonin	Lung, Breast
Prostaglandins, Erythropoietin	Kidney

Markers for specific tumors are given in the next Table:

Markers for Specific Tumors		
Tumor	Marker	
	Alpha-Fetoprotein	HCG
Testicular Tumors		
Choriocarcinoma	-	+
Embryonal Carcinoma	+	

Other tumors may contain syncytiotrophoblastic cells that secrete HCG.

Alpha-fetoprotein or HCG or both are present in 80 to 90 percent of patients with stage 3 testicular cancer.

Lactic dehydrogenase(LDH) levels are elevated in about 60 percent of patients with stage 3 testicular cancer; serial LDH may be helpful in detecting recurrence of cancer.

S. Bakerman

CANDIDA ANTIBODY, SERUM
SPECIMEN: Red top tube, separate and then freeze serum.
REFERENCE RANGE: Negative. Results most indicative of invasive candidiasis
are as follows: conversion from a negative to a positive reaction; a fourfold or
greater rise in antibody titer; a single titer of 1:8 or more, an increasing
number of precipitin bands (Penn, R.L. et al., Arch. Intern. Med. 143,
1215-1220, 1983).
METHOD: Counterimmunoelectrophoresis(CIE), immunodiffusion(ID) and latex
agglutination(LA). The ID test is not quantitative; the LA and CIE tests may be
quantitated.
INTERPRETATION: Candidiasis is caused by several candida species; the common
pathogen is Candida albicans.
    Candida organisms normally occur in the mouth, gastrointestinal tract and
the vagina of some normal individuals. Disease occurs in immunocompromised
patients, especially those with depressed T-cell function; newborn infants,
particularly those bottle-fed; diabetes; the urinary bladder of a catheterized
patient; debilitating illnesses, e.g., cancer, tuberculosis, etc; patients
treated with broad-spectrum antibiotics sufficient to reduce normal bacterial
flora and permit overgrowth of the fungi; vaginal candidiasis in females taking
oral contraceptives or during pregnancy; candidial endocarditis (valvular)
following cardiac surgery, in drug addicts and in patients with long-term venous
catheters.
    The sensitivity of the tests for candida antibody is about 50 percent;
false positive results are obtained in patients with bacterial endocarditis,
Torulopsis glabrata infection, and in other patients (Penn, R.L. et al, Arch.
Intern. Med. 143, 1215-1220, 1983).
    Diagnosis of Candida infections of the skin and mucous membrane may be made
by obtaining smears or scrapings on glass slides; add a drop of saline solution
or 10 percent NaOH or KOH. Stains are usually not necessary; however, Gram,
Wright, Ziehl-Neelsen, Giemsa, Papanicolaou, periodic acid-Schiff(PAS) or
methenamine silver method may be used.
    The routine isolation medium is Sabouraud's agar.

CARBAMAZEPINE (TEGRETOL)

SPECIMEN: Red top tube, separate serum and refrigerate; or green (heparin) top tube, separate plasma and refrigerate. Reject if serum frozen on cells or left standing on cells for several days.

REFERENCE RANGE: Therapeutic: 2-10mcg/ml; Toxic: >12mcg/ml. Time to Obtain Serum Specimens (Steady State): 2-4 weeks (see dosage for initiation of therapy below); Half-life: chronic dosing, adults and children, 5 to 27 hours.

METHOD: EMIT; HPLC; GLC; Fluorescence Polarization(Abbott); SLFIA(AMES).

INTERPRETATION: Carbamazepine (Tegretol) is used for the prophylaxis of grand mal (primary generalized) seizures, other seizure disorders, including simple and complex partial seizures, and for the relief of pain associated with trigeminal neuralgia (tic douloureux). It has no effect on absence attacks (petit mal).

Carbamazepine is given by mouth and is slowly absorbed from the G.I. tract; it reaches a peak in the serum in 2 to 24 hours; about 3/4 of the drug is bound to albumin. Carbamazepine is metabolized (99%) by the liver. The clearance and half-life decrease with time so that it is advisable to initiate therapy gradually over two to four weeks; dose schedule and time to obtain blood specimens are as follows:

Dose Schedule and Blood Specimens	
Dose	Blood Specimens
Adults:	
Week 1: 2 to 5mg/kg/day (2 to 4 dose/day)	Blood specimen obtained during weeks 3 or 4, several days after starting
Week 2: 4 to 10mg/kg/day (2 to 4 dose/day)	maintenance dose. Optimally, obtain blood specimen at start, middle and
Week 3: (Maintenance Dose) 7 to 15mg/kg/day (2 to 4 dose/day)	end (trough) of a dosing period; otherwise, obtain specimen at end of dosing period.
Children:	
Week 1: 4 to 7mg/kg/day (3 or 4 dose/day)	Blood specimen obtained during weeks 3 or 4, several days after starting
Week 2: 8 to 14mg/kg/day (3 or 4 dose/day)	maintenance dose; optimally, obtain blood specimen at start, middle and
Week 3: (Maintenance Dose) 15 to 20mg/kg/day (3 or 4 dose/day)	end (trough) of a dosing period; otherwise, obtain specimen at end of dosing period.

There is variability in carbamazepine absorption and the trough concentration may not represent the lowest drug level during the dosing interval; therefore, multiple determinations may be necessary in some patients.

The time to peak concentration is about 3 hours after an oral dose when patients are on chronic therapy.

The rate of clearance of carbamazepine in the presence of phenobarbital and phenytoin is increased.

Aplastic anemia has been reported to occur in patients on carbamazepine; therefore routine blood counts should be performed at 2-4 months intervals.

CARBON DIOXIDE CONTENT (CO_2 CONTENT)

SPECIMEN: Red top tube, separate serum
REFERENCE RANGE: 24-30mmol/liter
INTERPRETATION: The carbon dioxide content is the sum of concentration of bicarbonate and carbonic acid as given by the following equation:

$$CO_2 \text{ Content} = [HCO_3^-] + [H_2CO_3]$$
$$[H_2CO_3] = PCO_2 \times 0.03$$

The $[HCO_3^-]$ is about 20 times the $[H_2CO_3]$; therefore the CO_2 content approximates the value of the $[HCO_3^-]$. CO_2 content is decreased in metabolic acidosis and increased in metabolic alkalosis. The CO_2 content is mildly increased in chronic respiratory acidosis with compensation. The CO_2 content is mildly decreased in chronic respiratory alkalosis with compensation.

CARBON MONOXIDE (CARBOXYHEMOGLOBIN)

SPECIMEN: Grey (oxalate fluoride) top tube
REFERENCE RANGE: Rural Nonsmokers: 0.4 to 0.7%; Urban Nonsmokers: 1 to 2%; Smokers: 5 to 6%; Toxic: >15% Saturation. The half elimination time is 240 minutes breathing room air; it is 40 minutes when the patient is breathing 100 percent oxygen. To convert conventional units in % to international units, multiply conventional units by 0.01.
METHOD: Spectrophotometric; commercial instrument: IL182
INTERPRETATION: Carbon monoxide combines with hemoglobin as follows:

CO + Hemoglobin ⟶ CO Hemoglobin (Carboxy hemoglobin)

The correlations of CO in the atmosphere, saturation of CO in blood and symptoms, are given in the next Table:

Carbon Monoxide Levels and Symptoms		
CO in Atmosphere (%)	% Saturation CO in Blood	Symptoms
to 0.01%	to 10%	None
0.01% to 0.1%	10% to 50%	Headache, Tendency to Collapse
0.1% to 1.00%	> 50%	Respiratory Failure and Death

Severe symptoms develop in almost all patients when CO saturation rises above 20% saturation. Levels above 40% saturation are fatal if not treated immediately. Metabolic acidosis is an ominous sign. If carbon monoxide poisoning is suspected, take a blood specimen and treat patient with 100% oxygen at a high flow rate; give whole blood transfusion (Jackson, D.L. and Menges, H., JAMA 243, 772-774, 1980).

Carbon monoxide is produced wherever organic matter is burned; cigarettes, automobile exhaust and fires are the most frequent sources of exposure to toxic gas. Carbon monoxide is tightly bound to hemoglobin; carbon monoxide has an avidity for hemoglobin more than 200 times that of oxygen. The major pathophysiologic disturbances in patients with carbon monoxide poisoning is that of tissue hypoxia.

CARCINOEMBRYONIC ANTIGEN (CEA)

SPECIMEN: Lavender top tube (EDTA); separate plasma and place in plastic vial.

REFERENCE RANGE: Non-Smokers, <2.5ng/ml; Smokers, <4.0ng/ml.

METHOD: Enzyme-linked immunosorbent assay (ELISA), and RIA, Abbott; RIA, Roche. The antibodies in the Abbott and Roche kits measure different CEA antigen determinants and may give different values. Heparin may give spuriously elevated values with the Roche method.

INTERPRETATION: Serum carcinoembryonic antigen concentration is most useful for monitoring therapy with anti-neoplastic drugs and following surgical removal of neoplasms; it is not a screening test for cancer. Conditions that have been associated with an increase in serum concentration of carcinoembryonic antigen and the approximate sensitivities of CEA in these conditions are given in the next Table:

<table>
<tr><td colspan="4" align="center">Conditions Associated with Increase in Serum Concentration
of Carcinoembryonic Antigen</td></tr>
<tr><td>Carcinomas:</td><td>Sensitivities (%)</td><td></td><td>Sensitivities</td></tr>
<tr><td>Colon and Rectum</td><td>70-80</td><td>Non-Malignant Conditions:</td><td>(%)</td></tr>
<tr><td>Pancreas</td><td>60-90</td><td>Pulmonary Emphysema</td><td>20-50</td></tr>
<tr><td>Lung</td><td>65-75</td><td>Active Ulcerative Colitis</td><td>10-25</td></tr>
<tr><td>Stomach</td><td>30-60</td><td>Alcoholic Cirrhosis</td><td>25-70</td></tr>
<tr><td>Breast</td><td>50-65</td><td>Cholecystitis</td><td>6-20</td></tr>
<tr><td>Other Carcinomas</td><td>20-50</td><td>Rectal Polyps</td><td>4-20</td></tr>
<tr><td>Non-Carcinoma Malignancy</td><td></td><td>Benign Breast Disease</td><td>4-15</td></tr>
</table>

CEA levels are elevated in smokers and former smokers. CEA levels are elevated in renal failure (Branstetter, R.D. et al., Ann. Intern. Med. 91, 867-868, 1979).

The CEA level is relatively high in patients with well differentiated and widely metastasized carcinoma of the colon and rectum as compared to poorly differentiated or localized disease (Goslin, R. et al., Am. J. Med. 71, 246-253, 1981).

CEA levels have been used as a prognostic factor in advanced colorectal carcinoma; patients with an initially normal level of CEA versus those with an abnormal level of CEA had median survivals of 23 and 9.2 months respectively. (Kemeny, N. and Brawn, D.W., Am. J. Med. 74, 786-794, 1983.)

CEA levels are increased in almost 40 percent of patients with ovarian cancer (Bast, R.C. et al., N. Engl. J. Med. 309, 883-887, 1983).

CEA and Metastases: The CEA assay complements the use of liver scan for detecting liver metastases. In a study by McCarthy, W.H. and Hoffer, P.B. (JAMA 236, 1023-1027, 1976), 56 percent of patients who had liver metastases had an elevated CEA while 80 percent had a positive liver scan. If both CEA assays and liver scans gave positive results, the probability of metastases was virtually 100 percent and if both tests had negative results, the probability of liver involvement was approximately 1 percent.

CAROTENE

SPECIMEN: Patient must be fasting and receive no vitamin supplement or foods containing vitamin A or carotene for 24 hours before testing for younger age group (0-6 mo.); 48 hours in the older group (older than 6 months). Red or green (heparin) top tube; protect from light by wrapping in foil; separate serum or plasma and continue to protect from light. The specimen may be stored in the freezer (-20°C) for four days.

REFERENCE RANGE:

Change of Serum Carotene with Age		
	Range	
Age	(mg/dl)	(umol/liter)
0-6 months	0-40	0-0.75
6 months-Adult	40-180	0.75-3.4

Meites, S., Ed., Pediatric Clinical Chemistry, Am. Assoc. Clin. Chem., 1725 K. Street, N.W., Washington, D.C. 20006.

METHOD: Spectrophotometric

INTERPRETATION: Causes of low serum carotene are given in the next Table:

Causes of Low Serum Carotene
Liver Disease
Fat Malabsorption Syndromes
Lack of Carotenoids in the Diet
High Fever

The concentration of serum carotene is affected most by hepatic function.

Vitamin A is fat soluble and is absorbed with lipid; malabsorption of vitamin A may occur in any condition in which steatorrhea (pathological increase in stool fat) occurs such as pancreatic insufficiency, deficiency of bile and impairment of intestinal absorption.

CATECHOLAMINES, FRACTIONATION (NOREPINEPHRINE, EPINEPHRINE AND DOPAMINE),PLASMA, FREE

SPECIMEN: Prepare the patient prior to obtaining the blood specimen as follows: Place a heparinized catheter into a vein and reassure the patient regarding the procedure. After the patient has remained in the supine position in nonstimulating surroundings for at least 30 minutes, draw blood through the indwelling catheter into a green top tube (heparin anticoagulant). Use minimum tourniquet time. Mix gently by inversion; do not shake. Immediately place tube in an ice bath. Centrifuge tube in the cold. Immediately, separate the plasma and freeze plasma. For evaluation of orthostatic (postural) hypotension, the patient should be instructed to stand and a second specimen is obtained after 10 minutes.

REFERENCE RANGE: (Mayo Medical Laboratories, Interpretative Data for Laboratory Tests)

Fraction	Mean (pg/ml)	95% Range (pg/ml)	Observed Range (pg/ml)
Norepinephrine			
Supine	220	100-400	70-750
Standing	500	300-900	200-1700
Epinephrine			
Supine	30	<70	<110
Standing	50	<100	<140
Dopamine	<30	No postural change	

INTERPRETATION: The conversion of norepinephrine to epinephrine occurs in the adrenal medulla. Throughout most of the sympathetic nervous system, the paraganglion system and the central nervous system, norepinephrine is synthesized and not epinephrine. In the adrenal medulla, norepinephrine and epinephrine are synthesized. Fractionation of catecholamines may be useful in differentiating extra-adrenal pheochromocytoma from adrenal pheochromocytoma; in extra-adrenal pheochromocytoma, norepinephrine may be significantly elevated as compared to epinephrine. However, the results of fractionation of catecholamines in some patients with pheochromocytoma are not compatible with this premise.

CATECHOLAMINES, FREE, URINE

SPECIMEN: 24 hour urine: There is a diurnal variation in catecholamines.
Instruct the patient to void at 8:00 A.M. and discard the specimen. Add 25ml of
6N HCl to container prior to collection. Then, collect all urine including the
8:00 A.M. specimen at the end of the 24 hour collection period. Refrigerate jug
as each specimen is collected. Following collection, add 6N HCl to pH 1 to 2
(Do not use boric acid). Catecholamines are unstable in urine at alkaline pH.
Record 24 hour urine volume. In urine collected at a pH of less than 3 and
stored at 4°C, the catecholamines are stable for at least one week. Forward
100ml aliquot of 24 hour urine for analysis.

Patient should be drug free if possible. Methyldopa (Aldomet), drug free,
1 week; apresoline-drug free, 72 hours; no quinidine, epinephrine or
norepinephrine related drugs, i.e., neosynephrine nose drops. Interfering
substances, include L-dopa and methenamine mandelate, should be discontinued 48
hours prior to collection of specimen. Kidney function test dyes interfere.

REFERENCE RANGE:

Normal Values of Urinary Catecholamines	
Age	Normal Values (micrograms/24 hours)
Birth to 1 year	4 - 20
1 to 5 years	5 - 40
6 to 15 years	5 - 80
Over 15 years	20 - 100
Adults	Up to 100

Vorrhess, M.L.: "Urinary Catecholamine Excretion by Healthy Children", Ped. 39,
252-257, 1967.

METHOD: Extraction followed by HPLC reverse phase chromatography (Moyer,
T.P., Tyce, G.N., Jianp, N.S. and Sheps, S.G., "Analysis for urinary
catecholamines - liquid chromatography with amperometric detection: methodology
and clinical interpretation of results," Clin. Chem. 25, 256-263, 1979).

INTERPRETATION: Free catecholamines are elevated in patients with
pheochromocytoma and neuroblastoma.

Urinary catecholamines (epinephrine, norepinephrine and dopamine) are
excreted as glucuronide or sulfate metabolites plus free catecholamines. Total
catecholamines have a large range of normal which may obscure the diagnosis of a
minimally excreting tumor. In addition, dietary catecholamines occur in
conjugated form. Therefore, urine free or unconjugated catecholamines is the
preferred test. The sensitivity of this test is 100 percent during periods of
hypertension; false negative results are observed in patients with
pheochromocytoma who are normotensive during specimen collection. There is a 3
percent false positive rate. A pheochromocytoma is highly likely if:

Free Epinephrine plus Free Norepinephrine $\geq$ 100mcg/24 hours

or

Free Dopamine Excretion > 400mcg/24 hours.

CATECHOLAMINES, PLASMA, TOTAL

SPECIMEN: Prepare the patient prior to obtaining the blood specimen as
follows: Place a heparinized catheter into a vein and reassure the patient
regarding the procedure. After the patient has remained in the supine position
in nonstimulating surroundings for at least 30 minutes, draw blood through the
indwelling catheter into a green top tube (heparin anticoagulant). Use minimum
tourniquet time. Mix gently by inversion; do not shake. Immediately place tube
in an ice bath. Centrifuge tube in the cold and separate the plasma; freeze
plasma immediately.

REFERENCE RANGE: 2 to 32ng/ml

METHOD: RIA

INTERPRETATION: The catecholamines (epinephrine, norepinephrine and dopamine)
are elevated in pheochromocytoma and neuroblastoma.

CATECHOLAMINES, TOTAL
(INCLUDES UNCONJUGATED AND CONJUGATED), URINE

SPECIMEN: 24 hour urine: There is a diurnal variation in catecholamines. Instruct the patient to void at 8:00 A.M. and discard the specimen. Add 25ml of 6N HCl to container prior to collection. Then, collect all urine including the 8:00 A.M. specimen at the end of the 24 hour collection period. Refrigerate container as each specimen is collected. Following collection, add 6N HCl to pH 1 to 2 (Do not use boric acid). Catecholamines are unstable in urine at alkaline pH. Record 24 hour urine volume. In urine collected at a pH of less than 3 and stored at 4°C, the catecholamines are stable for at least one week. Forward 100ml aliquot of 24 hour urine for analysis.

Patient should be drug free if possible. Methyldopa (Aldomet) drug free 1 week. Apresoline-drug free 72 hours. No quinidine, epinephrine or norepinephrine related drugs, i.e., neosynephrine nose drops. Kidney function test dyes interfere.

REFERENCE RANGE: The normal urinary contents of catecholamines (expressed as mcg/24 hours) are lower in children than in adults. The reference range is given in the next Table:

Reference Range of Catecholamines	
Age (years)	Total (mcg/24 hours)
<1	Up to 20
1-5	Up to 40
6-15	Up to 80
>15	Up to 100
Adult	Up to 270

Meites, Samuel in Pediatric Clinical Chemistry, Am. Ass. for Clin. Chem. 2nd Ed. 1981, 1725 K Street, N.W., Wash. D.C. 20006.

METHOD: The total urinary catecholamine amine content (norepinephrine, epinephrine, dopamine) of urine (free and conjugated) is usually determined fluorometrically by the trihydroxindole method.

INTERPRETATION: Urinary catecholamines are elevated in patients with pheochromotocytoma (especially during episodic hypertensive episode), neuroblastomas (malignant neoplasm that can appear at any time from birth to six years of age). They are also elevated in individuals undergoing vigorous excercise and in patients with myasthenia gravis and progressive muscular dystrophy.

CEREBROSPINAL FLUID (CSF)

SPECIMEN: 5 or more ml of CSF

REFERENCE RANGE: Cell count: lymphocytes and monocyte 0-10/cu mm; infant, 1-20/cu mm; neutrophils and erythrocytes are absent; gram stain; cytology; CIE; latex agglutination; cultures; cryptococcus (India Ink Preparation); protein, 15-45 mg/dl; albumin/globulin ratio, 3:1; glucose, 50-85 mg/dl; chloride, 120-130 mEq/liter; pressure (mm H_2O), 50-200; LDH <40; lactate, <20mg/dl

METHOD: Prior to lumbar puncture, examine eyegrounds for evidence of papilledema. In the presence of papilledema, the mortality rate from lumbar puncture may be about 0.3 percent (Marshall, J., Brit. J. Hosp. Med. 3, 216, 1970). Potential problems and complications of lumbar puncture are as follows: herniation of uncus or cerebellar tonsils; progression of paralysis with spinal tumor; hematoma in patients with clotting defects; meningitis in presence of sepsis; asphyxiation of infants due to excessive constraint or tracheal obstruction caused by pushing the head forward; and introduction of infection (Krieg, A.F., Clinical Diagnosis and Management, Henry, J.B. ed. Saunders, Philadelphia, 1979, Vol. 1, page 637).

Lumbar puncture is performed between L3 and L4; in infants and children, puncture is done between L4 and L5. The patient lies on his side with back flexed as much as possible to "open up" the space. Mark the site for puncture, sterilize skin and anesthetize skin and underlying tissue. Place needle with stylet in place at midline, direct slightly cephaled and push slowly; at about 2-3 cm, remove stylet, check for flow of spinal fluid and before advancing further, replace stylet. If the needle strikes bone, withdraw and redirect. A "give" indicates that the needle is in the spinal canal.

A manometric reading is obtained (opening pressure) when spinal fluid begins to flow. If the tap is bloody, drain 2 to 3 ml; if it clears, then a blood vessel was ruptured during puncture. Collect three tubes, each filled to 2 to 3 ml: one for chemical determinations (protein, glucose, chloride, LDH, lactate); one for cell count, differential and serology; and the final tube for microbiological studies (India Ink preparation for cryptococci, gram stain, latex agglutination tests, CIE, bacterial, fungal, and viral cultures). Differentiation of Traumatic Tap from Pathologic Bleeding: Traumatic tap is differentiated from pathologic bleeding by the following: The traumatic tap shows non-homogeneous mixing in the manometer; the traumatic tap shows visible clearing between the first and subsequent tubes and the erythrocyte count decreases in a similar manner; a very bloody tap will clot on standing; xanthochromia (pink to orange to yellow color in the supernatant of centrifuged CSF) is found in pathologic bleeding. The specimen must be centrifuged within one hour after collection to avoid false positives.

INTERPRETATION: Typical CSF findings are shown in the next Table:

			Typical CSF Findings in Some Conditions		
Condition	Pressure mm CSF	Gross Appearance	Cells (cu. mm.)	Protein (mg/dl)	Glucose (mg/dl)
Normal	50-200	Clear, Colorless	0-10 Lymphs & Monocytes	Under 45	2/3 Blood glucose 50-80
Acute Bact. Meningitis	200-500 Increased	Turbid, May Clot	100-10,000; Polyps	50-500	Absent or very low
Tuberculous Meningitis	Increased 200-500	Turbid, Pellicle Common	10-500, chiefly Lymphs	50-500	Often under 40
Aseptic or Viral Meningitis	Increased 200-500	Clear or Slightly Turbid	10-500, chiefly Lymphs	45-200	Normal
Multiple Sclerosis	Normal	Clear, Colorless	Normal or 10-50 Lymphs	Normal or 45-100	Normal
Cerebral Thrombosis	Usually normal to slightly inc.	Clear	Usually Normal	Normal or 45-100	Normal
Cerebral Hemorrhage	Usually Normal	Bloody or Xanthochromic	Increased RBC	Increased 45-100	Normal
Subarachnoid Hemorrhage	Increased 200-500	Bloody or Xanthochromic	Increased RBC	Increased 50-1000	Normal
Brain Tumor	Usually Increased 200-500	Clear or Xantho-chromic	Normal to 50 Lymphs	Normal or Slightly Increased	Normal or Slightly Decreased

CIE = Counterimmunoelectrophoresis

113

Cerebrospinal Fluid   S. Bakerman
(CSF)

Diagnosis of Infectious Diseases: Tests of the cerebrospinal fluid (CSF) for
the diagnosis of infectious diseases are given in the next Table:

Tests of the CSF for the Diagnosis of Infectious Diseases
Conventional Laboratory Methods:
    Cellular, Protein and Glucose Content
    Gram Stain
    Culture with Isolation of Organisms
    India Ink Preparation for Cryptococci
Detection of Specific Bacterial Antigens:
    Latex Agglutination (LA)
    Counterimmunoelectrophoresis (CIE)
Rapid and Non-specific Methods:
    Lactic Acid
    Lactate Dehydrogenase (LDH)
    C-Reactive Protein (C-RP)

Bacterial and Viral Meningitis: The sensitivity and specificity of various CSF
laboratory studies for bacterial meningitis versus viral meningitis are given in
the next Table; (Corrall, C.J. et al. J. Pediat. 99, (3), 365-369, 1981):

Sensitivity and Specificity (%)
Screening Tests for Bacterial Meningitis versus Viral Meningitis

Screening Test (CSF)	Sensitivity (%)	Specificity (%)
WBC >500 cells/cu mm	74	94
Polymorphonuclear Leukocytes >200 cells/cu mm	91	84
Protein >100 mg/dl	74	94
Glucose <40 mg/dl	78	100
Gram Stain	74	100
Lactic Acid >35 mg/dl (3.9 mmol/liter)	100	100
C-Reactive Protein	100	94
LDH > 40	High	High

Rapid and specific diagnosis of bacterial meningitis is done by latex
agglutination (LA) tests and counterimmunoelectrophoresis (CIE) for assay of
antigens of microorganisms; the tests, available as of May, 1984, are listed in
the next Table:

Latex Agglutination (LA) and Counterimmunoelectrophoresis (CIE)
for Bacterial Antigens

Antigen	Latex Agglutination	Counterimmunoelectrophoresis
Neisseria Meningitidis	Yes	Yes
Hemophilus Influenzae, Polyvalent (a-d)	Yes	Yes
Streptococcus Pneumoniae	Yes	Yes
Listeria Monocytogenes		Yes
Group B Streptococcus	Yes	Yes

Bacteria Causing Meningitis: The organisms most commonly responsible for
bacterial meningitis, by age-group, are given in the next Table:

Usual Organisms, by Age, in Bacterial Meningitis

Age	Organisms
Newborns	Gram Negative Bacilli, Group B Streptococci, Listeria Monocytogenes
Children, >2 months	Haemophilus Influenzae, Type b
Healthy Adults	Neisseria Meningitidis (Meningococci)
Elderly	Streptococcus Pneumoniae (Pneumococci)

The approximate incidence, in percent, of different microorganisms causing
meningitis in the age range from two days to 10 years is as follows: H.
influenzae, Type b (67%), N. meningitidis (16%), Str. pneumoniae (9%), Group B
streptococci (5%), Listeria monocytogenes (4%). H. influenzae, type b, is the
most common causative organism in meningitis; the approximate incidence in
percent of different microorganisms associated with mortality with meningitis is
as follows: Streptococcus pneumoniae (20%), H. influenza, type b (10%),
Neisseria meningitidis (5%). Bacteremia occurs in 40 to 42 percent of patients
with meningitis.

The organisms that most commonly cause neonatal meningitis are Escherichia
coli and Group B streptococci. In the first week of life, infections are
likely to be caused by bacteria colonizing the maternal birth canal, and from
organisms transmitted from people caring for the newborn infant (usually via the

hands) and from organisms flourishing in the humidifying apparatus that is used in neonatal intensive care units.

It usually has been assumed that maternally derived IgG bactericidal antibody was protective to the infant where H. influenzae pneumococci and meningococci were concerned. Infection by organisms other than gram-negative bacilli become relatively common from about the third month of life when the maternal antibodies have disappeared and the infant's own production is still inadequate.

In children, H. influenzae, Type b, is the most common cause of bacterial meningitis. Meningitis in adults is rarely caused by H. influenzae; the most common causes of meningitis in adults are meningococci and pneumococci. Bacterial meningitis may occur secondary to pneumonia, endocarditis, osteomyelitis or extension of infections of the sinuses and ears.

Opportunistic Infections of the CNS: There are specific conditions that have a striking association with infection of the CNS by particular organisms. Some of these conditions are associated with impaired immunity . This is largely the result of the widespread use of immunosuppressive drugs such as cytotoxic drugs and corticosteroids to treat malignant diseases and connective tissue disorders and after organ transplantation. These patients and those with congenital or acquired immune deficiency are particularly prone to serious infections. The microorganisms involved are many times not pathogenic to normal persons; these infections are termed opportunistic. Conditions that have a striking association with infection of the central nervous system by particular organisms are shown in the next Table:

Conditions That Are Associated with Infection of the CNS by Particular Organisms	
Condition	Organism
CNS Surgery	Gram-Negative Bacilli
	Staphylococcus Aureus
Multiple Antibiotic Therapy in Seriously Ill Patients	Fungi
Leukopenia	Gram-Negative Bacilli
	Fungi
Leukemia, Lymphomas Cytotoxic Drug or Corticosteroid Administration	Viruses Fungi Listeria Nocardia
Treated Acute Lymphoblastic Leukemia in Childhood	Measles Virus
Splenectomy in Childhood and Adolescence	Streptococcus Species (Pneumococci) H. Influenzae

Culture: Spinal fluid should be cultured for bacteria, fungi, and tuberculosis. Neisseria meningitidis is cultured on chocolate and blood agar plates in a 3-5% CO_2 atmosphere. Incubate 2-3 ml to 37°C for 24 hours and incubate in a second set of cultures.

TUBERCULOUS MENINGITIS: The diagnosis of tuberculous meningitis requires demonstration of acid-fast organisms on smears of CSF or growth of mycobacteria in CSF cultures. The demonstration of T.B. on smears occurs infrequently; growth on culture media may require 8 weeks.

Mardh, P.A., et al (Lancet 1, 367, 1983) identified tuberculostearic acid, a characteristic constituent of mycobacteria, using gas chromatography and mass spectrometry, in CSF of a single patient within two days of clinical onset of tuberculous meningitis. The diagnosis was subsequently confirmed by detection of mycobacteria in the CSF.

MULTIPLE SCLEROSIS: See MYELIN BASIC PROTEIN and PROTEIN ELECTROPHORESIS

PITUITARY TUMORS: See PITUITARY PANEL for hormones in the CSF in patients with pituitary tumor and suprasellar extension.

CEREBROSPINAL FLUID (CSF) PROTEIN

SPECIMEN: 1ml CSF

REFERENCE RANGE: Premature Newborn: 15-130 mg/dl; Full-Term Newborn: 40-120 mg/dl; <1 month: 20-80 mg/dl; Adult Total: 15 to 45mg/dl; Percentage: Prealbumin, 7.0%; Albumin: 49%; Globulins: Alpha-1, 7.0%; Alpha-2, 8.6%; Beta: 19%; Gamma: 9.3%. The principal immunoglobulin in CSF is IgG. Normally, the concentration of CSF IgG is less than 8mg/dl while that of IgA and IgM is less than 2mg/dl. Albumin is 26.0mg/dl or less.

METHOD: Trichloroacetic Acid precipitation of protein with light scattering at 340nm with DuPont ACA; nephelometric assay of albumin and IgG; electroimmunodiffusion; coomassie blue dye-binding method.

INTERPRETATION: An increase in CSF protein occurs in the conditions listed in the next Table:

Increase in CSF Protein
Multiple Sclerosis
Postinfective Polyneuropathy (Possible Guillain-Barré Syndrome)
Spinal Block (e.g. Neoplasms in Spinal Canal)
Minor Degrees of Spinal Cord Compression e.g. Cervical Spondylosis; Recent Prolapse of an Intervertebral Disc
Meningitis (Bacterial, 100 to 500mg/dl; Viral, usually <100mg/dl)
Encephalitis
Cerebral Abscess
Cerebral Infarction
Neurosyphilis
Intracranial Venous Sinus Thrombosis

The sensitivities of three methods in known cases of multiple sclerosis are given in the next Table; (Markowitz, H. and Kokmen, E., Mayo clin. Proc. 58, 273-274, April 1983):

Sensitivities of Methods in Multiple Sclerosis		
Measurement	Upper Limit of Normal	Sensitivity (%)
CSF IgG	7.1mg/dl	25
CSF IgG/Albumin	0.26	38
CSF IgG Index	0.77	58

$$\text{CSF IgG Index} = \frac{\text{CSF IgG x Serum Albumin}}{\text{CSF Albumin x Serum IgG}}$$

The most sensitive method for detection of CSF protein abnormalities is measurement of the CSF IgG Index; this method is superior to the measurement of CSF IgG/albumin ratio or measurement of CSF IgG only (Hershey, L.A. and Trotter, J. L., Ann. Neurol. 8, 426-434, 1980).

The CSF IgG index is not specific for multiple sclerosis; in twelve percent of patients with other neurologic diseases (excluding multiple sclerosis and infections or inflammatory diseases), the CSF IgG index was elevated (Markowitz and Kokmen, 1983).

Detection of increased myelin basic protein and detection of oligoclonal bands on CSF protein electrophoresis are other methods used to aid in the diagnosis of multiple sclerosis.

CERULOPLASMIN

SPECIMEN: Red top cube; patient fasting; send frozen in plastic vial on dry ice.

REFERENCE RANGE:

Change of Serum Ceruloplasmin with Age	
Age	Serum Ceruloplasmin (mg/dl)
Newborn	2 - 15
Two	30 - 55
Ten	20 - 45
Young Adults	20 - 45
Adults	25 - 45

To convert conventional units in mg/dl to international units in mg/liter, multiply conventional units by 10.0.

METHOD: Ceruloplasmin catalyzes the oxidation of p-phenylenediamine by molecular oxygen with the formation of a blue-violet color.

INTERPRETATION: The causes of decreased serum ceruloplasmin are given in the next Table:

Causes of Decreased Serum Ceruloplasmin
Wilson's Disease
Severe Copper Deficiency that Accompanies
Total Parenteral Nutrition
Menke's Steely Hair Disease

Ceruloplasmin is decreased in most but not all patients with Wilson's disease (hepatolenticular degeneration). Wilson's disease is an uncommon inborn error of copper metabolism inherited as an autosomal- recessive trait. In this disease, there is excessive deposition of copper in the liver, basal ganglia, cerebral cortex, kidney and cornea. Symptomatic patients usually show pigmented cornea rings and the presence of this ring is an important aid in diagnosis. The Kayser-Fleischer ring is a golden-brown, greenish or brownish-yellow, or bronze discoloration in the zone of Descemets membrane in the limbi region of the cornea.

The defect in Wilson's disease may be a defect in the normal binding of copper or ceruloplasmin. Almost all of the circulating copper is normally bound to ceruloplasmin; this bound copper is known as the indirect-reacting copper. The remaining copper (2% to 5%) is loosely bound to serum albumin; this is known as the direct reacting fraction of serum copper. Patients with Wilson's disease have decreased total copper in their serum primarily because of the decrease in indirect reacting copper. This reflects a deficient level of ceruloplasmin. However, there is an increase in direct reacting copper which is loosely bound; this loosely bound copper is readily dissociated and deposited in tissue.

The laboratory diagnosis of Wilson's disease should be suspected in patients below the age of 30 who have chronic idiopathic hepatitis. The biochemical findings in Wilson's disease are listed in the next Table:

Biochemical Findings in Wilson's Disease
Decreased Ceruloplasmin in Serum
Decreased Total Copper in Serum
Decreased Indirect Reacting Copper
Increased Direct Reacting Copper
Increased Copper in Urine
Increased Copper Deposited in Liver

The causes of elevated serum ceruloplasmin are given in the next Table:

Causes of Elevated Serum Ceruloplasmin
Oral Contraceptives
First Trimester of Pregnancy
Primary Biliary Cirrhosis
Infections

High serum levels of ceruloplasmin may cause a greenish cast to plasma.

117

CHLAMYDIA ANTIBODIES (PSITTACOSIS ANTIBODIES)

SPECIMEN: Red top tube, separate and refrigerate serum; obtain acute and convalescent serum samples; the convalescent serum specimen should be obtained 2-3 weeks after onset.

REFERENCE RANGE: Presence of antibody indicates chlamydial infection in the past. A fourfold or greater rise in antibody titer between acute and convalescent phase sera indicates recent infection.

Method: Immunofluorescence. The test system consists of a slide preparation of McCoy cells infected with lymphogranuloma venereum (LGV). The heat-inactivated serially diluted serum specimen is added to the slide, Then, a drop of fluorescein-labeled-anti-IgG is added, and the slide preparation is incubated and then examined with a fluorescence light microscope. The highest dilution with faint but definite fluorescence is the titer reported.

INTERPRETATION: (see CHLADYMIA CULTURE) The lymphogranuloma venereum antigen used in this test cannot differentiate between C. trachomatis, psittacosis and lymphogranuloma venereum. The test is not useful for diagnosis of oculogenital syndrome because of the high incidence of background antibodies in the normal population (Mayo Medical Laboratories, Interpretive Handbook, pg. 47, 1983).

CHLAMYDIA CULTURE

SPECIMEN: Chlamydial organisms are labile in-vitro; use special transport media; the specimen must be kept cold; transport to laboratory as soon as possible. Use sterile swab to obtain specimen from suspected site. Adults: Urethra, cervix or eye. Infants: Nasopharynx, throat, sputum or eye. Extract swab into appropriate transport media, sucrose phosphate, without antibiotics. Discard swab and cap the vial containing transport media.

REFERENCE RANGE: Negative

METHOD: Because C. trachomatis is an obligate intracellular parasite, it cannot be grown on artifical media; tissue culture facilities are required. Cultivation of specimen in cycloheximide-treated McCoy cells with subsequent staining with iodide at 48 to 72 hours postinoculation and observation for inclusions present in the cytoplasm of the McCoy cells which are specific for C. trachomatis; C. psittaci inclusions do not stain with iodide; both C. trachomatis and C. psittaci inclusions stain with Giemsa stain.

INTERPRETATION: Chlamydiae are Gram-negative intracellular bacteria; there are two species: C. trachomatis and C. psittaci. C. trachomatis is the causative organism in ocular and urogenital disease. C. psittici causes respiratory, urogenital and systemic infections in a variety of animals; man is incidentally infected.

C. Trachomatis: C. trachomatis may be transmitted sexually; the type or stage is given in the next Table:

Chlamydia Trachomatis: Sexually Transmitted	
Type or Stage	Comments
Urethritis or Cervicitis	"Urethral syndrome" with sterile pyuria in sexually active women.
Pelvic Inflammatory Disease	Cervicitis may lead to invasion with endometritis and salpingitis and perihepatitis.
Epididymitis	C. trachomatis is a major cause of epididymitis in young, sexually active men.
Oculogenital Syndrome	Autoinoculation of the conjunctiva from a genital focus.
Proctitis	Occurs in homosexual men practicing anal intercourse.
Neonatal: Ophthalmia Pneumonia	About 10% of infants born to infected women will be infected with C. Trachomatis (Schachter, J., Lancet, 2, 377, 1979); these infants are at risk for conjunctivitis and pneumonitis. Give erythromycin ophthalmic ointment to prevent conjunctivitis; neonatal pneumonia can be treated with systemic erythromycin for 14 days.
Lymphogranuloma Venereum	Caused by certain types of C. trachomatis

The sexually transmitted diseases caused by C. trachomatis in men, women and infants are given in the next Table; (Syva Monitor, 2, No. 1, 1-4, Jan./Feb. 1984):

Chlamydia Trachomatis in Men, Women and Infants			
Men and Women	Women	Men	Infants
Lymphogranuloma Venereum	Cervicitis	Urethritis	Inclusion Conjunctivitis
Inclusion Conjunctivitis	Salpingitis	Epididymitis	Pneumonia
	Urethritis-Urethral	Reiter's Syndrome	Rhinitis
Otitis Media	Syndrome	(Suspected)	Otitis Media
Proctitis	Bartholinitis		Vaginitis (Suspected)
Pharyngitis	Perihepatitis		
(Suspected)	Endometritis		

The treatment of C. trachomatis sexually transmitted infections is given in The Medical Letter (24 (Issue 605), 29-34, March 19, 1982).

Venereal Disease: C. trachomatis is one of the most common genital infections in pregnant women. It causes a spectrum of diseases like that of Neisseria gonorrhoeae with urethritis, cervicitis, epididymitis, proctitis, pelvic inflammatory disease, oculogenital syndrome and ophthalmia. C. trachomatis is the single most important cause of pelvic inflammatory disease and its resulting infertility. The estimated annual incidence of C. trachomatis veneral infections is three to four million; of the patients with Neisseria gonorrhoeae, about 20 percent of men and 30 percent of women are coinfected with C. trachomatis. The infection is asymptomatic in about 70 percent of women.

C. trachomatis is treated with tetracycline and erythromycin.

Cytological examination using Papanicolaou staining is a useful screening test for C. trachomatis infection (Medly, G., The Lancet, 1, 1449, June 25, 1983).

Diseases of Infancy: C. trachomatis is an important cause of neonatal infection; it can be transmitted to newborns by their passage through an infected birth canal. The rate of transmission from infected mother is 40 to 60 percent. It has been linked with perinatal morbidity and mortality (Martin, D.H. et al., JAMA 247, 1585, 1982). There is little excess morbidity during the first year of life in infants who receive treatment for chlamydial infection (Hammersclag, M.R. et al., Pediatric Infectious Disease 1, 395-401, 1982). The signs and symptoms relate to nasal obstruction, tachypnea and cough; the organisms can be isolated from the throat, nasopharynx, eye, lungs, urethra or vagina.

C. trachomatis causes trachoma, the leading cause of preventable blindness in the world; almost all cases of trachoma are found in North Africa, the Mideast and Southeast Asia.

Psittacosis: C. psittaci causes psittacosis, a pneumonia derived from infected birds.

References: April, 1983 Edition of British Medical Bulletin, 39, No.2, pgs. 107-208; Schachter, J., N. Engl. J. Med. 298, 428-435, 490-495, 540-549, 1978; Holmes, K.K., JAMA 245, 1718-1723, 1981.

S. Bakerman

CHLAMYDIA TRACHOMATIS, ANTIGEN, ENZYME IMMUNOASSAY

<u>SPECIMEN</u>: Place the cervical or urethral swab in a transport tube; the tube contains phosphate buffered saline to keep the organisms in solution.
<u>REFERENCE RANGE</u>: Negative
<u>METHOD</u>: Enzyme immunoassay (EIA-Abbott Lab). Since the C. trachomatis organism survives <u>only</u> inside living cells, a special solution is used to dissolve the cell membranes and liberate the organism; the assay is illustrated in the next figure:

Principle of the Abbott C.Trachomatis Enzyme Immunoassay

Release of C. Trachomatis from Cells

Reaction of C. Trachomatis with Beads

Reaction with Antibody

Reaction of Antibody with Enzyme-Labeled Antibody
E = Horseradish Peroxidase

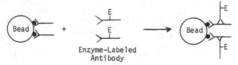

Detection
O-Phenylenediamine + H_2O_2 $\xrightarrow[\text{Peroxidase}]{\text{Horseradish}}$ Colored Product (492nm)

<u>INTERPRETATION</u>: Chlamydia trachomatis is the <u>predominant venereal</u> <u>disease-causing organism in the United States.</u>

The enzyme immunoassay test has a <u>sensitivity</u> of 88 percent and a <u>specificity</u> of 97 to 98 percent. This <u>method will</u> detect <u>both live and dead</u> <u>organisms</u>; therefore, dead organisms following successful therapy may yield a positive test result.

This method has been used to detect ocular trachoma, the single most important cause of preventable blindness in the United States.
<u>Ref.</u>: Medical News Section of JAMA <u>250</u>, 2257-2259, Nov. 4, 1983.

CHLAMYDIA TRACHOMATIS ANTIGEN, FLUORESCENCE IMMUNOASSAY

SPECIMEN: A swab is taken of the endocervical or urethral area. The specimen is processed according to the directions of Syva Corporation (Microtrak). The swab is rolled on a specially prepared slide; the slide is fixed in acetone or alcohol.

REFERENCE RANGE: Negative

METHOD: Fluorescent microscope is required. A test kit is available from Syva Corporation (Microtrak). Organisms are detected by adding a fluorescein-labeled monoclonal antibody to the slide preparation. The slide preparation is incubated at room temperature or 37°C for 15 to 30 minutes and then rinsed, mounted and viewed with the fluorescent microscope. C. trachomatis organisms are stained green.

INTERPRETATION: Chlamydia trachomatis is the predominant venereal disease-causing organism in the United States.

The Syva Microtrak test has a sensitivity of 91 percent and a specificity of 99 percent. Dead organisms may be detected by this method.

Ref.: Medical News Section of JAMA 250, 2257-2259, Nov. 4, 1983; Tam, M.R. et al., N. Engl. J. Med. 310, 1146-1150, May 3, 1984).

CHLORAMPHENICOL (CHLOROMYCETIN)   Chloramphenicol

SPECIMEN: Red top tube, separate serum

REFERENCE RANGE: Therapeutic: 10-20 mcg/ml; Toxic: >25 mcg/ml.

Half-Life: Adults and children: 1.5-5 hours;
          Premature infants (one to two days old): 24-48 hours;
          Premature infants (13 to 23 days old): 8-15 hours.

METHOD: HPLC; GLC

INTERPRETATION: Chloramphenicol is a broad-spectrum antibiotic which is used in infants when severe infections for which less potentially dangerous drugs are ineffective or contraindicated. Repeated courses and concurrent therapy should be avoided. The indications for its use are as follows: anaerobic infections, especially with Bacteroides fragilis; typhoid fever; Haemophilus influenzae meningitis; pneumococcal and meningococcal meningitis in patients allergic to penicillin; brain abscess; and rickettsial infections.

The pediatric dose is 25-100 mg/kg/day, orally or I.V. in 4 doses; the adult dose is 0.25-0.75 gm every six hours, 50 mg/kg/day.

The most important toxic effect of chloramphenicol is bone marrow suppression; this effect is dose related, especially with serum concentrations of 25 mcg/ml or more (Wilson, W.R. and Cockerill, F.R., Mayo Clin. Proc. 58, 92-98, 1983). Complete blood count should be obtained every two days to recognize reticulocytopenia, leukopenia, thrombocytopenia and anemia.

Chloramphenicol can cause the "gray" syndrome in premature infants; in this syndrome, there is cardiovascular collapse leading often to death. Chloramphenicol is converted to chloramphenicol-glucuronide by glucuronyl transferase in the liver and then excreted rapidly (80 to 90 percent of the dose) in the urine by the kidney. In the neonate, conversion of chloramphenicol to the glucuronide is slow because the activity of the enzyme, glucuronyl transferase, is low; thus, chloramphenicol may accumulate.

CHLORIDE, SERUM

SPECIMEN: Red top tube or green top tube (heparin)

REFERENCE RANGE: Premature: 95-110mEq/liter; full-term: 96-106mEq/liter; infant: 96-106mEq/liter; child and adult: 95-105mEq/liter.

METHOD: Coulometric measurement

INTERPRETATION: The causes of <u>increased</u> and <u>decreased</u> serum chloride are shown in the next Table:

Causes of Increased and Decreased Serum Chloride	
Increased Serum Chloride (>106mEq/liter)	Decreased Serum Chloride (<95mEq/liter)
Hyperchloremic Metabolic Acidosis	Metabolic Alkalosis
Respiratory Alkalosis	Respiratory Acidosis (chronic)
Urinary Loss-Renal Disease:	Congestive Heart Failure
Pyelonephritis	Overhydration
Polycystic Renal Disease	Inappropriate ADH
Obstructive Uropathy	Overtreatment with Hypotonic
Renal Tubular Acidosis	Solutions
Severe Dehydration,	Salt Depletion
i.e., Diabetes Mellitus	Burns
Diabetes Insipidus	Addison's Disease
I.V. Saline	

<u>Increased Serum Chloride</u>: Increased serum chloride occurs in <u>hyperchloremic metabolic acidosis</u>; anion gap is normal. In metabolic acidosis, $[HCO_3^-]$ is decreased; in hyperchloric metabolic acidosis serum chloride increases as $[HCO_3^-]$ decreases in order to maintain electroneutrality. In respiratory alkalosis hydrogen ions from intracellular sources enter the extracellular fluid and bicarbonate moves into the red blood cells in exchange for chloride, thus minimizing extracellular alkalosis.

<u>Decreased Serum Chloride</u>: In metabolic alkalosis, $[HCO_3^-]$ is increased; in order to maintain electroneutrality, chloride decreases. In respiratory acidosis, chloride excretion is a necessary concomitant of renal compensation for respiratory acidosis.

CHLORIDE, URINE

SPECIMEN: 10ml aliquot of 24 hour urine; preserved with 10ml of glacial acetic acid, indicate total volume.

REFERENCE RANGE: 110-250mEq/liter

METHOD: Coulometric measurement

INTERPRETATION: The level of urinary chloride may be used in the differential diagnosis of metabolic alkalosis (Harrington, J.T. and Cohen, J.J., N. Engl. J. Med. 293, 1241-1248, 1975). The common causes of metabolic alkalosis are two-fold: (1) those associated with <u>loss of hydrogen ion</u> and extracellular fluid volume (chloride responsive type) and (2) those associated with pathologically <u>excessive circulating levels of mineralocorticoid hormone</u> (chloride resistant type).

Differential Diagnosis of Metabolic Alkalosis	
Chloride Responsive (Urine Chloride <10mEq/liter)	Chloride Resistant (Urine Chloride >20mEq/liter)
Diuretic Therapy	Adrenal Disorders
Gastrointestinal Causes	Hyperaldosteronism
Vomiting	Cushing's Syndrome
Nasogastric Suction	Pituitary, Adrenal, Ectopic ACTH
Chloride-Wasting Diarrhea	Exogenous Steroid
Villous Adenoma-Colon	Gluco- or Mineralcorticoid
Rapid Correction of Chronic	Licorice Ingestion (Glycyrrhizic acid
Hypercapnea	has mineralocorticoid-like activity)
Carbenicillin	Carbenoxalone
Reduced Dietary Intake of	Barttler's Syndrome
Chloride	Alkali Ingestion

The measurement of urinary chloride is valuable in differentiation of causes of persistent metabolic alkalosis. The <u>most common causes of persistent metabolic alkalosis</u>, e.g., loss of gastric juice and diuretic induced, are due to chloride depletion and are responsive to the administration of chloride adequate to replace body stores.

In hyperadrenocorticism, persistent alkalosis results not from the loss of chloride but from indirect stimulation of renal bicarbonate reabsorption.

CHLOROMYCETIN (see CHLORAMPHENICOL)

CHOLECYSTITIS, ACUTE, PANEL

The test panel for acute cholecystitis is given in the next Table:

Test Panel for Acute Cholecystitis

Test	Test Result
Screening Tests:	
White Cell Count (WBC) and Differential	Leukocytosis with a Shift to the Left
Serum Bilirubin	Elevated (Mildly)
Serum Alkaline Phosphatase	Elevated (Mildly)
Serum Glutamic Oxalacetic Transaminase (SGOT)	May be elevated
Serum Amylase	Elevated with Associated Pancreatitis
Confirmatory Tests:	
Flat Plate of Right Upper Quadrant of Abdomen	10 to 30 Percent of Stones are Radiopaque
Ultrasonography	Diagnostic Accuracy in Detection of Gallbladder Calculi Superior to Oral Cholangiography
Oral Cholangiography	Complements Ultrasonography
Cholescintigraphy	Preferred Method for Diagnosis of Acute Cholecystitis; I.V. Tc99m Diisopropyl Iminodiacetic Acid-Visualization of Common Bile Duct without Visualization of Gallbladder is characteristic of Acute Cholecystitis
Computed Tomography	Selected Instances; Expensive

Refer to Radiological Examination of Gallbladder: Jacobson, H.G. and Stern, W.Z., JAMA 250:2977-2982, Dec. 2, 1983.

Laboratory studies usually show an increase in white blood cells with a shift to the left. Serum bilirubin and alkaline phosphatase are mildly elevated. The presence of jaundice (bilirubin > 3 mg/dl) suggests that stone or stones are present in the common bile duct. The SGOT may be elevated if there is ascending cholangitis; serum amylase and lipase should be checked for evidence of associated pancreatitis.

S. Bakerman

CHOLESTEROL, TOTAL
SPECIMEN: Red top tube
REFERENCE RANGE: Hypercholesterolemia values, for selecting men and women at moderate risk (75th to 90th percentiles) and high risk (>90th percentile) requiring treatment, are given in the next Table; (Consensus Conference, JAMA 253, 2080-2086, 1985):

Hypercholesterol Values for Selecting Men and Women at Risk for Treatment		
Age (Years)	Moderate Risk(mg/dl)	High Risk (mg/dl)
20-29	>200	>220
30-39	>220	>240
40 and Over	>240	>260

To convert conventional units in mg/dl to international units in mmol/liter, multiply conventional units by 0.02586.
METHOD: Enzyme assay
INTERPRETATION: The causes of elevated serum cholesterol are given in the next Table:

Causes of Elevated Serum Cholesterol
Lipoproteinemias (Types II, III, V)
Cholestasis, Intra- and Extrahepatic
Nephrotic Syndrome
Hypothyroidism
Oral Contraceptives
Normal Pregnancy
Acute Intermittent Porphyria (AIP)
Macroglobulinemia

A marked elevation of high density lipoprotein (HDL) cholesterol can cause the value for total cholesterol to exceed the upper limit of normal. As an example, a 28 year old female had a total cholesterol of 290 mg/dl; a level above the upper limit of normal; however, the HDL was 85 mg/dl. Thus, the elevated level of serum cholesterol was caused, in part, by high level of HDL-cholesterol (Land, E.C., personal communication.)

The causes of decreased serum cholesterol are given in the next Table:

Causes of Decreased Serum Cholesterol
Liver Disease
Malabsorption
Malnutrition
Hyperthyroidism
Anemia
Abetalipoproteinemia
Tangier's Disease

Abetalipoproteinemia: The demonstration of acanthocytes in fresh blood smears plus a serum cholesterol of less than 50mg/dl is virtually diagnostic of abetalipoproteinemia. The diagnosis can be confirmed by the absence (on electrophoresis) of all three lipoproteins that contain apolipoprotein B (LDL, VLDL and chylomicrons).

Clinically, abetalipoproteinemia is manifested by steatorrhea, retinitis pigmentosa, and ataxic neuropathic disease. The mode of transmission is that of autosomal recessive with a gene frequency of 1 in 20,000.
Tangier's Disease: This disease is characterized by reduced serum cholesterol and normal or raised triglycerides. The basic defect is a defect in the synthesis of HDL. In these patients there is marked accumulations of cholesterol in many tissues in the body, e.g., spleen, lymph nodes, intestinal mucosa and blood vessels. It may be that HDL mobilizes or transports cholesterol from peripheral tissues back to the liver where it can be excreted. The tonsils are orange or yellowish gray coloration.

CHOLINESTERASE (ACETYL-)
RED BLOOD CELL

SPECIMEN: Green (heparin) top tube
REFERENCE RANGE: Method dependent; units, U/liter
METHOD: Red blood cells are lysed and incubated with a buffered substrate
containing acetylcholine:

$$\text{Acetylcholine} + H_2O \xrightarrow{\text{Acetyl-Cholinesterase}} \text{Acetate} + \text{Choline}$$

The acetic acid that is liberated decreases the pH of the solution; the
rate of change of the pH is used as an index of acetylcholinesterase activity.
INTERPRETATION: Red cell cholinesterase is inhibited by insecticides; the
toxic effect is detected longer in the red cells than it is in plasma or serum.
When erythrocyte enzyme activity decreases to 25% of pre-exposure values, the
subject develops clinical illness.

Acetylcholinesterase is found principally in the red blood cell, at motor
end plates of skeletal muscle and in the central nervous system.

Decreased red blood cell cholinesterase is a useful adjunct in making the
diagnosis of paroxysmal nocturnal hemoglobinuria(PNH).

CHOLINESTERASE (PSEUDO-),
DIBUCAINE INHIBITION

Cholinesterase
Dibucaine Inhib.

SPECIMEN: Red top tube, separate and freeze serum; stable 6 months. The
patient should not have taken muscle relaxant or anticholinergic drugs within 24
hours.
REFERENCE RANGE: Inhibition of pseudocholinesterase activity by dibucaine.
Normal test value: 70-90% inhibition; Heterozygote: 30-60% inhibition;
Homozygote: less than 30% inhibition.
METHOD: This test is done to detect homozygous or heterozygous "atypical"
cholinesterase variant. The atypical variant is inhibited less strongly by
positively charged inhibitors, e.g., dibucaine, than normal cholinesterase;
dibucaine is a local anaesthetic.
INTERPRETATION: The atypical variant is autosomal recessively transmitted;
the carrier rate is 1/20; the number of couples at risk is 1/400, and the
incidence of the homozygote is 1/1600. If the level of inhibition is less than
70%, the patient should be considered to have an atypical pseudocholinesterase
variant and the administration of succinylcholine or similar type drugs may pose
a risk.

Patients with atypical variant have low pseudocholinesterase activity;
therefore, it is probably not necessary to perform dibucaine inhibition if the
uninhibited level is normal.

CHOLINESTERASE (PSEUDO-), SERUM OR PLASMA

SPECIMEN: Red or green (heparin) top tube; the specimens are stable at room temperature for 24 hours and for at least two weeks at 4°C.

REFERENCE RANGE: Cholinesterase activity is low at birth and for the first six months of life, increases to values 30% to 50% above adult values until about age five years, then gradually decreases until adult concentrations are reached at puberty (Hill, J.G. in Pediatric Clinical Chemistry, Meites, S., ed., 2nd edition, 1981, pg. 153, Am. Assoc. Clin. Chem. 1725 K. Street, N.W., Washington D.C. 20006). Values depend on method. If possible, measure cholinesterase prior to anticipated exposure to pesticides.

METHOD: Hydrolysis of various choline esters including acetylcholine.

INTERPRETATION: There are two types of cholinesterase in the blood. There is true or acetylcholinesterase which is found principally in the red blood cell, at motor end plates of skeletal muscle and in the central nervous tissue; there is the cholinesterase (pseudocholinesterase) which is found in plasma and in smooth muscle, liver and adipocytes.

Acetylcholinesterase has a well-defined physiological role at the motor end-plate; it acts on acetylcholine, which is an important mediator of nerve conduction.

The measurement of pseudocholinesterase is very useful in (a) monitoring exposure to organophosphate pesticides (b) identification of hypersensitivity to succinylcholine.

(a) Effect of Insecticides on Blood Cholinesterase Level: Cholinesterase activity levels can be used to detect the presence of organophosphate or carbomate insecticides; these insecticides inhibit the enzyme cholinesterase.

The reaction catalyzed by cholinesterase is as follows:

$$\text{Acetylcholine} + H_2O \xrightarrow{\text{Acetyl-Cholinesterase}} \text{Acetate} + \text{Choline}$$

The synthetic insecticides, e.g. organophosphates and carbomates, inhibit cholinesterase by combining with the enzyme; the organophosphates combine irreversibly with this enzyme. Both acetylcholinesterase and pseudocholinesterase are inhibited by organophosphate or carbomate insecticides. Pseudocholinesterase (serum) is depressed before acetylcholinesterase (red cells) but returns to normal before red cell cholinesterase.

When the action of cholinesterase in inhibited, acetylcholine accumulates. The signs and symptoms of poisoning by organophosphate or carbomate are similar to a continuous stimulation of the parasympathetic nervous system (Milby, T.D., J. Am. Med. Ass. 28, 2131-2133, 1971).

Acute poisoning causes almost complete loss of cholinesterase activity; chronic poisoning may cause a 25 percent to 50 percent depression of cholinesterase activity. Plasma cholinesterase regenerates at about 25 percent in the first week and returns to normal in about 6 weeks.

(b) Effect of the Muscle Relaxant Suxamethonium (Succinyldicholine; Anectine) on Persons with Inherited Abnormality or Reduced Level of Cholinesterase Activity Secondary to Insecticides: Persons with an inherited abnormality in cholinesterase activity and those with reduced levels of cholinesterase activity secondary to synthetic insecticides may show marked sensitivity when exposed to the short-acting muscle relaxant, suxamethonium (succinyldicholine).

Suxamethonium is used during tracheal intubation; the anaesthesia is injected I.V. Suxamethonium mediates its effects by its action on the myoneural junction; 90 to 95% of the drug is destroyed within a minute after injection. It is metabolized to the inactive form, succinylmonocholine, as shown in the next reaction:

$$\underset{\text{(Active Form)}}{\text{Succinyldicholine}} + H_2O \xrightarrow[\text{(Myoneural Junction)}]{\text{Cholinesterase}} \underset{\text{(Inactive Form)}}{\text{Succinylmonocholine} + \text{Choline}}$$

When there is an abnormality of cholinesterase, the action of succinyl-dicholine is prolonged; prolonged effects of succinyldicholine may cause paralysis of respiratory muscles and apnea. However, patients with apnea lasting many hours are readily and safely maintained with mechanical ventilation.

(c) Liver Function: Pseudocholinesterase is decreased in patients with liver disease but it is not a reliable index of parenchymal liver cell damage.

Conditions that are associated with depressed pseudocholinesterase levels are listed in the next Table:

Depressed Pseudocholinesterase Levels
Insecticide Poisioning
Genetic Variants
Liver Disease
Metastatic Carcinoma
Malnutrition
Anemias
Acute Infections
Myocardial Infarction
Dermatomyositis

Chorionic
Gonadotropin(HCG)

CHORIONIC GONADOTROPIN (see HUMAN CHORIONIC GONADOTROPIN, HCG)

Chorionic
Villus Biopsy

CHORIONIC VILLUS BIOPSY

SPECIMEN: Chorionic villus biopsy samples can be performed between the 8th and 12th weeks of gestation.

A catheter is inserted through the vagina and cervix into the uterus under direct observation with the aid of high resolution ultrasound. Tissue is obtained by snipping off a sample of chorionic villus or by suction (Simoni, G. et al., Human Genetics 63, 349-357, 1983; Ford, J.H. and Jahnke, A.B. The Lancet 2, pgs. 1491-1492, Dec. 24 and 31, 1983; Gregson, N.M. and Seabright, M.,, The Lancet 2, pg. 1491, Dec. 24 and 31, 1983).

INTERPRETATION: Chorionic villus biopsy can provide a larger sample of fetal tissues than amniocentesis; results are obtained faster and the procedures can be done much earlier in a pregnancy.

Chorionic villi protrude from a membrane called the chorion which surrounds the developing fetus and becomes part of the placenta. The chorion is not an anatomical part of the fetus; however, it is fetal rather than maternal in origin.

The sample is used for chromosome, enzyme and other analyses.

The fetal-loss rate in chorionic villus sampling averages about 12 percent, a rate far above the rate of midtrimester amniocentesis. It may be advisable to reserve this procedure for those conditions with relatively high genetic risks, such as the hemoglobinopathies (Hecht, F. et al., N. Engl. J. Med. 310, 1388, May 24, 1984).

Clostridium
Difficile Toxin

CLOSTRIDIUM DIFFICILE TOXIN ASSAY

SPECIMEN: Stool; not rectal swab; deliver immediately to laboratory in container with lid; refrigerate; if necessary, forward on dry ice to outside laboratory.

REFERENCE RANGE: Negative

METHOD: Tissue culture cytotoxicity assay. Stool extract is inoculated into monolayer cell culture (McCoy) and observed for cytopathic effect. Toxin is identified by neutralization with C. sordelli antitoxin. C. difficile produces two toxins, cytotoxin and enterotoxin.

INTERPRETATION: Pseudomembranous colitis (PMC), secondary to antibiotic use, has proved to be caused by an overgrowth of C. difficile, which produces an exotoxin that in turn leads to an enterocolitis with diarrhea.

Many antimicrobials have been shown to be capable of causing pseudomembranous colitis including ampicillin, cephalosporins, sulfamethoxazole and trimethoprim, tetracycline, and chloramphenicol. The clinical signs and symptoms usually develop within 4 weeks of antibiotic administration. Large-volume, watery diarrhea occurs, often with fever, leukocytosis, abdominal cramps and tenderness. Proctoscopic examination reveals a pseudomembrane (similar to diphtheritic membrane), composed of exudative, raised plaques with skip areas of edematous and hyperemic mucosa (Bartlett, J.G., Rev. Infect. Dis. 1, 530-539,1979).

It is thought that neonatal enterocolitis(NEC) is a result of local and systemic invasion of the damaged intestinal mucosa by microflora in a stressed neonate. An etiologic role of C. difficile has been invoked (Han, V.K.M. et al., Pediatrics 71, 935-941, 1983).

CLOT LYSIS TIME

SPECIMEN: Blue (Citrate) top tube with liquid anticoagulant; place tube on ice and transport to coagulation laboratory immediately. Do not freeze.
REFERENCE RANGE: Intact clot for 48 hours
METHOD: One ml of whole blood is placed into each of two 10 x 75mm test tubes; #1 tube is allowed to clot in a 37°C water bath; #2 tube is allowed to clot in the refrigerator. A normal control is tested at the same time. Observe at intervals for 24 and 48 hours for degeneration or disappearance of the clot.
INTERPRETATION: With excessive fibrinolysis, fragmentation of clot occurs; this test is useful only in severe hyperfibrinolysis.

CLOT RETRACTION TIME

SPECIMEN: Red top tube
REFERENCE RANGE: Clot begins to retract 30 to 60 minutes after collection.
METHOD: Clot retraction depends on intact platelets and the presence of divalent cations. The contractile protein of platelets, thrombasthenin, may be involved in this reaction. One ml of blood is placed into each of three 10 x 75mm test tubes and placed into a 37°C waterbath. The tubes are examined after two hours. If there is retraction in any of the tubes, the result is recorded as "positive". With normal blood, the clot begins to retract 30 to 60 minutes after collection.
INTERPRETATION: The clot retraction time is a measure of platelet function; if there is poor platelet function, clot retraction will be poor or may fail to occur. Clot retraction is abnormal in the following conditions:

Abnormal Clot Retraction Time
Low Platelet Count
Aspirin Therapy
Glanzmann's Disease
Disseminated Intravascular Coagulation (DIC)
Polycythemia
Severe Hemophiliac States

CLOT UREA SOLUBILITY

SPECIMEN: Blue (citrate) top tube, filled to capacity.
REFERENCE RANGE: Negative; in the absence of factor XIII, fibrin polymers will not become cross-linked and will be dissociated by urea into soluble fibrin monomers.
METHOD: Citrated plasma is recalcified and the fibrin clot is suspended in 5M urea solution at 37°C for 24 hours.
INTERPRETATION: This is a simple screening test for severe factor XIII (fibrin stabilizing factor) deficiency. Factor XIII deficiency is a rare inherited coagulation disorder manifested by umbilical cord bleeding at birth, delayed resorption of subcutaneous hematomas and poor wound healing.
  A positive test result indicates severe homozygous factor XIII deficiency; a negative test does not exclude mild or moderately severe factor XIII deficiency.

CLOTTING TIME, ACTIVATED

SPECIMEN: Obtain 3 ml. of blood in a <u>plastic</u> syringe. No anticoagulant. A clean venipuncture is required because contamination with tissue thromboplastin alters the clotting time.

REFERENCE RANGE: 1.5-4.5 minutes. The reference ranges vary with the laboratory.

METHOD: Timing of the test begins when one ml of the blood is placed in a glass tube which contains ground glass or celite which act as <u>activator</u>. It is preferable to perform the test at room temperature at the bedside. Tilt the tube at specific time intervals. The clotting time is the time interval from the start (when one ml of blood is placed in the glass tube) to the time that the blood has clotted.

INTERPRETATION: The activated clotting time is used to <u>monitor heparin therapy</u>. It is useful in <u>monitoring heparin therapy</u> given during extracorporeal circulation. Since the clotting time can be monitored at the bedside, heparin or its antagonists may be given to induce changes in the desired direction.

This test is not sensitive to deficiencies of blood coagulation factors except when these factors are very low (e.g., 1% of normal).

CLOTTING TIME (WHOLE BLOOD CLOTTING TIME) LEE-WHITE

SPECIMEN: Obtain 3 ml of blood in a <u>plastic</u> syringe. No anticoagulant. A clean venipuncture is required because contamination with tissue thromboplastin can even cause a severe hemophiliac to have a normal clotting time.

REFERENCE RANGE: 5-10 minutes using 13x10 mm acid washed Pyrex glass tubes. The reference ranges vary with the laboratory.

METHOD: <u>Timing</u> of the test begins when one ml of the blood is placed in the <u>three</u> glass tubes (13x100mm acid washed Pyrex). If the tubes are kept at 37°C, clotting time is enhanced; this may obscure minimal abnormalities. It is preferable to perform the test at room temperature and <u>in the room where the blood is drawn</u>; transporting the tubes causes agitation of the specimen and temperature change. Tilt tube number 1 every 30 seconds until it clots. Then, tube number 2 is tilted until it clots. Then, tube 3 is tilted until it clots. The clotting time is the time interval from the start (when one ml of blood is placed in first glass tube) to the time that blood in the third tube has clotted.

INTERPRETATION: <u>This test has been used to monitor heparin therapy</u> and <u>whole blood clotting system</u>. Patient should <u>not</u> receive heparin therapy within 3 hours of performing the test.

The <u>Lee-White clotting time</u> is <u>not</u> recommended for evaluating heparin activity because it <u>lacks reproducibility</u>. The test that is <u>recommended</u> for monitoring heparin therapy is the <u>partial thromboplastin time(PTT)</u>.

The sensitivity of the Lee-White clotting time is very low for detecting deficiences of blood coagulation factors; the level of many factors must fall below 1% of normal before the clotting time is prolonged. The Lee-White clotting time is prolonged when fibrinogen falls under 50 mg/dl, prothrombin under 30% of normal and factors VIII and IX under 2% of normal. This test is completely insensitive to platelet factor 3 deficiency.

A prolonged clotting time indicates that there is a clotting abnormality; <u>a normal clotting time does not guarantee that clotting is normal</u>.

COAGULATION FACTOR VIII ASSAY

SPECIMEN: Blue (citrate) top tube; tube must be filled to capacity. If multiple studies are being done, draw coagulation studies last; otherwise, draw 1-2ml into another vacutainer, discard and collect factor assay. This procedure avoids contamination of specimen with tissue thromboplastins. Separate plasma and refrigerate. Sample must be frozen if determination is not done immediately.

REFERENCE RANGE: 50-150 percent of normal.

METHOD: Specific human substrates using the partial thromboplastin time(PTT).

INTERPRETATION: Factor VIII is deficient in the conditions listed in the next Table:

Causes of Factor VIII Deficiency
Classic Hemophilia
von Willebrand's Disease
Acute Disseminated Intravascular Coagulopathy (DIC)

Prolonged bleeding times are characteristic of von Willebrand's but are normal in classic hemophilia.

Factor VIII-von Willebrand's factor assay allows for definitive diagnosis; this factor is diminished in von Willebrand's disease.

COAGULATION PANEL

COAGULATION PROFILES	
Profile	Tests
Pre-Operative Coagulation Panel	Prothrombin Time(PT)
	Partial Thromboplastin Time(PTT)
	Platelet Count
	Bleeding Time
Abnormal Bleeding Panel	Prothrombin Time(PT)
	Partial Thromboplastin Time(PTT)
	Platelet Count
	Bleeding Time
	Thrombin Time
	Fibrinogen
	Tourniquet Test
Easy Bruisability Panel	Prothrombin Time(PT)
	Partial Thromboplastin Time(PTT)
	Platelet Count
	Bleeding Time
	Tourniquet Test
	Platelet Function Tests
	Adhesiveness
	Aggregation
	Peripheral Smear for Megathrombocytes
Fibrinolysis Panel	Euglobulin Lysis
	Thrombin Clotting Time,
	Protamine Sulfate Corrected
Disseminated Intravascular Coagulation(DIC) (see Partial Thromboplastin Time)	Prothrombin Time (PT)
	Partial Thromboplastin Time(PTT)
	Fibrinogen
	Platelet Count
	Fibrin Degradation Products(FDP)
	Thrombin Time(TT)
	Euglobulin Lysis Time
	Peripheral Smear
von Willebrand's Panel	Partial Thromboplastin Time(PTT)
	Factor VIII Panel
	Factor VIII Related Antigen
	Bleeding Time
	Platelet Count
	Platelet Aggregation, Ristocetin
Thrombotic Tendency Panel	Partial Thromboplastin Time(PTT)
	Antithrombin III
Qualitative Platelet Disorders	Platelet Count
	Platelet Factor 3
	Bleeding Time
	Tourniquet Time
	Platelet Function Tests
	Adhesiveness
	Aggregation

Bleeding: The tests that should be done on bleeding patients are given in the next Table:

Tests on Bleeding Patients
Platelet Count
Prothrombin Time(PT)
Partial Thromboplastin Time(PTT)
Fibrinogen
Thrombin Clotting Time
Bleeding Time
Clot Urea Solubility
Clot Retraction Time

SCREENING TESTS IN CERTAIN COMMON DISORDERS OF COAGULATION (Cartwright, G.E., Diagnostic Laboratory Hematology, Grune and Stratton, New York, 1968, 4th Ed., pg. 356; Aledort, L.M., Diagnosis, pgs. 87-90, Jan. 1984):

Screening Tests in Certain Common Disorders of Coagulation

Disorder	Bleeding Time	Prothrombin Time (PT)	Partial Thromboplastin Time (PTT)	Plasma Thrombin Time
Hemophilia	Normal	Normal	Prolonged	Normal
von Willebrand's Disease	Prolonged	Normal	Prolonged	Normal
Deficiency of Vitamin K Dependent Factors	Normal	Prolonged	Usually Prolonged	Usually Normal
Defibrination Syndrome	Normal or Prolonged	Prolonged	Normal or Prolonged	Prolonged

Deficiency of Vitamin-K Dependent Factors: Deficiency of the vitamin-K dependent factors, prothrombin, factors VII, IX and X, occurs in coumarin toxicity, liver disease, malabsorption syndromes, hemorrhagic disease of newborns and vitamin K deficiency from any cause.

Hemostatic Findings in Various Conditions

Laboratory Measurement	Pregnancy	Hepatic Cirrhosis	Vit. C Def.	HDN	Heparin Therapy	ITP	Thrombas-thenia
Platelet Count (150,000-350,000/mm^3)	375,000/mm^3	N or ↓	N	N	N	↓	N
Bleeding Time (3 to 9 min)	2 minutes	N or ↑	↑	N	N	↑	↑
Capillary Fragility	N	N or ↑	↑	N	N	↑	↑
PTT (30 sec.)	23 seconds	N or ↑	N	N or ↑	↑	N	N
PT (11 sec.)	10 seconds	↑	N	↑	↑	N	N
Fibrinogen (0.15-0.35gm/dl)	430mg/dl	N or ↓	N	N	N	N	N

Hemostatic Findings in Various Conditions

Test	Thrombocytopenia	Aspirin Ingestion	Glanzmann's Thrombasthenia	Factor Deficiency
Bleeding Time	↑	↑	↑	N
Platelet Count	↓	N	N	N
PTT(a)	N	N	N	↑ or N
PT(b)	N	N	N	N or ↑
Thrombin Time (TT)	N	N	N	N
Fibrinogen	N	N	N	N
Clot Urea Solubility	N	N	N	N
Blood Clot Retraction	N	N	Abnormal	N

(a)Measures Factors 12, 11, 9, 8, 10, 5, 2, 1
(b)Measures Factors 10, 7, 5, 2, 1

For acutely ill patients with liver dysfunction, it is occasionally difficult to decide whether liver dysfunction and/or DIC is present; decrease in both factors V and VIII indicates DIC; decrease in factor V and normal factor VIII indicates liver disease only (Chaplinski, T.J., personal communication).

Function and Hemostatic Tests

Function	Test
Vascular Integrity	Bleeding Time
	Tourniquet Test
Platelets	Platelet Count; Bleeding Time; Clot Retraction
	Tourniquet Test: Platelet Aggregometry
Coagulation	Activated Partial Thromboplastin Time (PTT)
	Prothrombin Time(PT); Fibrinogen;
	Whole Blood Clot Lysis; Anti-Thrombin III
Fibrinolytic Studies	Euglobulin Lysis; Fibrin Split Products
	Thrombin Clotting Time

Coagulation Disorders: Characterized by joint, soft tissue and organ bleeding.
Defects in Platelet Function: Characterized by purpura (petechiae and ecchymoses) of the skin and hemorrhage from oral mucous membranes and other tissues.

COAGULATION TESTS

A list of coagulation tests and the clinical indications for these tests are given in the next Table:

Coagulation Tests and Clinical Indications	
Test	Clinical Indications
Partial Thromboplastin Time(PTT) [Blue (Citrate) Top Tube]	Monitor Heparin Therapy Liver Disease Disseminated Intravascular Coagulopathy(DIC) Factor Deficiency
Prothrombin Time(PT) [Blue (Citrate) Top Tube]	Monitor Oral Anticoagulants Vitamin K Deficiency Factor Deficiency Liver Disease Disseminated Intravascular Coagulopathy(DIC)
Anti-Thrombin III [Blue (Citrate) Top Tube]	Thrombosis or Risk of Thrombosis
Bleeding Time (Bedside Test)	Platelet Function
Clot Lysis Time [Blue (citrate) top tube]	Severe Hyperfibrinolysis
Clot Retraction Time (Red Top)	Platelet Function
Clot Urea Solubility [Blue (Citrate) Top Tube]	Severe Factor XIII Deficiency
Clotting Time, Activated (Plastic Syringe, Bedside)	Monitor Heparin Therapy Severe Factor Deficiency
Clotting Time, Lee-White (Plastic Syringe, Bedside)	Monitor Heparin Therapy Severe Factor Deficiency
Euglobulin Lysis Time [Blue (Citrate) Top Tube]	Monitor Urokinase and Streptokinase Therapy Evaluation of Abnormal Fibrinolysis
Factor VIII [Blue (Citrate) Top Tube]	Classic Hemophilia von Willebrand's Disease Disseminated Intravascular Coagulopathy (DIC)
Fibrin Split Products(FSP) (Red Top)	Disseminated Intravascular Coagulopathy (DIC)
Plasminogen [Blue (Citrate) Top Tube]	Disseminated Intravascular Coagulopathy (DIC)
Platelet Aggregation [Blue (Citrate) Top Tube]	Platelet Function
Thrombin Clotting Time [Blue (Citrate) Top Tube]	Disseminated Intravascular Coagulopathy (DIC)
Thrombin Clotting Time, Protamine Sulfate Corrected [Blue (Citrate) Top Tube]	Fibrinlysis
Tourniquet Test (Bedside)	Vascular Abnormality Decrease in Platelets Qualitative Defect in Platelets

COCAINE ("SNOW")
SPECIMEN: Random urine specimen
REFERENCE RANGE: Negative
METHOD: Thin-layer chromatography
INTERPRETATION: Cocaine is a stimulant derived from the coca bush, a plant growing in Chile, Bolivia and Peru.

In large doses, it produces violent stimulation, hallucinations and ecstatic effects. Similar to amphetamines, severe depressions occur as the effects of the drug decreases; thus, the user tends to continue its use.

Acute cocaine poisoning occurs in "body packers" or following massive intravenous infusion; "body packers" attempt to smuggle cocaine into the country in their gastrointestinal tracts wrapped in foil or rubber devices such as condoms. The drug packages may be successfully passed with purgation or surgery (McCarron, M.M. and Wood, J.D., JAMA 250, 1417-1420, Sept. 16, 1983). The clinical findings in acute cocaine poisoning are as follows: acute agitation; diaphoresis; ventricular dysrhythmia with hypotension; grand mal seizures leading to lactic acidosis; severe respiratory acidosis due to hypoventilation and respiratory arrest and metabolic acidosis due to the seizures.

The teatment is as follows: Treat seizures with diazepam, I.V. Treat acidosis with bicarbonate and ventilation (Jonsson, S. et al., Am. J. Med. 75, 1061-1064, Dec. 1983).

COCCIDIOIDES ANTIBODY, SERUM AND CSF
SPECIMEN: Red top tube, separate and then freeze serum. CSF specimens: 1.0ml of spinal fluid.
REFERENCE RANGE: Negative
METHOD: Complement fixation(CF); latex agglutination(LA); immunodiffusion (ID); tube precipitin(TP), counterimmunoelectrophoresis(CIE).

Antibody (IgG) is usually quantitated by CF. Titers of 1:2 and 1:4 are found early in the disease; titers of 1:16 to 1:32 are highly suggestive and over 1:32 are indicative of disease. Increasing titers are diagnostic.
INTERPRETATION: The change with time in skin test, complement fixation test(IgG) and tube precipitin(IgM) test following infection with Coccidioides immitis is shown in the next Figure; (Penn, R.L. et al., Arch. Intern Med. 143, 1215-1220, 1983):

Skin Test, Complement Fixation Test (IgG) and Tube Precipitin (IgM) Test
Following Infection with Coccidioides Immitis

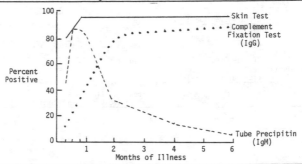

The TP(IgM) test is positive in about 90 percent of patients during the second and third weeks of illness.

Any titer of CF antibodies in CSF is indicative of meningeal infection.

Coccidioidomycosis is most prevalent in the Far West and Southwest of the United States and is particularly common in the San Joaquin Valley ("valley fever" or "San Joaquin fever") of California.

Coccidioides immitis in tissue sections appears as a thick-walled, nonbudding spherule, 20 to 60 micrometer in diameter, often filled with endospores.

Most patients with coccidioidomycosis are asymptomatic; infections usually involve the lungs and may progress to affect any or all organs of the body.

Two weeks following inhalation of the spores, the patient may develop symptoms; chest pain may be prominent. Within a day or two, a rash may be seen.

CODEINE

SPECIMEN: Random urine specimen
REFERENCE RANGE: The lethal dose for adults is 0.5-1g.
METHOD: Thin-layer chromatography
INTERPRETATION: Accidental overdose with codeine-containing cough medicines occurs in children. Overdose produce the following symptoms: somnolence, ataxia, miosis, vomiting, rash and itching of the skin.

COLD AGGLUTININ TITER

SPECIMEN: Red top tube; keep specimen at room temperature and transport to the laboratory immediately. The blood should be incubated preferably at 37°C or alternately at room temperature and allowed to clot before the serum is separated. Do not refrigerate prior to separation of serum from red cells. Refrigerate serum.
REFERENCE RANGE: Titers less than 1:32 dilution.
METHOD: Two fold serial dilutions of serum are prepared. Add an aliquot of serum to human group O erythrocytes and incubate at 0 to 5°C for one hour. Positive test is agglutination of red blood cells. The agglutination is reversible when warmed to 37°C.

Patients with cold agglutinins develop anti-I antibodies; these antibodies react with the I antigen on adult human Type O red cells at 4°C.

Antibiotic therapy may interfere with antibody formation.
INTERPRETATION: The cold agglutinin titer is a measure of antibody titer (anti-I) in the serum of conditions listed in the next Table:

Causes of Positive Cold Agglutinin Test
Mycoplasma Pneumoniae Infections
Viral Infections, such as EB Virus and Cytomegalovirus
Lymphoma
Chronic Liver Disease
Staphylococcemia
Tuberculosis
Malaria
Infectious Mononucleosis
Idiopathic

Mycoplasma Pneumoniae: A four-fold increase in antibody titer is found in 55 to 65 percent of patients with M. pneumoniae infections; 90 percent of patients with rising or falling titers are severely affected or have prolonged illness. In primary atypical pneumonia, cold agglutinins are detected one week after onset, peak at 12 to 25 days, and rapidly fall after day 30. Mycoplasma pneumoniae is probably the only mycoplasma consistently shown to cause disease in man. Culture and identification of this organism require three to four weeks. High titer of cold agglutinin antibodies is associated with M. pneumoniae infections.

Most cold agglutinins are autoagglutinins; these may cause problems in blood grouping or cross-matching.
Cold agglutinin disease(CAD): Cold agglutinin disease(CAD) is associated with a hemolytic process, characterized as follows:
Red Cell Pathology: Clumping of red cells; apparent increase in MCV by some automated particle counters due to clumping; usually moderate normochromic normocytic anemia;
Sedimentation Rate: Markedly increased at room temperature due to RBC agglutination.
Other Findings in Acute Attacks of CAD: Hemoglobinuria; hemosiderinuria, ahaptoglobinemia; increased unconjugated bilirubin.

COMA PANEL

The differential diagnosis of <u>coma</u> and <u>stupor</u> is given in the next Table; (Nicholson, D.P., Medical Clinics of North America <u>67</u>:1279-1293, Nov. 1983):

<u>Differential Diagnosis of Coma and Stupor</u>

<u>Central Nervous Disease:</u>
  Intracranial Disease
  Injury
  Infection
  Subdural Hematoma
<u>Metabolic:</u>
  Diabetes: Hyperosmolar Coma; Diabetic Ketoacidosis
  Uremia
  Hepatic Encephalopathy
  Anoxic Encephalopathy (Include CO Poisoning and Inert Gas)
  Hypothermia
  Hypothyroidism(Myxedema Coma)
<u>Electrolyte and Acid-Base Derangements:</u>
  Hyponatremia
  Hypernatremia
  Hypercalcemia
  Hypo-or hyperkalemia
  Hypomagnesemia
  Respiratory Acidosis
  Respiratory Alkalosis
<u>Drugs and Toxic Agents:</u>
  <u>Sedatives:</u> Alcohol, Phencyclidine(PCP), Tricyclics and Opiates.
  <u>Metabolic Acidosis with Increased Anion Gap:</u> Salicylates,
    Methyl Alcohol, Ethylene Glycol, Paraldehyde
  <u>Enzyme Inhibitors:</u> Heavy Metals, Organic Phosphates and Cyanide

Laboratory studies in coma of unknown cause are given in the next Table:

<u>Laboratory Studies in Coma</u>

<u>Complete Blood Count(CBC)</u>
<u>Blood Chemistries</u>
  Glucose
  Calcium
  Magnesium
  Liver Function Tests
    SGOT
    Bilirubin, Total and Direct
  Electrolytes: Sodium, Potassium, Chloride, CO_2 Content, Calculate
           Anion Gap
  Arterial Blood Gases: pH, PCO_2, PO_2
  Acetone
  Creatine Phosphokinase(CPK)
  Blood Urea Nitrogen(BUN)
  Creatinine
  Osmolality
<u>Urinalysis, including Osmolality</u>
<u>Drug Screen(Toxicology)</u>
<u>Volatiles</u>
<u>Chest X-Ray</u>
<u>Radiographic Examination of Any Injuries</u>
<u>Computed Tomography for Head Injury or CVA</u>

COMPLEMENT C3

<u>SPECIMEN:</u> Red top tube, separate serum; store frozen in plastic tubes. On
standing at room temperature, C3 breaks down into split products which can give
falsely high values.

<u>REFERENCE RANGE:</u> 83-177mg/dl. To convert <u>conventional</u> units in mg/dl to
<u>international</u> units in g/liter, <u>multiply conventional units by 0.01.</u>

<u>METHOD:</u> Radial immunodiffusion; nephelometry

<u>INTERPRETATION:</u> C3 Complement is <u>decreased</u> in the serum in conditions listed
in the next Table (Madaio, M.P. and Harrington, J.T., N. Engl. J. Med. <u>309</u>,
1299-1302, Nov. 24, 1983):

Decreased C3 Complement in Serum

Systemic Diseases:
    Systemic Lupus Erythematosus (<u>Focal</u>, approx. 75 percent;
        <u>Diffuse</u>, approx. 90 percent)
    Subacute Bacterial Endocarditis (approx. 90 percent)
    "Shunt" Nephritis (approx. 90 percent)
    Cryoglobulinemia (approx. 85 percent)
Renal Diseases:
    Acute Poststreptococcal Glomerulonephritis (approx. 90 percent)
    Membranoproliferative Glomerulonephritis
        Type I (approx. 50 to 80 percent)
        Type II (approx. 80 to 90 percent)
Other:
    Gram-Negative Septicemia
    Fungal Diseases, such as Cryptococcal Septicemia

Conditions that cause <u>acute nephritis</u>, but associated with <u>normal serum</u>
complement, are listed in the next Table (Madaio, M.P. and Harrington, J.T., N.
Engl. J. Med. <u>309</u>, 1299-1302, Nov. 24, 1983):

Conditions Causing Acute Nephritis but Associated with Normal Serum Complement

Systemic Diseases:
    Polyarteritis Nodosa Group
    Hypersensitivity Vasculitis
    Wegener's Granulomatosis
    Henoch-Schönlein Purpura
    Goodpasture's Syndrome
    Visceral Abscess
Renal Diseases:
    IgG-IgA Nephropathy
    Idiopathic Rapidly Progressive Glomerulonephritis
    Anti-Glomerular Basement Membrane Disease
    Immune-Complex Disease
    Negative Immunofluorescence Findings

C3 is an acute phase reactant.

COMPLEMENT, C1Q BINDING ASSAY

Complement,
C1Q Binding Assay

<u>SPECIMEN:</u> Red top vacutainer, separate the serum and freeze immediately; ship
frozen in dry ice.

<u>REFERENCE RANGE:</u> 0-13% Bound = Normal; 13-16% Bound = Borderline;
>16% Bound = Abnormal (Scripps-BioScience Reference Laboratory).

<u>METHOD:</u> The C1q binding assay measures the binding sites on immune complexes
available to radiolabeled C1q, a sub-unit of the first component of complement.
If immune complexes are present, the complexes will bind the radioactive C1q;
polyethylene glycol (PEG) is used to precipitate the complex. Free C1q is not
precipitated by PEG. The percentage of total added C1q that is precipitated is
calculated and expressed as "% Bound" (Scripps-BioScience Reference Laboratory)

<u>INTERPRETATION:</u> Some diseases reported to be associated with <u>circulating
immune complexes</u> are given in the Table in IMMUNE COMPLEX PROFILE.

COMPLEMENT PROFILE, C3, C4, FACTOR B (C3PA)

<u>SPECIMEN</u>: Red top tube, separate serum and refrigerate serum; <u>or</u> lavender (purple) top tube, separate plasma and refrigerate. If the specimen is shipped, freeze and send frozen.

<u>REFERENCE RANGE</u>: C3, 90-200mg/dl; C4, 15-45mg/dl; factor B(C3PA), 17-40mg/dl.

<u>METHOD</u>: Immunodiffusion or immunoprecipitin or laser-nephelometric light scattering.

<u>INTERPRETATION</u>: Complement is the term for a system of <u>sequentially reacting</u> <u>serum proteins</u> which participates in <u>pathogenic processes</u>; the complement sequence leads to <u>inflammatory injury</u>. Complement pathways are illustrated in the next Figure:

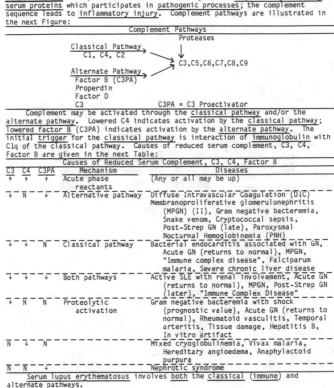

Complement Pathways

Complement may be activated through the <u>classical pathway</u> and/or the alternate pathway. Lowered C4 indicates activation by the <u>classical pathway</u>; lowered factor B (C3PA) indicates activation by the <u>alternate pathway</u>. The initial <u>trigger</u> for the classical pathway is interaction of immunoglobulin with Clq of the classical pathway. Causes of reduced serum complement, C3, C4, Factor B are given in the next Table:

C3	C4	C3PA	Mechanism	Diseases
+	+	+	Acute phase reactants	(Any or all may be up)
+	N	+	Alternative pathway	Diffuse Intravascular Coagulation (DIC) Membranoproliferative glomerulonephritis (MPGN) (II), Gram negative bacteremia, Snake venom, Cryptococcal sepsis, Post-Strep GN (late), Paroxysmal Nocturnal Hemoglobinemia (PNH)
+	+	N	Classical pathway	Bacterial endocarditis associated with GN, Acute GN (returns to normal), MPGN, "Immune complex disease", Falciparum malaria, Severe chronic liver disease
+	+	+	Both pathways	Active SLE with renal involvement, Acute GN (returns to normal), MPGN, Post-Strep GN (later), "Immune Complex Disease"
+	N	N	Proteolytic activation	Gram negative bacteremia with shock (prognostic value), Acute GN (returns to normal), Rheumatoid vasculitis, Temporal arteritis, Tissue damage, Hepatitis B, In vitro artifact
N	+	N		Mixed cryoglobulinemia, Vivax malaria, Hereditary angioedema, Anaphylactoid purpura
N	N	+		Nephrotic syndrome

Causes of Reduced Serum Complement, C3, C4, Factor B

Serum lupus erythematosus involves <u>both</u> the <u>classical</u> (immune) and alternate pathways.

COMPLEMENT, TOTAL (CH50 or CH100)

SPECIMEN: Red top tube; collect specimen on ice and separate serum from clot with minimum centrifugation time; refrigerate or freeze specimen. It is essential that serum be fresh or stored at -75°C for reliable results. Fluids other than blood, such as, pleural fluid and synovial fluid, should be treated in a similar manner, that is, maintain on ice and freeze.

REFERENCE RANGE: CH50 Test: Serum: 35-55 CH50 units/ml; synovial fluid: 20-40 CH50 units/ml; pleural fluid: 10-20 CH50 units/ml. CH100 Test: Serum: 58-114 units/ml.

METHOD: CH50 Test: This test uses sheep red blood cells coated with antibody. Addition of complement (serum or synovial fluid or pleural fluid) in the presence of Mg^{++} and Ca^{++} causes lysis of the coated cells. The amount of hemolysis is evaluated quantitatively and compared with known controls.

CH100 Test: This is a radial diffusion test in agarose gel for quantitating hemolytic complement levels. The clinical specimen to be assayed for complement activity is delivered into a well cut in a uniform layer of buffered agarose gel which is impregnated with sheep erythrocytes which have been previously sensitized with specific rabbit antibody. As complement diffuses radially through the gel from the well cytolysis of the sensitized erythrocytes occurs leaving a clear zone of hemolysis around the well. After a specified time under controlled temperature the diameters of the cleared zones are measured, diameter being proportional to the concentration of active complement. Unknown values are determined from a reference curve which is constructed by plotting the log of the concentrations (CH100 units/ml) of reference calibrators against the diameters of their respective zones of hemolysis. (Meloy Laboratories, 6715 Electronic Drive, Springfield, Va. 22151 [800-336-4555]).

INTERPRETATION: These tests are done to detect complement deficiencies. All of the numbered components of complement are required for red blood cell lysis. Normal CH50 or CH100 values indicate all components present in normal levels; these tests do not measure the alternate pathway (non-immune) activation of complement.

COMPOUND S (SEE 11-DEOXYCORTISOL)

Compound S

CONNECTIVE TISSUE DISEASE PANEL

Connective tissue diseases are listed in the next Table:

Connective Tissue Diseases
Rheumatoid Arthritis(RA)
Related and Associated Syndromes to R.A.
Felty's Syndrome
Sjögren's Syndrome
Juvenile Rheumatoid Arthritis (Still's Disease)
Reiter's Syndrome
Ankylosing Spondylitis (Marie-Strumpell)
Systemic Lupus Erythematosus(SLE)
Scleroderma and Progressive Systemic Sclerosis(PSS)
Polymyositis
Mixed Connective Tissue Disease(MCTD)

LABORATORY TESTS IN CONNECTIVE TISSUE DISEASES: The laboratory tests in connective tissue diseases may be classified as follows:

Laboratory Tests in Connective Tissue Diseases
(a)   Non-immunological Laboratory Tests
(b)   Immunological Laboratory Tests

NON-IMMUNOLOGICAL LABORATORY TESTS: The non-immunological laboratory tests are those that are often done in screening; these tests are listed in the next Table:

Non-immunological Laboratory Tests in Connective Tissue Diseases	
Test	Remarks
Markers of Inflammation:	
Sedimentation Rate	Elevated
C-Reactive Protein	Elevated
White Blood Cell Count(WBC)	Decreased
	May have lymphocytosis
Red Blood Cell Count(RBC)	Decreased
RBC Indices	Anemia, normocytic, normochromic
Platelets	Decreased
Coombs	
Renal Function Tests:	
BUN, Creatinine, Creatinine	May be abnormal
Clearance	

Markers of Inflammation: Fever, WBC, erythrocyte sedimentation rate(ESR), and C-reactive protein(CRP) increase with inflammation; fever, WBC, and ESR may be affected by conditions other than inflammation. However, CRP is said to be specific for inflammation and/or tissue necrosis, such as rheumatoid arthritis. These patients often have a lymphocytosis.

The white blood cell count may be decreased due to the presence of anti-leukocyte antibodies.

Red Blood Cell Count(RBC) and RBC Indices: The red blood cell count is often decreased possibly due to antierythrocyte antibodies; normochromic, normocytic anemia is a common finding in connective tissue diseases.

Platelets: Platelets may be decreased due to antiplatelet antibodies.

Renal Function Tests: The BUN and serum creatinine may be elevated in connective tissue diseases (except polymyositis) reflecting renal disease.

IMMUNOLOGICAL LABORATORY TESTS: The immunological laboratory tests in connective tissue diseases are listed in the next Table:

Immunological Tests in Connective Tissue Diseases
(1)   Antinuclear Antibodies
(a)   Fluorescence-Titer; Pattern
(b)   Specific Antinuclear Antibodies
(2)   Complement Profile
(3)   Immune Complexes
(4)   Rheumatoid Factor(RF)
(5)   Cryoglobulins

RHEUMATOID ARTHRITIS: There are no laboratory tests that are specific for rheumatoid arthritis; the laboratory tests that should be considered in work-up of a patient for suspected rheumatoid arthritis are listed in the next Table:

Laboratory Profile in Rheumatoid Arthritis

Complete Blood Count(CBC): RBC; WBC; Indices; Platelets
Erythrocyte Sedimentation Rate(ESR); C-Reactive Protein
Rheumatoid Factor(RF)
Antinuclear Antibodies(ANA)
Examination of Synovial Fluid

Laboratory Findings in Rheumatoid Arthritis(RA): The laboratory findings in rheumatoid arthritis are listed in the next Table:

Laboratory Findings in Rheumatoid Arthritis(RA)

Finding	Remarks
Anemia	Normocytic, Normochromic
Lymphocytosis	25% of Patients
Erythrocyte Sedimentation Rate(ESR)	Elevated; Degree of Elevation Roughly Parallels Disease Activity
C-Reactive Protein	Elevated
Rheumatoid Factor(RF)	Positive in 85% of Patients with Active Disease
Antinuclear Antibodies(ANA)	20% to 70% of Patients
Antibodies to Histones	20% of Patients
Synovial Fluid	Turbid, Poor Mucin Clot: Lysosomal Enzymes may Depolymerize Synovial Hyaluronate. WBC's: 5000 to 20,000/cu mm, 2/3 pmn's. Rheumatoid Factor(RF);Complement; Immune Complexes

Typically, the RF is absent when the clinical signs and symptoms of rheumatoid arthritis first appear. Patients in which RF is demonstrable early in the course of rheumatoid arthritis have a greater risk of developing articular destruction and having sustained disabling disease.

SYNDROMES RELATED TO OR ASSOCIATED WITH ARTHRITIS: Syndromes related to or associated with arthritis are listed in the next Table:

Syndromes and Arthritis

Condition	Syndrome
(1) Felty's Syndrome	Rheumatoid Arthritis, splenomegaly and leucocytopenia; high RF titers
(2) Sjögren's Syndrome (Keratoconjunctivitis Sicca, Sicca = Dry; Xerostomia Xero = dry)	May occur alone or associated with other connective tissue diseases. 90% occur in women, usually middle-aged. Heavy infiltration of lymphocytes in salivary, lacrimal and other secretory glands: Keratoconjunctivitis Sicca, pharyngitis sicca and parotid gland enlargement are the chief symptoms. Autoantibodies: RF, 90%; ANA, 70%; antisalivary, 60%-70% (Diagnostic); very high incidence of SS-A (occurs in 70% of patients), SS-B (occurs in 60% of patients) and Nucleolar RNA; Native DNA antibody is absent.
(3) Juvenile Rheumatoid Arthritis (Still's Disease)	Pathology identical to Adult-Onset R.A. Low Frequency of RF. Three Clinical Forms: (a) Systemic - High fever, skin rash, lymphadenopathy, pleurisy, pericarditis. (b) Pauciarticular - Few large joints. (c) Polyarticular - Multiple small joint involvement. Associated with HLA, B-27.
(4) Reiter's Syndrome	Triad: Arthritis, Urethritis, Conjunctivitis; Usually RF and ANA neg.; Associated with HLA, B-27.
(5) Ankylosing Spondylitis (Marie-Strumpell)	Erosive sacroiliitis may progress to include entire vertebral column and may result in ankylosis and immobilization; shoulders, hips and small peripheral joints may be involved. Associated with HLA, B-27.

LUPUS ERYTHEMATOSUS: Lupus erythematosus is a disease of the connective tissue
affecting about 500,000 people per year. Its cause is unknown. In this
disease, the body forms too many antibodies and the antibodies attack healthy
tissue.

Criteria for diagnosis of systemic lupus erythematosus(SLE) are given in
the next Table; (JAMA 48, 622, Aug. 13, 1982):

Criteria for Diagnosis of Systemic Lupus Erythematosus(SLE)
1. Malar Rash
2. Discoid Lupus
3. Photosensitivity
4. Oral Ulcers
5. Arthritis
6. Proteinuria Greater than 0.5 g/day, or Cellular Casts
7. Seizures or Psychosis
8. Pleuritis or Pericarditis
9. Hemolytic Anemia or Leukopenia or Lymphopenia or Thrombocytopenia
10. Antibody to DNA or Sm Antigen or the Presence of LE Cells or a Biologically False-Positive Serologic Test Result for Syphilis
11. Positive Fluorescence Antinuclear Antibody Test Results

A patient must have at least 4 of the 11 conditions to be classified as
having SLE. The sensitivity is 96 percent and the specificity is 96 percent.
Systemic Lupus Erythematosus(SLE): The laboratory tests results that should be
considered in the work-up of patients with SLE are listed in the next Table:

Laboratory Profile in Systemic Lupus Erythematosus (SLE)
Non-Immunological Tests:
Complete Blood Count(CBC): RBC; WBC; Indices
Erythrocyte Sedimentation Rate(ESR)
C-Reactive Protein
Urinalysis
Blood Chemistries (SMA 12/60-especially Total Protein and BUN or Creatinine)
Immunological Tests:
Antinuclear Antibodies(ANA)
Anti-Double-Stranded-DNA
Anti-Sm Antibodies
Complement Profile
Immune Complexes

Hemolytic Anemia (5%) or Leucopenia (<4000/cu mm) (15%) or Thrombocytopenia
(<100,000/cu mm) (5%) or Any Combination of These: Cell specific antibodies may
be found in patients with lupus; erythrocyte antibodies may be responsible for
leukopenia and platelet antibodies may be responsible for autoimmune
thrombocytopenia.
Proteinuria >3.5g/day: Proteinuria occurs in about two-thirds of patients with
systemic lupus erythematosus; proteinuria >3.5g/24 hr. occurs in about 20% of
patients.

The incidence of renal disease in patients with lupus erythematosus is
about 75%; the most common cause of death in these patients is renal failure.
Renal lesions in lupus include focal glomerulonephritis, diffuse proliferative
glomerulonephritis and membranous glomerulonephritis. Cellular casts develop in
association with disease.
Antinuclear Antibodies by Immunofluorescence: There are four fluorescent
patterns by immunofluoresence, three of which occur in lupus erythematosus; the
patterns are the diffuse, homogeneous pattern; rim pattern and speckled pattern.
The rim pattern is relatively specific for lupus.

Antinuclear antibodies in SLE are shown in the next Table; (Tan, E.M.,
Hospital Practice, pgs. 79-84, Jan. 1983):

Antinuclear Antibodies in Systemic Lupus Erythematosus (SLE)

Antibodies to:	Incidence
Native DNA	Occurs in 50-60% of Patients at Significant Titers.
DNP	Occurs in Up to 70% of Patients usually at Titers of >1:10,000 by Hemagglutination.
Sm	Occurs in 30% of Patients usually at Titers of 1:40 to 1:640 by Hemagglutination.
Histones	Occurs in Up to 60% of SLE Patients and 95% of Drug-Induced Lupus Patients
SS-A	30-40%
SS-B	15%
RNP	30-40%
PCNA	<5%

Anti-Double-Stranded DNA: High levels of antibody to native double-stranded deoxyribonucleic acid (DNA) is found in 40 percent to 60 percent of serum of patients with active systemic lupus erythematosus (SLE) but uncommonly in other diseases. There is good correlation between active SLE (particularly lupus nephritis) and elevated levels of anti-DNA antibodies; exacerbations of SLE is associated with a rim or homogeneous pattern of immunofluorescence. Inactive SLE usually is accompanied by low or absent serum levels of anti-DNA antibodies. Low levels of antibody to double-stranded DNA are found in a number of other connective tissue diseases.

Anti-Sm Antibodies: Antibodies to Sm antigens are considered almost specific of SLE, and they are found in 20 to 30 percent of patients with this disease. The presence of anti-Sm antibody is associated with a much higher incidence of vasculitis, resulting in peculiar visceral manifestations, which can be poorly responsive to therapy (Beaufils, M., et al., Am.J.Med. 74, 201-216, Feb. 1983).

Serum Complement in Lupus Erythematosus: Serum complement levels are decreased in systemic lupus erythematosus; the components that are depressed involve the Classical Pathway C-1, C-4, C-2, C-3 and the Common Pathway.

The presence of high binding capacity of antibodies to native DNA and low C-3 was 100 percent correct in predicting the diagnosis of SLE (Weinstein, A., et al., Amer. J. Med. 74, 206-216, 1983).

Other Immunological Tests and Immunological Parameters in Systemic Lupus Erythematosus: Immunological tests other than those used for routine diagnostic purposes, are listed in the next Table:

Other Immunological Tests and Immunological Parameters in Systemic Lupus Erythematosus

(1) L.E. Cell Phenomena
(2) Rheumatoid Factor (30%)
(3) Biological False Positive - VDRL - (10-20%)
(4) Anti-Platelet Antibodies
(5) Anti-Thyroglobulin Antibodies
(6) Anti-Cytoplasmic Antibodies (Anti-Ribosomal Antibodies are most Frequent)

Drug Induced Lupus Erythematosis: Drugs that can cause an illness with some of the features of SLE are shown in the next Table:

Drug Induced Lupus Erythematosis

Hydralazine (Apresoline) - Antihypertensive Agent
Procainamide - Used in Treatment of Cardiac Arrhythmias
Isoniazid; Practolol; Hydantoins; Chlorpromazine;
D-Penicillamine; Nitrofurantoin

There are differences between drug-induced disease and spontaneous SLE; some similarities and differences are listed in the next Table:

Similarities and Differences: Drug-Induced Disease Versus Spontaneous SLE

Systemic Lupus Erythematosis (SLE)	Drug-Induced Disease
Immune Complex Type Renal Disease is Common	Low Incidence of Immune Complex Type Renal Disease
Antibodies to Native (DS) DNA and Sm (Sm = Smith)	Antibodies to DS-DNA and Sm are Absent; ANA to Single-Stranded DNA and to DNP
Chronic - Primarily Females; Usually 20-40 Years of Age	Usually reversible - Male and Female; any Age

Procainamide: About 30% of patients who take therapeutic doses of procainamide for a prolonged time will develop a reversible lupus-like syndrome with arthralgias, myalgias, fever, pleuritis and pericarditis. Symptoms may develop as early as two weeks or as late as two years after starting treatment. Rechallenging procainamide-lupus patients with procaiamide will cause a recurrence or remission of symptoms of procainamide lupus. These observations support the hypothesis that the aromatic amino group of procainamide is important for the induction of procainamide lupus and that acetylation of this amino group blocks this effect of procainamide (Kluger, J. et al., Ann Int. Med. 95, 18-23, 1981).

SCLERODERMA AND PROGRESSIVE SYSTEMIC SCLEROSIS (PSS): The immunological findings in PSS are listed in the next Table:

Immunological Findings in PSS

Test	Remarks
Fluorescent Anti-Nuclear Antibody (FANA)	Positive in almost 80% of patients but titers are usually low. Usually speckled pattern, but may be homogeneous or rim, or nucleolar; the nucleolar pattern is associated with PSS.
Specific Immunological Tests: Antibodies to Nucleolar RNA (occurs in 40% to 50% of patients at titers varying from 1:100 to 1:1000 by immunofluorescence) and Antibodies to Scl-1 (occurs in 10% to 20% of patients)	Usually detected by hemagglutination or immunodiffusion
Rheumatoid Factor	Present in about 25% of patients
Cryoglobulin	Immune Complex

CREST: CREST is a mnemonic for the characteristics of this syndrome: calcinosis cutis, Raynaud's phenomenon, esophageal motility abnormalities, sclerodactyly and telangiectasia. CREST is a variant of scleroderma or progressive systemic sclerosis(PSS). CREST patients differ from patients with PSS in that they do not have widespread involvement and the skin changes are confined to the hands and face. Antibodies to centromeric chromatin are found in 80 to 90% of patients with the CREST syndrome.

POLYMYOSITIS: The laboratory diagnosis is based on three tests; these are listed in the next Table:

Laboratory Diagnosis of Polymyositis
(a) Muscle-Enzyme Studies
(b) Electromyograph (EMG)
(c) Muscle Biopsy

(a) Muscle-Enzyme Studies: Serum creatine phosphokinase(CPK) is elevated in patients with polymyositis.
(b) Electromyograph(EMG): This test gives positive results when involved muscles are examined.
(c) Muscle Biopsy: This is pathognomonic for polymyositis; however, it may not be necessary if the clinical picture, enzyme studies and EMG give adequate evidence for diagnosis. Muscle biopsy shows initially intense perivascular inflammation.

Immunological Findings in Polymyositis: Some patients with polymyositis have findings similar to that observed in "connective tissue" diseases; these are given in the next Table:

Immunological Findings in Polymyositis

Finding	Remarks
Antinuclear Antibodies	Pattern variable; speckled; positive in 1/3 of patients
PM-1	Antibodies to PM-1 antigen occur in 50% of polymyositis patients and in 10% of dermatomyositis patients at titers of 1:2 or 1:4 by immunodiffusion (Tan, E.M. Hospital Practice, pgs. 79-84, Jan.1983).
Rheumatoid Factor	Positive in occasional patient

MIXED CONNECTIVE TISSUE DISEASE(MCTD): Mixed connective tissue disease has a
"mixed" clinical picture suggesting symptoms of systemic lupus erythematosus,
scleroderma, and polymyositis. All of these patients have an unusually high
titer of an antinuclear antibody with a specificity for a nuclear ribonucleo-
protein(RNP) antigen. Antibodies to RNP occur in 95%-100% of patients; there is
absence of other antinuclear antibodies (Tan, E.M., Hospital Practice, pgs.
79-84, Jan. 1983). Patients are often responsive to corticosteroid therapy.
     Almost 3/4 of these patients have muscle pain, muscle tenderness and
weakness with abnormal electromyograms consistent with inflammatory myositis.
     The laboratory findings in MCTD are given in the next Table:

Laboratory Findings in MCTD	
Finding	Remarks
Sedimentation rate	Elevated
Anemia	Moderate
Coombs Positive Hemolytic Anemia	Rare
Leukopenia	Moderate
Thrombocytopenia	Rare
CPK	Elevated when muscle disease present
Fluorescent Antinuclear Antibodies Patterns	Titers are High
	Low dilutions, mixed patterns; high dilutions, speckled pattern
RNP Antibody	High titers of RNP antibody are found in MCTD (95%-100% of patients at titers >10,000) and SLE
Rheumatoid Factor	Found in 50% of patients.

COOMBS, DIRECT (ANTI-HUMAN GLOBULIN TEST, DIRECT ANTIGLOBULIN TEST)

SPECIMEN: Purple (EDTA) top tube, or 2 drops blood in saline micro-capillary tube; include history of recent and past pregnancy and drug therapy.
REFERENCE RANGE: Negative
METHOD: The anti-human globulin test is used to detect antibodies on red cells and is based on the principle that anti-human globulin antibodies induce agglutination of erythrocytes coated with globulins. The reaction is illustrated in the next Figure:

Coombs Test, Direct

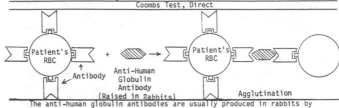

The anti-human globulin antibodies are usually produced in rabbits by immunization. These antibodies are directed against human immunoglobulins (mainly IgG) and complement (mainly C3).
INTERPRETATION: This test is used to answer the question: Are IgG and/or complement bound to red cell membrane? The direct Coombs test is positive in the following conditions:

Positive Direct Coombs Test

Hemolytic Disease of the Newborn (HDN)
Transfusion Reactions, Hemolytic
Autoimmune Hemolytic Anemias
Drug Induced: Methyl Dopa (Aldomet); Penicillin;
   Cephalosporins (Keflin); Quinidine, Insulin, Sulfonamides; Phenacetin
Warm Autoimmune Hemolytic Anemia
Cold Autoimmune Hemolytic Anemia
Paroxysmal Cold Hemoglobinuria

Antibodies directed against human immunoglobulins (and occasionally complement) coating red blood cells are as follows: HDN, hemolytic transfusion reactions and antibodies induced by drugs, such as, methyl dopa (Aldomet), penicillin, cephalosporins (Keflin).

Antibodies directed against complement coating red blood cells are: Warm autoimmune hemolytic anemia; cold agglutinin disease; paroxysmal cold hemoglobinuria; antibodies induced by drugs, such as, Quinidine, insulin, sulfonamides and phenacetin.
Hemolytic Disease of the Newborn (HDN): HDN is due to Rh(-) mother and Rh(+) fetus and occurs as follows: The mother, on exposure to Rh(+) fetal red blood cells, develops antibodies to the Rh(+) antigens; the antibodies pass from the maternal blood into the fetus. In the fetus, the antibodies combine with the Rh antigens of the fetal red blood cells; complement may combine with the antibody. Macrophages or lymphocytes attach to the Fc region of the immunoglobulin of the red blood cells. Sphering and then lysis of the red cell occurs; antibody-coated red cells may also be destroyed by polymorphonuclear leukocytes or by phagocytes. These effects are illustrated in the next Figure:

Lysis of Red Blood Cells in Erythroblastosis Fetalis

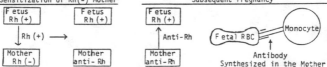

Diagnosis of erythroblastosis fetalis in the newborn is made as follows: mother, Rh(-); infant, Rh(+); Coombs test(+), the jaundice is due to Rh incompatibility.
   Erythroblastosis fetalis can be prevented by passively administered antibody which can interfere with the induction of antibody formation.

COOMBS, INDIRECT (ANTIBODY SCREEN, INDIRECT COOMBS, SELECTIGEN)

SPECIMEN: Red top tube, separate serum; do not use serum separator tube; include history of recent and past transfusions, pregnancy and drug therapy.
REFERENCE RANGE: Negative
METHOD: The indirect Coombs test is used to detect antibodies in sera; the sera, suspected of containing antibody, is reacted with red cells having known antigens on their surfaces. The reaction is illustrated in the next Figure:

Coombs Test, Indirect

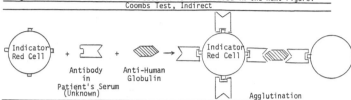

INTERPRETATION: This test is used to detect red cell antibodies in patient's serum in order to evaluate possible causes of hemolysis. This test is performed routinely with compatibility testing. This test is positive in the following conditions: antibody specific (previous transfusion) to red blood cell antigens, platelet antigens, leukocyte antigens, etc.; autoantibody, non-specific in acquired hemolytic anemia.

COPPER, LIVER

SPECIMEN: Needle or wedge biopsy; refrigerate. Forward to reference laboratory in special metal-free plastic container; it is not necessary to refrigerate during shipment.
REFERENCE RANGE: 10-35mcg/gram dry weight of liver. To convert conventional units in mcg to international units in micromol/liter, multiply conventional units by 0.01574.
METHOD: Atomic absorption
INTERPRETATION: Copper, in liver, is increased in the conditions listed in the next Table:

Increase in Copper in the Liver
Hepatolenticular Degeneration (Wilson's Disease)
Primary Biliary Cirrhosis

Characteristics of hepatolenticular degeneration are described(see COPPER, SERUM).
Primary Biliary Cirrhosis: The clinical features are listed in the next Table:

Clinical Features of Primary Biliary Cirrhosis
Middle-aged Females
Insidious Onset
Pruritis, Skin Pigmentation, Jaundice
Skin Xanthomata, Xanthelasmata
Hepatomegaly, Splenomegaly
Steatorrhea (Fat in Stool)
Signs and Symptoms of Portal Hypertension
Associated Diseases (Sicca Syndrome, Thyroiditis, etc.)
Death in Liver Failure

About 90 percent of the patients are women. The disease starts insidiously. Pruritis may be the presenting complaint, and usually precedes jaundice. Pruritis is usually attributed to raised bile acid concentrations. Hepatosplenomegaly is not usually a presenting finding. Steatorrhea is caused by defective bile acid excretion and thus poor fat and fat-soluble vitamin absorption leading to vitamin K deficiency and vitamin D deficiency . As the disease progresses, portal hypertension may develop.

Primary biliary cirrhosis often occurs in association with other autoimmune conditions such as Hashimoto's thyroiditis or Sjögren's syndrome.
Diagnosis of Primary Biliary Cirrhosis: Laboratory findings in primary biliary cirrhosis are those of biliary obstruction, e.g. serum alkaline phosphatase increased, serum bilirubin increased, serum cholesterol increased, serum IgM values usually increased and positive serum mitochondrial antibody test.

COPPER, SERUM

SPECIMEN: Red top tube, separate serum; refrigerate at 4°C.

REFERENCE RANGE:

AGE	Serum Copper(mcg/dl)
Newborn	10-27
Two	95-190
Ten	70-165
Young Adults	65-150
Adults	85-150

To convert conventional units in mcg/dl to international units in micromol/liter, multiply conventional units by 0.1574.

METHOD: Atomic absorption spectrophotometry

INTERPRETATION: Serum copper is decreased in the conditions listed in the next Table:

Causes of Decreased Serum Copper
Hepatolenticular Degeneration (Wilson's Disease)
Diarrhea
Malnutrition
Parenteral Nutrition with Cu Deficient Solutions
Malabsorption
Menke's Syndrome ("kinky" or "steely" hair syndrome)

Hepatolenticular Degeneration (Wilson's Disease); (Cartwright, G.E., "Diagnosis of Treatable Wilson's Disease," N. Engl. J. Med. 298, 1347-1350, 1978): Wilson's disease (hepatolenticular degeneration) is an uncommon inborn error of copper metabolism inherited as an autosomal-recessive trait. In this disease, there is excessive deposition of copper in the liver, basal ganglia, cerebral cortex, kidney, and cornea. The classic findings are hepatic, neurologic and renal disease; other common manifestations include hemolytic anemia, arthritis, coagulopathies, and endocrinologic disorders including hypoparathyoidsm (Carpenter, T.O. et al., N. Engl. J. Med. 309, 873-877, 1983). Symptomatic patients usually show pigmented cornea rings and the presence of this ring is an important aid in diagnosis. The Kayser-Fleischer ring is a golden-brown, greenish, or brownish- yellow, or bronze discoloration in the zone of Descemet's membrane in the limbi region of the cornea.

The defect in Wilson's disease may be a defect in the normal binding of copper to ceruloplasmin. Almost all the circulating copper is normally bound to ceruloplasmin; the bound copper is known as the indirect-reacting copper. The remaining copper (2% to 5%) is loosely bound to serum albumin; this is known as the direct reacting fraction of serum copper. Patients with Wilson's disease have decreased total copper in their serum primarily because of the decrease in indirect reacting copper. This reflects a deficient level of ceruloplasmin. However, there is an increase in direct reacting copper which is loosely bound; this loosely bound copper is readily dissociated and deposited in tissue.

The laboratory diagnosis of Wilson's disease should be suspected in patients below the age of 30 who have chronic idiopathic hepatitis. The biochemical findings in Wilson's disease are listed in the next Table:

Biochemical Findings in Wilson's Disease
Decreased Ceruloplasmin in Serum
Decreased Total Copper in Serum
Decreased Indirect Reacting Copper
Increased Direct Reacting Copper
Increased Copper in Urine
Increased Copper Deposited in Liver

The diagnosis of Wilson's disease is aided by the demonstration of decreased levels of the enzyme, ceruloplasmin, and copper in the serum and increased levels of copper in the urine.

A serum ceruloplasmin level of less than 20mg/dl and a urinary copper level in excess of 100mcg/24 hour urine is compatible with the diagnosis of Wilson's disease.

Copper is increased in the serum in the conditions listed in the next Table:

Causes of Increased Serum Copper
Spraying of Grapes with a Copper Sulfate
Fungicide (Bordeau Mixture)
Dialysis Secondary to Use of Copper Tubing
Ingestion of Copper Sulfate
Copper in Drinking Water
Acute and Chronic Diseases
Malignant Diseases
Hemochromatosis
Biliary Cirrhosis
Thyrotoxicosis
Various Infections
Patients on Oral Contraceptives
Patients Taking Estrogens
Pregnancy

COPPER, URINE

SPECIMEN: Collect 24 hour urine in acid washed-containers. Use plastic containers (borosilicate, polyethylene or polypropylene); add 10% HCl solution to the container and allow to "soak" for 10 minutes; rinse with five volumes of tap water and then five volumes of deionized or distilled water. The patient should urinate at 8:00 A.M. and the urine is discarded. Then, urine is collected for 24 hours including the next day 8:00 A.M. specimen. Indicate 24 hour volume. A 50ml aliquot is used for analysis.

REFERENCE RANGE: 15-100mcg/24 hour collection. To convert conventional units in mcg/24 hour to international units in micromol/day, multiply conventional units by 0.01574.

METHOD: Atomic absorption spectroscopy

INTERPRETATION: Copper is increased in the urine in the following conditions:

Causes of Increased Urine Copper
Hepatolenticular Degeneration (Wilson's Disease)
Primary Biliary Cirrhosis
Nephrotic Syndrome

A discussion of hepatolenticular degeneration is given under COPPER, SERUM.

CORTISOL, PLASMA (COMPOUND F, HYDROCORTISONE)

SPECIMEN: Green top tube (lithium heparin), separate plasma and freeze plasma, or red top tube, separate serum and freeze serum.

REFERENCE RANGE: A.M.: 7 to 25mcg/dl; P.M.: 2 to 9mcg/dl. To convert conventional units in mcg/dl to international units in nmol/liter, multiply conventioanl units by 27.59.

METHOD: RIA

INTERPRETATION: The change in plasma cortisol in adrenal disorders is given in the next Table:

| Change in Plasma Cortisol in Adrenal Disorders ||
Disorder	Plasma Cortisol
Cushing's Syndrome	Increased
Adrenal Insufficiency	Decreased
Adrenogenital Syndromes	Decreased
Secondary Adrenal Insufficiency	Decreased

In Cushing's syndrome there is loss of diurnal variation of plasma cortisol; elevated levels of plasma cortisol are maintained around the clock. Blood should be drawn at 8:00 A.M. and 4:00 P.M. to evaluate diurnal variation. The normal diurnal variation of plasma cortisol and the change in Cushing's syndrome is shown in the next Figure:

Normal Diurnal Variation of Plasma Cortisol and Change in Cushing's Syndrome

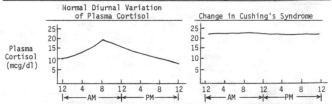

ACTH Infusion Test: ACTH infusion test is done to measure the functional reserve of the adrenal cortex. The manner in which the test may be done is as follows (Melby, J.C., N. Engl. J. Med. 285, 735-739, 1971): Obtain a blood specimen for cortisol assay; inject 0.25 mg of alpha, 1-24, corticotropin (synthetic unit of ACTH), I.V.; obtain blood specimen at 0, 30, 60, 90 and 120 minutes for cortisol assay. The results are given in the next Figure (Melby, 1971):

Effect of ACTH Injected Intravenously on Plasma Cortisol Concentrations in Normal Subjects and Patients with Addison's Disease (Melby, 1971)

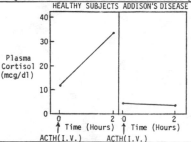

Dexamethasone Test for Diagnosis of Cushing's Syndrome: Dexamethasone is a fluorinated steroid that has about 30 times the potency of cortisol. Dexamethasone, 1mg, is taken orally at midnight and a fasting blood cortisol is obtained at 8:00 A.M.; a value of blood cortisol greater than 5mcg/dl is compatible with Cushing's syndrome. A negative test (<5mcg/dl) is found in less than 2 percent of the cases of Cushing's syndrome (Crapo, L., Metabolism 28,

955-977, 1979). However, abnormal results (>5mcg/dl) are often found in patients under stress for any reason, including at least 30 percent of patients in a hospital who are acutely ill (Connolly, C.K. et al., Br. Med. J. 2, 665-667, 1968). (See also, Dexamethasone Suppression Test as a Diagnostic Aid in Depression.) The tests that are used to answer the question, "Does the patient have Cushing's syndrome?," are listed in the next Table:

Does the Patient Have Cushing's Syndrome? - Tests	
Test	Result in Cushing's Syndrome
Plasma Cortisol	Increased
Urinary 17-OH Corticosteroids, 17-Keto-Steroids and 17-Ketogenic Steroids	Increased
Urinary Free Cortisol	Increased
Diurnal Variation of Plasma Cortisol	Loss of Diurnal Variation
Dexamethasone, 1mg at Midnight	8 A.M. Cortisol Greater than 5 mcg/dl Compatible with Cushing's

CORTISOL, URINE, FREE

Cortisol, Urine, Free

SPECIMEN: 20ml aliquot of 24 hour urine containing 10ml glacial acetic acid.
REFERENCE RANGE: 4 months-10 years: 35-176mcg/g creatinine; 11-20 years: 1-44mcg/g creatinine; adult: 13-60mcg/g creatinine.
METHOD: RIA
INTERPRETATION: Urinary free cortisol is the non-conjugated cortisol excreted in the urine; it reflects the concentration of the free cortisol in the serum, that is, cortisol not bound to transcortin. This test is the most sensitive, specific, and cost-effective approach for initial screening in hospitalized patients suspected of having Cushing's syndrome. (See Hospital Practice, pgs. 37-43, June, 1983).

CIE

COUNTERIMMUNOELECTROPHORESIS(CIE) (SEE BACTERIAL ANTIGENS)

"C"-PEPTIDE, INSULIN (CONNECTING PEPTIDE, INSULIN)   "C"-Peptide

SPECIMEN: Red top tube, separate serum; place serum in a separate plastic vial and freeze immediately.
REFERENCE RANGE: 1.5 to 10.0ng/ml; To convert from ng/ml to picomoles/ml, divide ng/ml by 3.56 (or multiply by 0.28). To convert from picomoles/ml to ng/ml, multiply picomoles/ml by 3.56.
METHOD: RIA
INTERPRETATION: The "C" peptide is the peptide that connects the A and B chains of insulin as illustrated in the next Figure:

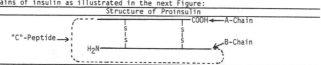
Structure of Proinsulin

Assay of "C" peptide is done for the following reasons:
(1) Differential diagnosis of insulinoma versus factitious hypoglycemia: In insulinoma, "C" peptide and insulin are increased while in factitious hypoglycemia, insulin levels are increased and "C" peptide levels, which reflect endogenous insulin secretion, are decreased.
(2) Insulin Secretion in Diabetics: Almost all diabetics who receive insulin develop serum antibodies after several months. Since commercial RIA insulin determinations employ heterologous antibodies, immunoassay of insulin in these individuals is virtually useless. The measurement of C-peptide is the best way to assess beta cell function. Since C-peptide and insulin are secreted in equimolar amounts, quantitation of C-peptide reflects insulin secretion; C-peptide does not contain common antigenic determinants with administered insulin; proinsulin does. Since the half-life of C-peptide is over three times longer than insulin, (14 minutes versus 4 minutes), normally there is more C-peptide than insulin in the blood.

Patients with insulin-dependent diabetes mellitus (type I) usually have no or very low concentrations of C-peptide; patients with non-insulin dependent diabetes mellitus (type II) tend to have normal or elevated C-peptide levels.

C-REACTIVE PROTEIN (CRP)

SPECIMEN: Red top tube, separate serum.

REFERENCE RANGE: None detected

METHOD: Latex agglutination, radial immunodiffusion (RID), electroimmuno-diffusion (EID), nephelometry, enzyme-linked immunosorbent assays (ELISA)

INTERPRETATION: Causes of elevation of serum levels of "C"-reactive protein are listed in the next Table; (Gewurz, H., Hosp. Pract. pgs. 67-81, June, 1982):

Causes of Elevation of Serum "C"-Reactive Protein

Screening for Organic Disease
    General screening aid for inflammatory diseases, infections, and neoplastic diseases or tissue injury

Detection and Evaluation of Inflammatory Disorders (>CRP)
    Rheumatoid arthritis
    Seronegative arthritides (e.g., Reiter's syndrome)
    Rheumatic fever
    Vasculitic syndromes (e.g., hypersensitivity vasculitis)
    Inflammatory bowel disease

Detection and Management of Infections (>CRP)
    Neonatal infections
    Postoperative infections
    Intercurrent infections in leukemia
    Bacterial infections in systemic lupus erythematosus
    Pyelonephritis

Detection and Evaluation of Tissue Injury and Neoplasia (>CRP)
    Myocardial infarction and embolism
    Transplant rejection
    Certain tumors (e.g., Burkitt's lymphoma)

Aid in Differential Diagnosis
    Systemic lupus erythematosus versus rheumatoid arthritis and other arthritides (< SLE)
    Crohn's disease versus ulcerative colitis (< Crohn's)
    Pyelonephritis versus cystitis (> pyelonephritis)
    Bacterial versus viral infections (> bacterial)
    Acute bronchitis versus asthma (< asthma)

"C"-reactive protein(CRP) is an acute reacting protein which may increase dramatically(up to a thousandfold) in inflammatory conditions; it's nonspecific. CRP levels follow the course of the acute phase. It can be used as a supplement to or complement to erythrocyte sedimentation rate(ESR) studies. The conditions that elevate ESR usually cause elevation of CRP but CRP levels more closely approximate the degree of ongoing tissue damage. CRP responds quickly to inflammation(6-10 hrs); it has a short half-life(5-7 hrs). The "C" of CRP was derived from earlier studies which demonstrated that serum from acutely ill individuals contained a substance that precipitated with the C form(termed C-polysaccharide) of the cell wall of pneumococci. CRP is synthesized in the liver. CRP was found to be elevated in the serum of patients with bacterial meningitis but not in patients with viral meningitis or meningoencephalitis. CRP may be useful in mOnitoring the course of bacterial meningitis; levels characteristically return to normal within seven days in bacterial meningitis in the absence of complications (Peltola, H.O., Lancet 1, 980, 1982).

The change in the serum level of CRP before and following cholecystectomy is shown in the next Figure; (Gewurz, H. Hosp. Pract., pgs. 67-81, June, 1982):

Change in Serum Level of CRP Before and Following Choleystectomy

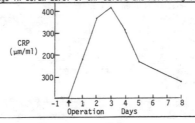

CRP in Acute Pyelonephritis and Acute Cystitis: CRP is increased in patients
with pyelonephritis and normal in patients with cystitis as shown in the next
Figure; (Morley, J.J. and Kushner, I., "Serum C-Reactive Protein Levels in
Disease," Ann. N. Y. Acad. Sci. 389, 406-418, 1982):

Serum C-Reactive Protein (CRP) in Acute Pyelonephritis and Acute Cystitis

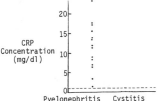

Serum CRP levels in patients with pyelonephritis are greater than 1mg/dl;
serum CRP levels in patients with cystitis are less than 1mg/dl. In this study,
only one patient with cystitis and three patients with pyelonephritis were
children; otherwise, the patients were adults.

In another study, serum CRP levels were measured in children (ages 5 to 17
years) with pyelonephritis and cystitis. All children with pyelonephritis had
serum levels of CRP greater than 2.5mg/dl; all children with cystitis had serum
levels of CRP less than 1mg/dl (Jodal, V., Lindberg, U. and Lincoln, K., "Level
Diagnosis of Symptomatic Urinary Tract Infection in Childhood", Acta Paediatr.
Scand. 64, 201-208, 1975).

Prediction of chorioamnionitis in Premature Rupture of Membranes: The CRP
has a sensitivity of 88 percent and a specificity of 96 percent for predicting
chorioamnionitis (Hawrylyshyn, P. et al., Am. J. Obstet. Gynecol. 147, 240-246,
1983).

CREATINE PHOSPHOKINASE (CPK)

SPECIMEN: Red top tube, separate serum; freeze if analysis delayed.
REFERENCE RANGE: Newborn: Up to 10x Adult Value, depending on trauma at birth; 4 Days: 3x Adult Value; >6 Weeks: Adult Value; Adult Male: 35-230 U/liter; Adult Female: 20-210 U/liter
METHOD: CPK is assayed using two consecutive kinase reactions followed by a dehydrogenase reaction as the indicator reaction as shown below:

$$\text{Creatine Phosphate (CP) + ADP} \xrightarrow[\text{Phosphokinase (CPK)}]{\text{Creatine}} \text{Creatine (C) + ATP}$$

$$\text{ATP + Glucose} \xrightarrow[\text{(HK)}]{\text{Hexokinase}} \text{Glucose-6-P + ADP}$$

$$\text{Glucose-6-P + NADP}^+ \xrightarrow[\text{Dehydrogenase (G6PD)}]{\text{Glucose-6-P}} \text{6-Phosphogluconate + NADPH + H}^+ \quad (340\text{nm})$$

INTERPRETATION: Creatine phosphokinase is found in relatively high concentration in three tissues, the heart, skeletal muscle and the brain; conditions affecting these tissues that are associated with elevated creatine phosphokinase are listed in the next Table:

Conditions Affecting Heart and Skeletal Muscle Associated with Elevated Serum Creatine Phosphokinase	
Heart	**Skeletal Muscle**
Acute Myocardial Infarction	Vigorous Exercise
Acute Myocarditis	Intramuscular Injections
	Skeletal Muscle Diseases; e.g.,
	Polymyositis; Dermatomyositis; Muscular
	Dystrophies of All Types
	Malignant Hyperpyrexia (Hyperthermia)
	Cerebral Vascular Accidents
	Convulsions
	Trauma
	Surgery
	Alcoholism
	Reye's Syndrome
	Head Injury
	Hypothyroidism

Creatine phosphokinase is elevated in the serum 3-4 hours following myocardial infarction, reaches its peak of activity before 36 hours and is followed by a sharp drop to normal levels by the second to fourth day. An intramuscular injection or injury prior to obtaining the serum may result in the elevation of the serum CPK; approximately one-third of patients develop elevation of serum CPK after intramuscular injection; elevation varies up to 10 times the upper limit of normal.

This enzyme is raised in progressive muscular dystrophy and other primary diseases (dystrophies, not atrophies) of skeletal muscle. Also, it may be elevated in cerebral infarcts. The enzyme is not elevated in liver disease or in pulmonary infarction.

Occasionally, it may be necessary to distinguish the origin of elevated serum creatine phosphokinase, e.g., heart vs. skeletal muscle. This can be done by serum CPK isoenzyme studies.

CPK of brain origin (CPK-BB) does not ordinarily appear in the serum even in patients with cerebral vascular accidents or head injury. It is CPK-MM of skeletal muscle origin that appears in the serum in these conditions.

Creatine phosphokinase tends to be decreased in hyperthyroidism.

CREATINE PHOSPHOKINASE ISOENZYMES
(CPK ISOENZYMES)

SPECIMEN: Red top tube; separate serum; freeze if assay is not done within 48 hours; the specimen should be rejected if there is gross hemolysis.

REFERENCE RANGE: CPK-MM: 94-100%; CPI-MB (Heart): 0-6%; CPK-BB (Brain): 0%.

METHOD: Agarose electrophoresis

INTERPRETATION: CPK isoenzymes are most often done to document an acute myocardial infarction. Optimally, a minimum of three serum specimens should be obtained for analysis for CPK-MB at admission and at 12 and 24 hours after onset of symptoms of acute myocardial infarction. Negative results of analysis for CPK-MB in samples obtained before 12 hours or after 24 hours should not be used to exclude the diagnosis of acute myocardial infarction. A total CPK value within the normal range is not a reliable screening test to exclude analysis of CPK-MB. (Irvin, R.G., Coff, F.R. and Roe, C.R., Arch. Intern. Med. 140, 329-334, 1980; Wagner, G.S., Arch. Int. Med. 140, 317-319, 1980; Fisher, M.L. et al, JAMA, 249, 393-394, 1983.)

Following acute myocardial infarction, CPK-MB begins to rise in 2 to 12 hours, reaches its peak in 12 to 40 hours and returns to normal in 24 to 72 hours. The change in CPK-MB is summarized in the next Table:

Change in CPK-MB Following Acute Myocardial Infarction		
Beginning of Increase	Maximum	Return to Normal
2 - 12 hours	12 - 40 hours	24 - 72 hours

Irvin et al., Arch. Intern. Med. 140, 329-334, 1980.

The causes of elevation of CPK-MB are given in the next Figure:

Causes of Elevation of CPK-MB	
Pattern	Causes

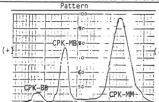

Causes:
Acute Myocardial Infarction
Acute Pericarditis with Myocardial Involvement
Polymyositis
Reye's Syndrome
Malignant Hyperpyrexia
Carbon Monoxide Hemoglobin
Duchenne's Muscular Dystrophy

CPK-MB is increased in conditions other than acute myocardial infarction. CPK-MB is elevated in the serum of patients with pericarditis when there is a concomitant underlying myocarditis. (Marmor, A. et al. Arch. Intern. Med. 139, 819-820, 1979). About 15 to 20 percent of patients with polymyositis have involvement of cardiac muscle and increase in CPK-MB. Practically all patients with Duchenne's muscular dystrophy have elevation of CPK-MB and CPK-MM.

Macro-CPK is formed from either an immune complex consisting of autoantibody to CPK-MM and CPK-BB or from mitochondrial CPK. On electrophoresis, macro-CPK usually migrates between CPK-MM and CPK-MB, occasionally migrates anodally to CPK-MM or with CPK-MM or CPK-BB. Macro-CPK leads to incorrect results when CPK-MB is separated by column, e.g., DuPont ACA (Pudek, M.R. et al., Clin. Chem. 28, 1400, 1982). Ectopic production of CPK-MB was found in a patient with metastatic carcinoma of the colon (Annesley, T.M. et al., Am. J. Clin. Pathol. 79, 255-259, 1983).

Strenuous exercise may cause elevation of CPK-MB such as observed in marathon runners (Ohman, E.M. et al., Brit. Med. J. 285, 1523-1526, 1982).

Elevation of serum CPK-BB is uncommon; conditions associated with elevation of serum CPK-BB include infarction of the colon, Reye's syndrome, malignant hyperpyrexia, renal dialysis patients, cerebrovascular accidents and massive brain injury. CPK-BB has been found in the serum of some patients with active metastatic disease (Zweig, M.H. et al., Clin. Chem. 25, 1190-1191, 1979); it is reported to be raised in 90 percent of patients with metastatic carcinoma of the prostate (Silverman et al., Clin Chem. 25, 1432-1435, 1979).

CREATINE, SERUM

SPECIMEN: Red top tube, separate serum; freeze if analysis delayed

REFERENCE RANGE: Male: 0.2-0.7mg/dl; Female: 0.35-0.95mg/dl. Normal levels are higher in children and during pregnancy. To convert conventional units in mg/dl to international units in micromol/liter, multiply conventional units by 76.25.

METHOD: Creatine is converted to creatinine in acid solution; then creatinine is measured by the Jaffe reaction. The difference in the creatinine content before and after treatment with acid gives the creatine content.

INTERPRETATION: Creatine is formed primarily in the liver and is transported to the muscles where it is phosphorylated; creatine phosphate acts as a storage depot for muscle energy. The muscle enzyme, creatine phosphokinase(CPK), catalyzes the reaction between creatine and creatine phosphate.

$$\text{Creatine} + P \xrightarrow{\text{CPK}} \text{Creatine Phosphate}$$

Creatine is increased in the conditions listed in the next Table:

Increased Serum Creatine
Skeletal Muscle Necrosis or Atrophy:
Trauma
Muscular Dystrophies
Poliomyelitis
Amyotrophic Lateral Sclerosis
Dermatomyositis
Myasthenia Gravis
Starvation
Endocrine Disorders:
Hyperthyroidism
Acromegaly
Other Conditions:
Acute Rheumatoid Arthritis
Systemic Lupus Erythematosus
Leukemia
Infections
Burns

This test has been largely replaced by CPK measurements to reflect skeletal muscle necrosis and dystrophies.

CREATINE, URINE

SPECIMEN: 24 hour urine, preserve with thymol or toluene; record 24 hour volume; obtain aliquot and freeze for prolonged storage.

REFERENCE RANGE: Male: 0-40mg/day; Female: 0-80mg/day; increased in pregnancy. To convert conventional units in mg/day to international units in micromol/day, multiply conventional units by 7.625.

METHOD: See SERUM CREATINE

INTERPRETATION: Creatinuria is found in myopathies, infections, starvation, impaired carbohydrate metabolism and hyperthyroidism; See SERUM CREATINE.

CREATININE CLEARANCE

SPECIMEN: Red top tube, separate serum and determine creatinine. Obtain 24 hour urine, record volume and measure urine creatinine. Instruct patient on 24 hour urine collection as follows: Void at 8:00 A.M. and discard specimen. Collect all urine during the next 24 hours including the 8:00 A.M. specimen the next morning. Keep specimen on ice during collection. Transport specimen to laboratory immediately. Measure 24 hour urine volume.

REFERENCE RANGE: Male: 85-125ml/min; Female: 75-115ml/min.

METHOD: Alkaline picrate

INTERPRETATION: Creatinine clearance reflects glomerular filtration rate. The equation for the calculation of creatinine clearance is as follows:

$$\text{Clearance } \frac{ml}{min} = \frac{\text{Urine Volume (ml/min) x Urine Creatinine}}{\text{Serum Creatinine}} \times \frac{1.73}{A}$$

where A = body surface area.

The relationship between plasma creatinine and creatinine clearance is illustrated in the next Figure; (Davidsohn, I. and Henry, J.B., Clinical Diagnosis by Laboratory Methods, Saunders, 1979, p. 140):

Relationship of Plasma Creatinine to Creatinine Clearance

(Creatinine Clearance)

Serum creatinine increases as creatinine clearance decreases.

From this Figure, it can be seen that the serum creatinine level may be within normal range when the glomerular filtrate rate is significantly reduced. Normal creatinine clearance is 90ml/min to 130ml/min. Stages of renal failure may be arbitrarily related to creatinine clearance as follows: Renal impairment, >50ml/min; renal insufficiency, 20ml/min to 50ml/min; renal failure, <20ml/min and as shown in the next Table:

Stages of Renal Failure and Creatinine Clearance

Stage	Creatinine Clearance
Renal Impairment	>50ml/min
Renal Insufficiency	20ml/min to 50ml/min
Renal Failure, Uremia	5ml/min to 20ml/min

A two hour creatinine clearance test has been found to be useful in intensive care units when urine was collected using an indwelling Foley catheter. Accurate timing during the specimen collection is absolutely necessary. (Wilson, R.F. and Soullier, G., Crit. Care Med. 8, 281-284, 1980).

The value of creatinine clearance is particularly useful when the dose of a drug depends on glomerular filtrate.

Creatinine clearance may be estimated from the serum creatinine level alone by use of the following formula:

$$\text{Creatinine Clearance} = \frac{98 - (4/5)(\text{Age}-20)}{\text{Serum Creatinine}}$$

This formula is empirically derived; it is useful over an age range from 20 to 80 years (Jeliffe, R.W., Ann. Intern. Med. 79, 604, 1973), and it is useful to calculate medication schedules at the bedside.

With age, muscle mass decreases, and thus serum creatinine may not reflect creatinine clearance. Creatinine clearance decreases with age; the decrease in creatinine clearance with age is probably due to a lowered glomerular filtration rate caused by a reduction in renal plasma flow.

In pregnancy, there is an increase in renal blood flow and glomerular filtration rate (GFR). The normal values for BUN and serum creatinine are significantly lower; a serum creatinine concentration of 1.2mg/dl indicates an approximate 50 percent reduction in GFR in pregnant patient.

Cefoxitin falsely elevates creatinine values; creatinine clearance measurements are unreliable while patients are receiving cefoxitin.

CREATININE, SERUM

SPECIMEN: Red top tube, separate serum.
REFERENCE RANGE: Newborn: 0.8 to 1.4mg/dl; Infant: 0.7-1.7mg/dl; Adult:
0.7-1.5mg/dl. To convert conventional units in mg/dl to international units in
micromol/liter, multiply conventional units by 88.40.
METHOD: Alkaline picrate (Jaffé reaction)
Enzymatic Method: (Boehringer's Creatinase Enzymatic Kit)

$$Creatinine + H_2O \xrightarrow{Creatinase} Creatine$$

$$Creatine + ATP \xrightarrow[CPK]{Creatine Kinase} Creatine Phosphate + ADP$$

$$ADP + Phosphoenolpyruvate(PEP) \xrightarrow[(CPK)]{Pyruvate Kinase} ATP + Pyruvate$$

$$Pyruvate + NADH + H^+ \xrightarrow[(LDH)]{Lactate Dehydrogenase} Lactate + NAD^+$$

INTERPRETATION: Serum creatinine is increased in renal disease, muscle
necrosis and hypovolemia.

Serum creatinine is useful in evaluation of glomerular function. However,
the serum level is not sensitive to early renal damage and responds more slowly
than the BUN to hemodialysis during treatment of renal failure. The serum
creatinine together with serum BUN is used to differentiate pre-renal, renal and
post-renal (obstructive) azotemia since an elevated urea with only slight to
moderate elevation of creatinine suggests pre-renal or post-renal azotemia. The
BUN/creatinine ratio in different conditions is shown in the next Table:

BUN/Creatinine Ratio	
State	Ratio
Normal	10/1
Dehydration	15 to 20/1
Prerenal	> 10/1
Renal Disease	10/1

Creatinine is formed in muscles from creatine and creatine phosphate.
Creatine is formed primarily in the liver and is transported to the muscles
where it is phosphorylated; creatine phosphate acts as a storage depot for
muscle energy. The muscle enzyme, creatine phosphokinase(CPK) catalyzes the
reaction between creatine phosphate and creatine.

$$Creatine \ phosphate \xrightarrow{CPK} Creatine + P$$

The metabolic end product of creatine metabolism is creatinine, a cyclic
anhydride of creatine. Determination of creatinine concentrations in serum is
the most commonly used clinical method for measuring glomerular filtration rate.
The plasma creatinine concentration is increased when the glomerular filtration
is decreased. Serum creatinine varies with subjects age, body weight and sex.

Creatinine clearance may be estimated from the serum creatinine level alone
by use of the following formula:

$$Creatinine \ Clearance = \frac{98 - (4/5)(Age-20)}{Serum \ Creatinine}$$

This formula is empirically derived; it is useful over an age range from 20
to 80 years (Jeliffe, R.W., Ann. Intern. Med. 79, 604, 1973), and it is useful
to calculate medication schedules at the bedside.

The daily production rate of creatinine is approximately 15 mg per kg; with
complete renal shut-down, the serum level will rise at a rate of 1 to 2 mg/dl
per day. If the rate of rise is less, residual renal function is present. If
the rate of rise is greater than 3 mg/dl per day, there is muscle disease, such
as rhabdomyolysis or severe catabolism (Goldstein, M., Med. Clin. N. Am. 67,
1325-1341, Nov. 1983).

Serum creatinine is <u>low</u> in subjects with <u>relatively small muscle mass,</u>
<u>cachetic patients, amputees, patients with muscle disease</u> and <u>some infants</u> and
<u>children</u> and <u>older persons</u>; <u>older persons</u> have a <u>decreased muscle mass</u> and a
<u>decreased rate of creatinine production.</u> A serum creatinine level that would
usually be considered normal does <u>not</u> rule out the presence of impaired renal
function.
<u>Cefoxitin Interference in Creatinine Assays</u>: Some cephalosporins, notably
cefoxitin and cephalothin, falsely elevate creatinine values in the assay
commonly used by hospital laboratories, that is, the alkaline picrate (Jaffe
reaction); this interference is observed in all alkaline picrate methods
including the widely used Beckman's kinetic Jaffe reaction (Steinback, G. et
al., Clin. Chem. <u>29</u>, 1700-1701, 1983). At peak cefoxitin levels, this
elevation may be 1.5 to 8.5 times the true serum creatinine. In patients with
normal renal function, the serum creatinine assay is reliable two to four hours
after the dose; in mild to moderate renal failure, six to eight hours. In
severe renal failure, the serum creatinine determination is unreliable. To
prevent interference during treatment with cefoxitin or cephalothin, order serum
creatinine assays to be drawn <u>immediately prior to the dose.</u> <u>Creatinine</u>
<u>clearance measurements are unreliable under all conditions while patients are</u>
<u>receiving cefoxitin</u> and the alkaline picrate method is used.
The enzymatic method for the determination of creatinine is not influenced
by Cefoxitin (Steinbach, G. et al, Clin. Chem. 29, 1700-1701, 1983).
Cephapirin, cefazolin, cefamandole and moxalactam have no effect on the
assay. (References: Sahh, A.J. and Koch, T.R. JAMA, <u>247</u>,205-206, 1982; Kirby,
M.G. et al., Clinical Chemistry <u>28</u>,1,981, 1982.)
<u>Diabetic Ketoacidosis and Elevated Creatinine</u>: Falsely elevated serum creatinine
values occur in diabetic ketoacidosis when the alkaline picrate reaction is used
for assay of creatinine (Nanji, A.A. and Campbell, D.J., Clin. Biochem. <u>14</u>,
91-93, 1981).

CRYOGLOBULINS, QUALITATIVE   Cryoglobulins
<u>SPECIMEN</u>: Red top tube; keep specimen at 37°C until centrifugation and
separation of cells.
<u>REFERENCE RANGE</u>: Negative
<u>METHOD</u>: Incubate at 4°C; positive cryoglobulins are analyzed by
immunoelectrophoresis.
<u>INTERPRETATION</u>: Cold precipitable cryoglobulins may be present in macroglobu-
linemia, myeloma, chronic lymphatic leukemia, lupus & other autoimmune diseases.
Cryoglobulins are serum globulins, that precipitate at lower temperatures
and redissolve upon warming to 37° C. The proteins in cryoglobulinemias are
immunoglobulins and other proteins. Immunoproliferative, proliferative and
autoimmune disorders are the most frequent diseases associated with
cryoglobulinemia. Brouet, J.C. et al, (Am. J. Med. <u>57</u>, 775, 1974), classified
patients with cryoglobulins into three types:
Type I - Isolated monoclonal immunoglobulins including IgG, IgM, IgA, and
Bence-Jones.
Type II - Mixed cryoglobulins with a monoclonal component possessing
antibody activity towards polyclonal IgG; combinations included IgM-IgG; IgG-IgG
and IgA-IgG.
Type III - Mixed Polyclonal Cryoglobulins. These are composed of one or
more classes of polyclonal immunoglobulins and sometimes non-immunoglobulin
molecules such as complement or lipoprotein.
<u>Laboratory Findings</u>: Some laboratory findings in mixed cryoglobulenemia are
given in the next Table; (Gorevic, P.D., Am. J. Med. <u>69</u>, 287-308, 1980):

Laboratory Findings in Mixed Cryoglobulinemia
Increased Erythrocyte Sedimentation Rate(ESR)
Rheumatoid Factor
Serum Protein Electrophoresis: Increased Gamma Globulin
Increased Immunoglobulins(IgM or IgA or IgG)
Decreased Serum Complement(C4, C2, C3, CH50, CH100)
Anemia (Hematocrit<35%)
Liver Function Test Abnormalities: Elevated Serum Alkaline Phosphatase; Elevated Serum Transaminases
Proteinuria and Hematuria

CRYPTOCOCCAL ANTIGEN TITER (CSF OR SERUM)

SPECIMEN: Cerebral spinal fluid (CSF); or red top tube, separate serum; or urine.

REFERENCE RANGE: Negative

METHOD: Latex agglutination(LA) kit; commercial sources for Cryptococcus Latex Agglutination Test Reagents: International Diagnostic Technology, 989 S. Stone Street Ave., Rockville, Maryland 20880; Wampole Corporation, Half Acre Road, Cranbury, NJ 08512, (609) 665-1100, Technical service (800) 257-9525, Technical information (301) 279-7050; Meridian, Inc., P.O. Box 44216, Cincinnati, OH 45244, (513) 474-2175. The latter agglutination(LA) methods must include a control to identify nonspecific agglutination reactions caused by the presence of rheumatoid factor.

INTERPRETATION: Cryptococcal antigen in body fluids indicates active disease. A negative test does not exclude a diagnosis of cryptococcus; it may be necessary to examine more than one specimen. Occasionally patients who are serologically negative for cryptococcal antigen will have cutaneous lesions, sputums, or well-isolated pulmonary nodules that yield positive C. neoformans cultures. In cases of proved cryptococcal meningitis, LA has detected cryptococcal antigen in 94 percent of CSF and in 70 percent of the serum samples (Bennett, J.E. in Klastersky, J. (ed.), Infections in Cancer Patients, New York, Raven Press, pp. 131-139, 1982).

Increasing titers reflect progressive disease, while declining titers indicate response to chemotherapy and recovery. Failure of CSF or serum antigen titers to fall suggests inadequate therapy. However, in some treated patients, titers remain positive at a low level for an indefinite period of time during which the fungus is no longer viable.

Cryptococcus neoformans usually enters by the pulmonary route; it is the most common fungus to involve the CSF.

The usual laboratory findings in the CSF in patients with cryptococcal meningitis are as follows: increased pressure, increased protein, decreased glucose, variable nature and number of inflammatory cells. Organisms may be visualized following concentration of the specimen by centrifugation, membrane filtration or cytocentrifugation; stain with Papanicolaou, Wright or Giemsa stain.

CRYPTOCOCCUS NEOFORMANS ANTIBODY

SPECIMEN: Red top tube, separate and then freeze serum.

REFERENCE RANGE: Negative

METHOD: See comment below

INTERPRETATION: Cryptococcal antibody testing is of limited usefulness in the diagnosis and prognosis of cryptococcal infection (Penn, R.L. et al., Arch. Intern. Med. 143, 1215-1220, 1983).

CULTURES

Recommended number of cultures, to detect or rule out the presence of a pathogen, is shown in the next Table; (Fluornoy, D.J., Clinical Microbiology Newsletter, 4, No.7, 50-51, 1982):

Specimen	Organisms	Recommended Number of Cultures
Sputum (not saliva)	Acid Fast Bacilli	A series of 3-6 consecutive early a.m. specimens, one/day; duplicate specimens collected on same day are pooled.
	Bacteria	A series of 2 consecutive early a.m. specimens, one/day
	Fungi	Same as acid-fast bacilli
Blood	Bacteria	A total of 3-4 blood culture bottle sets (2 bottles/set, 5ml blood/bottle) collected in a 24 hr.period, 3 different venipunctures at different sites, preferably at 1 hr. intervals; accuracy is greater than 90%.
Urine	Bacteria	Women: 2 consecutive early a.m. clean-catch, midstream specimens, one/day; Men: 1 early a.m. clean-catch, midstream specimen; accuracy greater than 90%.
Stool	Bacteria	A series of 3 consecutive specimens, one/day
	Ova and Parasites	A series of 3 specimens collected 2-3 days apart, one/day; if these tests are negative, catharsis or sigmoidoscopy may be necessary.
Wound	Bacteria	1-2 specimens

If the pathogen is found in the first of a series of cultures, it may not be necessary to complete the series.

CYANIDE, BLOOD   Cyanide

SPECIMEN: Lavender(EDTA) top tube or grey (oxalate, fluoride) top tube. Use whole blood as specimen

REFERENCE RANGE: < 0.2microgram/ml

Units: Traditional Units to International Units: $\frac{microgram}{ml} \times 38.5 = \frac{micromol}{liter}$

METHOD: Gas-liquid chromatography, electron capture detector (Valentour, J.C. et al., Anal. Chem. 46, 924-925, 1974); pyridine-barbituric acid colorimetric technique after microdiffusion.

INTERPRETATION: Cyanide as a gas (like carbon monoxide and hydrogen sulfide gases) can cause death in a few minutes; when salts of cyanide are ingested, death may take one hour. Fires involving synthetic polymers may result in toxic levels of cyanide in firemen and other people who may be exposed. It acts by blocking cellular respiration by inactivating cytochrome oxidase. Oxygen cannot be utilized and venous blood may appear bright red.

Well publicized events of cyanide poisoning have occurred. In 1978, mass poisoning occurred in Jonestown, Guyana, following ingestion of grape Kool-Aid mixed with cyanide. In 1982, cyanide was placed into the over-the-counter capsules of Tylenol and several people died following ingestion of these capsules (Lifschultz, B.D. and Donoghue, E.R., Forensic Pathology Check Sample, 25, No.1, 1983).

Cyanide causes a high anion gap metabolic acidosis. High serum levels of ketones only occur in association with diabetes mellitus, alcoholic ketoacidosis, ingestion of propyl alcohol and cyanide intoxication.

Treat patients with hydroxycobalamin, 50 mg per kg. or approximately 3000 ampules or 3 liters; a stronger solution of hydoxycobalamine (10 mg per ml) is to be marketed.

CYCLIC AMP, NEPHROGENOUS

SPECIMEN: Plasma Specimen: Lavender(EDTA) top tube; transport specimen to laboratory immediately; separate plasma and freeze immediately. Use this specimen for determination of plasma AMP and creatinine. Do not perform determination of cyclic AMP, nephrogenous if the glomerular filtration rate is less than 25ml/min.

Urinary Specimen: Timed two hour urine specimen; plastic container with no preservatives; measure volume and then freeze the urine specimen immediately. Use this specimen for determination of urinary cAMP and creatinine.

REFERENCE RANGE: 0.34 to 2.7nmol/100ml GF (GF = glomerular filtrate) see Broadus, A.E., Nephron., 23, 136-141, 1979.

METHOD: Radioreactor assay. A RIA kit is available from Immuno Nuclear Corp., P.O.Box 285, Stillwater, MN 55082.

INTERPRETATION: Nephrogenous cyclic AMP is calculated as follows:

$$\begin{array}{cccc} \text{Total cAMP} & = & \text{Serum Creatinine} & X & \text{Urinary cAMP} \\ \text{(nmol/100ml GF)} & & \text{(mg/dl)} & & \text{(nmol/mg Creatinine)} \end{array}$$

$$\begin{array}{cccc} \text{Nephrogeneous cAMP} & = & \text{Total Urinary cAMP} & - & \text{Plasma cAMP;} & \text{GF = Glomerular} \\ \text{(nmol/100ml GF)} & & \text{(nmol/100ml GF)} & & \text{(nmol/dl)} & \text{Filtrate} \end{array}$$

The cAMP found in the urine of normal persons is derived from two sources: 50% to 75% is derived by glomerular filtration of plasma and the remaining 25% to 50% is synthesized in the renal cortex of the kidney and excreted by renal tubular cells; this is illustrated in the next Figure; (Broadus, A.E., Nephron, 23, 136-141, 1979):

Renal Clearance and Sources of cAMP in Human Urine

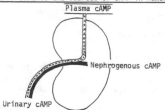

Plasma cAMP

Nephrogenous cAMP

Urinary cAMP

Parathyroid hormone stimulates adenyl cyclase in the renal cortex, thus converting ATP to cAMP. Measurement of urinary cyclic AMP is an index of parathyroid function.

Causes of increased and decreased renal cAMP are given in the next Table:

Causes of Increased and Decreased Renal cAMP	
Increased Renal cAMP	Decreased Renal cAMP
Primary Hyperparathyroidism	Chronic Hypoparathyroidism
Some Patients with Vitamin D	Pseudohypoparathyroidism
Deficiency and Osteomalacia	Some Patients with Hypercalcemia other
Patients with Calcium Urolithiasis	than Primary Hyperparathyroidism
and Hypercalciuria	Some Patients with Malignancy-Associated
Some Patients with Malignancy-	Hypercalcemia
Associated Hypercalcemia	

The assay of nephrogenous cAMP in response to intravenous or oral calcium administration is useful in diagnosis of patients with subtle primary hyperparathyroidism or intermittent hypercalcemia (Broadus, A.E., Nephron, 23, 136-141, 1979).

Seventy percent of patients with hyperparathyroidism excrete increased amounts of cAMP in their urine. Some patients with vitamin D deficiency with osteomalacia may excrete excess cAMP; this reverts to normal with vitamin D therapy. Patients with calcium urolithiasis associated with hypercalciuria have increased urinary cAMP; when these patients are given an oral load of calcium, their urinary cAMP falls toward normal. An oral calcium load does not depress elevated cAMP of patients with primary hyperparathyroidism. Some patients with malignancy associated hypercalcemia may excrete increased cAMP in their urine.

Hypoparathyroid patients excrete about one-third the normal amount of cAMP.
Patients with cancer-associated hypercalcemia may have elevated urinary
cAMP or suppressed urinary cAMP. Tumors of patients with elevated urinary cAMP
produce a nonparathyroid hormone peptide that mimics certain aspects of PTH
hormonal activity, such as increased urinary cyclic AMP, but not other aspects
of PTH activity (Stewart, A.F. et al., N. Engl. J. Med. 303:1377-1383, 1980).

Broadus, A.E. et al., J. Clin. Invest. 60:771-783, 1977; Pak, C.Y.C.,
Hamet, P. and Sands, H. (eds.), "Advances in Cyclic Nucleotide Research" 12:,
N.Y., Raven Press, pgs. 393-403, 1980; Broadus, A.E., Recent Progress in Hormone
Research 37:667-701, 1981.

CYCLIC AMP, URINE

SPECIMEN: Timed two hour urine specimen; use plastic container with no
preservative; measure volume and then freeze the urine specimen immediately.

REFERENCE RANGE:

Traditional Units (microgram/g. creat.)	Conversion Factor	International Units (nmol/mmol creat.)
2.9-5.6	113.1	330-630

METHOD: Radioreceptor or radioimmunoassay(RIA). A RIA kit is available from
Immuno Nuclear Corp., P.O. Box 285, Stillwater, MN 55082.

INTERPRETATION:

$$\text{Total cAMP (nmol/100ml GF)} = \text{Serum Creatinine (mg/dl)} \times \text{Urinary cAMP (nmol/mg Creatinine)}$$

The cAMP found in the urine of normal persons is derived from two sources:
50% to 75% is derived by glomerular filtration of plasma and the remaining 25%
to 50% is synthesized in the renal cortex of the kidney and excreted by renal
tubular cells; this is illustrated in the next Figure; (Broadus, A.E., Nephron,
23, 136-141, 1979):

Renal Clearance and Sources of cAMP in Human Urine

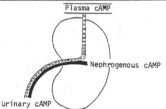

Parathyroid hormone stimulates adenyl cyclase in the renal cortex, thus
converting ATP to cAMP. Measurement of urinary cyclic AMP is an index of
parathyroid function.

See CYCLIC AMP, NEPHROGENOUS, for further discussion.

CYTOMEGALOVIRUS (CMV) ANTIBODIES (IgM AND IgG)

SPECIMEN: Red top tube, separate serum; for <u>congenital infections</u>, obtain <u>serially</u> collected sera from both the mother and infant. In postnatal and for other patients, obtain <u>serially</u> collected sera.

REFERENCE RANGE: Two antibodies: IgM and IgG. The <u>presence</u> of <u>IgM</u> or a fourfold or greater rise in IgG titer indicates recent infection.

METHOD: Indirect fluorescent antibody test (IFAT), kits commercially available; anticomplement immunofluorescence test (ACIF).

INTERPRETATION: There are <u>two types</u> of antibodies: IgM and IgG. The <u>presence</u> of <u>IgM</u> indicates <u>current infection</u>; IgM antibodies persist for one to two months. CMV is commonly <u>transmitted in-utero</u> (<u>congenital</u> CMV <u>infection</u>) or by <u>blood transfusions</u>.

Congenital CMV: Congenital cytomegalovirus infection is probably the <u>most common</u> viral infection acquired <u>in-utero</u> by the fetus; in the United States, the incidence is one to two percent. CMV may also be <u>transmitted</u> to infants following <u>blood transfusions</u>.

If a woman who has had a cytomegalovirus(CMV) infection has a <u>recurrence</u> of the infection <u>during pregnancy</u>, the chance that her <u>baby</u> will have a harmful congenital <u>CMV infection is much less than</u> it is if a woman has <u>her first CMV infection during pregnancy</u> (Medearis, D.N., N. Engl. J. Med. <u>306</u>, 985-987, 1982). Stagno et al (N. Engl. J. Med. <u>306</u>, 945-949, 1982) found that 52 percent of the babies born after a primary infection during gestation had congenital infections; however, not more than 6 percent of the babies born after recurrent infections had congenital infections (Medearis, D.N. 1982). <u>The most severe effects on the fetus occurs during the second trimester.</u>

Unlike <u>rubella</u>, the presence of <u>antibodies</u> to CMV in the <u>mother</u> does <u>not</u> protect the fetus against infection. The <u>absence</u> of antibodies in the mother rules out congenital infection. IgG antibodies may be transferred from the mother to the infant by placental transfer; these antibodies will decay over 2 to 3 months. A <u>rising</u> (four-fold) or <u>persistent</u> titer of IgG antibodies in the infant's serum is suggestive of a <u>congenital infection</u> or a 1:64 titer or more is also suggestive of infection.

The <u>presence</u> of <u>IgM</u> antibodies in the infant indicates <u>current infection</u> since IgM antibodies, unlike IgG, do not cross the placental barrier.

Most infants with congenital CMV infections are asymptomatic or have mild hepatomegaly with moderately abnormal liver function tests and jaundice.

Postnatal and CMV Infections in Adults: The presence of IgM antibodies or a four-fold or greater rise in antibody titers between acute and convalescent sera is indicative of infection. About 80% of adults over the age of 35 have IgG antibodies.

Other Diagnostic Procedures: A definitive diagnosis of CMV can be made by culturing the virus from urine or throat specimens. A histologic study of stained urine sediments helps to make a presumptive diagnosis in 25 to 50 percent of symptomatic cases; the finding of single large red round intranuclear inclusions surrounded by a halo, giving the cell the appearance of an owl's eye, is distinctive for CMV infection.

DEHYDROEPIANDROSTERONE SULFATE(DHEA-S)

SPECIMEN: Red top tube, separate serum; freeze serum as soon as possible

REFERENCE RANGE:

Change of DHEA-S with Age			
Age	Traditional Units (ng/ml)	Conversion Factor	International Units (micromol/liter)
Newborn	1670-3640	0.002714	4.5-9.9
Children	100-600	"	0.3-1.6
Pre-Pubertal Male	2000-3350	"	5.4-9.1
Female(Premenopausal)	820-3380	"	2.2-9.2
Female(Postmenopausal)	100-610	"	0.3-1.7
Pregnancy(Term)	230-1170	"	0.6-3.2

METHOD: RIA

INTERPRETATION: DHEA-S is a good indicator of adrenal function; testosterone is a good indicator of ovarian function. A marked elevation of DHEA-S (60-340mcg/dl) is usually an indication of an adrenal abnormality (tumor or emzymatic). DHEA-S measurements have replaced the more cumbersome and less accurate measurements of urinary 17-ketosteroids.

The most common indication for measurement of DHEA-S is in the evaluation of patients with hirsutism.

11-Deoxycortisol
(Compound S)

11-DEOXYCORTISOL (COMPOUND S)

SPECIMEN: Green(heparin) top tube; transport to laboratory immediately and separate plasma and freeze plasma.

REFERENCE RANGE: 0-2mcg/dl; post-metyrapone: greater than 10mcg/dl. To convert traditional units in mcg/dl to international units in nmol/liter, multiply traditional units by 28.86.

METHOD: RIA

INTERPRETATION: Assay of 11-deoxycortisol (compound S) is done in the work-up of patients with possible 11-beta hydroxylase deficiency (a virilizing adrenal hyperplasia) and in evaluating pituitary reserve following metyrapone. The causes of increased plasma 11-deoxycortisol (compound S) are listed in the next Table:

Causes of Increased Plasma 11-Deoxycortisol
11-Beta Hydroxylase Deficiency (Virilizing Adrenal Hyperplasia)
Post-Metyrapone
Other: Eclampsia, Stress, Pancreatitis

The causes of decreased plasma 11-deoxycortisol (Compound S) are listed in the next Table:

Causes of Decreased Plasma 11-Deoxycortisol
Hypofunction of Anterior Pituitary
Addison's Disease

The metabolic block in 11-beta-hydroxylase deficiency and in post-metyrapone are illustrated in the next Figure:

Metabolic Block: 11-Beta Hydroxylase Deficiency and Post-Metyrapone

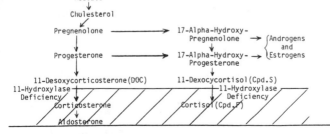

1-Beta Hydroxylase Deficiency: The sex characteristics, electrolytes and blood pressure in 21-, 11- and 17-hydroxylase deficiency are shown in the next Table:

Sex Characteristics, Electrolytes and Blood Pressure in Adrenogenital Syndromes

Characteristic	Hydroxylase Deficiency		
	21-	11-	17-
Sex	Excess Sex Steroids	Excess Sex Steroids	Deficiency of Sex Steroids
Electrolytes:			
Sodium Retention	No	Yes	Yes
Sodium Loss	Yes	No	No
Potassium Loss	No	Yes	Yes
Hypertension	No	Yes	Yes

In 11-hydroxylase deficiency, plasma 11-deoxycortisol is increased because the metabolic block is between 11-desoxycortisol and cortisol. Note that the metabolic block occurs in the hydroxylation of DOC; failure to hydroxylate DOC results in overproduction of this steroid. DOC is a powerful mineralocorticoid; thus, patients with this syndrome have clinical signs, symptoms and laboratory values compatible with hypermineralocorticism with hypertension, sodium retention, potassium depletion, increased extracellular volume; renin production is inhibited because of the volume expansion. As in all adrenogenital syndromes, blood cortisol is low but ACTH is increased; urinary 17-ketosteroids are increased and the patient exhibits masculinization; little or no aldosterone is found in the urine. The urinary 17-OH-corticosteroids, as measured by the Porter-Silber method, are increased because 11-desoxycortisol(Cpd.S) is increased.

Post-Metyrapone: Metyrapone(Metopirone, SU-4885) inhibits the conversion of 11-deoxycortisol to cortisol by blocking the 11-hydroxylase enzyme. In normal subjects, the resulting decrease in plasma cortisol triggers an increase in ACTH secretion, with a consequent increase in adrenal steroid synthesis and release of the steroid, 11-deoxycortisol, proximal to the enzymatic block; 11-deoxycortisol is measured directly or indirectly as a 17-OHCS in the urine (see METOPIRONE TEST).

Metyrapone(Metopirone) is used to differentiate adrenal hyperplasia from adrenal tumor and is used to measure pituitary reserve - Also see METOPIRONE TEST. Perform the metyrapone as follows:

Metyrapone Test

Day	Time	Metyrapone	Blood Specimen
1	0800	Give orally 300mg/sq.meter (up to 570mg)	11-Deoxycortisol (Normal: 0-2mcg/dl)
1	1200 Noon	"	--
1	1600	"	--
1	2000	"	--
1	2400	"	--
2	0400	"	--
2	0800	---	11-Deoxycortisol (Normal: Greater than 10mcg/dl)

Following the administration of metyrapone (300mg/sq.meter, up to 750mg) every four hours for six doses starting at 0800, plasma 11-deoxycortisol level is obtained at that time; another plasma level is obtained 24 hours later (0800 on day 2). In normal subjects, the 11-deoxycortisol rises from less than 2mcg/dl to well over 10mcg/dl; in patients with diminished ACTH reserve, the post-metyrapone value is less than 8mcg/dl.

DEPAKENE (see VALPROIC ACID)

DEXAMETHASONE SUPPRESSION TEST
(CUSHING'S SYNDROME VERSUS NORMAL)

SPECIMEN: The dexamethasone suppression test for the diagnosis of Cushing's syndrome is as follows: Dexamethasone, 1mg, is taken orally by the patient between 11:00 P.M. and midnight; a fasting blood cortisol is obtained the next morning at 8:00 A.M, in green (heparin) top tube, separate plasma, or red top tube, separate serum. Store and ship frozen.

REFERENCE RANGE: Suppression by dexamethasone in normal individuals of blood cortisol is less than 5mcg/dl.

METHOD: RIA

INTERPRETATION: A value of blood cortisol greater than 5mcg/dl is compatible with Cushing's syndrome. A negative test (<5mcg/dl) is found in less than 2 percent of the cases of Cushing's syndrome (Crapo, L., Metabolism 28, 955-977, 1979). However, abnormal results (>5mcg/dl) are often found in patients under stress for any reason, including at least 30 percent of patients in a hospital who are acutely ill (Connolly, C.K. et al., Br. Med. J. 2, 665-667, 1968).

   A protocol for the work-up of a patient with suspected Cushing's syndrome is given in the next Figure:

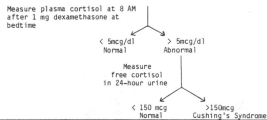

Cushing's Syndrome?

Measure plasma cortisol at 8 AM after 1 mg dexamethasone at bedtime

< 5mcg/dl Normal   > 5mcg/dl Abnormal

Measure free cortisol in 24-hour urine

< 150 mcg Normal   >150mcg Cushing's Syndrome

   At least three conditions, depression, obesity and alcoholism may produce results similar to that seen in Cushing's syndrome.

   Urinary free cortisol is the best single test for differentiation of normal from Cushing's syndrome.

   Suppression test using high dose dexamethasone has been used to differentiate adenoma from hyperplasia. Eight mg/day of dexamethasone tends to suppress urinary excretion of steroids of patients with hyperplasia but not that of patients with adenoma.

DEXAMETHASONE SUPPRESSION TEST
AS DIAGNOSTIC AID IN DEPRESSION

SPECIMEN: A protocol for the dexamethasone suppression test for endogenous depression is given in the next Table:

Protocol for the Dexamethasone Suppression Test on Inpatients			
	Time	Specimen	Normal Value
Day 1	11:00 P.M.	Specimen for Blood Cortisol	<5mcg/dl
	One mg Dexamethasone		
	(11:00 P.M. to 12:00 P.M.)		
Day 2	4:00 P.M.	Specimen for Blood Cortisol	<5mcg/dl
	11:00 P.M.	Specimen for Blood Cortisol	<5mcg/dl

The dexamethasone suppression test may be used on an outpatient basis. Only a 4 P.M. post-dexamethasone serum cortisol is obtained.

Collect blood in a green top tube (heparin), separate plasma; or red top tube, separate serum; store serum or plasma frozen.

REFERENCE RANGE: Blood cortisol: <5mcg/dl
METHOD: RIA

INTERPRETATION: When 1mg of dexamethasone is given orally at 11:00 P.M. to 12:00 Midnight, normal subjects maintain suppressed plasma cortisol concentrations for at least 24 hours; patients with endogenous depression may suppress their plasma cortisol concentrations temporarily but fail to maintain plasma cortisol suppression below 5mcg/dl for 24 hours. The abnormal escape of plasma cortisol concentrations usually occurs by 4 P.M. and may be observed as early as 8:00 A.M. The use of this test identifies depression with a sensitivity of 67% and a specificity of 96%. The sensitivity and specificity is dependent on the depressed population (Shapiro, M.F. et al., Arch. Intern. Med. 143, 2085-2088, 1983). Normalization of the results of the test occurs with clinical recovery; when the results of the test fail to normalize at discharge, patients are at high risk for early relapse.

Caution should be exercised with regard to the routine application of the dexamethasone suppression test for depression (Hirschfeld, R.M.A. et al., JAMA 250, 2172-2175, 1983; Health and Public Policy Committee, American College of Physicians, Ann. Intern. Med. 100, 307-308, 1984).

Conditions and drugs that cause false positive and false negative results in the dexamethasone suppression test for endogenous depression are given in the next Table; (Ritchie, J.C. and Carroll, B.J., Lab. World, pgs. 24-29, Sept. 1981):

Conditions and Drugs Causing False Positive and False Negative Results in the Dexamethasone Suppression Test for Endogeneous Depression	
Conditions	Drugs
False Positive Tests	False Positive Tests
Major Physical Illness	Phenytoin
Trauma, Fever, Nausea, Dehydration	Barbiturates
Temporal Lobe Disease	Meprobamate
Pregnancy (or high dosage estrogens)	Methaqualone
Cushing's Disease	Glutethimide
Unstable Diabetes	Methyprylon
Malnutrition	Carbamazepine
Anorexia Nervosa (<80% ideal weight)	Reserpine (?)
Heavy Alcohol Use	
Alcohol Withdrawal (acute <10 days)	
E.C.T. (on Post dex. day)	False Negative Tests
Other Endocrine Disease?	High-dose Benzodiazepines
False Negative Tests	(>25 mg/day of diazepam)
Addison's Disease	Cyproheptadine
Corticosteroid Therapy	Synthetic Steroid Therapy
Hypopituitarism	

The basis of the dexamethasone suppression test is the regulation of neuroendrocrine function by the limbic system and hypothalamus of the brain. (The limbic system includes the hippocampus, cingulate gyrus, the fornix, mammillary bodies and the anterior part of the thalamus). These areas are thought to be the sites of the pathology in the depression disorders.

DIABETIC KETOACIDOSIS (DKA) PANEL

Diabetic ketoacidosis(DKA) has a mortality of 5% to 15%; of those patients who die, one-quarter die within 6 hours of admission to the hospital, nearly half die within 12 hours and three quarters die within the first 24 hours. It is the most common cause of death below the age of 20 years in diabetics. DKA occurs at all ages, often in people previously unrecognized as diabetic and in insulin dependent diabetics. Disorders that accompany or precipitate DKA include bronchopneumonia, pyelonephritis, alcoholism, septicemia, gastroenteritis, myocardial infarction, pancreatitis and cerebral vascular disease.

Diagnosis of DKA: The diagnosis of DKA is made through determination of glucose, ketones, acid-base status, and electrolytes. The differential diagnosis of anion-gap metabolic acidosis includes: ketoacidosis: diabetes, alcohol, starvation, lactic acidosis; chronic renal failure; and drugs: salicylates, ethylene glycol, methanol and paraldehyde. The mean laboratory values in patients with DKA on admission to the hospital and a schedule for monitoring laboratory tests for the first six hours are given in the Table: (Foster, D.W. and McGarry, J.D., N. Engl. J. Med. 309, 159-169, 1983):

Mean Laboratory Values in Patients with Diabetic Ketoacidosis(DKA)
on Hospital Admission and Schedule for Monitoring for First Six Hours

Summary of Laboratory Data in Diabetic Ketoacidosis		
Laboratory Data	Average Concentration (Initial)	Schedule for Monitoring
Glucose(Lab) (80-120mg/dl)	500	Every Hour For 4-6 Hours
Glucose (Dextrostix;Chemstrip)	500	Every 0.5 Hour for 4-6 Hours
Urine: Glucose	+3-4	Every 2 Hours
Ketones	+2-4	Every 4 Hours
Serum Osmolality(285-295mOsm/l)	313	Every Hour For 4-6 Hours
Arterial Blood Gases:	-	Every Hour For 4-6 Hours
pH (7.36-7.44)	7.1	
PCO$_2$ (32-45 mm)	20-35	
HCO$_3$(24-29 mEq/l)	5-10	
Base Excess (-3 to +3)	-20	
Electrolytes:		Every Hour For 4-6 Hours
K$^+$ (3.5-5.0 mEq/l)	5.3	If K$^+$<3.0, every 0.5 hrs.
Na$^+$ (135-145 mEq/l)	133	
Cl$^-$ (95-110 mEq/l)	100	
CO$_2$ Content (24-30 mEq/l)	5-10	
Anion Gap:		
[Na$^+$-(Cl$^-$+HCO$_3$)]<15mmol/l	25-35	Every Hour for 4-6 Hours
Ketones: Acetoacetate(<0.3mM)	4.0	Every Hour for 4-6 Hours
Acetone(<0.2mM)	4-5	
Beta-Hydroxybutyrate(<0.5mM)	12	
Lactic Acid(0.5-1.66mmol/l)	3.5	As Appropriate
BUN (5-20 mg/dl)	27	As Appropriate
Phosphorus:Adult:(2.3-4.3mg/dl)	4	Every 2 Hours
Adult: (2.3-4.3 mg/dl)	5	Every 2 Hours
Urine Output		Continuously
Electrocardiogram(ECG)		Every 4 Hours
Pulmonary Capillary Wedge Pressures		As Appropriate

Other Tests: Vital signs (B.P., pulse and respirations) must be carefully monitored; the patient is often hypotensive with low B.P., and the pulse is increased. The respiratory rate is increased secondary to metabolic acidosis (Kussmaul respirations). Evaluation of weight loss may give some concept of dehydration. The state of mentation should be carefully monitored.

Other tests that should be done initially include the following: blood and urine culture, urine analysis, serum creatinine, calcium and magnesium, chest x-ray, and complete blood count(CBC) with differential.

Specimens: Collect arterial blood for blood gases in a heparinized syringe (on ice). Use either arterial or venous blood for lactate; collect in a grey (fluoride/oxalate) top tube (on ice). Collect blood for culture in special tubes. Collect all other blood specimens in a red top tube. These specimens are collected every hour for four to six hours. Collect urine for culture and analysis.

Glucose: The average concentration of plasma glucose in patients with DKA on admission to the hospital is about 500 mg/dl; however, levels range from nearly normal to a thousand or more. Blood-glucose test strips (glucose oxidase) are used to determine blood glucose values in the physician's office, emergency room or at the bedside.

Urinalysis: Glucose and ketone bodies are measured in the urine by dip-stick methodology; this is done at the bedside.

Osmolality: The average serum osmolality is 310-315 mOsm/liter when the serum glucose is 500 mg/dl; the serum osmolality may be obtained by direct measurement or calculated from the formula:

$$mOsm/kgH_2O = 2Na^+ + \frac{Glu}{18} + \frac{BUN}{2.8}; \text{ Sodium (mEq/l), Glucose and BUN (mg/dl)}$$

Each 100 mg/dl of glucose contributes 5.5 to the serum osmolality.

Arterial Blood Gases(pH, pCO_2, HCO_3^-). The average value for arterial blood pH is 7.1. At this pH, the HCO_3^- is 12 mEq/liter when the pCO_2 is 40 mmHg. The brainstem centers that regulate alveolar ventilation are stimulated by acidified extracellular fluid. Increased ventilation lowers pCO_2. Maximal respiratory compensation occurs within 24 to 36 hours. In metabolic acidosis with respiratory compensation (24 to 36 hours):

Expected pCO_2(mmHg)=(1.5 HCO_3^- + 8) $\pm$ 2.
If HCO_3^-=10 mEq/liter, then pCO_2=23(21 to 25)mmHg.

Base Excess: The base excess is about -17 when the blood pH is 7.1 and the pCO_2 is 40 mmHg; the base excess is -23 when the pH is 7.1 and the pCO_2 is 20 mmHg.

Electrolytes (K^+, Na^+, Cl^-, CO_2 Content):
Potassium: The serum potassium and electrocardiogram(ECG) are used to monitor serum potassium. The average value for serum potassium is 5.3 mEq/liter; potassium is normal or elevated in 80 to 90% of patients with DKA. The serum potassium does not reflect the total body potassium. In DKA, there is an osmotic diuresis with total body deficit of 3-5 mmol/kg and as much as 10 mmol/kg; however, there is a shift of K^+ from the cells as hydrogen ion enters the cell in exchange for intracellular K^+. A low-serum potassium indicates profound total body depletion and replacement (with K^+ as KCl or K phosphate) must be done as soon as possible.

Sodium: The average value for serum sodium is 133 mEq/liter; glucose draws water into the extracellular space and thus decreases the sodium concentration. Other causes of low sodium include vomiting, diarrhea, and dilution secondary to water intake. Serum sodium may be spuriously depressed due to severe hypertriglyceridemia; this occurs when the popular Beckman Astra or other electrolyte analyzers are used that dilute the serum sample prior to analysis.

Chloride: Serum chloride is usually within normal range; there is a normochloremic metabolic acidosis (anion gap).

Carbon Dioxide (CO_2) Content: The CO_2 content is the sum of the HCO_3^- concentration plus the PCO_2; normally, the concentration of HCO_3^- is twenty times that of the concentration of PCO_2. Thus, the CO_2 content reflects the concentration of the HCO_3^-.

Anion Gap: The anion gap is increased to 25 to 35 mmol/liter. The increase in the anion gap corresponds to the quantity of ketones present in plasma (usually 10 to 20 mEq/liter). The anion gap is calculated from the formula:
[Na^+-(Cl^-+HCO_3^-)]; when the average initial values for Na^+, Cl^- and HCO_3 found in DKA are substituted in this equation: [133-(100+5)], a value of 28 mmol/liter is obtained.

Ketone Bodies: Metabolic acidosis is caused primarily by the accumulation of ketone bodies, acetone and acetoacetate and beta-hydroxybutyrate. The nitroprusside test measures acetoacetate and acetone but not beta-hydroxybutyrate; nitroprusside tests are 15-20 times more sensitive to acetoacetate than to acetone.

Occasionally, the nitroprusside test is negative in the presence of severe ketoacidosis because ketones are in the form of beta-hydroxybutyrate; this situation occurs in conditions associated with hypoxemia, i.e., hypovolemia; hypotension, low arterial pO_2 or association of DKA with alcoholism.

A screening test for ketone bodies in blood is done as follows: Dilute serum with an equal volume of water or saline consecutively, 1:1; 1:2; 1:4; etc. With the nitroprusside reaction, significant ketonemia is indicated if the serum ketones are positive at a dilution of 1:8 or greater.

Lactic Acid: About one-third of patients with DKA have lactic acidosis.

BUN and Creatinine: There is moderate volume depletion in DKA with serum BUN in the range of 25 to 30 mg/dl; serum creatinine is usually normal; however, acetoacetic acid interferes with the assay of serum creatinine; at a serum acetoacetic acid concentration of 8 to 10 mmol/liter, the serum creatinine is falsely elevated by 3 to 4 mg/dl by the Astra methodology which uses the alkaline picrate method.

Phosphorus: During DKA there is a shift of phosphorus from tissues to the extracellular compartment and excessive loss of phosphorus in the urine. The initial serum phosphorus level is usually normal but may be elevated or low. During therapy, the serum phosphorus falls progressively.

Prior to phosphorus replacment therapy, it is necessary to measure both the serum calcium level and BUN or creatinine to exclude the possibility of renal disease.

Electrocardiogram(ECG): The ECG is a helpful guide to potassium levels at the bedside, but it is less reliable than direct measurements of potassium in the serum. In hypokalemia, there is a low T wave, the presence of a U wave and a depression of the ST segment. In hyperkalemia, there is a tall T wave, a wide QRS interval and decrease or absence of the P wave.

Pulmonary Capillary Wedge Pressure(PCWP): Pulmonary capillary wedge pressure is a useful and reliable indicator of left ventricular dynamics and pulmonary congestion (>21 mmHg). In severe DKA, PCWP may be increased.

Vital Signs (Blood Pressure, Pulse, Respirations): The degree of volume depletion may be guaged by evaluating vital signs; when volume depletion is severe, there is hypotension, marked tachycardia and altered mentation. Volume contraction can be life threatening, with development of myocardial infarction, stroke or irreversible shock from underperfusion.

Chest X-Ray: Chest X-ray is done to detect pulmonary infection and pulmonary congestion.

Other: The CBC is normal except for a leukocytosis which is associated with ketoacidosis.

Treatment of DKA Mellitus: Principles of treatment of DKA are listed in the next Table:

Principles of Treatment of Diabetic Ketoacidosis(DKA)
Correct Fluid and Electrolyte Disturbances
Give Sodium, Potassium and Glucose
Give Insulin to Restore and Maintain Intermediary Metabolism in a Normal State, e.g., metabolize Ketone Bodies.
Optional: Sodium Bicarbonate and Potassium Phosphate

It is estimated that a 70 kg man has a deficit of 5-7 liters of water, 300-450 mEq of sodium, and 200 to 400 mEq of potassium.

Fluids are given to correct fluid deficits and lower the glucose level; the concentration of ketones and beta-hydroxybutyrate do not decrease when fluids are given without insulin.

Start I.V. with normal saline at 10-20 ml/kg/hour (70 kg. Adult: 700-1400 ml/hr; 30 kg. Child: 300-600 ml/hr). For an adult, the average fluid deficit is 3 to 5 liters; about 1 liter per hour is given for the first two or three hours. If hypernatremia occurs during therapy, use half-normal saline. Fluid replacement for patients with congestive heart failure or pulmonary edema and DKA must be done with great care.

Evaluation of Initial Laboratory Results: Laboratory results should be available within an hour from the time that the specimens are obtained.

If the HCO_3^- is less than 10 mEq/liter, calculate the HCO_3^- necessary to correct to plasma HCO_3^- to 15 mEq/liter over a 6 hour period by using the following formula:

[15-Measured HCO_3^-] (Wt in kg.) (0.6) = mEq/liter of HCO_3^- required.

0.6 is the bicarbonate volume of distribution (60%). If the value for the measured HCO_3^- is not available, substitute value of carbon dioxide content for the measured HCO_3^-.

Example: 30 kg. child; measured HCO_3^- = 5 mEq/liter
[15-5][30][0.6] = 180 mEq of HCO_3^- required.

Example: 70 kg. adult; measured HCO_3^- = 5 mEq/liter
[15-5][70][0.6] = 420 mEq of HCO_3^- required.

Prepare HCO_3 solution using stock solutions on ward.

If HCO_3^- is required, add HCO_3^- to half-normal saline to provide a solution that approaches isotonicity and infuse over 4-6 hours. Never give HCO_3^- by I.V. push. The maximal allowable concentration of bicarbonate is 0.5 normal which is equal to 80 mEq/liter. Stop infusion of bicarbonate when the pH reaches 7.2 to avoid rebound metabolic alkalosis as ketone bodies are metabolized.

Start insulin therapy using arm other than that used for saline infusion. Prepare insulin solution for initial I.V. infusion: Use U100 (100 units/ml) regular insulin. Child: Give, I.V., 0.1 U/kg both as an initial bolus and as the hourly infusion rate. Adult: Give, I.V., 10 units as an initial bolus and then at the rate of 5 to 10 units per hour. The insulin solution should be prepared for the first 6 hours of therapy. Use normal saline to dilute insulin to a maximum of 50 ml. The insulin is given "piggyback" or via a Y-shaped tubing (proximal port) into the I.V. line. The insulin solution should be delivered by a pump or other method that produces a constant rate of flow. Insulin is continued until ketosis is cleared.

The level of insulin in normal individuals is 5-40 mcU/ml in the fasting state. Infusion of regular insulin to raise the insulin level to 50 mcU/ml suppresses hepatic glucose output by 70 to 75 percent; at 100 mcU/ml, almost complete suppression of glucose output by the liver occurs and stimulation of peripheral glucose uptake is 70 to 75% of maximum. At plasma insulin levels of 50 to 100 mcU/ml, lipolysis is completely blocked. "Low dose" insulin, as described in this protocol, raises the plasma insulin level to 150 to 200 mcU/ml.

Thirty to 45 minutes prior to the termination of the IV insulin infusion, probably between 6-8 hours after initiation of therapy, subcutaneous or IM insulin must be administered at an adequate dose, probably 1/4 U/kg, to insure adequate coverage of glucose metabolism.

Potassium Therapy: As already mentioned, serum potassium is usually elevated initially and no treatment with potassium is necessary. After therapy with fluids and insulin has started, potassium concentration in plasma decreases as potassium returns to the cell as acidosis is reversed. Administration of insulin in the presence of glucose will also facilitate potassium uptake. Ordinarily, potassium therapy is not necessary until two to four hours after therapy has begun; then, therapy with potassium is started.

However, if serum potassium is normal, add 20 to 30 mEq/liter of KCl(1 mEq of KCl equals 75 mg of KCl); if serum potassium is below 3.5 mEq/liter, initially give 30 to 50 mEq/liter. If hypokalemia is severe, 100 mEq/liter of KCl (or K phosphate) may be required.

Glucose Levels During Therapy: As fluids and insulin are given, plasma glucose levels begin to fall. The expected rate of fall of plasma glucose is 75 to 100 mg/dl per hour. The plasma glucose level must not be allowed to fall below 250 mg/dl during the first four to six hours of insulin therapy. If this occurs, glucose should be infused to avoid hypoglycemia. It is necessary to continue insulin infusion even after glucose levels have fallen in order to eliminate ketone bodies; maintain glucose around 200 mg/dl. In general, hyperglycemia is corrected in 4 to 8 hours. Glucose levels must not fall too rapidly or cerebral edema and neurological complications will result.

Ketone Levels During Therapy: Plasma glucose levels fall before ketone bodies disappear; it is necessary to continue insulin infusion when acidosis or ketosis are still present even though glucose has fallen to 250 mg/dl. In general, ketosis is corrected in 10 to 20 hours. Ketonuria may persist for 24 to 48 hours after ketone bodies have cleared from the blood.

Chloride Levels During Therapy: Hyperchloremia normally develops during therapy and therapy using Ringer's lactate should be instituted.

Osmolality During Therapy: The fall in serum osmolality should be controlled at a rate not to exceed 10 mOsm/hour.

Monitoring Laboratory Data After First Four to Six Hours: Blood gases electrolytes, osmolality, ketones, and glucose should be assayed every 2-4 hours. Blood gases should be obtained at 12, 24, and 36 hours even if acidosis is apparently corrected.

Complications: Complications of DKA are as follows: shock, infection, arterial thrombosis and cerebral edema in children. Cerebral edema is usually fatal; if the patient survives, permanent brain damage may result.

Cerebral edema may be due to a more rapid fall in blood glucose than in
intracerebral glucose; this may be due to too rapid rehydration and/or by use of
low-dose subcutaneous insulin instead of continuous infusion. DKA is often
precipitated by infection or other factors listed in the introduction. A
diligent search for such a precipitating event is indicated.
General References: Barrett, E.J. and DeFronzo, R.A., Hospital Practice,
89-104, April 1984; Foster, D.W. and McGarry, J.D., N. Engl. J. Med. 309,
159-169, 1983; Tunbridge, W.M.G., The Lancet, 2, 569-572, 1981.

DIAZEPAM (VALIUM)   Diazepam (Valium)
SPECIMEN: Red top tube, separate serum
REFERENCE RANGE: Therapeutic range: 100-1500ng/ml;
Toxic Range: 3,000-14,000ng/ml
METHOD: Qualitative: Thin layer chromatography. Quantitative: Gas-liquid
chromatography; ultraviolet spectrometry.
INTERPRETATION: Diazepam (Valium) is a benzodiazepine and is used for the
treatment of anxiety and insomnia. Diazepam (Valium) is prescribed for oral
administration more often than any other drug in the United States.
     Diazepam acts rapidly when taken as a single dose; it is absorbed rapidly
from the gastrointestinal tract, and since it is soluble in lipids, it quickly
reaches therapeutic concentrations (serum therapeutic range, 10-1500ng/dl) in
the central nervous system. Diazepam is metabolized in the liver by oxidation
and has active metabolites; it is "long-acting" with a half-life greater than 24
hours. It may be more hazardous for the elderly and for patients with liver
disease.
     In the blood, Valium is strongly bound to protein (85%-95%). Death is rare
in overdose when diazepam is taken alone; when death occurs when diazepam is
taken, the actual cause of death is more likely attributable to the other drugs
ingested (Finkle, B.S. et al., JAMA 242, 429-434, 1979; Jatlow, P. et al., Am.
J. Clin. Path. 72, 571-577, 1979).

S. Bakerman

DIGOXIN

SPECIMEN: Red top tube; remove serum and freeze if greater than 24 hours; Reject specimen if not clotted. Blood specimen must be drawn 6-8 hours after administration of the last dose.

REFERENCE RANGE: Therapeutic range: 0.5-2ng/ml; Toxic: >2ng/ml; potentially, >1.5ng/ml; in patients whose plasma potassium concentrations are below normal, plasma digoxin concentrations between 1.3 and 2.0ng/ml may be associated with toxicity. Half-Life: 1.6 days (normal renal and hepatic function); 5.0 days (markedly reduced renal and hepatic function).

The time to steady state is 1 to 2 weeks; the time to steady state may be prolonged with either renal or hepatic dysfunction. Therapeutic monitoring is done after steady state has been achieved.

METHOD: RIA; EMIT; Fluorescence Polarization (Abbott). There is a natural substance in premature and newborn babies that reacts like digoxin in several commercial kits used for assay of digoxin; the substance is not detected in infants older than two months.

INTERPRETATION: Digoxin is given by mouth or intravenously. Digoxin is used in the treatment of congestive heart failure, atrial fibrillation and supraventricular tachyarrhythmias. Congestive Heart Failure: When pathology of the left ventricular myocardium is involved, digitalis glycosides should be used. When the causes of congestive heart failure are pulmonary, endocrine, hypertension, mechanical or muscle replacement, digitalis will be slightly effective or ineffective.

The usual (oral) dose of digoxin is 0.25mg digoxin tablets; patients with reduced renal function take 0.125mg tablets or less. Firm recommendations about modification of dose of digoxin in patients with uremia have been devised (British National Formulary, London: British Medical Association and the Pharmaceutical Society of Great Britain 11, 1983). However, consider the measurements of digoxin in uremia as described below.

Serum Digoxin in Uremia: Blood samples from uremic patients, who are not receiving digoxin, may give false values for digoxin; two-thirds of blood specimens were positive for digoxin when measured by immunoassay methods (Graves, S.W. et al., Ann. Intern. Med. 99, 604-608, 1983).

In another study, acute renal failure with uremia was found to be associated with a two-fold rise in serum digoxin concentration during a nine-day period after the last dose of the drug, when measured by immunoassay. The apparent increase in serum digoxin may have been caused by a decrease in the apparent volume of distribution of digoxin and its metabolites secondary to renal failure, or the occurrence of an endogenous digoxin-like substance or polar digoxin metabolites (Craver, J.L. et al., Ann. Intern. Med. 98, 483-484, 1983). In uremia, it may be that substances are produced that have both immunological activity and physiological activity similar to digoxin (Editorial, The Lancet 2, pgs. 1463-1464, Dec. 24 and 31, 1983).

Therapy is usually begun with or without a loading dose and therapeutic serum levels are maintained on a long-term basis after plasma-tissue equilibrium. The time course of distribution and elimination of a single oral dose of digoxin is shown in the next Figure; (Fenster, P.E. and Ewy, G.A., Resident and Staff Physician, 2s-15s, Dec. 1981):

Time Course of Distribution and Elimination of a Single Oral Dose of Digoxin

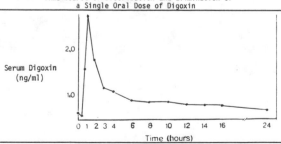

S. Bakerman   Digoxin

As with other drugs, after administration of a single dose, drug levels increase and then decline. The relatively high concentration of digoxin during the first six hours after oral dose reflects the relatively slow distribution of digoxin from the central compartment to the peripheral compartment. The serum level to determine peak activity should be obtained after the serum digoxin has had time to equilibrate with the tissues, that is, 8 to 24 hours after the oral dose. Digoxin is excreted mainly unchanged in the urine.

Toxicity: The most effective therapeutic dose of digoxin is a large fraction of the toxic dose, meaning that the therapeutic index is very low with only a very narrow difference between therapeutic and toxic dosages.

Digoxin toxicity dropped from 14% to 6% after the assay of digoxin was introduced (Koch-Weser, J., Therapeutic Drug Monitoring 3, 3-16, 1981). The toxic reaction to digoxin has a reported 20% mortality (Smith, T.W., Am. J. Med. 58, 470-476, 1975).

The features of digitalis toxicity are given in the next Table; (Anderson, G.J., Geriatrics, pgs. 57-65, June, 1980):

Features of Digitalis Toxicity	
Precipitating or potentiating factors	
Hypokalemia	Impaired excretion
Hypercalcemia	Severe heart disease
Hypoxia	Hypomagnesemia
Overdose	Hypothyroidism
	Drug interactions:
	Quinidine
	Spironolactone
	Verapamil
Clinical signs and symptoms	
Anorexia	Abdominal pain
Nausea	Psychosis
Vomiting	Mesenteric venous occlusion
Diarrhea	Weight loss
Headache	Scotomas
Depression	Blue-yellow vision
Congestive heart failure	
Incidence of cardiac arrhythmias	
Ventricular arrhythmias	55-85%
Atrioventricular (AV) block	20-40%
Atrial arrhythmias	20-25%
Junctional arrhythmias	12-18%
Sinoatrial arrhythmias	5-15%
AV dissociation	10-20%

Clinical Features of Digoxin Toxicity: The systems that are most commonly involved are the gastrointestinal tract and the central nervous system.

ECG Manifestations of Digitalis Toxicity: Between 30% and 70% of patients with ECG manifestations of digitalis toxicity will not have any clinical signs or symptoms; ECG changes occur in 85% to 95% of patients with digitalis toxicity. The changes are listed in the previous Table. Ventricular arrhythmias are the most prevalent ECG finding and are observed in about 60% of patients.

Precipitating or Potentiating Factors:

Hypokalemia: Hypokalemia potentiates or precipitates digitalis toxicity because the potassium ion competitively inhibits the binding of cardiac glycosides to the cardiac membrane.

Hypercalcemia: An increase in calcium concentration potentiates digitalis toxicity and induces abnormal automaticity through its effects on membranes.

Quinidine Effect on Digoxin: The simultaneous administration of quinidine and digoxin results in increased digoxin blood level. Therefore, digoxin dose should be decreased by one-third to one-half when quinidine is given.

Treatment: Absorption from the gastrointestinal tract should be prevented with either cholestyramine or colestipol. Restore body stores of potassium; mild intoxication can be corrected with oral potassium. The cardiac aspects of digoxin toxicity are treated with lidocaine or phenytoin. In a recent development, digoxin toxicity was treated with digoxin-specific Fab antibody fragments (Smith, T.W. et al., N. Engl. J. Med. 307, 1357-1362, 1982).

DILANTIN (SEE PHENYTOIN)

DIRECT ANTIGLOBULIN TEST (SEE COOMBS, DIRECT)

DISOPYRAMIDE (NORPACE)
SPECIMEN: <u>Red</u> top tube, separate serum; or <u>green</u> (heparin) top tube, separate plasma.
  Spuriously low plasma concentrations have been reported when blood is collected in certain commercial rubber-stoppered collection tubes. Use all-glass tubes or test any commercial collecting tube thoroughly before use.
REFERENCE RANGE: Range: Therapeutic: 2.0-5.0mcg/ml. Time to Obtain First Serum Specimen (Steady State): 36-48 hours. Peak: 2-3 hours. Sampling Time: Just before next dose. Half-Life: 4 to 10 hours. Toxicity: >7mcg/ml.
METHOD: EMIT, GLC, HPLC
INTERPRETATION: Disopyramide is given by mouth. Disopyramide is used in the treatment of <u>ventricular arrhythmias</u>. <u>Guidelines</u> for <u>therapy</u> and serum specimen monitoring are given in the next Table:

Guidelines for Therapy and Monitoring	
Therapy	Monitoring
Loading Dose: 300mg	36-48 hours and at least 6 hours after last dose
Maintenance Dose: 100-150mg every 6 hours	

  Loading dose decreased to 200mg if the patient weighs less than 50kg.
Dosage is modified in the conditions listed in the next Table:

Modification of Dosage of Disopyramide		
Condition	Effect	Modification of Dosage
Renal Failure	Half-Life Longer; Serum Conc. Higher	Decrease by increasing dosing interval
Acute Myocardial Infarction	Half-Life Longer; Volume of Distribution Higher	Decrease by decreasing maintenance dose
Congestive Heart Failure	Avoid disopyramide unless failure is caused by arrhythmias.	
Hypokalemia		Higher Doses Are Necessary

DISSEMINATED INTRAVASCULAR COAGULOPATHY (DIC), TEST PROFILE
  In DIC, the <u>coagulation factors</u> and platelets are <u>consumed</u> by <u>intravascular coagulation</u>. The laboratory tests and the test results in DIC are shown in the next Table; (Corash, L., Primary Care 7, 423-438, 1980):

Test Results in Disseminated Intravascular Coagulopathy(DIC)	
Test	Test Results
Prothrombin Time(PT)	Prolonged
Partial Thromboplastin Time(PTT)	Prolonged
Plasma Fibrinogen	Decreased
Platelet Count	Decreased [below 70,000 per cubic millimeter]
Serum Fibrin Degradation Products(FDP)	Increased
Thrombin Time(TT)	Prolonged
Euglobulin Lysis Time	Shortened
Peripheral Smear	Fragmented Red Cells

  The PT, fibrinogen level and platelet count are <u>screening</u> tests; confirmatory tests are thrombin time, euglobulin lysis time and fibrin degradation product(FDP) titer. The combination of <u>prolonged prothrombin time</u>, <u>hypofibrinogenemia</u>, <u>thrombocytopenia</u> and <u>abnormality</u> of <u>one</u> of the <u>confirmatory tests</u> is strong evidence for the diagnosis.
Prothrombin Time(PT): The PT is <u>prolonged</u> due to decreases in the level of factors II, V, IX and fibrinogen.
Partial Thromboplastin Time: The PTT may be paradoxically <u>short</u> due to the presence of <u>activated factors</u> or <u>mildly prolonged</u> due to depletion of factors.

Plasma Fibrinogen: Plasma fibrinogen is decreased because the catabolic rate exceeds the synthetic rate. However, fibrinogen is an acute reacting protein and may be elevated prior to the onset of DIC; following onset of DIC, the level of fibrinogen may not fall below the normal range.

Platelet Count: Thrombocytopenia is found in DIC.

Fibrin Degradation Products(FDP): The excessive activation of thrombin leads to overactivation of the fibrinolytic system and to increased production of fibrin degradation products. This is the test that tends to confirm the diagnosis of DIC although other conditions give rise to FDP.

Thrombin Time: The thrombin time is prolonged because of hypofibrinogenemia (usually at levels of fibrinogen less than 75 mg/dl) and because of circulating anticoagulants in the form of fibrin degradation products.

Euglobulin Lysis Time: This is a reliable and easily formed test for increased fibrinolytic activity; the euglobulin lysis test is usually shortened to less than 120 minutes; however this test is not often available in clinical laboratories.

Peripheral Smear: Beside thrombocytopenia, there is fragmentation of red blood cells which occurs in about 50 percent of patients.

Other diseases may be associated with fragmented erythrocytes; these diseases include thrombotic thrombocytopenia purpura, myelophthistic marrow disorder, acute leukemia and beta-thalassemia-hemoglobulin E disease (Visudhiphan, S., et al., N. Engl. J. Med., 309, 113, 1983; Chaplinski, T.J., personal communication).

Bacterial infection, particularly gram-negative septicemia, is the most common cause of DIC; other common causes are malignancy, trauma and surgery. Clinical conditions complicated by DIC are given in the next Table; (Preston, F.E., Brit. J. Hosp. Med. pgs. 129-137, Aug. 1982):

Clinical Conditions Associated With DIC	
Mechanism	Clinical Condition
Infection	Bacterial especially Gram-Negative Sepsis and Meningococcemia, Viral, Protozoal, Rickettsial especially Rocky Mountain Spotted Fever
Neoplastic	Mucin-Secreting Adenocarcinoma, Acute Leukemia especially Promyelocytic Leukemia, Carcinoma of the Prostate, Lung and Other Organs
Tissue Damage	Trauma, Surgery especially Prostatic Surgery, Heat Stroke, Burns, Dissecting Aneurysms
Obstetric	Obstetrical Complications: Abruptio Placentae, Amniotic Fluid Embolism, Retained Fetal Products, Eclampsia
Immunological	Immune-Complex Disorders, Allograph Rejection, Incompatible Blood Transfusion, Anaphylaxis
Metabolic	Diabetic Ketoacidosis
Miscellaneous	Shock, Snake Bite, Cyanotic Congenital Heart Disease, Fat Embolism, Severe Liver Disease, Cavernous Hemangioma

Hemorrhage and thrombosis are the major criteria for beginning therapy. Thrombosis in arteries and veins of major vessels and small vessels of the digits, skin and organs occurs secondary to formation of fibrin clots. Hemorrhage occurs secondary to depletion of clotting factors and platelets. Therapy is first directed toward the underlying disease process. Heparin (starting dose: 50 U per kg over six hours) is the treatment of choice when additional treatment is necessary. Heparin may be monitored by whole blood clotting time, which should be twice the control value one hour before the next dose. Follow the fibrinogen concentration and platelet count which are usually the first laboratory tests to show improvement. Heparin therapy is most often helpful in patients with liver dysfunction plus DIC or cancer plus DIC.

Severe Liver Disease: It is possible to determine if DIC is present in patients with liver disease by measuring factors V and VIII; decrease in both factors indicates DIC; decrease in factor V and normal factor VIII indicates liver disease only (Chaplinski, T.J., personal communication).

DRUG SCREEN (TOXICOLOGY)

SPECIMEN: 30ml of random urine; red top tube, separate serum.
REFERENCE RANGE: Negative
METHOD: Spot tests, thin-layer chromatography and EMIT
INTERPRETATION: The substances that are most commonly <u>found</u> in emergency room toxicology specimens are shown in the next Table:

Substances in Emergency Room Toxicology Specimens	
Classification	Drug
Ethanol	
Analgesics	Salicylates, acetaminophen (Tylenol), codeine, propoxyphene (Darvocet), morphine, heroin
Tranquilizers	Benzodiazepines: diazepam (Valium), librium, dalmane and others
	Phenothiazines: Chlorpromazine (Thorazine), thioridazine (Mellaril), prochlorperazine (Compazine), trifluoperazine (Stelazine) and others
Antidepressants	Tricyclic antidepressants: Amitriptyline, nortriptyline, imipramine, desipramine, doxepin
Stimulants	Amphetamines
Barbiturates	Phenobarbital, pentobarbital, secobarbital, amobarbital
Hypnotics other than barbiturates	Methaqualone, ethchlorvynol, glutethimide
Anticonvulsants	Phenytoin (Dilantin)
Hallucinogens	Lysergic acid diethylamide (LSD), phencyclidine (PCP)
Volatiles	Ethylene glycol, isopropanol

The drugs that are <u>detected</u> by Toxi-Screen using thin-layer chromatography are given in the next Table:

Qualitative Drug Analysis	Qualitative Drug Analysis	Qualitative Drug Analysis
Acetaminophen (Tylenol etc)	Hydrocortisone	Procainamide (Pronestyl)
Amphetamine	Imipramine (Tofranil)	Propoxyphene (Darvon)
Barbiturates	Ipecac	Propranolol (Inderal)
Caffeine	Lidocaine (Xylocaine)	Pseudoephedrine (Sudafed)
Chloridiazepoxide (Librium)	MDA	Pyrilamine (Triaminic)
Chlorpromazine (Thorazine)	Meperidine (Demerol)	Quinidine
Chlorprothixene (Taractan)	Meprobamate (Miltown)	Quinine
Cimetidine (Tagamet)	Mescaline	Spironolactone (Aldactone)
Cocaine	Methadone (Dolophine)	Strychnine
Codeine	Methamphetamine	Tegretol (Carbamazepine)
Dextromethorphan	Methaqualone (Quaalude)	Terpin Hydrate
Diazepam (Valium)	Methocarbamol (Robaxin)	Thioridazine (Mellaril)
Dimenhydrinate (Dramamine)	Morphine	Thiothixene (Navane)
Diphenhydramine (Benadryl)	Nicotine	Triamterene (Dyrenium,
Disopyramide (Norpace)	Orphenadrine (Norflex)	Dyazide)
Doxepin (Sinequan)	Oxazepam (Serax)	Tricyclic Antidepressants
Doxylamine (Bendectin)	Pentazocine (Talwin)	Trifluoperazine
Ephedrine (Tedral)	Phenacetin (Empirin)	(Stelazine)
Erythromycin	Phencyclidine (Angel Dust)	Triflupromazine (Vesprin)
Ethchlorvynol (Placidyl)	Pheniramine, Related Drugs	Trimeprazine
Flurazepam (Dalmane)	Phenolphthalein	Trimethobenzamide (Tigan)
Glutethimide (Doriden)	Phenylpropanolamine	Trimethoprim (Bactrim,
Heroin	Phenytoin (Dilantin)	Septra)
		Tripelennamine
		(Pyribenzamine)

Alcohols (ethyl alcohol, ethylene glycol and isopropanol) and aspirin are not detected in drug screens.

A useful <u>screening test</u> for <u>volatiles</u> is <u>serum osmolality</u>. The contribution to serum osmolality of each 100 mg/dl of ethanol, ethylene glycol and methanol is given in the next Table:

Contribution to Serum Osmolality (Each 100mg/dl of Volatile Substance)	
Substance	Contribution
Ethanol	21.7
Methanol	31.0
Ethylene Glycol	16.3

Thus, if serum osmolality is normal, the patient does not have ethanol, methanol or ethylene glycol toxicity.

Besides drug screen and serum osmolality, other useful laboratory tests are as follows: electrolytes, Na^+, K^+, Cl^-, CO_2 content; glucose, creatinine, urea, arterial blood gases, urinalysis, and liver function tests.

The causes of death by accidental poisoning, in British children under 10, arranged in decreasing order of incidence are given in the next Table; (Fraser, N.C., Br. Med. J. 28, 1595-1598, 1980).

Accidental Poisoning by Medicinals and Nonmedicinals	
Medicinals	Nonmedicinals
Salicylates	Lead
Tricyclic antidepressants	Corrosives
Barbiturates	Petroleum products
Opiates and analogues	Paraquat
Digoxin	Alcohols
Quinine	Nicotine
Strychnine	

Treatment Using Emetics: Syrup of ipecac may be administered in the emergency room or at home. If syrup of ipecac is not available, use household liquid detergents e.g., Ivory Liquid, Palmolive Liquid, Dove Liquid, Dawn Liquid. Do not use laundry detergents or electric dishwashing detergents. Add 3 tablespoons of liquid detergent in 8 ounces of water. The guidelines for use of emetics are as follows; (Mack, R.B., N.C. Med. Journal 44, 297-298, 1983).
1. Emesis is indicated in a potentially poisoned patient;
2. No contraindications to emesis are present, e.g., coma, ingestion of caustic caustic, etc.;
3. The patient is more than 12 months of age;
4. Access to syrup of ipecac is not possible within a reasonable time frame;
5. Use only the type of liquid dishwashing detergent mentioned. Salt water is contraindicated as an emetic because of the danger of hypernatremia.

Gastric Lavage: Gastric lavage may be done if the drugs were ingested within the previous four hours (eight hours for aspirin). Lavage is not recommended in an awake patient in whom the use of activated charcoal is more appropriate. Gastric lavage tends to be overused.

Charcoal: The use of charcoal for gastrointestinal clearance of drugs has been reviewed (Levy, G., N. Engl. J. Med. 307, 676-678, 1982; see also Letters to the Editor, N. Engl. J. Med. 308, 156-157, 1983).

Poison Treatment: Commonly described antidotes are given in the next Table; (Arena, J.M., Clin. Symp. 30, 44, 1978):

Antidotes	
Poison	Antidote
Acetaminophen	N-Acetyl Cysteine
Anticholinergics	Possibly Physostigmine
Carbon Monoxide	100% Oxygen
Cyanide	Lilly Kit: Amyl Nitrite; I.V. Sodium Nitrate; I.V. Sodium Thiosulfate
Ethylene Glycol	Ethyl Alcohol
Iron	Deferoxamine
Lead	Penicillamine; BAL; EDTA
Methemoglobinemia caused by Acetanilid, Aniline Derivatives, Nitrites, Chlorates, Nitrobenzene	Methylene Blue(1%), infuse over 30 minutes
Methyl Alcohol	Ethyl Alcohol
Narcotics (Opiates)	Naloxone
Phenothiazines	Benadryl
Organophosphates	Atropine; Pralidoxime (Protopam, 2-PAM Chloride, 1 gram I.V.)
Tricyclics	Bicarbonate; Physostigmine
Warfarin	Vitamin K

Physostigmine can cause seizures; it should be used only in life-threatening situations.

Poison Control Centers: Information on treatment of poisons is available from poison control centers:
Pittsburgh Poison Control Center: 1-412-681-6669;
Duke University Poison Control Center: 1-919-684-8111, extension 3957;
East Carolina University Poison Control Center: 1-919-757-4461

D-XYLOSE (see XYLOSE)

ENDOCRINE EMERGENCIES

A list of conditions that may present as endocrine emergencies is given in the next Table; (Daggett, P., Brit. J. Hosp. Med. pgs. 38-43, Jan. 1979):

Conditions that May Present as Endocrine Emergencies
Diabetes Mellitus
Ketoacidosis
Hyperosmolar Coma
Lactic Acidosis
Hypoglycemia
Thyroid Gland
Myxedema Coma
Thyroid Crisis(Thyroid Storm)
Adrenal Glands
Addisonian Crisis
Iatrogenic Steroid Deficiency
Acute Hypercortisolism
Pheochromocytoma
Pituitary Gland and Hypothalamus
Hypopituitary Coma
Diabetes Insipidus
Inappropriate ADH Activity
Parathyroid Glands (and Other Causes of Hypocalcemia and Hypercalcemia):
Hypocalcemia <7 mg/dl
Hypercalcemia >13 mg/dl

Diabetes mellitus is the only common endocrine emergency. Endocrine emergencies are usually secondary to a precipitating cause such as <u>infection</u>, trauma or surgery.

The list of conditions that may present as endocrine emergencies and the laboratory tests and findings are given in the next Table:

Conditions that May Present as Endocrine Emergencies and the Laboratory Tests and Findings	
Condition	Laboratory Tests and Findings
Diabetes Mellitus	

Ketoacidosis	Serum Glucose; Usually in Range of 300 mg/dl to 800 mg/dl; Measure every 1 to 2 hours.
	 Ketone Bodies; Elevated in Blood and Urine.
	 Blood Gases:
	   pH: Usually Below 7.1
	   HCO_3: Usually Less Than 10
	   pCO_2: Usually Low Due to Compensation
	 Other:
	   Hemoglobin: Usually High due to Dehydration
	   WBC's: May Be Raised Even in Absence of Infection

Hyperosmolar Coma	Plasma Glucose often 900 mg/dl or Greater. Plasma Osmolality Usually 350-400 mosmol/kg.
Lactic Acidosis	Plasma Lactate is Very High (Upper Limit of Normal, 1.4 mmol/liter). Increased Anion Gap. pH: Usually below 7.0.
Hypoglycemia	Blood Glucose is Low
Thyroid Gland:	
Myxedema Coma	Decreased T-4, Free T-4, and Elevated TSH Hyponatremia (Dilutional) Hypercholesterolemia Macrocytosis With or Without a Low Hemoglobin Acidosis: Increased pCO_2 and Hypoxia Secondary to Alveolar Hypoventilation
Thyroid Crisis (Thyroid Storm)	Elevated T-4, and Free T-4

Table Continued:

Conditions that May Present as Endocrine Emergencies and the Laboratory Tests and Findings	
Condition	Laboratory Tests and Findings
Adrenal Gland:	
Addisonian Crisis	Low Plasma Cortisol (10 ml of Heparinized Blood) and High Plasma ACTH (20 ml of Heparinized Blood Separated at Once and Freeze Plasma Immediately) Low Serum Sodium High Serum Potassium
Iatrogenic Steroid Deficiency	
Acute Hypercortisolism	Hypokalemic Alkalosis Elevated Plasma Cortisol
Pheochromocytoma	Phenolamine Test (Use when Diagnosis has not previously been made): Give 5 mg Phentolamine I.V.; a Positive Response is a Fall of at Least 35 mm Hg in Systolic b.p. Within Two Minutes of I.V. Injection.
Pituitary Gland and Hypothalamus:	
Hypopituitary Coma	Decreased Plamsa T-4, Decreased Cortisol, Decreased TSH and Decreased ACTH Skull X-Ray Hypoglycemia Dilutional Hyponatremia
Diabetes Insipidus	Simultaneous Plasma and Urine Osmolality; Increased Plasma Osmolality (Usually Around 350 mosmol/kg) and Decreased Urine Osmolality (Usually Around 150 mosmol/kg) Large Urine Volume
Inappropriate ADH Activity	Simultaneous Plasma and Urine Osmolality Decreased Plasma Osmolality and Increased Urine Osmolality Dilutional Hyponatremia
Parathyroid Gland (and Other Causes of Hypocalcemia and Hypercalcemia)	
Hypocalcemia	Serum Calcium <7 mg/dl
Hypercalcemia	Serum Calcium >13 mg/dl

S. Bakerman

EOSINOPHIL COUNT, TOTAL

<u>SPECIMEN</u>: Lavender(EDTA) top tube; deliver specimen to laboratory within one
hour; stable up to 4 hours in the refrigerator.
<u>REFERENCE RANGE</u>: Newborn (<24 hours old): 20-850/cu mm; <u>1 day to 1 year old</u>:
50-700/cu mm; 1 year to adult: 0-450/cu mm.
<u>METHOD</u>: Manual Method: Unopette (Becton, Dickinson and Co.) technique; the
red cells are lysed and eosinophils only stain bright orange-red. <u>Automatic
Methods</u>: (1) <u>Selective cytochemical stains</u> in a liquid milieu; sensing device
measures light scatter and light absorption (Technicon D/90). (2) <u>Stained blood
smears</u>; automated microscope with computerized morphologic and tintorial
criteria for cell identification (Hematrak). (3) Unstained cells are classified
by phase microscopy on the basis of <u>size</u> and <u>refractive index</u> in a liquid
milieu; classified on basis of size and density (Coulter Electronics).
<u>INTERPRETATION</u>: The causes of increased blood eosinophils are listed in the
next Table; (Editorial, The Lancet <u>1</u>, 1417-1418, June 25, 1983):

Causes of Increased Blood Eosinophils

Parasitic Diseases: Especially with tissue invasion.
 Trichinosis; Visceral Larva Migrans (Toxocara Cunis or
 T. Cati), and Strongyloides
Allergic Diseases: Bronchial Asthma and Seasonal Rhinitis (Hay Fever)
Skin Disorders: Atopic Dermatitis; Eczema; Acute Urticarial Reactions;
 Pemphigus
Pulmonary Eosinophilias: Loeffler's Syndrome; Pulmonary Infiltration
 with Eosinophilia(PIE Syndrome); Tropical Pulmonary Eosinophilia
 caused by Microfilariae.
Infectious Diseases in Immunodeficient Children
Certain Vasculitic and Granulomatous Diseases
Connective Tissue Diseases such as Polyarteritis Nodosa, Lupus
 Erythematosus and Eosinophilic Fasciitis
Many Types of Drug Eruptions
Malignant Tumors: Bronchogenic Carcinoma, Hodgkin's Disease
 and T-Cell Leukemias
Hypereosinophilic Syndrome(HES)

The characteristics of HES are discussed in an editorial (The Lancet <u>1</u>,
1417-1418, June 25, 1983).

EPILEPTIC ATTACK PANEL

A protocol for examination of a patient with possible epilepsy is shown in
the next Figure; (Scott, A.K., Brit. Med. J. <u>288</u>, 986-987, March 31, 1984):

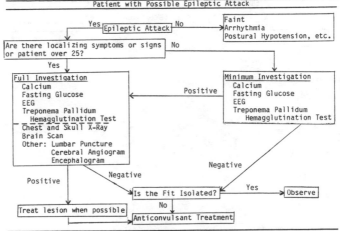

Patient with Possible Epileptic Attack

182

EPSTEIN-BARR VIRUS(EBV) ANTIBODIES, SERUM

SPECIMEN: Red top tube, separate serum and refrigerate serum.

REFERENCE RANGE: Three antibodies to the Epstein-Barr virus are measured; these are antibodies to viral capsid antigen, (anti-VCA, IgM and IgG) and antibody to Epstein-Barr nuclear antigen(EBNA). The reference ranges are as follows:

Antibodies to Epstein-Barr Virus	
Antibody	Reference Range
Anti-Viral Capsid Antigen(VCA)IgM	< 1:10
Anti-Viral Capsid Antigen(VCA)IgG	< 1:10
Anti-EB Virus Nuclear Antigen(EBNA)	< 1:5

METHOD: Indirect immunofluorescence for anti-VCA(IgG or IgM); anti-EBNA by anti-complement immunofluorescence.

INTERPRETATION: Epstein-Barr virus(EBV) is the cause of infectious mononucleosis. Tests of antibodies to Epstein-Barr virus are recommended only when a screening procedure, such as Monospot Test or heterophil absorption are negative and the diagnosis of infectious mononucleosis is suspected. The Monospot test and heterophil tests are negative in about 10% of adult patients with infectious mononucleosis; the changes in the different antibodies in infectious mononucleosis are shown in the next Figure:

Changes in Heterophil Antibodies and Epstein-Barr Virus Antibodies During Course of Infectious Mononucleosis

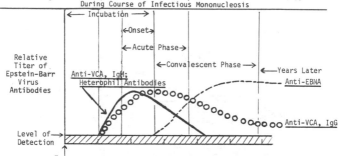

During the course of infectious mononucleosis, heterophil antibodies are detected in up to 90 percent of patients while antibodies to viral capsid antigen (anti-VCA, IgM and IgG) and antibody to Epstein-Barr nuclear antigen (anti-EBNA) are detected in 100 percent of patients. (Capsid is the protein coat that surrounds the nucleic acid, DNA or RNA, core of viruses.)

Both anti-VCA, IgM and anti-VCA, IgG reach high titers during the acute phase of infectious mononucleosis. Anti-VCA, IgM antibodies peak about the second week of illness and then, gradually disappear within one or two months, that is, soon after the disease subsides. Anti-VCA, IgG antibodies decrease to lower levels which persist for many years. Presence of anti-VCA, IgM, antibodies indicate recent, primary infection with EBV. A four-fold increase in anti-VCA, IgG antibodies is diagnostic of active disease; a single titer of 1:320 is strongly suggestive and a titer of 1:640 is definitive of active or recent EBV infection. The specific tests for the diagnosis of infectious mononucleosis are given in the next Table; (McCarthy, L.R., Clin. Microbiol., Newsletter, 6, 17-20, Feb. 1, 1984):

Epstein-Barr Virus Specific Tests for the Diagnosis of Infectious Mononucleosis	
EBV Test	Comment
IgM anti-VAC	Antibody formed early in course of disease, disappears in convalescent. Positive test indicates infection.
IgG anti-VCA	Antibody formed during acute illness; antibody remains elevated for prolonged duration. Fourfold increase in titer diagnostic. Formation of IgG may be delayed, and not present in early stage of acute infection.
Anti-EBNA	Antibody formed during convalescence. Presence of antibody indicates past disease.
Virus Culture Throat	Positive during acute disease; intermittent/continuous shedding up to three months after disease.
Leukocyte Culture	Positive during acute disease. Cultures may be positive during convalescence.

Anti-EBNA antibodies may begin to appear in the fourth to eighth week but are usually delayed for about six months after onset of illness. Anti-EB nuclear antibodies persist for life.

The diagnosis of infectious mononucleosis is usually straight forward; the "classic" findings in infectious mononucleosis are as follows: malaise, fever, sore throat or pharyngitis, lymphadenopathy, mild to moderate hepatitis, splenomegaly, and positive Monospot or heterophil serology. Approximately 80% of patients present with pharyngitis and/or lymphadenopathy; the remaining 20% have fever without either pharyngitis or lymphadenopathy. About 50% of patients have splenomegaly.

The atypical lymphocytes are not diagnostic of infectious mononucleosis; they are also seen in other viral infections, such as hepatitis, mumps, rubella, rubeola, varicella, Herpes simplex, H. zoster, and cytomegalovirus infections, in drug-induced hepatitis and in toxoplasmosis. The atypical lymphocytes are T-cells.

The diagnosis of infectious mononucleosis may be difficult when the clinical features are atypical or absent and the Monospot and heterophil tests are negative. Then, the acute phase of infectious mononucleosis may be confirmed by positive antibody to EB virus capsid antigen(VCA), IgM or raised levels of IgG. The absence of antibody EBV-associated nuclear antigen (anti-EBNA) indicates lack of previous contact with EBV. The atypical and confusing clinical findings has been reported in patients, aged 40 to 78; these patients may present with predominant hepatic involvement; (Horwitz, C.A. et al., Medicine, 62, 256-262, 1983; Editorial, The Lancet, pg. 143, January 21, 1984).

Evidence of Epstein-Barr virus(EBV) infection is present in almost every case of Burkitt's lymphoma in Africa; about 20% of sporadic or non-African Burkitt-like tumors are also associated with EBV. EBV is also associated with undifferentiated carcinoma of the nasopharynx, and a cytomegalovirus-like disease in renal transplant recipients and in post-transfusion syndrome. These conditions are associated with high levels of EBV antibodies.

ERYTHROCYTE SEDIMENTATION RATE (ESR)

SPECIMEN: Lavender (EDTA) top tube

REFERENCE RANGE: Newborn: 0-2mm/hr; Neonates and children: 3-13mm/hr. Post-adolescent male (less than 40 years): 1-15mm/hr; Post-adolescent female (less than 40 years): 1-20mm/hr. The ESR (Westergren Method) varies with age; the maximum normal ESR at a given age is calculated using the formulas: Men: Age in years/2; Women: (Age in Years + 10)/2 (Miller, A. et al., Br. Med. J. 286, 266, 1983).

METHOD: The erythrocyte sedimentation rate (ESR) is the measurement, under standard conditions, of the rate of settling of erythrocytes in anticoagulated blood. Westergren sedimentation; Wintrobe; zeta method.

INTERPRETATION: The ESR is used as a marker of tissue inflammation, some causes of elevation of ESR are given in the next Table: (Lascari, A.D., Pediatric Clinics of North America, 19, 1113-1121, Nov. 1972):

Causes of Elevation of ESR	
Infections	**Collagen Diseases**
Majority of bacterial infections	Rheumatic fever
Infectious hepatitis	Rheumatoid arthritis
Cat scratch disease	Lupus erythematosus
Post-perfusion syndrome	Dermatomyositis
Primary atypical pneumonia	Scleroderma
Tuberculosis	Systemic vasculitis
Secondary syphilis	Henoch-Schönlein purpura
Leptospirosis	Mediterranean fever
Systemic fungal infections	**Renal**
Hematologic and Neoplastic	Acute glomerulonephritis
Severe anemia	Chronic glomerulonephritis
Leukemia	with renal failure
Lymphoma	Nephrosis
Metastatic tumors	Pyelonephritis
Chronic granulomatous disease	Hemolytic-uremic syndrome
Gastrointestinal	**Miscellaneous**
Ulcerative colitis	Hypothyroidism
Regional ileitis	Thyroiditis
Acute pancreatitis	Sarcoid
Lupoid hepatitis	Infantile cortical hyperostosis
Cholecystitis	Surgery, burns
Peritonitis	Drug-hypersensitivity reactions

The ESR has a relatively high sensitivity and low specificity. Changes in ESR may provide information concerning the activity of a disease process and serial readings may reflect therapeutic response. Normal results do not exclude serious illness.

An increased ESR is caused by enhanced erythrocyte aggregation; this is caused by increased levels of asymmetrical macro-molecules, principally fibrinogen and the globular proteins in the blood. Normally, red blood cells have a negative charge on their surface; the like charges on red cells cause these cells to repel each other. Plasma proteins, especially fibrinogen, tend to adhere to the red cell membrane and neutralize the surface charges and makes the cells more likely to aggregate forming stacks or rouleaux. The aggregated (stacked) cells have a higher ratio of mass to surface area than single cells and, therefore will fall out from the plasma more readily.

Conditions that alter ESR by interfering with the formation of rouleaux are as follows: acanthocytosis and poikilocytosis. Anemia may be associated with elevated ESR levels. In iron-deficiency anemia, there is reduction in the ability of smaller red cells to sediment; this may be compensated by the accelerating effect of an increased proportion of plasma.

The ESR is decreased in the following conditions: polycythemia vera; some forms of macroglobulinemia (e.g., hyperviscosity syndrome); and sickle cell anemia. An ESR of 20 mm/hr in a sickler may equate to a value of 100 mm/hr in a normal individual and is a signal of possible acute infection (Thomas, R.E., Personal Communication).

In a study by Kirkeby, A.K. and Leren, P. (Nordisk Medicin 48, 1193, 1952) of patients with ESRs of more than 100mm, 24% had pneumonia, 17% had chronic renal disease and 19% had malignancy. (Zacharski, L.R. and Kyle, R.A., JAMA 202, 264-266, 1967; Gibbs, D., Brit. J. Hosp. Med., 493-496, May, 1982; Cunha, B.A., Diagnosis, pgs. 62-69, Feb. 1983).

ESTRADIOL

SPECIMEN: Red top tube, separate serum. The serum is stable at room temperature for 1 week, at refrigerator temperature for 1 month, in a frost-free freezer for 1 year and in a non-defrosting freezer for 3 years. Lavender(EDTA) top tube, separate plasma.

REFERENCE RANGE:

Reference Range for Estradiol	
	pg/ml
Female	
Prepubertal, Normal	4.0-12.0
Ovulating, Normal	
Early Follicular	30-100
Late Follicular	100-400
Luteal Phase	50-150
Pregnant, Normal	Up to 35,000
Human Menopausal	
Gonadotropin Treatment:	
Therapeutic Range	350-750
Postmenopausal or Castrate	5.0-18.0
On Oral Contraceptives	Under 50
Male Normal	
Prepubertal	2.0-8.0
Adult	10-60

METHOD: RIA

INTERPRETATION: Estradiol is most often measured in the evaluation of amenorrhea; other clinical applications of serum estradiol measurements include precocious puberty in females; ovarian induction; and gynecomastia in males.

Ammenorrhea: If estradiol is less than 30pg/ml, obtain serum FSH levels in order to differentiate primary from secondary ovarian failure.

Precocious Puberty in Females: Elevation of serum estradiol suggests estrogen secreting tumor or true precocious puberty associated with elevation of serum gonadotropins(FSH, LH) which are in the adult female range.

Ovulation Induction: Serum estradiol is measured during induction of ovulation (Larsen, S. and Honore, E., Fertil. Steril. 33, 378, 1980; Smith, D.H. et al., Fertil. Steril. 33, 387,1980).

Gynecomastia in Males: Elevated estradiol levels have been observed in patients with cirrhosis and tumors of the liver and gonadal tumors.

ESTRIOL, SERUM

SPECIMEN: Red top tube, separate serum and freeze; specimen must not remain at room temperature for more than one hour.

REFERENCE RANGE: The change in estriol during pregnancy is shown in the next Figure:

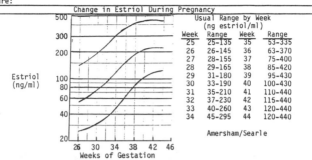

Change in Estriol During Pregnancy

Estriol (ng/ml)

Weeks of Gestation

Usual Range by Week (ng estriol/ml)

Week	Range	Week	Range
25	25-135	35	53-335
26	26-145	36	63-370
27	28-155	37	75-400
28	29-165	38	85-420
29	31-180	39	95-430
30	33-190	40	100-430
31	35-210	41	110-440
32	37-230	42	115-440
33	40-260	43	120-440
34	45-295	44	120-440

Amersham/Searle

Estriol increases until term, reaches a plateau at the 40th week of gestation and then declines.

METHOD: RIA

INTERPRETATION: Estriol is the main estrogen in the blood and urine of pregnant females. Serum estriol determinations reflect the clinical status of the fetus and are used to evaluate fetal distress and placental function.

Guidelines for Interpretation of Results:

1. The usual range quoted is useful for preliminary evaluation of a single level; however, a single estriol determination is of limited use, except to rule out a serious problem.
2. Values below the usual range are abnormal and should be followed up.
3. Values at the low end of the usual range may indicate a problem pregnancy and should be followed with serial serum estriol determinations to determine the trend of estriol values.
4. In patients with low levels, repeat testing may document a persistent low level indicative of a problem pregnancy. However, if values increase during pregnancy, the outcome is generally good.
5. Stable or falling values may indicate a problem pregnancy.

Diabetic pregnancies: Frequent estriol assays are mandatory in evaluating diabetic pregnancies; a decline of 40 percent or more compared to the mean of the three preceding determination is considered to be a predictor of fetal distress in diabetic pregnancies; biweekly assays, during week 32 to 34 and daily assays from the 34th week to the time of delivery, is recommended (Distler, W. et al., Am. J. Obstet. Gynecol. 130, 424-431, 1978).

Pre-Eclampsia; (Hypertension plus proteinuria and edema in pregnancy): The peri-natal mortality in pre-eclampsia is 29 percent and pre-ecalampsia is associated with intrauterine growth retardation(IUGR). Serial serum estriol during the last four to six weeks of gestation in cases of IUGR usually fail to show the surge in estriol levels typical of normal pregnancy; in general, the decline in estriol production in cases of IUGR is less severe or rapid than that associated with maternal diabetes (Rock, R.C., Laboratory Medicine 15, 95-97, 1984).

Severe Toxemia: Estriol levels are usually low. Impairment of renal function may give false normal levels.

Prolonged Gestation: Serum estriol reaches a plateau at 40th week of gestation and declines thereafter. Serial estriol measurements may help to detect the postdate gestation in which the fetus is at risk from the postmaturity syndrome (Rock, R.C., Laboratory Medicine 15, 95-97, 1984).

ESTROGEN RECEPTOR ASSAY WITH
PROGESTERONE RECEPTOR ASSAY

SPECIMEN: Estrogen-receptors are very labile at room temperature; therefore, the tumor tissue should be packed in ice, without direct contact, as soon as it is removed. The tissue must <u>not</u> be placed in <u>formalin.</u>

A biopsy specimen (at least one gram) should be placed in a container surrounded by ice in the operating room. The specimen should contain viable tumor tissue and not hemorrhagic or necrotic tissue; a separate similar specimen from "normal" breast, to be used as control, should be obtained from an adjacent area. The specimens should be transported immediately to the pathology laboratory where the surgical pathologist should trim the tissue and obtain a small specimen for microscopic examination. The trimmed specimen should be weighed and then cut into 5mm³ pieces. If the assay is not done on the same day, then the specimen should be immediately frozen in liquid nitrogen (-196°C) or dry ice (-70°C); liquid nitrogen is preferred.

REFERENCE RANGE: Negative: Binding is undetectable or less than 4 femtograms/dgm. Positive: Estrogen or progesterone binding of 4 femtograms/dgm.

METHOD: The tissue is homogenized and the soluble cytoplasmic protein is separated by centrifugation. Radiolabeled estriol is added to the supernatant; the unbound reagent (radiolabeled estriol) is separated from bound reagent by density gradient, charcoal, electrophoresis, etc. If estrogen-receptors are present, the tracer will be found in the protein fraction.

INTERPRETATION: General Statements on Estrogen Receptors in Breast Carcinomas are listed in the next Table; (McGuire, W.L. Cancer 36, (Suppl), 634-644, 1975; Seger, M.A., BAMC Progress Notes, 21, No. 8, pgs. 14-18, Nov., 1977; Clark, G.M. et al., N. Engl. J. Med. 309, 1343-1347, 1983):

Estrogen-Receptors in Breast Cancer

(1) Fifty to Sixty Percent of Estrogen-Positive Tumors Responded to Endocrine Therapy; Only 5 Percent of Estrogen-Negative Tumors Responded (response to therapy was defined as reduction in size of at least half of the measurable lesions by more than 50 percent.)

(2) There was No Correlation Between Presence of Estrogen-Receptors and:
    (a) Histologic Type of Tumor
    (b) Clinical Stage
    (c) Positive and Negative Axillary Nodes

(3) Patients with Low Estrogen-Receptor Levels have <u>Recurrence Sooner</u> than Patients with Moderate Levels; There is no Significant Difference Between Patients with High Estrogen Levels and Moderate Estrogen Levels.

(4) Estrogen-Receptor Values from Primary or Metastatic Tumors would Predict Equally Well.

The presence of the estrogen-receptor is predictive of longer disease-free period and longer survival than patients whose tumors are negative regardless of their treatment (Hubay, C.A. et al., Breast Cancer Res. Treat. 1, 77-82, 1981).

The presence of estrogen receptor often occurs in the post-menopausal patient.

About <u>three-fourths</u> of <u>breast tumors</u> containing <u>both estrogen-receptor</u> and <u>progesterone-receptor</u> respond to various endocrine therapies. (Edwards, D.P., et al., Biochim Biophys Acta 560, 457-486, 1979). <u>Both receptor assays</u> should be performed on breast cancers obtained at time of surgical procedures. The presence of <u>either</u> estrogen receptors <u>or</u> progesterone-receptors is <u>positively</u> correlated with disease-free survival when analyzed separately, whether or not the adjuvant regimen included an endocrine component. <u>Disease-free survival is directly related to</u> the amount of <u>progesterone-receptor</u> in the tumor, that is, the higher the progesterone-receptor level, the longer the disease-free survival interval. Determination of the <u>progesterone-receptor</u> <u>concentration</u> is of <u>equal</u> or <u>greater value</u> than determination of the estrogen-receptor concentration for predicting the <u>disease-free survival</u> of patients with breast cancer (Clark, G.M. et al., N. Engl. J. Med. 309, 1343-1347, 1983).

Treatment of Breast Carcinomas: Endocrine and/or chemotherapy are used to complement surgery. A protocol for treatment of pre- and post-menopausal patients with metastatic breast carcinoma is shown in the next Figure; (Allegra, J.C., Seminars in Oncology 10, No. 4, Suppl. 4, 25-28, Dec. 1983):

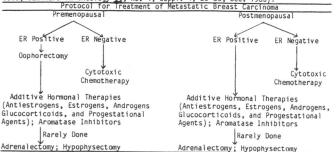

Protocol for Treatment of Metastatic Breast Carcinoma

Premenopausal

ER Positive → Oophorectomy → Additive Hormonal Therapies (Antiestrogens, Estrogens, Androgens Glucocorticoids, and Progestational Agents); Aromatase Inhibitors → Rarely Done → Adrenalectomy; Hypophysectomy

ER Negative → Cytotoxic Chemotherapy

Postmenopausal

ER Positive → Additive Hormonal Therapies (Antiestrogens, Estrogens, Androgens, Glucocorticoids, and Progestational Agents); Aromatase Inhibitors → Rarely Done → Adrenalectomy; Hypophysectomy

ER Negative → Cytotoxic Chemotherapy

ETHANOL (ALCOHOL) BLOOD

SPECIMEN: Do not use alcohol swab to clean venipuncture site; use iodine; visible hemolysis must be treated with 6% TCA. Red top tube, separate serum.
REFERENCE RANGE: Negative
METHOD: Enzymatic using alcohol dehydrogenase; gas-liquid chromatography.
INTERPRETATION: Blood alcohol level and probable toxic condition are given in the next Table:

Toxic Effects of Ethyl Alcohol			
Blood Ethyl Alcohol Level (mg/dl)	Milliosmolality of Ethanol	Serum Milliosmolality	Condition
100 $\overset{50}{-}$ 150	22 $\overset{11}{-}$ 33	312 $\overset{301}{-}$ 323	Subject drinking but may not be intoxicated
150 - 500	33 - 110	323 - 400	Intoxication
500 -	110 -	400 -	Generally comatose, often die

Ethanol is absorbed through the stomach wall (20-25%) and the remaining is absorbed in the small intestines. The peak level in the blood usually occurs in about 1 hour after ingestion; food in the stomach decreases the rate of absorption. Once the blood alcohol has peaked, the hourly elimination from blood varies from 4 to 40 mg/dl/hour; the female elimination rate is about 1.25 times that for the males.

Ethanol distributes itself between the blood and gas phase of the lungs (according to Henry's Law: the amount of gas dissolved is proportional to the pressure of the gas) and the ratio of the concentrations of alcohol in the blood and the alveolar air is a constant (k = 2100). Thus, the alcohol concentration of the alveolar air is an accurate reflection of the concentration of alcohol in the blood. Actually, breath alcohol determinations tend to underestimate the blood concentration of ethanol (Lovell, W.S., Science 178, 264-272, 1972).

Measurement of serum osmolality may be used as a screening test for alcohol in that each 100mg/dl contributes 22mosm/liter to the serum osmolality. If the normal serum osmolality is 290mosm/liter, 100mg/dl of ethanol in the serum will yield a serum osmolality of 312mosm/liter. A laboratory decision tree to differentiate an unknown alcohol, based on serum osmolality is given in the next Figure; (Glasser, L. et al., Am. J. Clin. Pathol. 60, 695-699, 1973):

Laboratory Decision Tree for Differentiation of an Unknown Alcohol

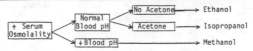

About 95 percent of the alcohol absorbed into the blood is oxidized in the liver to acetaldehyde and then to acetic acid.

There are two laboratory tests that serve as "markers" for alcoholism; these are the enzyme gamma-glutamyl-transpeptidase (GGTP) which tends to be elevated in the alcoholic and mean corpuscular volume (MCV) of red blood cells which also tends to be elevated even without folate deficiency.

The laboratory tests that tend to be abnormal in acute alcoholism are as follows: increased triglycerides; acid-base changes: respiratory alkalosis, metabolic acidosis; hypokalemia; hypomagnesemia; hypophosphatemia; and hyperuricemia.

Urine samples have an alcohol content 1.3 times that of blood.

ETHOSUXIMIDE (ZARONTIN)
SPECIMEN: Red top tube, separate serum; or green (heparin) top tube, separate plasma. Reject: hemolyzed specimen or serum frozen on cells.
REFERENCE RANGE: Therapeutic: 40-100mcg/ml. Time to Obtain First Serum Specimen (Steady State): Children: 6 days, Adults: 12 days. Trough: Immediately prior to the next dose. Time to Peak Concentration: 2-4 hours after an oral capsule; less for syrup. Half-Life: Children: 30 hours, Adults: 60 hours. Toxic: >100mcg/ml.
METHOD: EMIT, GLC, HPLC
INTERPRETATION: Ethosuximide is given by mouth; it used in the treatment of patients with petit mal seizures; petit mal attacks are characterized by transient episodes of loss of awareness without convulsive movements. Guidelines for therapy and serum specimen monitoring are given in the next Table:

Guidelines for Therapy and Monitoring	
Therapy	Monitoring
Children (<11 years): 15-40mg/kg/day	Children: 6 days
Adults: 15-30mg/kg/day	Adults: 12 days
	Maintenance Monitoring: 4 to 6 month intervals provided therapeutic response is good

Around puberty, a lower dose is usually required to achieve the same serum level.

Ethosuximide is oxidized to inactive form in the liver; 10-20% is excreted unchanged by the kidneys.

ETHYLENE GLYCOL, SERUM OR URINE

SPECIMEN: Red top tube or random urine; store serum or urine at 4°C.
REFERENCE RANGE: Normally not present in serum or urine.
Toxicity: If sampled shortly after ingestion, ethylene glycol >20mg/dl in serum.
METHOD: Gas-liquid chromatography; Sunshine, Methodology for Analytic Toxicology, CRC Press, 1975; Spectrophotometric assay; Eckfeldt, J.H. and Light, R.T., Clin. Chem. 26, 1278-1280 (1980).
INTERPRETATON: Ethylene glycol is the major ingredient in antifreeze. Small amounts of ethylene glycol can be ingested without ill effects; the minimum lethal dose of ethylene glycol is approximately 100 ml.

Blood and urine levels of ethylene glycol during hemodialysis and therapy with oral ethanol are given in the next Figure; (Peterson, C.D. et al., N. Engl. J. Med. 304, 21-23, 1981):

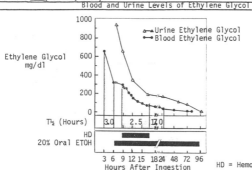

Blood and Urine Levels of Ethylene Glycol

The half-lives of ethylene glycol are as follows: No therapy, 3.0 hours; hemodialysis, 2.5 hours; oral ethanol, 17.0 hours. Ethanol is used to treat ethylene glycol toxicity; ethanol competes with ethylene glycol for metabolic sites in the liver.

Most laboratories do not offer gas-liquid chromatography on a "stat" basis. However, a metabolic acidosis with an increased anion gap and a substantial discrepancy between the measured and calculated osmolality exists for ethylene glycol ingestion.

The increase in serum osmolality for each 100mg/dl of ethylene glycol is 16.3mosm/liter; the normal range for serum osmolality is 285-295mosm/liter. Therefore, 100mg/dl of ethylene glycol will yield a serum osmolality of about 306mosm/liter.

The metabolism of ethylene glycol may help to explain acidosis, and acute renal failure. The pathway for metabolism of ethylene glycol is shown in the next Figure; (Levinsky, N.G., Discussant in Case Records of the Mass. Gen. Hosp., N. Engl. J. Med. 301, 650-657, 1979):

Metabolism of Ethylene Glycol

ethylene glycol
↓
glycoaldehyde
↓
glycolic acid
↓
glycine ← glyoxylic acid
↓   ↓
hippurate   oxalic acid

Acidosis is due to direct acidifying effect of acid breakdown products of ethylene glycol and to the effects of the toxic metabolic products resulting in production of lactic acid. Oxalate and hippurate crystals may be seen in the renal collecting system. Appearance of oxalate crystals in the urine strongly suggests the diagnosis.

The clinical symptoms of ethylene glycol toxicity are as follows: During the first 12 hours, CNS symptoms predominate; in the first hour or so, the patients may appear drunk; coma and convulsions, hypertension and leukocytosis may follow during the first 12 hours. The cerebrospinal fluid shows leukocytosis and xanthochromia. Over the next 12 to 24 hours, the patient may develop hypertension and cardiopulmonary failure with evidence of cardiac enlargement, congestive heart failure, pulmonary edema, and bronchopneumonia. Other effects include acute oliguric renal failure and myopathy.

EUGLOBULIN LYSIS TIME

SPECIMEN: Blue (citrate) top tube with liquid anticoagulant; place tube on ice and transport to coagulation laboratory immediately. Do not freeze. Test must be performed immediately. The specimen is centrifuged in the cold and the plasma removed.

REFERENCE RANGE: 120-240 minutes

METHOD: Plasma is treated with acetic acid and refrigerated for 15 minutes or until precipitate forms; it is centrifuged in the cold. The precipitate (euglobulins) contain fibrinogen, plasminogen, plasminogen activator and plasmin. The euglobulins fraction of plasma is that portion which precipitates at low pH and decreased ionic strength. The supernatant is decanted and saline-borate buffer is added to dissolve the precipitate. Thrombin is added to the solution and the resultant clot is incubated at 37°C and checked after 10, 20 and 30 minutes and every subsequent half-hour period. Time of lysis is recorded.

INTERPRETATION: This test is mainly a measure of plasminogen and plasminogen activator; the fibrinolytic system is illustrated in the next Figure:

Fibrinolytic System

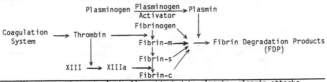

Plasminogen is proteolytically activated to plasmin; plasmin attacks fibrinogen and fibrin to produce fibrin degradation products(FDP).

This test is used in monitoring urokinase and streptokinase therapy; (Schafer, A.I., Patient Care, pgs. 87-115, Nov. 30, 1982). Urokinase is a pharmacologic activator of plasminogen, directly stimulating the conversion of plasminogen to plasmin. Streptokinase acts indirectly initially combining with plasminogen to form a streptokinase-plasminogen complex. This complex then activates plasminogen to form plasmin. Thus, the end result is the same with either drug-increased generation of plasmin. Aminocaproic acid (Amicar) neutralizes the effects of streptokinase or urokinase.

This test is also used in evaluation of abnormal fibrinolysis.

A shortened lysis time indicates activation of the fibrinolytic system.

EXCRETION FRACTION OF FILTERED SODIUM

SPECIMEN: Urine: 10ml aliquot of random urine specimen; no preservative is
necessary; this urine specimen will be used for assay of sodium and creatinine.
Blood: Red top tube, separate serum; this specimen will be used for the assay of
sodium and creatinine.
REFERENCE RANGE: See interpretation below.
METHOD: Flame photometry or ion-specific electrodes for sodium; alkaline
picrate is the usual method for creatinine.
INTERPRETATION: This is a sensitive and specific test for acute tubular
necrosis (Espinel, C.H. and Gregory, A.W., Clin. Nephrol. 13, 73-77, 1980). The
measurements that are done are as follows: Serum and urine sodium, serum and
urine creatinine; no timed specimen is necessary. The excretion fraction of
filtered sodium is given by the next equation:

$$\text{Excretion Fraction of Filtered Sodium} = \frac{\text{Urine Na}^+ \times \text{Plasma Creatinine}}{\text{Plasma Na}^+ \times \text{Urine Creatinine}} \times 100$$

The results of the test for excretion fraction of the filtered sodium
($F_{E_{Na}}$ Test) for patients with acute tubular necrosis and those with prerenal
azotemia are illustrated in the next Figure:

Excretion Fraction of Filtered Sodium ($F_{E_{Na}}$ Test) for Patients with
Acute Tubular Necrosis and Patients with Prerenal Azotemia

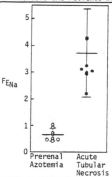

The two conditions are clearly distinguished by the $F_{E_{Na}}$ values.
Patients with acute tubular necrosis present with an $F_{E_{Na}}$ of more than 2.
Patients with prerenal azotemia present with an $F_{E_{Na}}$ of less than 1.
Levels of $F_{E_{Na}}$ in different conditions are shown in the next Table:

Levels of $F_{E_{Na}}$ in Different Conditions	
Low $F_{E_{Na}}$	High $F_{E_{Na}}$
Prerenal Azotemia	Acute Tubular Necrosis
Acute Glomerulonephritis	Urinary Obstruction
Hepatorenal Syndrome	Chronic Uremia
	Diuretics

In acute glomerulonephritis and hepatorenal syndrome, the renal tubule
reabsorbs sodium avidly.
In acute tubular necrosis, functioning nephron units excrete a large
fraction of the filtered sodium. In chronic uremia, sodium hemostasis is
maintained by a reduction in tubular sodium reabsorption. Inhibition of sodium
reabsorption by diuretics results in a high $F_{E_{Na}}$.

EYE CULTURE AND SENSITIVITY

<u>SPECIMEN:</u> Swab and Culturette; do <u>not</u> refrigerate.
<u>REFERENCE RANGE:</u> <u>Normal flora:</u> The most common are Staphylococcus epidermidis and S. Aureus.
<u>METHOD:</u> Usual culture techniques.
<u>INTERPRETATION:</u> The causes of ocular infections are given in the next Table; (Barza, M. and Baum, J., Medical Clinics of North America, <u>67</u>, 131-152, 1983):

Causes of Ocular Infections

Site	Organisms
Eyelids - Blepharitis	Staphylococci Epidermidis
Stye (Infection of Superficial Sebaceous glands of the Eyelid)	Staphylococci Epidermidis
Chalazion (Infection of a Large Sebacious Gland of the Eyelid)	Staphylococci Epidermidis
Impetigo	Staphylococci Aureus and/or Streptococcus Pyogenes
Other Types of Bacterial Blepharitis (Rare)	Moraxella Lacunata Escherichia Coli Pseudomonas Species Proteus Species
Spread from Other Areas	S. Aureus, Streptococcus Pyogenes, H. Influenzae in Younger Children, Pseudomonas Aeruginosa
Candidal Blepharitis	Primarily in Immunosuppressed Patients
Conjunctivitis	
Viral	Adenoviruses, Types 8 and 19 (Epidemic Keratoconjunctivitis, EKC) Herpes Simplex Virus (HSV)
Bacterial	Streptococcus Pneumoniae Streptococcus Pyogenes Hemophilus Influenzae Staphylococcus Aureus Staphylococcus Epidermidis Enteric Bacilli and Moraxella Niesseria Gonorrhoeae Chlamydial Infection
Conjunctivitis in Neonates	Chlamydial, Gonococcus and Staphylococcus

FACTOR VIII ASSAY (SEE COAGULATION FACTOR VIII ASSAY)

FASTING, DIET, EXERCISE AND LABORATORY TESTS

The laboratory tests that are commonly done and that require that the blood specimen be drawn in the <u>fasting state or consideration given to diet</u> are as follows:

1. Serum <u>glucose</u>, <u>triglycerides</u> and <u>potassium</u> are <u>elevated</u> after a meal.
2. Serum <u>organic phosphorus</u> is <u>decreased</u> after a meal; phosphorus follows glucose into the cells.
3. <u>Serum Lipids and Electrolytes</u>: In circumstances where there are large amounts of triglyceride in blood, such as in patients with diabetes, nephrotic syndrome, acute pancreatitis, Type IV or V hyperlipidemias, the <u>electrolytes</u> may be spuriously <u>low</u>; in these states, the lipid volume is <u>expanded</u> and the water volume is <u>contracted</u>. In hyperproteinemia, such as in multiple myeloma, the <u>volume</u> occupied by <u>protein</u> is <u>expanded</u> and the water volume is contracted; thus, electrolytes may be spuriously low. <u>The electrolytes are only present in the aqueous of plasma.</u>

For example if in 1000 ml of plasma, there is 800 ml of water and 200 ml of lipid, and the measured plasma sodium concentration is 120 mmol/liter, then the <u>true plasma sodium is 150 mmol/liter; this is illustrated in the next Figure:</u>

Sodium Ion in Hyperlipidemia

Measured Sodium Ion = 120 mmol/liter
However, the 120 mmol is actually in 800 ml.
Therefore:

$$\frac{120 \text{ mmol}}{800 \text{ ml}} = \frac{X}{1000 \text{ ml}}$$

$$X = \frac{600}{4} = 150 \text{ mmol in } 1000 \text{ ml}$$

The results may be spurious when using ion selective electrodes(ISE) electrolyte analyzers, such as Technicon SMAC and the Beckman Astra, which <u>dilute</u> the plasma <u>before</u> it is assayed by the electrodes. However, electrolyte analyzers such as the Nova Biomedical, Nova 1, which <u>do not</u> dilute plasma, will give a more meaningful result.

When the electrode is used without diluting the sample (direct potentiometry) lipids would not be expected to interfere with the value because the activity of sodium only in the water phase is detected (Ladenson, J.H. et al., Ann. Int. Med. <u>95</u>, 707-708, 1981.)

4. A <u>highly unsaturated diet</u> will yield an increase in <u>serum cholesterol</u>.
5. <u>Hyperbilirubinemia of Fasting</u>: Following a 48 hour fast, the total serum bilirubin concentration increases progressively during the second day of the fast; it averaged 240 percent for normal subjects (Barrett, P.V.D., JAMA <u>217</u>, 1349-1353, 1971).
6. A diet <u>extremely high in protein</u> will result in a <u>slight elevation in serum</u> and urine urea(BUN), urate and ammonia.

<u>EXERCISE</u>:

<u>Exercise</u> causes <u>increases</u> in the activities of <u>muscle enzymes</u>; the changes in enzyme activity at various times following one hour of strenuous exercise are shown in the next Table; (King, S. et al., Clin. Chem. Acta, <u>72</u>, 211, 1976):

Changes in Enzyme Activity of Muscle Enzymes After Exercise for One Hour

Enzyme	Percent Increase				
	1 Hour	5 Hours	11 Hours	29 Hours	43 Hours
Creatine Phosphokinase(CPK)	40	80	120	80	30
Serum Glutamate Oxalacetate Transaminase(SGOT)	20	40	20	20	2
Lactate Dehydrogenase(LDH)	20	20	20	10	2

FAT, FECES (see FECAL FAT)

FAT MALABSORPTION PANEL

The causes of fat malabsorption are given in the next Table:

Causes of Fat Malabsorption
(1) Deficiency of Pancreatic Digestive Enzymes: 　　Chronic Pancreatitis 　　Cystic Fibrosis 　　Pancreatic Carcinoma 　　Pancreatic Resection
(2) Impairment of Intestinal Absorption: 　　Celiac Disease (Coeliac Disease, Non-Tropical Sprue, 　　　Gluten-Sensitive Enteropathy) 　　Rare Causes: 　　　Tropical Sprue 　　　Abetalipoproteinemia 　　　Lymphangiectasis 　　　Intestinal Lipodystrophy 　　　Amyloidosis 　　　Lymphoma 　　　Surgical Loss of Functional Bowel
(3) Deficiency of Bile: 　　Extrahepatic Bile Duct Obstruction 　　Intrahepatic Disease 　　Cholecystocolonic Fistula

The causes of steatorrhea in 47 elderly patients are given in the next Table (Price, H.L., Gazzard, B.G., Dawson, A.M. "Steatorrhea in the Elderly", Brit. Med. J., 1, pp. 1582-1584, June 18, 1977):

Cause	Age <65	Age >65	Total
Celiac Disease	12	4	16
Pancreatic Insufficiency			
Carcinoma	4	0	4
Other	3	7	10
Postgastrectomy	6	2	8
Jejunal Diverticula	1	1	2
Tropical Sprue	0	2	2

The common causes of steatorrhea are celiac disease and pancreatic insufficiency; other causes of steatorrhea are "collagen" disease, diabetes mellitus, scleroderma and Whipple's disease; all these occurred in patients less than age 65. Patients with conditions associated with deficiency of bile were not included in the population examined by the investigators.

Tests for Malabsorption and Steatorrhea: Tests for malabsorption are as follows:

(a) Fecal Fat (72 Hour Collection): More than 5g/24 hours represents steatorrhea.

(b) Microscopic Examination of the Stool Fat: Sudan III or Oil Red O are used to stain fat globules; the sensitivity is 72% and the specificity is 95%; (Bin, T.L. et al., J. Clin. Pathol. 36, 1362-1366, 1983).

(c) Trypsin, Immunoreactive, Serum: Serum immunoreactive trypsin is decreased in pancreatic insufficiency with associated steatorrhea; serum immuno-reactive trypsin is normal in pancreatic insufficiency without steatorrhea and in steatorrhea with normal pancreatic function; (Jacobson, D.G., N. Engl. J. Med. 310, 1307-1309, May 17, 1984).

(d) Serum Carotene: (See Carotene)

(e) Glycerol Tri [1-^{14}C] Oleate: The measurement of $^{14}CO_2$ breath excretion, after the administration of a 60g fat meal containing glycerol tri [1-^{14}C] oleate, is useful in evaluating fat malabsorption. Subjects with fat malabsorption have reduced respiratory excretion of $^{14}CO_2$ (West, P.S. et al., Brit.Med.J. 282, 1501-1504, May 9, 1981). The simplicity and convenience of breath analysis make it an attractive alternative to analysis of fecal fat excretion in screening for fat malabsorption.

A two-stage triolein breath test may be used to differentiate pancreatic insufficiency from other causes of malabsorption (Goff, J.S., Gastroenterology 83, 44-46, 1982). Patients are given labeled triolein before and after ingestion of exogenous pancreatic enzymes. Labeled CO_2 excretion increases following ingestion of pancreatic enzymes when steatorrhea is secondary to pancreatic disease.

(f) <u>Double-Labeled Schilling Test:</u> Patients are given equal proportions of: Co-58-labeled cobalamin bound to intrinsic factor. A lack of trypsin leads to an inability to absorb cobalamin bound to R protein. Therefore, the ratio of urinary ^{58}Co to urinary ^{57}Co is a measure of exocrine pancreatic function. A low ratio indicates exocrine pancreatic insufficiency.

(g) <u>D-Xylose Absorption Test:</u> (see Xylose)

(h) <u>Sweat Electrolytes for Cystic Fibrosis:</u> (see Sweat Test)

(i) <u>Intestinal Biopsy for Intestinal Disorders</u>

(j) <u>Trypsin in Stool</u>

(k) <u>Glucose Tolerance Test for Pancreatic Deficiency</u>

(l) <u>Breath Test for Lactose Intolerance</u>

Laboratory Findings in Malabsorption: Laboratory findings reflecting malabsorption are given in the next Table:

Laboratory Findings Reflecting Malabsorption
Fat Malabsorption
Vitamin D Deficiency
Serum Calcium Decreased
Secondary Hyperparathyroidism, Alkaline Phosphatase Increased,
(Osteomalacia)
Serum Cholesterol Decreased
Folic Acid Deficiency
Megaloblastic Anemia
Lactate Dehydrogenase (LDH) Increased
Protein Malabsorption
Total Protein Decreased
Albumin Decreased
Vitamin K Deficiency
Prolonged Prothrombin Time (PT)
Vitamin A Deficiency
Serum Carotene Decreased
Retinol Binding Protein Decreased

Deficiency of fat soluble vitamins (A, D, E and K) may accompany malabsorption.

FEBRILE AGGLUTININS (TYPHOID, PARATYPHOID, BRUCELLA)

SPECIMEN: Collect acute phase specimen. Use a red top tube, separate serum and freeze serum. Then, collect convalescent-phase sera, ten to fourteen days later. Use red top tube, separate serum and freeze.

REFERENCE RANGE: Less than fourfold increase in titer between the acute and convalescent serum titers.

METHOD: Slide and tube agglutination tests

INTERPRETATION: Tests for febrile agglutinins have been widely used in the past in screening patients with fever of unknown origin (FUO). However, the need for these screening tests has diminished significantly in recent years for the following reasons: sensitivity and specificity often low; improved understanding of these diseases; decreased incidence of these diseases; improved ability to culture the causative organisms; and greater sensitivity and specificity of selected antibody tests. The antigens commonly used in this test to detect antibodies are listed in the next Table:

Antigens Commmonly Used in Tests for Febrile Agglutinins	
Infections	Antigens
Salmonella	Typhoid O and H Paratyphoid A and B
Rickettsial Infection	Proteus OXK, OX2, OX19
Brucellosis	Brucella

Salmonella Infections: This test is useful when organisms in parenteral tissue invade the blood stream and stimulate the production of antibody; as a corollary, if blood invasion does not occur, there will be no antibody production.

The antibody response that characteristically occurs in salmonella infection is as follows: Antibodies to O Antigen: Titers of O agglutinins in patients with Salmonella disease are elevated in 50% of patients by the end of the first week and in 90 to 95% of patients by the fourth week peaking in the sixth week and falling or disappearing within six to 12 months. Antibodies to H Antigen: Titers of H agglutinins in patients with Salmonella disease are detected later, peak later, and remain elevated for several years. Antibody to H antigen may be detected for years following immunization.

Specificity is poor in that antigens of salmonellae are shared by enterobacteriaceae, and there may be nonspecific stimulation of Salmonella agglutinins by febrile disease of other cause.

The most appropriate test for Salmonellae infection is culture.

Rickettsial Infection: The basis of this test is that Proteus antigens are shared by R. rickettsii, R. prowazekii, R. mooseri and R. tsutsugamushi and infections by these Rickettsial agents often elicit antibodies that react with Proteus antigens. This is a very insensitive test in that infection by Rickettsial agents may not elicit antibodies that react with Proteus antigens and antibodies may be produced by other infections e.g. Proteus, spirochetal, brucellosis, tularemia and others.

Brucellosis: This test is more sensitive than the two previous tests but is less specific in that other conditions produce antibodies that react with Brucella antigen.

Ref. Fuchs, P.C., Medical Laboratory Observer, pgs 15-17, April, 1983.

FECAL EXAMINATION FOR OCCULT BLOOD

SPECIMEN: Hemoccult slide test for occult blood (Smith Kline Corp.):
(1) Collect a very small stool specimen on tip of wooden applicator.
(2) Apply thin smear of specimen inside the circle.
(3) Close cover; dispose of applicator.
(4) Allow specimen to dry completely.
(5) Open perforated window in back of slide.
(6) Apply two or three drops of developing solution to slide opposite specimen.
(7) Read results after 30 seconds.

REFERENCE RANGE: Positive: Trace of blue. Negative: No detectable blue color. Hemoccult begins to turn positive in the presence of about 5.0mg of hemoglobin per gram of feces. Some specimens may turn from positive to negative after storage for two days.

METHOD: A stool specimen is applied to guaiac-impregnated filter paper. In the presence of hemoglobin and hydrogen peroxide the reaction occurs as follows:

$$Guaiac + H_2O_2 \xrightarrow{\text{Heme}} Blue\ Color + H_2O$$

The heme moiety of hemoglobin has pseudoperoxidase activity and catalyzes the oxidation of guaiac by hydrogen peroxide to a blue color.

Methods, other than the guaiac procedure, have been developed (Welch, C.L. and Young, D.S., Clin. Chem. 29 2022-2025, 1983; Schwartz, S. et al., Clin. Chem. 12, 2061-2067, 1983).

INTERPRETATION: Colorectal cancer is one of the most common tumors; it is the second leading cause of cancer death in females and the third leading cause of cancer death in males in the United States. One hundred and twenty-thousand new cases are diagnosed yearly. The test for occult blood in the feces is done to help detect colorectal cancer.

FECAL FAT, SEMI-QUANTITATIVE (SUDAN III STAIN, FECAL FAT STAIN , FATTY ACID, NEUTRAL FAT)

SPECIMEN: Fresh random stool (semi-quantitative) or 72 hour stool (quantitative)

REFERENCE RANGE: Up to 5 grams/24 hours.

METHOD: Semi-Quantitative: Sudan III or Oil Red O. This test has a sensitivity of 72% and specificity of 95%; (Bin, T.L. et al., J. Clin. Pathol. 36, 1362-1366, 1983). Quantitative: Saponification, acidification, extraction of lipids with petroleum ether, vaporization of ether and weigh residue.

INTERPRETATION: This test is done in the work-up of patients for possible steatorrhea; steatorrhea refers to a pathological increase in stool fat. The major causes of steatorrhea are listed in the next Table:

Causes of Steatorrhea
(1) Deficiency of Pancreatic Digestive Enzymes:
Chronic Pancreatitis
Cystic Fibrosis
Pancreatic Carcinoma
Pancreatic Resection
(2) Deficiency of Bile in Intestinal Lumen:
Extrahepatic Bile Duct Obstruction
Intrahepatic Disease, e.g., Primary Biliary Cirrhosis
Cholecystocolonic Fistula
Intestinal Stasis Syndrome
(3) Impairment of Intestinal Absorption:
Celiac Disease (Nontropical Sprue, Gluten-Sensitive Enteropathy)
Postgastrectomy
Tropical Sprue
Lymphangiectasis
Intestinal Lipodystrophy
Amyloidosis
Lymphoma
Surgical Loss of Functional Bowel

The major causes of steatorrhea are: (1) deficiency of pancreatic digestive enzymes, (2) deficiency of bile and (3) impairment of intestinal absorption.

FECAL LEUKOCYTES (see STOOL LEUKOCYTES)

FERRIC CHLORIDE TEST FOR AMINOACIDURIA

SPECIMEN: Random urine; an early morning specimen is preferred to reduce variations due to diet. If not analyzed immediately, the urine should be acidified to a pH less than 4; if necessary, freeze at -20°C for no longer than one week.

REFERENCE RANGE: Negative, that is, no color change.

METHOD: Add 10% FeCl$_3$ by drops to 1-2ml of urine

INTERPRETATION: The ferric chloride test is non-specific and gives color reactions with metabolites from amino acid disorders, other metabolites and drugs as shown in the next Table:

Ferric Chloride Test in Urine	
Disorder and Urinary Product	Color Change
Phenylketonuria (PKU)	
Phenylpyruvic Acid	Blue or Blue-Green, Fades to Yellow
Tyrosinosis(emia)	
p-Hydroxyphenylpyruvic Acid	Green, Fades in Seconds
Alkaptonuria	
Homogentisic Acid	Blue or Green, Fades Quickly
Maple Syrup Urine Disease	
Alpha-Ketoisovaleric Acid	Blue
Alpha-Ketoisocaproic Acid	Blue
Alpha-Keto-Beta-Methyl Valeric Acid	Blue
Histidinemia	
Imidazole Pyruvic Acid	Green or Blue-Green
Diabetics, Alcoholics, Starvation	
Acetoacetic Acid	Red or Red Brown
Other Products:	
Bilirubin	Blue-Green
Ortho-Hydroxyphenyl Acetic Acid	Mauve
Ortho-Hydroxyphenyl Pyruvic Acid	Red
Alpha-Ketobutyric Acid	Purple; Fades to Red Brown
Pyruvic Acid	Deep Gold-Yellow or Green
Xanthurenic Acid	Deep Green, Later Brown
Drugs:	
Salicylates	Stable Purple
Aminosalicylic Acid	Red-Brown
Phenothiazine Derivatives	Purple Pink
Antipyrines and Acetophenetidines	Red
Phenol Derivatives	Violet
Cyanates	Red

FERRITIN, SERUM

SPECIMEN: Red top tube, separate serum.
REFERENCE RANGE: Females, 13-125ng/ml; males, 27-220ng/ml; children (6 months
to 15 years), 13-145ng/ml; infants, 13-220ng/ml; iron deficiency, <12ng/ml; iron
overload, >220ng/ml (Skikne, B.S. and Cook, J.D., Laboratory Management, pgs.
31-35, May 1981). To convert conventional units in ng/ml to international units
in mcg/liter, multiply conventional units by 1.00.
METHOD: RIA or enzyme immunoassay
INTERPRETATION: The conditions in which serum ferritin measurements are of
value are given in the next Table; (Skikne, B.S. and Cook, J.D., Laboratory
Management, pgs. 31-35, May, 1981):

Conditions in Which Serum Ferritin Measurements Are of Value
Detection of Iron Deficiency
Detection of Response and/or Endpoint to Oral Iron Therapy
Differentiation of Anemia of Chronic Disease from Iron Deficiency Anemia
Monitoring Iron Status of Patients with Chronic Renal Disease
Detection of Iron Overload
Monitoring the Rate of Iron Accumulation in Iron Overload
Monitoring the Response to Iron Chelation Therapy in Iron Overload
Determination of Iron Status of a Population and Response to Iron Fortification

The causes of altered serum ferritin are given in the next Table:

Causes of Altered Serum Ferritin	
Decreased	Increased
Iron Deficiency	Iron Overload:
	Hemochromatosis
	Transfusion Hemosiderosis
	Inflammation
	Chronic Liver Disease

Iron Deficiency Anemia: The first biochemical change in iron deficiency is a low
serum ferritin level; this occurs before iron is decreased and before
morphologic abnormalities appear in red blood cells. Iron deficiency and no
other disease is associated with a serum ferritin level less than 10ng/ml.
     Serum ferritin level is helpful in differentiating between iron deficiency
anemia and anemia of chronic inflammation, infection or chronic disease. In
iron deficiency anemia and in anemia associated with chronic conditions, serum
ferritin is low (below 10ng/ml). However, anemia associated with infection and
malignancy, serum ferritin is above 10ng/ml; a serum ferritin level lower than
50ng/ml in a patient with obvious inflammatory disease is a strong indicator of
iron deficiency.
     In the presence of liver disease, serum ferritin is increased; in iron
deficiency anemia plus liver disease, serum ferritin may be normal.
     In patients with iron deficiency anemia who are being treated with iron
orally, serum ferritin measurements are useful in monitoring the response to
therapy and in determining the time when iron should be discontinued; a normal
hemoglobin does not necessarily indicate that the body iron stores have been
replenished. Serum ferritin assays should be done at 3 to 4 week intervals
until serum ferritin rises above 50ng/ml (body iron stores of about 400mg).
(Skikne, B.S. and Cook, J.D., Laboratory Management, pgs. 31-35, May 1981.)
     Serum ferritin reflects iron stores if the patient does not have liver
disease or an acute inflammatory reaction that may produce spurious elevation of
serum ferritin concentration; 1.0ng/ml of serum ferritin is equivalent to 8mg of
storage iron. Total iron stores are 0.7 to 1.5g. About 15 to 20 percent of
iron is stored within ferritin. Serum ferritin does not reflect bone marrow
stores, examples: patients treated with iron; dialysis patients; uremia with
depression of erythropoiesis.
Hemochromatosis: Serum ferritin is usually elevated in patients with
hemochromatosis. During the development of hemochromatosis, iron is confined
mainly to liver parenchymal cells and in the absence of cellular damage,
relatively little ferritin is released.

Not all patients with hemochromatosis have elevated serum ferritin; in a study of homozygous patients with hemochromatosis 23 of 32 patients had elevated serum ferritin (Edwards, C.Q. et al., Ann. Intern. Med. 93, 519-525, 1980).

Transferrin saturation is a more sensitive test for hemochromatosis than is serum ferritin (Edwards, C.Q. et al., Ann. Intern. Med. 93, 519-525, 1980). The three tests that are used in the diagnosis of hemochromatosis are given in the next Table; (Gollan, J.L., Patient Care, pgs. 102-119, Sept. 30, 1982):

Tests in Diagnosis of Hemochromatosis			
Test	Abnormal Value	False Positive Frequency	False Negative Frequency
Serum Iron	>175mcg/dl	10%	24%
Transferrin Saturation	>50%	33%	0%
Serum Ferritin	>250ng/ml	2%	3%

Characteristics of Ferritin: Ferritin is the main storage molecule for iron. It is present in highest concentration in the reticuloendothelial cells of the liver and spleen and in the erythroblasts of bone marrow. It consists of two basic constituents - a protein shell (apoferritin) and a central core of iron in the ferric form existing as heterogenous ferric oxyhydroxide micelles with some phosphate.

FETAL HEMOGLOBIN (see HEMOGLOBIN F)

FETAL HEMOGLOBIN IN MATERNAL BLOOD; SCREEN:
THE ROSETTE METHOD

SPECIMEN: Purple (EDTA) top tube; for best results, examine fresh specimen immediately.
REFERENCE RANGE: Negative
METHOD: Ortho Diagnostics: Fetal Screen: A suspension of maternal red blood cells is incubated with anti-Rho(D) typing serum and then washed and incubated with an Rho(D)-positive indicator red blood cell in combination with a potentiating medium. If any Rho(D)-positive (fetal) red blood cells are present in the maternal sample, rosettes will be formed. Rosetted red blood cells are quickly and easily detected microscopically.
INTERPRETATION: This is a screening test for fetal-maternal hemorrhage. It is recommended that all Rho(D)-negative women who give birth to Rho(D)-positive babies be screened by rosetting technique (Taswell, H.F. and Reisner, R.F. Mayo Clin. Proc. 58, 342-343, 1983). This method easily detects a fetal-maternal hemorrhage of 15ml and can detect as little as 5ml of fetal red blood cells in maternal blood; there is a false-positive result of 1.5% and 100 sensitivity for detection of hemorrhages of more than 5ml.

FETAL HEMOGLOBIN STAIN (BETKE-KLEIHAUER STAIN)

SPECIMEN: Blood: Purple(EDTA) top tube; for best results, examine fresh specimen immediately; otherwise refrigerate. Amniotic Fluid: No preservative; send to laboratory for analysis immediately; do not refrigerate.

REFERENCE RANGE: Negative

METHOD: Dilute blood sample with saline, 1:1; if hematocrit is low, saline dilution is not necessary. Prepare thin film of the diluted capillary or fresh EDTA blood. Air dry slides and fix with 80% alcohol in order to precipitate hemoglobin within red cells. Treat slides with citric acid-phosphate buffer, pH 3.3; at this pH adult hemoglobins are soluble and are dissolved out of the red cells while HgF remains precipitated and is next stained with eosin. The cells that contained HgA appear as "ghosts" while the HgF cells stain red. Five hundred cells are counted under high dry magnification.

INTERPRETATION: This test is most often used as a quantitative measure of Rho(D) cells in the maternal circulation; use Fetal Hemoglobin Test as a screen. There are three different patterns; these are illustrated in the next Figure:

Patterns of HgF and HgA in Red Cells and Examples

Mother's Blood: Fetal-maternal hemorrhage; 50-90% of fetal RBC's contain HgF. Collect maternal blood immediately following delivery if newborn has anemia of newborn or when a mother is Rh negative or Du-negative. In the case of a full-term delivery, the red blood cells of the infant must be Rho(D) positive and the direct antiglobulin test (for anti-Rho(D)) negative for the mother to be a candidate for RhIG. The amount of fetal blood that has escaped into the maternal circulation is roughly calculated using the formula:

$$\text{Fetal Blood (ml)} = \% \text{ HgF cells} \times 50; \quad \% \text{ HgF} = \frac{\text{No. of Fetal Cells}}{\text{No. of Maternal Cells}} \times 100$$

$$\text{Vials of Rho(D) Immune Globulin Needed (RhIG)} = \frac{\text{ml of Fetal Blood}}{30}$$

One vial (300 microgram) of RhIG will suppress the immunization of 15ml of fetal red blood cells or 30 ml of whole blood hemorrhage; two vials will suppress the immunization of 2 x 15ml or 30ml, etc. Many blood bankers recommend doubling the calculated dose of RhIG since the method of calculating fetal bleed is not accurate and the results of undertreatment are serious but the effects of "overdose" are minor. If the calculated number is three or more, add one additional vial. A minidose vial has been standardized to suppress the immunogenic challenge of 2.5ml of Rho(D) positive red cells (Blood Component Therapy, a Physicians Handbook, Third edition, 1981, Am. Ass. Blood Banks, 1925 L Street, N.W. Suite 608, Wash. D.C. 20036).

Even distribution of HgA + HgF (Heterozygous Form)

Hereditary Persistance of HgF (2 Alpha Chains, 2 Gamma Chains):

Homozygous Form: 100 percent HgF

Heterozygous Form: Hemoglobin F (15 to 35%) and is evenly distributed throughout the red cells.

Uneven Distribution of HgA + HgF: Thalassemia minor and major, HgS-S disease, Fanconi Anemia and heriditary spherocytosis.

S. Bakerman

FIBRIN SPLIT PRODUCTS (FSP)

SPECIMEN: Obtain specimen before starting heparin therapy; red top tube, collect blood in dry syringe; inject blood into FSP sample tube. Immediately mix by inverting the tube gently several times; the blood will clot firmly within a few seconds. Do not shake the tube, transport to laboratory as soon as possible. Do not put specimen on ice. The FSP sample tube contains thrombin and proteolytic inhibitor.

REFERENCE RANGE: <10mcg/ml

METHOD: Fibrinogen fragments D and E antibody bound to latex particles (Burroughs-Wellcome)

INTERPRETATION: The conditions associated with increased fibrin split products are those associated with disseminated intravascular coagulopathy (DIC); see Disseminated Intravascular Coagulopathy(DIC) Profile.

FIBRINOGEN

SPECIMEN: Blue (citrate) top tube with liquid anticoagulant, fill tube to capacity; separate plasma; freeze plasma if determination is not done immediately.

REFERENCE RANGE: 200-400mg/dl. To convert conventional units in mg/dl to international units in g/liter, multiply conventional units by 0.01.

METHOD: Light scattering; gel diffusion; antisera; functional ability to form a clot.

INTERPRETATION: Fibrinogen is decreased in the plasma in the following conditions:

Causes of Decreased Plasma Fibrinogen	
Acquired Deficiencies:	
Liver Disease	
Disseminated Intravascular Coagulopathy(DIC)	
Clinical Conditions Associated with DIC:	
Mechanism	Clinical Condition
Infection	Bacterial especially Gram-Negative Sepsis and Meningococcemia, Viral, Protozoal, Rickettsial especially Rocky Mountain Spotted Fever
Neoplastic	Mucin-Secreting Adenocarcinoma, Acute Leukemia especially Promyelocytic Leukemia, Carcinoma of the Prostate, Lung and Other Organs
Tissue Damage	Trauma, Surgery especially Prostatic Surgery, Heat Stroke, Burns, Dissecting Aneurysms
Obstetric	Obstetrical Complications: Abruptio Placentae, Amniotic Fluid Embolism, Retained Fetal Products, Eclampsia
Immunological	Immune-Complex Disorders, Allograph Rejection, Incompatible Blood Transfusion, Anaphylaxis
Metabolic	Diabetic Ketoacidosis
Miscellaneous	Shock, Snake Bite, Cyanotic Congenital Heart Disease, Fat Embolism, Severe Liver Disease, Cavernous Hemangioma
Congenital Hypofibrinogenemias: Afibrinogenemia; Hypofibrinogenemia; Dysfibrinogenemia	

In hypofibrinogenemia, both PT and PTT are prolonged; the bleeding time is normal.

Hypofibrinogenemia can be distinguished from dysfibrinogenemia by RIA.

Fibrinogen is increased in the conditions listed in the next Table:

Causes of Increased Plasma Fibrinogen
Pregnancy
Oral Contraceptives
Acute Inflammation and Tissue Damage

Functionally, fibrinogen serves as the substrate of the proteolytic enzyme, thrombin, as illustrated in the next Figure:

Actions of Thrombin on Fibrinogen

Thrombin acts on fibrinogen and on factor XIII.

204

FLUORESCENT TREPONEMAL ANTIBODY-ABSORBED TEST (FTA-ABS) (see also SYPHILIS, TEST FOR ANTIBODIES)

<u>SPECIMEN:</u> Red top tube, separate serum
<u>REFERENCE RANGE:</u> Negative
<u>METHOD:</u> Antigen derived from T. pallidum is used.
<u>INTERPRETATION:</u> The FTA-ABS test is a test for both IgG and IgM <u>antibodies</u> in the serum of patients with syphilis; it is <u>not useful</u> for detection of antibodies in the CSF. This test is <u>most commonly used</u> to determine whether the results of a <u>non-treponomal test</u> are due to <u>syphilis</u> or <u>due to</u> a <u>condition</u> causing a <u>false-positive</u>. This test may <u>also</u> be used to <u>detect syphilis</u> in patients with <u>negative non-treponemal test</u> results but with clinical evidence of late syphilis. The reactivity of non-treponemal and treponemal tests in untreated syphilis is given in the next Figure; (Henry, J.B. [ed.] Todd-Sanford-Davidsohn, Clinical Diagnosis by Laboratory Methods, 16 ed., 1979, pg. 1890, Vol. II, W. B. Saunders Co., Phila. Pa):

Reactivity of Nontreponemal and Treponemal Tests in Untreated Syphilis

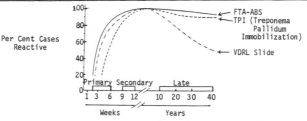

<u>Effect of Therapy:</u> The TPI and FTA-ABS test results remain positive following therapy in most patients (Schroeter, A.L. et al., JAMA <u>221</u>, 471, 1972). With successful therapy, the titer of VDRL will tend to fall and will become negative in at least two-thirds of patients if <u>treatment</u> is given in the <u>primary</u> or <u>secondary stages</u>. However, treatment during late syphilis is infrequently associated with reversion of a reactive VDRL to negative.

The TPI (treponema pallidum immobilization) test is the standard against which all other serologic tests for syphilis are measured. The test procedure is technically difficult and requires maintenance of "cultures" of live Treponema pallidum; few research laboratories still perform this test.

<u>False positive</u> reactions with the FTA-ABS test occur in the <u>conditions</u> listed in the next Table:

False Positive Reactions with the FTA-ABS Test
Increased or Abnormal Globulins
Lupus Erythematosus
Acute Genital Herpes
Pregnancy
Pinta, Yaws and Bejel

An IgM-specific FTA-ABS test is available but it is a relatively difficult test to perform.

<u>Ref.:</u> Wiesner, P.J., JAMA, <u>244</u>, 1147 (1980).
Schmidt, B.L., Sexually Transmitted Diseases <u>7</u>, 53-58 (1980).
Felman, Y.M. and Nikitas, J.A., Arch. Dermatol. <u>116</u>, 84-89 (1980).

S. Bakerman

FOLATE (FOLIC ACID), RED BLOOD CELL

SPECIMEN: Assay usually includes serum folate. Two blood specimens are required: red top tube and a lavender (purple)(EDTA) top tube for red cell folate and hematocrit. Red top tube, separate serum and freeze serum. Lavender top tube, mix by inverting at least six times; add 0.5 ml of EDTA blood to a vial containing 4.5 ml of freshly prepared 1% ascorbic acid solution; then freeze hemolysate.

REFERENCE RANGE: 200-700 ng/ml. To convert conventional units in ng/ml to international units in nmol/liter, multiply conventional units by 2.266.

METHOD: RIA

INTERPRETATION: RBC folate in conjunction with serum folate is a good index for diagnosing folate deficiency.

Serum folate values fluctuate significantly with diet; measurement of red cell folate more closely reflects tissue folate stores. Following dietary deprivation or poor absorption of folate, serum folate levels will be decreased within 3 weeks while the RBC folate, representing the storage form, will remain normal for 3-4 weeks.

See FOLATE (FOLIC ACID), SERUM for causes of folate deficiency.

The drug sulfasalazine (Azulfidine) is used in the treatment of ulcerative colitis; patients taking more than 2 grams of sulfasalazine daily for several months may develop subclinical folate depletion as reflected by red cell folate deficiency but not by serum folate deficiency (Longstreth, G. and Green, R., Arch. Int. Med. 143, 902-904, May, 1983).

FOLATE (FOLIC ACID), SERUM

SPECIMEN: Red top tube, separate serum. The test should not be ordered for patients who have recently received a radioisotope, methotrexate, or other folic acid antagonist. Strongly consider ordering vitamin B_{12} determination.

REFERENCE RANGE: 2-14ng/ml. To convert conventional units in ng/ml to international units in nmol/liter, multiply conventional units by 2.226.

METHOD: RIA

INTERPRETATION: The causes of folate deficiency are given in the next Table:

Causes of Folate Deficiency
Inadequate Intake:
Alcoholism
Nutritional Deficiencies
Relative Inadequate Intake:
Pregnancy
Hemolytic Anemia
Chronic Hemodialysis or Peritoneal Dialysis (Folate Dialyzable)
Inadequate Absorption:
Malabsorption Syndromes:
Tropical Sprue
Gluten-Sensitive Enteropathy (Non-Tropical Sprue)
Crohn's Disease
Lymphoma or Amyloidosis of Small Bowel
Diabetic Enteropathy
Intestinal Resections or Diversions
Interference with Folic Acid Metabolism
Alcohol
Drugs Blocking Action of Difolate Reductase:
Methotrexate, Trimethoprim, Pyrimethamine
Drugs Interfering by Unknown Means: Anticonvulsants
(Phenytoin), Ethanol, Antituberculous Drugs,
Oral Contraceptives

Folate deficiency is most commonly encountered in pregnancy and alcoholism; the requirement during pregnancy is about 4 times that of the normal requirement. Decreased red blood cell folate and normal serum folate occasionally occur with pure vitamin B_{12} deficiency (folate-trap hypothesis).

Folate deficiency is associated with megaloblastic anemia.

FOLLICLE STIMULATING HORMONE(FSH)

SPECIMEN: Red top tube, separate serum from cells as soon as possible; if serum is not analyzed immediately, store at -20°C. Repeated freezing and thawing should be avoided.

There are rapid cyclical changes in concentration; thus, <u>optimally</u>, three blood specimens should be drawn at 60 minute intervals; combine the serum of the three specimens.

REFERENCE RANGE: Units(IU/liter); Prepubertal: Up to 5. In the <u>female</u>, FSH concentrations start to increase at about age 10 or 11 and plateau at the time of menarche; in the <u>male</u>, FSH concentrations start to increase about 2 years later. <u>Female</u>: Changes in FSH during the <u>menstrual cycle</u> are given in the next Figure:

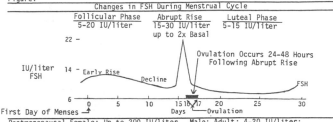

Changes in FSH During Menstrual Cycle

Follicular Phase	Abrupt Rise	Luteal Phase
5-20 IU/liter	15-30 IU/liter	5-15 IU/liter
	up to 2x Basal	

Ovulation Occurs 24-48 Hours Following Abrupt Rise

<u>Postmenopausal Female</u>: Up to 200 IU/liter. <u>Male</u>: <u>Adult</u>: 4-20 IU/liter; Castrate: Up to 200 IU/liter.

METHOD RIA

INTERPRETATION: FSH controls development of ovarian follicles and ova of the ovaries in the female and with the aid of LH, testosterone and sperm cells in the testis in the male.

FSH determinations are indicated in the work-up of patients for <u>hypogonadism</u> or <u>infertility</u>; determinations are done to determine whether these are due to primary failure of ovaries or testicles or failure secondary to pituitary failure. The test results, expected in <u>pituitary</u>, <u>ovarian</u> or testicular and end-organ failure, are given in the next Table:

FSH Levels in Pituitary, Ovarian or Testicular and End-Organ Failure

Failure	Serum FSH
Pituitary (Consider Prolactinoma)	Decreased
Ovarian (Consider Polycystic Ovarian Disease)	Increased
End-Organ	Normal

In isolated pituitary gonadotropin (FSH and LH) deficiency, there are no clinical signs before the age of puberty. Secondary sexual development fails to occur during adolescence. If isolated pituitary gonadotropin deficiency occurs in the female adult, amenorrhea develops.

Increased levels of FSH are found in conditions listed in the next Table:

Increased Serum Levels of FSH

Females	Males
Primary Ovarian Failure:	Primary Testicular Failure:
Menopause, Premature Menopause	Undescended Testis; Radiation
Ovariectomy	Castration
Ovarian Agenesis, Dysgenesis	Anorchia (Testicular Agenesis)
Genetic Disorders: Turner's Syndrome	Genetic Disorders: Klinefelter's Syndrome
	Sertoli-Cell-Only Syndrome

Decreased levels of FSH are found in conditions listed in the next Table:

Decreased Serum Levels of FSH

Females	Males
Medications: Estrogens, Testosterone,	Medications: Testosterone or Estrogen
Oral Contraceptives	Hypopituitarism
Hypopituitarism; Hypophysectomy	Hypophysectomy
Adrenal or Ovarian Neoplasm	Adrenal Neoplasm
Polycystic Ovarian Disease	
Various Menstrual Disorders	
Sheehan's Syndrome (Postpartum Pituitary Necrosis)	

FRACTIONAL EXCRETION OF FILTERED SODIUM
(SEE EXCRETION FRACTION OF FILTERED SODIUM)

FUNGAL PANEL, CEREBROSPINAL FLUID(CSF)

The fungi found in the CSF and their distinguishing features are given in the next Table; (Bigner, S.H., Am. Soc. Clin. Path. Check Sample, Cytopathology II, No.5, 1983):

Fungi in CSF and Their Distinguishing Features				
Organism	Shape and Size	Capsule	Bud Type	Usual Host
Cryptococcus Neoformans	Round Yeast, 2 to 15 Microns	Thick; Muco-polysaccharide; Stains with Alcian Blue and Mucicarmine	Thin-necked	Lymphoma or Steroids; occ. Healthy Individuals
Blastomyces Dermatididis	Round Yeast, Usually 1 to 5 Microns	None	Thick-necked	Healthy Individuals
Histoplasma Capsulatum	Round Yeast, 1 to 5 Microns	None	Thin-necked	Healthy Individuals
Candida Albicans	Round, Oval or Elongated, 2 to 5 Microns	None	Thin-necked	Lymphoma, Steroids or Cytotoxic Agents
Coccidioides Immitis	Round or Oval Endospores, Up to 60 Microns	None	Rarely seen	Healthy Individuals
Aspergillus	Septate Filaments 3 to 4 Microns	—	—	Antibiotics or Steroids

Cryptococcus neoformans is the most common fungus involving the human central nervous system.

FUNGAL SEROLOGICAL PANEL

The common invasive fungi are listed in the next Table; (Penn, R.L. et al., Arch. Intern Med. 143, 1215-1220, 1983):

Common Invasive Fungi	
Pathogenic Fungi	Opportunistic Fungi
Blastomyces Dermatididis	Aspergillus Species
Coccidioides Immitis	Candida Species
Cryptococcus Neoformans	
Histoplasma Capsulatum	

The pathogenic fungi can infect normal individuals; the opportunistic fungi usually cause infections in immunocompromised hosts or in patients with indwelling catheters.

The serologic tests that are most commonly done for diagnosis of invasive fungi are given in the next Table:

Serologic Tests for Invasive Fungi	
Fungi	Serologic Tests
Blastomyces Dermatidis	Unreliable
Coccidioides Immitis	Tube Precipitin(TP) Test for IgM Antibodies Complement Fixation(CF) Test for IgG Antibodies
Cryptococcus Neoformans	Latex Agglutination(LA) Test for Cryptococcal Antigen
Histoplasma Capsulatum	Complement Fixation(CF) Test for Antibodies
Aspergillus Species	Immunodiffusion(ID), CF, etc. for Antibodies for Noninvasive Aspergillosis
Candida Species	Immunodiffusion(ID), Latex Agglutination(LA) and Counterimmunoelectrophoresis(CIE) for Antibody

FREE ERTHROCYTE PROTOPORPHYRIN(FEP)
(SEE ZINC PROTOPORPHYRIN(ZPP), BLOOD)

GAMMA GLUTAMYL TRANSPEPTIDASE (GGTP)

SPECIMEN: Red top tube, separate serum

REFERENCE RANGE: Male: 15-85IU/liter; Female: 5-55IU/liter

METHOD: Substrates: Gamma-glutamyl-p-nitroanilide + Glycylglycine

INTERPRETATION: Some of the diseases that are associated with increased serum GGTP are given in the next Table:

Conditions Associated with Elevated Serum GGTP
Alcoholism
Liver Disease
Infectious Mononucleosis
Pancreatic Disease
Diabetes Mellitus (Boone et al, Am. J. Clin. Pathol. 61, 321-327, 1974)
Myocardial Infarction (Betro et al, AM. J. Clin. Pathol. 60, 679-683, 1973)
Congestive Heart Failure
Neurologic Disorders
Trauma
Nephrotic Syndrome
Chronic Renal Failure (occasionally)
Sepsis (Fang, M.H., et al., Gastroenterology 78, 592-597, 1980)
Drugs, such as Phenobarbital, other Barbituates and Dilantin (Diphenylhydantion), Antipyrine

Assay of gamma glutamyl transpeptidase is most often used to differentiate the source of an elevated serum alkaline phosphatase, e.g., liver or bone; gamma glutamyl transpeptide is not present in bone.

Serum gamma glutamyl transpeptidase (GGTP) activity is elevated in all forms of liver disease and in intra-hepatic or extra-hepatic biliary obstruction. It is a sensitive method to determine whether bone or liver is the source of increased serum alkaline phosphatase and has gained wide popularity for this purpose. Serum alkaline phosphatase is increased during the growth period because of increased osteoblastic activity associated with bone growth and during the last trimester of pregnancy because of the high content of alkaline phosphatase in the placenta. Serum GGPT is absent from the bone and the placenta. If alkaline phosphatase is high due to bone disease, serum GGTP is normal.

Serum GGTP is normal in the following conditions:

Normal Serum Gamma Glutamyl Transpeptidase (GGTP)
Children
Adolescents
Pregnancy
Bone Disease
Muscle Disease

Alcoholism: GGTP is the most sensitive enzyme used to detect liver damage from excessive alcohol intake.

Gamma-glutamy transpeptidase is situated on the smooth endoplasmic reticulum. Any substance, such as alcohol or barbiturates, which causes microsomal proliferation will cause an increase in serum gamma-glutamyl transpeptidase. One or two drinks may elevate serum gamma-glutamyl transpeptidase level.

Boone et al. (Boone, D.J., Tietz, N.W. and Weinstock, A., Ann. Clin. Lab. Sci. 7, 25-28, 1977) found that gamma-glutamyl transferase activity is useful in the assessment of alcohol-induced liver disease and for monitoring the progress of therapy as well as alledged abstention from alcohol in known alcoholics.

Lamy et al (Lamy, J., Baglin, M.C., Weill, J. and Aron, E. "Gamma-Glutamyl Transpeptidase Serique et Alcoholisme" Nov Presse Med. 4, 487-490, 1975) studied the change of GGTP activity in the serum of "heavy drinkers" and alcoholics during detoxification; the decrease in enzyme activity is shown in the next Figure:

Decrease in Gamma-Glutamyl Transferase Activity in Serum of
"Heavy Drinkers" and "Alcoholics" during Detoxification (Lamy et al., 1975)

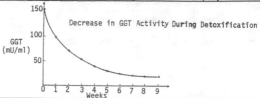

Decrease in GGT Activity During Detoxification

As seen in the previous Figure, Lamy (1975) found that after alcohol deprivation, the GGTP activity of alcoholics decreased in the first few days exponentially with a half-life of return to normal of 5 to 17 days.

Elevated serum gamma glutamyl transpeptidase activity in the presence of normal results of other liver function tests and in the absence of jaundice and hepatomegaly is apparently a valuable indicator of significant alcohol consumption in adolescents who consumed six or more drinks per day (Westwood et al., Pediat. 62, 560-562, 1978).

Conditions other than liver disease cause elevation of GGTP. Drugs such as the antiepileptics, phenobarbital and diphenylhydantoin, or other drugs which stimulate the smooth endoplasmic reticulum, cause elevation of GGTP (Berg, B. and Tryding, N., the Lancet, pg. 1162, May 23, 1981).

GASTRIN, SERUM

SPECIMEN: Fasting 8-12 hours overnight; red top tube; pour serum into plastic tube; freeze immediately.
REFERENCE RANGE: Normal: 0 to 10pg/ml; borderline: 100-200pg/ml; elevated: 200pg/ml or more. To convert conventional units in pg/ml to international units in ng/liter, multiply conventional units by 1.0.
METHOD: RIA
INTERPRETATION: Serum gastrin is elevated in the conditions listed in the next Table; (Essop, A.R., Segal, I., Ming, R., N. Engl. J. Med. 307, 192, July 15, 1982):

Causes of Elevated Serum Gastrin
Zollinger-Ellison Syndrome
Pernicious Anemia
Duodenal Ulcers
Pyloric Stenosis
Chronic Renal Failure
Atrophic Gastritis
Intestinal Resection
Antral G-cell Hyperplasia
Rheumatoid Arthritis
Ulcerative Colitis
Administration of Steroids and Calcium

A fasting gastrin level greater than 600pg/ml (normals and duodenal ulcer patients, less than 200pg/ml) with acid hypersecretion (greater than 15mEq/hour) is virtually diagnostic of Zollinger-Ellison syndrome.

The Zollinger-Ellison syndrome is characterized as follows:
(1) Presence of a non-beta cell tumor secreting gastrin (Gastrinoma)
(2) Marked secretion of acid from gastric parietal cells
(3) Single ulcers of gastric or duodenal mucosa in two-thirds of patients.

GASTROINTESTINAL HEMORRHAGE PANEL

Gastrointestinal hemorrhage panel is given in the next Table:

Gastrointestinal Hemorrhage Panel

Initial Laboratory Tests:
    Stool for Blood
    Complete Blood Count(CBC)
    Urinalysis
    Blood Urea Nitrogen(BUN)
    Electrolytes: Sodium, Potassium, Chloride, CO_2 Content
    Coagulation Screen:
        Prothrombin Time(PT)
        Partial Thromboplastin Time(PTT)
        Fibrinogen Level
        Platelet Count
    Liver Function Tests:
        Bilirubin, Total and Direct
        Serum Glutamic Oxalacetic Transaminase(SGOT)
        Serum Alkaline Phosphatase
    Serum Albumin
    Serum or Urine Amylase
    Gastric Aspirate for Blood
    Routine X-Ray Films of the Chest and Abdomen
    Electrocardiogram for Older Patients - Silent
        Infarct may Occur with Significant Gastrointestinal Bleeding
Specific Diagnosis:
    Nasogastric Aspiration to Rule-out Upper Gastrointestinal Hemorrhage
    Barium Enema for Diverticulosis
    Fiberoptic Colonoscopy
    Sigmoidoscopy
    Selective Angiography
    Sodium Pertechetate Tc99m Nuclear Scans for Ectopic Gastric Mucosa

Colacchio, T.A., Am. J. Surg. 143, 607-610, 1982.

Causes of gastrointestinal hemorrhage are given in the next Table:

Causes of Gastrointestinal Hemorrhage

Upper Gastrointestinal Hemorrhage: Peptic Ulcer Disease;
                Esophageal Varices; Gastritis;
                "Stress" Ulcerations
                Mallory-Weiss Syndrome
Lower Gastrointestinal Hemorrhage:
                Diverticular Bleeding
                Bleeding from Colonic Vascular Ectasias (Angiodysplasia)
                Colorectal Carcinoma
                Colonic Polyps
                Ischemic Colitis
                Inflammatory Bowel Diseases
- -
Gastrointestinal Bleeding in Infants and Children:
                Neonates:
                    Anorectal Fissure
                    Hemorrhagic Disease of the Newborn
                    Infectious Diarrhea
                    Volvulus
                    Vascular Malformations
- -
Infants up to Age Two: Infectious Diarrhea; Anal Fissure;
                    Intussusception; Meckel's Diverticulum; Peptic Ulcer;
                    Volvulus
Children, 2 to 12 Years: Infectious Diarrhea; Juvenile Polyp;
                    Peptic Ulcer; Esophageal Varices;
                    Inflammatory Bowel Disease; Meckel's Diverticulum;
                    Intestinal Duplication; Intussusception;
                    Vascular Malformation
Adolescents:
                    Peptic Ulcer
                    Inflammatory Bowel Disease
                    Esophageal Varices

GENETIC DISEASE AND RECOMBINANT DNA

Using techniques of recombinant DNA, genetic diseases have been diagnosed in-utero; these diseases are listed in the next Table; (Editorial, The Lancet, pgs, 1404-1405, Dec. 17, 1983):

Genetic Disease and Recombinant DNA	
Disease	Reference
Huntington's Chorea..........	Gusella, J.F. et al., Nature 306, 234-238, 1983
Phenylketonuria..............	Woo, S.L.C. et al., Nature 306, 151-155, 1983
Sickle Cell Anemia..........	Chang, J.C. and Kan, Y.W., N. Engl. J. Med. 307, 30-32, 1982; Orkin, S.H. et al., N. Engl. J. Med. 307, 32-36, 1982. Boehm, C.D. et al., N. Engl. J. Med. 308, 1054-1058, 1983)
Beta-Thalassemia.............	Pirastu, M. et al., N. Engl. J. Med. 309, 284-287, 1983
Alpha-1-Antitrypsin.......... Deficiency	Kidd, V.J. et al., Nature, 304, 230-234, 1983
Antithrombin III............. Deficiency	Prochownik, E.V. et al., N. Engl. J. Med. 308, 1549-1552, 1983
Sex-Linked Disorders,e.g.,.... Duchenne Muscular Dystrophy Hemophilia Lesch-Nyhan	Lau, Y-F. et al., The Lancet, pgs. 14-16, Jan. 7, 1984

Determination of Fetal Sex

The technique for antenatal diagnosis using recombinant DNA probe is shown in the next Figure; (Permutt, M.A. and Rotwein, P., Am. J. Med. 75, pgs. 1-7, Nov. 30, 1983):

Antenatal Diagnosis Using Recombinant DNA	
(1) Genomic DNA Digested with Restriction Endonuclease:	Discussion
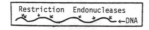	Analysis is performed on an amniotic fluid specimen; the cells are obtained following centrifugation. The cells are lysed and the nuclei are pelleted by centrifugation. The DNA is extracted and digested with restriction endonucleases.
(2) Gel Electrophoresis of DNA Fragments: 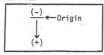	Gel electrophoresis is done on the DNA fragments; the DNA fragments are separated according to size with the large fragments remaining near the top and the small fragments of DNA migrating toward the bottom.
(3) Transfer of DNA to Nitrocellulose Filter by Blotting: 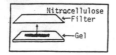	The double-stranded DNA in the gel is denatured with alkali, converting the double-stranded DNA to single-stranded. Then, the gel is overlaid with a nitrocellulose filter; absorbent paper towels are placed on the nitrocellulose filter and the single-stranded DNA is drawn to and bound to the nitrocellulose filter. Thus, a replica of the gel is obtained on the nitrocellulose filter.

Antenatal Diagnosis Using Recombinant DNA

(4) Hybridization of Filter
with P-32 Labeled Human   Discussion
Insulin Gene Probe:

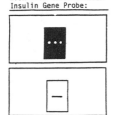

The single-stranded DNA is treated with P-32
labeled cloned gene probes that bind to
the gene by hydrogen bonding; this is
known as nucleic acid hybridization.
As an example, the gene probe for hemoglobin
is obtained as follows: reticulocytes
contain globin messenger RNA almost
exclusively. Using reverse
transcriptase, a globin complementary
DNA copy can be synthesized to make gene
specific probes.

Chorionic villi may be used, instead of amniotic fluid, as a source of
fetal cells.
     A simple and rapid method for identification of Y-DNA sequences for sex
determination has been reported; the procedure involves centrifugation of
amniotic fluid, heating the cell pellet with an alkaline solution, spotting the
DNA on two filters, and hybridizing the filters with two repeat-sequence probes;
(Law, Y-F. et al., The Lancet, pgs. 14-16, Jan. 7, 1984).

                                                                    Gentamicin
GENTAMICIN
SPECIMEN: Red top tube, separate serum and freeze. Obtain serum specimens as
follows:
(1) 24-48 hours after starting therapy if loading dose is not given.
(2) 5 to 30 minutes before I.V. gentamicin (trough).
(3) 30 minutes after a 30 minute I.V. infusion of gentamicin (peak).
I.M. administration: 30 minutes to one hour.
Heparinized tubes should not be used to collect specimens.
REFERENCE RANGE: Therapeutic: 4-10 mcg/ml; Toxic: >12mcg/ml; Peak Values:
4-10mcg/ml; Trough: <2mcg/ml. Time to Steady State: Adults(<30 years), 2.5-15
hours; Adults(>30 years), 7.5-75 hours; Children, 2.5-12.5 hours; Neonates,
10-45 hours. Time to steady state and time to peak concentration may be
significantly prolonged in patients with renal dysfunction.
METHOD: RIA, EMIT
     High concentrations of beta-lactam antibiotics inactivate aminoglycosides
(gentamicin, streptomycin, amikacin, tobramycin and kanamycin); to reduce this
interaction, specimens containing both classes of antibiotics should either be
assayed immediately using a rapid method or stored frozen.
INTERPRETATION: Give I.V. or I.M. Gentamicin is an aminoglycoside antibiotic
(gentamicin, streptomycin, amikacin, tobramycin and kanamycin) which is used
frequently in hospitals to treat patients who have serious gram-negative
bacterial infections, especially septicemia and staphylococcal infections,
untreatable with penicillins.
     The first serum level of gentamicin is obtained when gentamicin has reached
steady-state serum concentrations; steady-state is reached in 24 to 48 hours
after starting therapy if the patient has not received a loading dose (steady
state = 5-7 drug half-lives; half-live of gentamicin is 2 to 4 hours). The
following times should be recorded on the laboratory requisition form and on the
patient's chart:
  Trough Specimen Drawn _____(Time) (5 to 30 minutes before Gentamicin)
  Gentamicin Started   _____(Time)
  Gentamicin Completed _____(Time) (30 minutes I.V. infusion)
  Peak Specimen Drawn   _____(Time) (30 minutes after I.V. Gentamicin)

The <u>three</u> main <u>toxic</u> side effects of gentamicin are <u>ototoxicity</u>, <u>nephrotoxicity</u> and <u>neuromuscular blockage</u> (Smith, C.R. et al., N. Engl. J. Med. <u>302</u>, 1106, May 15, 1980); it is important to control the dose given by monitoring <u>peak</u> and <u>trough</u> levels of the drug, <u>particularly</u> in patients with any degree of <u>renal failure</u>. As <u>renal function declines</u>, <u>drug half- life increases</u> to up to 50 hours.

To <u>minimize risk</u> of <u>toxicity</u>, it has been recommended that <u>peak levels not exceed 10mcg/ml</u> and that <u>trough levels</u> should fall <u>between 1 and 2 mcg/ml</u>.

Gentamicin is eliminated exclusively by <u>renal excretion</u>; excessive serum concentrations may occur and lead to further <u>renal impairment</u>. The renal damage is to the renal proximal tubules and is usually reversible if discovered early.

<u>Ototoxicity</u> is usually due to vestibular damage and is often <u>not</u> reversible.

GLIADIN ANTIBODIES (SCREENING TEST FOR CELIAC DISEASE IN CHILDREN)

<u>SPECIMEN</u>: Red top tube, separate serum.
<u>REFERENCE RANGE</u>: Negative
<u>METHOD</u>: Fluorescent immunosorbent test for gliadin antibody; commercial kit may not be available.
<u>INTERPRETATION</u>: Assay for antibodies to gliadin has been examined in children with malabsorptive disorders in a prospective multicenter study. The test had a sensitivity of 100 percent and a specificity of 84 percent for the diagnosis of childhood celiac disease. Among the 16 percent of the patients with other malabsorptive disorders who had gliadin antibodies, 12 percent had a low titer and 3.5 percent showed moderate to high titers (Burgin-Wolff, A. et al., J. Pediatr. <u>102</u>, 655-660, May 1983)..

The results of this study indicate that assay of gliadin antibody is useful as a <u>screen</u> for celiac disease in children; <u>a negative test result excludes the diagnosis of celiac disease</u>. The <u>definitive</u> diagnosis of celiac disease includes a <u>mucosal biopsy</u> before and after a gluten free diet and again after rechallenge.

GLOMERULONEPHRITIS ACUTE, PANEL (see ACUTE GLOMERULONEPHRITIS)

GLUCOSE, BLOOD, SELF-MONITORING

SPECIMEN: Capillary blood obtained on finger-stick using automated, spring-operated lance; a commercially available lance is the Autolet.

REFERENCE RANGE: Infants and children: 60-115mg/dl; adults: 65-120mg/dl. To convert conventional units in mg/dl to international units in mmol/liter, multiply conventional units by 0.05551.

METHOD: A reagent strip is impregnated with glucose oxidase, reduced chromogen and peroxidase; in the presence of glucose, the following reaction takes place:

$$\text{Glucose} + O_2 + H_2O \xrightarrow{\text{Glucose Oxidase}} H_2O_2 + \text{Gluconic acid}$$

$$H_2O_2 + \text{Reduced Chromogen} \xrightarrow{\text{Peroxidase}} \text{Oxidized Chromogen} + H_2O$$

Reagent strips may be interpreted visually by comparing the color generated to a color chart or reagent strips may be read with a reflectance meter that generates a numerical value for blood glucose.

Visual Interpretation: There are two reagent test strips that are available; recommendations are given in the next Table; (Medical Letter, 25, 42-44, April 29, 1983):

Visual Interpretation: Reagent Test Strips and Recommendations			
Reagent Strip	Range of Measurements (nm)	Color Stability	Price*/ Strip
Chemistrip bG (Bio-Dynamics)	20 to 800	Good	0.55
Visidex (Ames)	20 to 800	Poor	0.55

*Suggested retail price, 1983.

Chemistrip bG and Visidex strips can be cut in half-lengthwise, thus cutting the cost in half. A timer is necessary since reading is done in about two minutes. The Chemistrip bG (Bio Dynamics) strips can also be read in a meter.

Visual interpretation is convenient; a range, rather than a precise number for blood glucose, is obtained. Diabetics treated with photocoagulation may have problems in properly interpreting strips.

Instruments: The reflectance meters for quantitatively measuring blood glucose are given in the next Table; (Medical Letter, 25, 42-44, April 29, 1983):

Reflectance Meters				
Instrument	Range (mg/dl)	Reagent Strips	Price/ Strip	Price of Reference Meter
Dextrometer (Ames)	10-399	Destrostix	$0.56	$169.00
Glucometer (Ames)	10-399	Destrostix	0.56	200.00
		Glucoscan	0.44	
Glucoscan II (Lifescan)	40-450	Glucoscan	0.44	250.00
		Dextrostix	0.56	
Stat-Tek (Bio-Dynamics)	50-350	Stat-Tek	0.70	375.00
Accu-Chek bG (Bio-Dynamics)	40-400	Chemstrip bG	0.55	265.00

These instruments weigh from 300 grams (Glucoscan II) to 1,100 grams.

INTERPRETATION: Some diabetics prefer testing blood to urine and urine testing may be insufficiently sensitive (Holman, R.R. and Turner, R.C., Lancet, 1, 469-474, 1977). Maintenance of normal glucose levels in diabetics is important in retarding the development of complications of diabetes mellitus (Peterson, C.M. et al. in Friedman, E.A. and L'Esperance, F.A., Jr. eds., Diabetic Renal-Retinal Syndrome 2, N.Y., Grune and Stratton, 1982, pg. 73). Poorly controlled diabetes is associated with higher incidence of major congenital anomalies in the offspring of pregnant women (Miller, E. et al., N. Engl. J. Med. 304, 1331-1334, 1981; Frienkel, N., N. Engl. J. Med. 304, 1357-1359, 1981). Close control of blood glucose in the pregnant diabetic has been shown to decrease infant mortality and maternal morbidity (Jovanovic, L. and Peterson, C.M., Diabetes Care 5, Suppl. 1, 24, 1982). The development of diabetic neuropathy has been related to the degree of hyperglycemia (Porte, D. et al., Am. J. Med. 70, 195-200, 1981). However, no improvement was noted in any measurement of nerve function associated with improved blood glucose control (Service et al., Mayo Clin. Proc. 58, 283-289, 1983). The benefits of self-monitoring of blood glucose have been reviewed (Bell, P.M. and Walshe, K., Brit. Med. J. 286, 1230-1231, April 16, 1983); arguments for randomized clinical trials are given (Barbosa, J., Arch. Intern. Med. 143, 1118-1119, 1983).

GLUCOSE, SERUM

SPECIMEN: Red top tube, separate serum; or grey top tube (oxalate and fluoride), separate plasma; glucose will decrease in samples left on the clot and in tubes other than fluoride prior to analysis. Fluoride inhibits glucolysis by blood cells.

REFERENCE RANGE: Newborn: Premature: 20-80mg/dl; full-term: 20-90mg/dl; infants and children: 60-115mg/dl, adults: 65-120mg/dl. To convert conventional units in mg/dl to international units in mmol/liter, multiply conventional units by 0.05551.

METHOD: Glucose oxidase or hexokinase method.

INTERPRETATION: The causes of hyperglycemia are given in the next Table:

Causes of Hyperglycemia
Diabetes Mellitus
Acute Pancreatitis
Endocrine Hyperfunctions:
Cushing's Disease; Pheochromocytoma; Acromegaly;
Hypothalamic Lesions; Carcinoid Syndrome;
Eosinophilic Adenoma of the Pituitary Gland
Hemochromatosis
Ataxia Telangiectasia
Drugs:
Anabolic Hormones; Epinephrine and Norepinephrine;
Benzothiadiazine Diuretics; Diphenylhydantoin
Note: Increased with I.V. Glucose; Stress; Non-Fasting Specimen

A blood glucose >120 mg/dl during the early hours following the onset of ischemic stroke is associated a poor neurologic outcome. A good recovery from hemispheric stroke was achieved by 36% of normoglycemic patients but only 14% of hyperglycemic; stroke-related death occurred with eight times the frequency in hyperglycemic patients. The mechanism by which hyperglycemia contributes to morbidity in ischemic brain damage is unknown (Longstreth, W.T. et al., N. Engl. J. Med. 308, 1378,1983; Ann. Int. Med. 98, 588, 1983; Pulsinelli, W.A. et al., Am. J. Med. 74, 540, 1983).

The causes of hypoglycemia are given in the section on Glucose Tolerance Test.

GLUCOSE TOLERANCE TEST,
DIAGNOSIS OF DIABETES MELLITUS IN ADULTS

REFERENCE RANGE: Fasting: 70-105mg/dl; 60 minutes: 120-170mg/dl; 90 minutes: 100-140mg/dl; 120 minutes, 70-120mg/dl. To convert conventional units in mg/dl to international units in mmol/liter, multiply conventional units by 0.05551.

INTERPRETATION: The biochemical diagnosis of diabetes mellitus in non-pregnant adults is given in the next Table; (National Diabetes Data Group, Diabetes 28, 1039-1057, 1979; Keen, H. et al., Diabetologia 26, 283-285, 1979):

Biochemical Diagnosis of Diabetes Mellitus in Non-Pregnant Adults

Classical Symptoms of Diabetes
(Polyuria, Polydipsia, Ketonuria, Rapid Weight Loss)
Plus
Random Plasma Glucose > 200mg/dl
} = Diabetes

↓ Overnight Fast (> 10 hours)

A.M. Blood Specimens for Glucose Assay

Plasma Glucose Values (mg/dl)
> 140 = Diabetes if this level obtained on more than one occasion; OGTT not required

---

Give Oral Glucose Load (75g)
↓ 60 Minutes
Blood Specimen for Glucose Assay   ≥ 200
↓ 120 Minutes
Blood Specimen for Glucose Assay   ≥ 200
} = Diabetes

GLUCOSE TOLERANCE TEST,
DIAGNOSIS OF DIABETES MELLITUS IN CHILDREN

REFERENCE RANGE: See below
INTERPRETATION: The biochemical diagnosis of diabetes mellitus in children is
given in the next Table; (National Diabetes Data Group, Diabetes 28 1039-1057,
1979; Keen, H. et al., Diabetologia 26, 283-285, 1979):

Biochemical Diagnosis of Diabetes Mellitus in Children

Classic Symptoms of Diabetes
(Polyuria, Polydipsoa, Ketonuria, Rapid Weight Loss)

Plus   } = Diabetes

Random Plasma Glucose > 200 mg/dl

Asymptomatic Subject:

	Plasma Glucose Values (mg/dl)
Overnight Fast (>10 hours)	
A.M. Blood Specimen for Glucose Assay	≥ 140
Give Oral Glucose Load	
(1.75g/kg ideal body weight up to a maximum of 75g)	
60 Minutes	
Blood Specimen for Glucose Assay	≥ 200
120 Minutes	
Blood Specimen for Glucose Assay	≥ 200

GLUCOSE TOLERANCE TEST,
DIAGNOSIS OF DIABETES MELLITUS IN PREGNANCY

REFERENCE RANGE: See Below
INTERPRETATION: The biochemical diagnosis of diabetes mellitus in pregnancy
is given in the next Table; (National Diabetes Data Group, Diabetes 28
1039-1057, 1979; Keen, H. et al., Diabetologia 26, 283-285, 1979):

Biochemical Diagnosis of Diabetes Mellitus in Pregnancy

	Plasma Glucose Values (mg/dl)
Overnight Fast (>10 hours)	
A.M. Blood Specimen for Glucose Assay	≥ 105
Give Glucose Load (100g)	
60 minutes	
Blood Specimen for Glucose Assay	≥ 190
120 minutes	
Blood Specimen for Glucose Assay	≥ 165
180 minutes	
Blood Specimen for Glucose Assay	> 145

Diabetes Mellitus in Pregnancy = Two or more of the plasma glucose values
elevated.

Close control of blood glucose in the pregnant diabetic has been shown to
decrease infant mortality and maternal morbidity (Jovanovic, L. and Peterson,
C.M., Diabetes Care 5, Suppl. 1, 24, 1982).

GLUCOSE AND "MIXED MEALS"(MM) TOLERANCE TESTS, HYPOGLYCEMIA

SPECIMEN: See Glucose, Blood.

REFERENCE RANGE: Healthy Males: Blood glucose levels, obtained by the glucose oxidase method, following an oral glucose administration, are shown in the next Table; (Sussman, K.E., Stimmler, L. and Birenboim, H., Diabetes 15, 1-4, 1966):

Blood Glucose Response to Oral Glucose Administration (100g)	
Time (Min.)	Serum Range (mg/dl)
Fast	63.6-105.2
15	63.9-140.5
30	76.9-176.3
45	88.2-169.2
60	85.6-164.4
90	64.2-172.6
120	70.6-137.0
150	76.9-133.7
180	53.9-141.7
240	64.7-105.5

In this study on healthy males, plasma glucose levels did not decline below 55 mg/dl during a 24 hour fast. However, 23 percent of the normal population during a glucose tolerance test exhibit blood glucose levels below 50 mg/dl between two and five hours (Cahill, G.F. and Soeldner, J.S., N. Engl. J. Med. 291, 905-906, 1974) and occasionally, values below 35 mg/dl may be noted in normal persons without symptoms (Park B.N., Kahn, C.B., Gleason, R.E. et al., Diabetes 21, 373, 1972).

Children: (Mietes, S. Ed., Pediatric Clinical Chemistry, Am. Assoc. Clin. Chem., 1725 K. Street, N. W., Washington D.C. 20006, pg. 286)

Blood Glucose Response to Oral Glucose Administration (1.75g/kg Body Weight) in Children			
Time Min.	Serum Glucose (mg/dl)	Insulin (Micro USP units/ml)	Phosphorus (mg/dl)
Fasting	56-96	5-40	3.2-4.9
30	91-185	36-110	2.0-4.4
60	66-164	22-124	1.8-3.6
90	68-148	17-105	1.6-3.6
120	66-122	6-84	1.8-4.2
180	47-99	2-46	2.0-4.6
240	61-93	3-32	2.7-4.3
300	63-86	5-37	2.9-4.4

METHOD: Glucose oxidase or hexokinase method.

INTERPRETATION: The causes of hypoglycemia are given in the next Table:

Causes of Hypoglycemia
Postprandial (Reactive) Hypoglycemia: Alimentary Hyperinsulinism, e.g., Gastrectomy, Gastrojejunostomy, Pyloroplasty or Vagotomy; Hereditary Fructose Intolerance; Galactosemia; Leucine Sensitivity; and Idiopathic.
Factitious Hypoglycemia: Insulin; Oral Hypoglycemic Agents
Falciparum Malaria (White, N.J. et al., N. Engl. J. Med. 309, 61-66, 1983)
Reye's Syndrome
Liver Disease (advanced)
Malnutrition
Renal Glycosuria
Islet Beta Cell Tumor (Insulinoma)
Malignancy (large tumors)
Hypoglycin Ingestion (Ackee fruit)
Neonatal Hypoglycemia
Dormandy's Syndrome (Familial Fructose and Galactose Intolerance)
Endocrine Hypofunctions: Anterior Pituitary; Addison's Disease
Enzyme Deficiencies: Glycogen Storage Diseases
Artifactual: Polycythemia Vera (Billington, C.J., JAMA 249, 774-775,1983)
Bacteria Contamination of Specimen; Failure to Separate Clot from Serum Promptly

Postprandial (Reactive) Hypoglycemia: Some causes of postprandial hypoglycemia
are shown in the previous Table; the most common cause is alimentary.
Hypoglycemia follows meals because of rapid gastric emptying with rapid
absorption of glucose and excessive insulin release. Glucose concentrations
decrease rapidly leading to hypoglycemia.
Oral Glucose Tolerance Test, "Mixed Meal" and "Reactive" Hypoglycemia: The use
of the oral glucose tolerance test(OGTT) for the evaluation of "reactive"
hypoglycemia should be discontinued for the following reasons: almost one-fourth
of the normal population develop hypoglycemia following OGTT without
experiencing symptoms of hypoglycemia; in patients with "reactive" hypoglycemia,
there is no consistent relationship between hypoglycemia and symptoms. When
subjects with symptoms suggestive of postprandial hypoglycemia are given a
"mixed meal," signs and symptoms of the postprandial syndrome develop but
chemical hypoglycemia does not occur. The "mixed meal," consisting of
carbohydrate, fat and protein, more closely approximates food intake that
precipitates postprandial symptoms.
     It is concluded that the cause of postprandial symptoms is not known and
certainly is not due to hypoglycemia. The term "reactive" hypoglycemia is
discarded because low blood glucose is not involved in the pathogenesis of this
condition; the disorder should be termed idiopathic postprandial syndrome
(Charles, M.A. et al., Diabetes 30, 465-470, 1981; Hogan, M.J. et al., Mayo
Clin. Proc. 58, 491-496, 1983).
Factitious Hypoglycemia: Factitious (made by or resulting from art; artificial;
produced by special causes) hypoglycemia may be produced by surreptitious
self-injection of insulin. The diagnosis of factitious hypoglycemia induced by
insulin is made by the simultaneous presence of the following:

Diagnosis of Factitious Hypoglycemia Induced by Insulin
Low Serum Glucose
High Immunoreactive Plasma Insulin
Decreased Plasma C-Peptide

     Factitious hypoglycemia may be induced by surreptitious taking of oral
hypoglycemic agents; the diagnosis of factitious hypoglycemia induced by oral
hypoglycemic agents is given in the next Table:

Diagnosis of Factitious Hypoglycemia Induced by Oral Hypoglycemic Agents
Low Serum Glucose
Increased Plasma Insulin
Increased Plasma C-Peptide
Detection of Oral Hypoglycemic Agents in Urine

     Low serum glucose, increased plasma insulin and increased plasma C-peptide
are also found in patients with insulinoma.
Hypoglycemia and Hyperinsulinemia in Falciparum Malaria: Hypoglycemia occurred
in 8 percent of patients with falciparum malaria. The hypoglycemia is induced
by quinine; quinine is the only available parenteral treatment for severe
chloroquine-resistant falciparum malaria. Quinine stimulates insulin secretion
which may partially account for the hypoglycemia. Other possible mechanisms
include large glucose requirements of the malaria parasites (White, N.J. et al.,
N. Engl. J. Med. 309, 61-66, 1983).

GLYCOHEMOGLOBIN (SEE HEMOGLOBIN A$_{1c}$)   Hgb A$_{1c}$

GONOCOCCAL ANTIGEN ASSAY
(GONOZYME, ABBOTT)

SPECIMEN: This assay is not used for specimens from the throat or anus. Swabs and transport media are provided by the manufacturer.

Cervix: Specimens from the cervix are obtained as follows: Use a sterile bivalve speculum; moisten the vaginal speculum with warm water before introduction. Do not use lubricating jelly because it may be lethal for gonococci. Wipe the cervix with sterile cotton swabs to remove vaginal secretions. Gently compress the cervix between the blades of the speculum to help to produce endocervical exudate. Rotate the swab in the cervix from 10 to 30 seconds to ensure adequate sampling and absorption by the swab. Avoid contamination by not allowing the swab to contact the vaginal walls.

Urethra in Males: Use a sterile wire swab to obtain specimen from anterior urethra by gently scraping the mucosa.

REFERENCE RANGE: Negative

METHOD: Enzyme immunoassay(EIA) by Abbott Diagnostics Division, North Chicago, Illinois. The test procedure is rapid, taking less than one hour. The swab specimen is incubated with beads; N. gonorrhoeae adheres to the bead. The bead is incubated with anti-gonococcal antibodies which react with gonococci on the bead.

Next, the bead is incubated with horseradish peroxidase-labelled sheep anti-rabbit globulins, which reacted with the antigen-antibody complex on the bead. After another incubation with an appropriate substrate the samples are read with a spectrophotometer to determine the relative quantity of N. gonorrhoeae antigens absorbed to the bead.

INTERPRETATION: The sensitivity and specificity of the EIA method, as compared to culture, are given in the next Table; (Data from Abbott Corp.; Danielson, D. et al., J. Clin. Pathol. 36, 674-677, 1983):

	Sensitivity and Specificity of EIA Method, as Compared to Culture	
	Percent	
Group	Sensitivity	Specificity
Males	98; 87	100; 94
Females	88; 91	98; 100

The Gonozyme (enzyme immunoassay, EIA) assay method takes 1 hour to obtain results; culture (incubation at 37° in a 5% CO_2 atmosphere on Thayer-Martin medium) takes 24 to 48 hours. The EIA method is not dependent upon viable organisms.

GONORRHEA CULTURE

SPECIMEN: Specimens may be obtained from cervix, vagina, urethral specimens from males, throat, or joint fluid. Specimens are obtained using a swab; the Culturette is convenient to use. Do not allow specimen to dry and avoid refrigeration.

Cervix: Specimens from the cervix are obtained as follows: Use a sterile bivalve speculum; moisten the vaginal speculum with warm water before introduction. Do not use lubricating jelly because it may be lethal for gonococci. Wipe the cervix with sterile cotton swabs to remove vaginal secretions. Gently compress the cervix between the blades of the speculum to help to produce endocervical exudate. Use a swab to obtain specimens for culture. The Gram Stain should not be done; the Gram stain is insensitive (38-69%) and cervical smear may be misleading because the normal cervix contains organisms that closely resemble N. gonorrhoeae. Diagnosis requires culture and a positive oxidase test.

Rectal Cultures: Rectal cultures should be done on all women at the same time cervical cultures are done because the yield is increased. It should be done on all patients that practice rectal sexual activity.

Urethra in Males: Specimens from the urethra in males are obtained as follows:
Urethral specimens should not be collected until at least one hour after
urination. Collect urethral discharge directly or from discharge obtained by
"milking" the urethra. If no discharge is available, insert an unmoistened thin
swab into the distal urethra for approximately 2cm and gently rotate it.
Another approach to obtaining a specimen in an asymptomatic male, which is not
embarrassing to the patient, is as follows: Collect a first void (not
mid-stream) urine specimen. Send to laboratory for "GC Gram stain and culture
of sediment." Gram stain should be done; a positive smear, i.e., typical Gram
negative diplococci within neutrophils is sufficient for office diagnosis;
confirmation should be obtained by culture.
Throat Culture: Depress tongue and expose pharynx. Swab posterior pharynx,
tonsils, and tonsillar fossae vigorously. Smears are insensitive and
nonspecific and should not be done.
     Specimens should be inoculated on selective medium in an enhanced CO_2
atmosphere for shipment to the laboratory. Transport: The best nutritive
transport system is the JEMBEC (John E. Martin Biological Environment Chamber)
which contains the appropriate media and a moisture activated CO_2 generating
tablet. Specimen must be inoculated on Martin-Lewis or Thayer-Martin or New
York City. New York City is the name of a media that supports the growth of the
gonorrheae organism. Neisseria gonorrhoeae is a very fragile bacteria and will
lose viability if allowed to dry. Plates for inoculation must be at room
temperature before use. Specimen must not be refrigerated. Plates are placed
in a CO_2 incubator.
REFERENCE RANGE: No N. gonorrheae organisms isolated.
METHOD: Neisseria gonorrhoeae is an aerobic, oxidase-positive, Gram negative
diplococcus. Specimens are almost always contaminated and thus must be
inoculated on selective media that will allow Neisseria to grow (chocolate agar
or other appropriate media) but containing antibiotics and antifungal agents
that inhibit the growth on contaminants. For instance, the modified
Thayer-Martin medium also contains vancomycin; cloistin, nystatin and
trimethoprim; the Martin-Lewis medium also contains vancomycin, cloistin,
trimethoprim and anisomycin (an antifungal agent). The JEMBEC system is the
recommended system to be used in the private office setting. Other systems
available are the Microcult system (Ames Company) and the Isocult system (SKF
Diagnostics). Bring the media to room temperature before inoculation.
     The swab is used to make a "Z" streak onto the surface of the medium; then
a needle is used to streak the inoculum.
     The plates are then incubated in a CO_2 incubator or a candle jar at 37°C;
moisture is maintained in the jar to prevent the medium from drying out.
     There are commercially available systems such as JEMBEC system and the
Gono-Pak system.
     The oxidase test is used to confirm N. Gonorrheae and is done on isolated
colonies; morphology (gram-negative diplococci) should be confirmed by Gram
stain. The organisms, Moraxella and N. meningitidis, are part of the normal
pharyngeal flora and morphologically resemble N. Gonorrhoeae; these organisms
are also oxidase-positive. Therefore, diagnosis must be confirmed by
carbohydrate fermentation pattern or by serologic methods.
INTERPRETATION: Cervical Specimens: Eighty to ninety percent of women with
endocervical gonorrheae have cultures positive for N. gonorrhoeae (Handsfield,
H.H., Hosp. Pract. 17, 99-116, 1982).
Urethral Specimens from Males: Gonococcal urethritis is usually accompanied by
urethral exudate.
Throat: N. gonorrhoeae has been cultured from the throat in 10 percent or more
of patients with genital gonorrheae; patients who participate in orogenital sex,
particularly male homosexuals, are at increased risk. Patients usually have no
symptoms is relation to the throat. Rarely, acute exudate pharyngotonsillitis
has been cause by gonoccoci and may be the only source of disseminated
gonococcal infection.

GRAM STAIN

SPECIMEN: Two specimens should be obtained; one for gram stain and the other for culture. Gram stained smears of exudates, abscesses or infected body fluids should be obtained; Gram stained smear of exudates should be obtained from all patients with acute urethritis, pelvic inflammatory disease, pneumonia (where there is productive cough), selected patients with infections of the skin or urinary tract.

For specimen collection, use sterile container, culturette; transport specimen to the laboratory as soon as possible and refrigerate.

REFERENCE RANGE: Depends on site of origin of specimen.

METHOD: Gram stain technique is done as follows: (Gulick, P. et al, Medical Clinics of North America, 67, 39-55, 1983):

Gram Stain Technique
1. Make a thin smear of the material onto a clean slide and allow the slide to air dry.
2. Fix the material onto the slide. This can be done by placing the slide on the slide warmer set at about 70°C or by passing the slide, right side up, through a Bunsen burner flame 3 or 4 times.
3. Overlay the smear with crystal violet solution for 1 minute.
4. Wash slide thoroughly with water.
5. Overlay the smear with Gram iodine for 1 minute.
6. Wash slide thoroughly with water.
7. Flood the surface of the slide with the decolorizer, acetone-alcohol, until no violet color washes off the slide. This usually requires about 10 seconds or less.
8. Wash slide thoroughly with water.
9. Overlay the smear with safranin counterstain for 1 minute.
10. Wash the slide thoroughly with water and allow the excess water to run off the slide, then allow the slide to air dry or blot dry with bibulous paper.
11. Examine the stained smear under low power to observe the nature of the specimen, then examine under 100 x (oil immersion) to observe the bacteria present. Gram-positive bacteria stain dark blue to purple and gram-negative bacteria stain pink-red; the nuclei of polymorphonuclear leukocytes should stain pink-red.

INTERPRETATION:

Throat Culture: Gram stained smears have been recommended as a method for the early diagnosis of streptococcal pharyngitis.

Vaginal Specimens: Gardnerella vaginitis and Candida albicans ("yeast vaginitis")

Cervical Specimens: Directions for obtaining cervical specimens are described under Gonorrheae Culture. One-half of culture-positive women have positive smears for gonococci. A positive smear for gonococci show gram-negative diplococci within polymorphonuclear leukocytes.

Urethritis in Males: Acute urethritis in males is classified as gonococcal and nongonococcal. Examine exudate by Gram stain; more than four polymorphonuclear leukocytes per oil immerson field in an average of five oil fields indicates urethritis. The presence of gram-negative diplococci within polymorphonuclear leukocytes indicates gonococcal infection.

GROSS CYSTIC DISEASE FLUID PROTEIN (GCDFP)

SPECIMEN: Red top tube, separate serum

REFERENCE RANGE: Normal Adult Female: 0-50ng/ml
 Normal Pregnant Female: 0-400ng/ml (Increased in Third Trimester)
 Benign Breast Diseases: 0-150ng/ml
 Metastatic Breast Carcinoma: Suspicious, 100-150ng/ml; Abnormal, >150ng/ml

METHOD: RIA

INTERPRETATION: Gross cystic disease fluid protein is a glycoprotein with a molecular weight of 15,000 daltons; it is found in the serum and is useful in following metastatic breast cancer, especially in early detection of recurrence before the clinical symptoms return. Assay of GCDFP is useful in about 1/3 of the cases of metastatic breast cancer.

Ref: Haagensen, D.E. et al., Ann. Surg. 185, 279-285, 1977; Haagensen, D.E. et al., Cancer, 42, 1646-1652, 1978; Haagensen, D.E. et al., J. Natl. Cancer Inst. 62, 239-247, 1979.

GROWTH HORMONE (GH), SERUM

SPECIMEN: Red top tube, separate serum. A.M. levels are obtained after an overnight fast.

REFERENCE RANGE: Children and adult, baseline, resting: 0-10ng/ml. To convert conventional units in ng/ml to international units in mcg/liter, multiply conventional units by 1.00.

METHOD: RIA

INTERPRETATION: Growth hormone assays are done to evaluate: (1) Deficiency of growth hormone in small children or (2) Excess of growth hormone such as occurs in gigantism or acromegaly.

(1) Growth hormone deficiency is usually idiopathic; the deficiency may be isolated or it may be accompanied by a deficiency of other pituitary hormones. Growth hormone is low in normal subjects; therefore, a variety of stimuli are used to evaluate impaired secretion; these are listed in the next Table:

Stimuli for Growth Hormone
Insulin
Arginine
L-Dopa
Glucagon
15 Minutes following Vigorous Exercise

Arginine-Insulin Stimulation Test: Arginine and insulin are often given in a piggy-back fashion in order to approach 100 percent sensitivity for growth-hormone stimulation test. In this test, arginine (0.5 gm/kg) is infused over 30 minutes; blood specimens are obtained at 15 minute intervals. At 60 minutes, give 0.075 U/kg regular insulin I.V. and collect blood specimens at 15, 30, 45 and 60 minutes. Observe the patient continuously for hypoglycemia; give glucose (50 percent) if the patient develops clinical signs and symptoms of hypoglycemia. As collected, each sample should be centrifuged and separated. A decrease in serum glucose of 50 percent or more is required for the insulin infusion test to be valid.

The details of the test are as follows: The patient is admitted to the hospital overnight for testing in the morning. Prepare for the test as follows: NPO after midnight. Obtain 10 percent solution of arginine hydrochloride from pharmacy (10 gms/dl); obtain sufficient arginine to give 0.5 gms per kg body weight. Have pharmacy prepare regular insulin so that 1 unit=1 ml; regular insulin is administered intravenously in a dosage of 0.05 to 0.1 unit per kg body weight. Have available a solution of 50% glucose in case the patient develops clinical symptoms and signs of hypoglycemia following I.V. insulin. In the A.M., start I.V. with normal saline solution and 3-way stopcock. Label red top vacutainer tubes as follows: -30, 0, +15, +30, +45, +60, +75, +90, +105, +120; these numbers correspond to minutes of the test. Follow the schedule for injection of arginine and insulin and drawing blood specimens as given in the next Figure:

Schedule for Arginine and Insulin Infusion and Time to Obtain
Serum Specimens for Growth Hormone Stimulation Tests

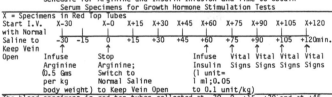

The blood specimens in red top tubes collected at -30, 0, +15, +30 and at +45 minutes are for growth hormone assay; the blood specimens in red top tubes collected at +60, +75, +90, +105 and +120 minutes are for growth hormone, glucose and cortisol assays.

Results of arginine-insulin stimulation test are shown in the next Figure:

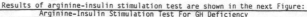

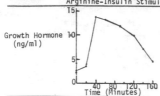

Growth Hormone (ng/ml)

Time (Minutes)

The results in this Figure show stimulation of growth hormone, indicating a normal response; subjects with growth deficiency show no stimulation of growth hormone during the arginine-insulin stimulation test.

L-Dopa Stimulation Test: The L-dopa (L-dihydroxyphenylalanine, levodopa) test is done as follows: Give L-dopa orally (less than 30 pounds, 125 mg; 30 to 70 pounds, 250 mg; > 70 pounds, 500 mg). Obtain serum specimens as follows: -15, 0, 30, 60, 90 and 120 minutes for growth hormone assay. Serum growth hormone normally peaks between 30 and 120 minutes.

(2) Growth Hormone Excess: Usually growth hormone assay per se is sufficient to indicate the diagnosis of acromegaly; growth hormone is almost always increased in acromegaly.

HAPTOGLOBIN
SPECIMEN: Red top tube; separate serum
REFERENCE RANGE: 25-180mg/dl; haptoglobin is absent in the first three months of life. To convert conventional units in mg/dl to international units in g/liter, multiply conventional units by 0.01.
METHOD: Radial immunodiffusion
INTERPRETATION: Haptoglobin is a serum protein that has the capacity to bind hemoglobin in vivo and in vitro. This test is generally used to help in the diagnosis of anemia. The changes of serum haptoglobin in disease are shown in the next Table:

Changes of Serum Haptoglobin in Disease	
Decreased (<30mg%)	Increased (>200mg%)
Hemolytic Anemia	Tissue Destruction:
Megaloblastic Anemia	Infections; Malignant Neoplasms;
Chronic Hepatocellular Disease	Collagen Diseases
Infectious Mononucleosis	Obstructive Jaundice
Toxoplasmosis	Inflammatory Reactions & Tissue Proliferation
Newborns	

Haptoglobin is a "reactive" protein that migrates with alpha-2 globulin band on protein electrophoresis.

HEAVY METAL SCREEN, URINE (ARSENIC, MERCURY, ANTIMONY, BISMUTH)
SPECIMEN: Random urine, refrigerate at 4°C
REFERENCE RANGE: Negative
METHOD: Reinsch test
INTERPRETATION: See ARSENIC, MERCURY

HEINZ BODY STAIN
SPECIMEN: Lavender (EDTA) top tube
REFERENCE RANGE: No Heinz bodies identified
METHOD: Oxidative denaturation of hemoglobin to Heinz bodies using acetylphenylhydrazine; staining of Heinz bodies using crystal violet.
INTERPRETATION: The three main conditions associated with the presence of Heinz bodies are given in the next Table:

Conditions Associated with Heinz Bodies
Exposure to Certain Chemicals or Drugs
Deficiency of one of the Reducing Systems of the Blood such as Glucose-6-Phosphate Dehydrogenase Deficiency
Presence of an Unstable Hemoglobin

Heinz bodies are insoluble inclusions of hemoglobin precipitated within the red blood cell.

HEMATOCRIT

SPECIMEN: Lavender(EDTA) top tube or microhematocrit (lavender) capillary tube containing EDTA. Specimens may be stored for 8 hours at room temperature or 24 hours in the refrigerator. Anticoagulants, other than EDTA may be used, e.g., heparin.

REFERENCE RANGE: Microhematocrit results average 3 percent higher than hematocrit obtained by calculation, that is, Hct = MCV x RBC's; trapping of leukocytes, platelets and plasma occurs with "spun" hematocrits.

Values of "spun" hematocrit are given in the next Table:

Normal Values of "Spun" Hematocrit	
Age	Percent
Birth	44-64
14-90 days	35-49
6 months-1 year	30-40
4-10 years	31-43
Adult, Male	42-52
Adult, Female	37-47

METHOD: Microhematocrit using capillary tube; calculated from determination from MCV and RBC's using Coulter counter or other automated counter from the relationship Hct = MCV x RBC's.

INTERPRETATION: Hematocrit is <u>decreased</u> in the conditions listed in the next Table:

Decrease in Hematocrit
Anemias (Irrespective of Cause)
Recovery Stage After Blood Loss

Hematocrit is <u>increased</u> in the conditions listed in the next Table:

Increase in Hematocrit
Polycythemia
Erythrocytosis of Dehydration
Hemoconcentration as in Shock associated with Trauma, Surgery and Burns
High Altitude

HEMOGLOBIN

SPECIMEN: Lavender(EDTA) top tube or microtube (lavender) which contains EDTA. The specimen is stable at room temperature for up to 8 hours and in the refrigerator for up to 24 hours. The anticoagulant, heparin (green top tube) may also be used.

REFERENCE RANGE: The reference range, with age, is given in the next Table:

Reference Range with Age	
Age	g/dl
Cord..................	14-20
Birth................	15.0-24.0
1 Week...............	13.0-20.0
1 Month..............	11.0-17.0
6 Months.............	10.5-14.5
1 Year...............	11.0-15.0
10 Years.............	11.0-16.0
15 Years (Male)......	14-18
(Female)......	12-16

At birth, hemoglobin values are very high (15-24 g/dl); this high hemoglobin concentration is attributable to the relatively low levels of oxygen in-utero. It decreases markedly and at <u>two to three months</u> of age, it reaches a value of about 10-14 g/dl. Then, hemoglobin slowly increases until it reaches the adult value at about age 15.

Fetal hemoglobin constitutes about <u>60 percent</u> of the hemoglobin at term; by 6 months of age, fetal hemoglobin constitutes <u>5 to 6 percent</u> of the total and <u>1 to 2 percent</u> by <u>adulthood</u>.

METHOD: The most popular method is the cyanmethemoglobin method.

INTERPRETATION: Decreased hemoglobin is caused by anemia of all types (see Anemia Panel).

The causes of <u>increased</u> hemoglobin are <u>polycythemia vera</u>, <u>secondary polycythemia</u>, vigorous exercise and high altitude.

HEMOGLOBIN A$_{1c}$ (GLYCOSYLATED HEMOGLOBIN)

SPECIMEN: The patient need not be fasting when the blood is drawn. Blood is collected in a grey top tube (oxalate as anticoagulant, fluoride as preservative); or lavender (EDTA) or green (heparin) top tubes. Blood is centrifuged; the red blood cells are separated and hemolyzed.

REFERENCE RANGE: Normal: 4.0-8.2%

METHOD: Cation-exchange column chromatography, such as Bio-Rad columns; hemoglobin F has been reported to co-chromatograph with HgA$_{1c}$; thus, conditions associated with elevated levels of hemoglobin F (thalassemias and certain hemoglobinopathies) are associated with "false" elevations of hemoglobin A$_{1c}$ (Goldstein, D.E. et al., Diabetes 31, Suppl. 3, 70-78, 1982).

INTERPRETATION: Glycosylated hemoglobin is a measure of chronic blood sugar control in the diabetic. Assay of hemoglobin A$_{1c}$ is useful as a means of monitoring carbohydrate control in diabetic patients and as a means of detecting diabetes.

When diabetic patients are carefully and optimally regulated, the levels of glycosylated hemoglobin begin to drop toward normal in from three to five weeks (Koenig et al, N. Engl. J. Med. 295, 417-420, 1976). Hemoglobin A$_{1c}$ can be measured at infrequent intervals in non-fasted patients to determine whether the patient's diabetes is well controlled; this monitoring will allow a more objective assessment of therapeutic efficacy (Koenig, R.J. and Cerami, A., Ann. Rev. Med., 11, 29-34, 1980; Tegos, C. and Beutler, E., Blood 56, 571-572, 1980).

The mean blood glucose concentration(MBG) can be calculated from the value of the HgA$_{1c}$ by the following equation:

$$MBG = 33.3(HgA_{1c}) - 86$$

However, an error of 1 percent in the measurement of glycosylated hemoglobin leads to an error of approximately 35 mg per deciliter in the blood glucose level.

Glycosylated hemoglobin assay provides information about the degree of long term control that is not otherwise obtainable in the usual clinical setting (Nathan, D.M. et al., N. Engl. J. Med. 310, 341-346, Feb. 9, 1984). The test is useful for identifying insulin-treated patients who are at risk of having serious hypoglycemic reactions; patients whose glycosylated hemoglobin levels are within the nondiabetic range have a high frequency of hypoglycemia (Goldstein, D.E., N. Engl. J. Med. 310, 384-385, Feb. 9, 1984).

Glycosylated hemoglobin cannot be used to monitor control in diabetic patients with chronic renal failure. Glycosylated hemoglobin is significantly lower in patients with chronic renal failure than in normal controls; the lowered levels in chronic renal failure are indicative of shortened erythrocyte survival secondary to chronic renal failure and are not due to hemodialysis (Freedman, D. et al., J. Clin. Path. 35, 737-739, 1982).

There is a significantly higher incidence of major congential anomalies in the offspring of women with elevated HgA$_{1c}$ values in early pregnancy as compared to women with normal levels of HgA$_{1c}$; thus, poorly controlled diabetes is associated with an increased risk of such anomalies (Miller, E. et al., N. Engl. J. Med. 304, 1331-1334, 1981; Freinkel. N., N. Engl. J. Med. 304, 1357-1359, 1981).

The mechanism that has been proposed for nonenzymatic glucosylation of proteins is shown in the next Figure; (Guthrow et al., Proc. Natl. Acad. Sci. 76, 4258-4261, 1979):

Nonenzymatic Glucosylation of Proteins

Glucose — Schiff Base — Ketoamine Structure — Glycosylated Protein

HEMOGLOBIN BART'S (ALPHA-THALASSEMIA SCREEN)

SPECIMEN: Lavender(Purple)(EDTA) top tube.

REFERENCE RANGE: <1 percent; hemoglobin Bart's may be found in the blood of normal infants; it disappears at about 3 months of age.

METHOD: Hemoglobin Bart's is quantitated by applying a freshly prepared hemolysate to a prepacked ready to use column (Isolab, Akron, Ohio). Hemoglobin Bart's is eluted in the first fraction; remaining hemoglobins are eluted with a second buffer. The absorbances at 415nm of each fraction are used to calculate the percent hemoglobin Bart's.

The migration of Bart's on electrophoresis at pH 8.6 on cellulose acetate is shown in the next Figure:

Electrophoresis of Hemoglobins on Cellulose Acetate at pH 8.6

Cathode (-) A_2 F A_1 Barts   H
Origin   C,E,O|αα| SD|G| αα αα   γγ   ββ (+)
                  |ΔΔ| γγ ββ   γγ   ββ [anode]

Carbonic
Annhydrase

INTERPRETATION: Alpha-thalassemia is a condition characterized by reduced synthesis of alpha chains. The normal hemoglobins that contain alpha chains are hemoglobin A(two alpha chains and two beta chains), hemoglobin A_2(two alpha chains and two delta chains) and hemoglobin F(two alpha chains and two gamma chains). Hemoglobin Bart's (four gamma chains) is formed when extra gamma chains of hemoglobin F combine in tetramers. Assay of hemoglobin Bart's is done on cord blood or blood of the newborn.

The characteristics of alpha-thalassemias are given in the next Table:

		Characteristics of Alpha Thalassemias	
		Hemoglobin Pattern	
Condition	Genotype and Risk	% Bart's in Cord Blood	Other
Normal	○○ ○○	<1%	
Heterozygous (Silent Carrier) (Alpha-Thal-2 Trait)	○○ (Silent Carrier x ● ○ Normal) (Risk = 1/2) One of 4 Alpha-Globulin Genes Deleted	0.8 to 3	Normal
Heterozygous (Alpha-Thal-1 Trait)	● ○ (Alpha-Thal. Trait x ● ○ Normal) (Risk = 1/2) Two of 4 Alpha-Globulin Genes Deleted	3 to 10	In Adults, Hemoglobin H Inclusion Bodies are Found in About 1 in 10^5 Red Cells. Differential Dx.: Iron Deficiency Anemia, Beta-Thal. or Decreased Alpha-Chain/Beta Chain Ratio (approx.0.6)
Heterozygous (Alpha-Thal-2 Trait)	○○ (Alpha-Thal. Trait x ● ● Normal) Two of 4 Alpha-Globulin Genes Deleted	3 to 10	
Hemoglobin H Disease	○○ (Alpha-Thal Trait x ● ● Silent Carrier) (Alpha-Thal Trait x Hemoglobin CS Heterozygote) (Risk = 1/4) Three of 4 Alpha-Globulin Genes Deleted	20 to 30; Level in Adults Variable	4-30% Hemoglobin H
Homozygous	● ● (Risk = 1/4) ● ● All Alpha-Globulin Genes Deleted	>80	No Hemoglobin A or F; Hemoglobin H Present

Hemoglobin H has four beta chains.

Clinical Severity:

Heterozygous (Silent Carrier) (Alpha-Thal-2 Trait): Benign condition; no hemolytic abnormalities; diagnosis not made reliably in adults.

Heterozygous (Alpha-Thal-1 Trait): Mild; very mild in Blacks; clinical picture similar to beta-thal trait with very mild anemia and microcytosis.

Hemoglobin H Disease: Variable; usually chronic anemia; MCV and MCH are decreased; hypochromia; reticulocytes = 4 to 5 percent.

Homozygous: Lethal; hydrops fetalis with hemoglobin Bart's.

HEMOGLOBIN ELECTROPHORESIS

SPECIMEN: Lavender (EDTA) top tube

REFERENCE RANGE:

Normal Values of Hemoglobin		
Normal Values:	Adult	Newborn
HbA$_1$	96-98.5%	20%
HbA$_2$	1.5-4%	2%
HbF	0-2.0%	80%

METHOD: Electrophoresis at alkaline pH(8.6) and acid pH using citrate buffer. The electrophoretic migration of the hemoglobins on cellulose acetate at pH 8.6 is shown in the next Figure:

Electrophoresis of Hemoglobins on Cellulose Acetate at pH 8.6

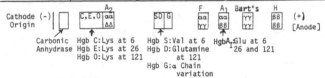

Carbonic   Hgb C:Lys at 6   Hgb S:Val at 6   HgbA₂:Glu at 6
Anhydrase Hgb E:Lys at 26   Hgb D:Glutamine   26 and 121
           Hgb O:Lys at 121   at 121
                               Hgb G:α Chain
                                  variation

The electrophoretic migration of the hemoglobins on citrate agar at pH 6.2 is shown in the next Figure:

Electrophoresis of Hemoglobins on Citrate Agar at pH 6.2

INTERPRETATION: Characteristics of hemoglobinopathies are given in the next Table:

Characteristics of Hemoglobinopathies						
Hemoglobinopathy	Hgb A$_1$	Hgb S	Hgb C	Hg A$_2$	Hgb F	Other
Normal	96-98.5%	--	--	1.5-4%	0-2.0%	
Sickle cell Trait	60-80%	20-40%	--	--		N
Sickle Cell Anemia	--	60-99%	--	--	1-40%	
C Trait	70%	--	30%	--	--	
C Disease	--	--	95%	--	5%	
Sickle C Disease	--	45-55%	45-55%	--	--	
Beta Thal. Minor (↑ Beta Chains)	100-A$_2$-F	--	--	N to ↑	N to 7	
Beta Thal. Major (↓ Beta Chains)	--	--	--	Low or↑	7 to 100	
Alpha Thal. Minor (↓ Alpha Chains)	--	--	--	N to Dec.	N to Dec.	Υ-4

Sickle Trait: This is a heterozygous state showing HbA$_1$, and HbS with a normal amount of HbA$_2$. The HbS concentration may vary from 20-50%.

Sickle Cell Anemia: This is a homozygous state showing almost exclusive HbS. A small amount of HbF (>15%) may also be present.
C Trait: This is a heterozygous state showing HgA$_1$, and HgS with a normal amount of HgA$_2$. The HgC concentration is usually about 30%.
C Disease: This is a homozygous state showing almost exclusively HbC plus up to 5% HbF.
Sickle C Disease: This is a heterozygous state showing HbS (45-55%) and HbC (45-55%).
Beta Thalassemia Minor: This state shows HbA$_1$ and HbA$_2$. The HbA$_2$ band is elevated to twice normal levels with a concentration of 5.8% in approximately 60% of the cases. HbA$_2$ values between 4-5% probably represent thalassemia minor, but repeated red cell morphologic abnormalities must be present for confirmation.
Beta Thalassemia Major: This condition shows HbF (70-90%), HbA$_1$, (2-20%) and HbA$_2$ (2-8%).
Beta Thalassemia-S Disease: This condition shows HbA$_1$ (0-25%), HbS (50-90%), HbA$_2$ (3-8%) and HbF (2-20%).
Beta Thalassemia-C Disease: This condition shows HbA$_1$ (0-25%), HbF (2-20%) and HbC (60-90%), HbA$_2$ cannot be measured.

HEMOGLOBIN F

Hemoglobin F

SPECIMEN: Lavender(Purple)(EDTA) top tube
REFERENCE RANGE: The level of fetal hemoglobin F with age is shown in the next Table:

| Fetal Hemoglobin with Age ||
Age	Hemoglobin F (Percent)
Newborn(Cord Blood)	50-85
1 Month	50-75
2 Months	25-60
3 Months	10-35
6 Months	8
1 Year	<2
Normal Adult	<2

METHOD: Hemoglobin F resists denaturation under alkaline conditions; most other hemoglobins are denatured at basic pH and are then precipitated with ammonium sulfate. Fetal hemoglobin remains soluble and is quantitated spectrophotometrically (NCCLS Proposed Standard: H13-P, Guidelines for the quantitative measurement of fetal hemoglobin by the alkali denaturation method).
INTERPRETATION: Hemoglobin F consists of two alpha chains and two gamma chains. Conditions associated with an increase in fetal hemoglobin are given in the next Table:

| Conditions Associated with Increased Fetal Hemoglobin ||
Condition	Hemoglobin F (Percent)
Thalassemia Major	10-90
Thalassemia Minor	2-12 Normal in more than 50 percent of patients; may be as high as 12 percent
HgS-Thalassemia	5-20
HgSS	1-40
HgC	<7
Hereditary Persistence of Fetal Hemoglobin:	
Homozygote	100
Heterozygote	20-30

Hemoglobin F is elevated in certain hemoglobinopathies and thalassemia. It may be elevated in acquired conditions, e.g., megaloblastic anemia, myelofibrosis, aplastic anemia, leukemias, erythroleukemia, refractory anemias, pregnancy and paroxysmal nocturnal hemoglobinuria.

S. Bakerman

HEMOLYTIC DISEASE OF THE NEWBORN(HDN) PANEL

Specimens and proposed tests for the work-up of infants with HDN are given in the next Table: (Komarmy, L. Pathologist, 266-267, April 1983):

Specimens and Proposed Tests for Infants with HDN			
	Test Group	Test	Specimen
Infant	Blood Bank	Blood Type	Red Top Tube
		Direct Coombs Test	Lavender(EDTA) Top Tube
		Crossmatch	
	Routine Hematology	Hemoglobin, Hematocrit,	Lavender(EDTA) Top Tube
		White Cell Count	Lavender(EDTA) Top Tube
		Wright Stained	Finger-Stick
		Peripheral Blood Smear	
	Arterial Blood Gases	pH, pO_2, pCO_2	
Mother	Blood Bank	Blood Type	Red Top Tube
		Antibody Screen	Red Top Tube
		Crossmatch	
	Routine Hematology	CBC with Indices	Lavender(EDTA) Top Tube
Father	Routine Hematology	CBC with Indices	Lavender(EDTA) Top Tube

A unit of uncrossmatched blood should be available for exchange transfusion.

HEMOSIDERIN, URINE

SPECIMEN: 30 ml of random urine specimen; no preservative. The specimen must reach the laboratory within one hour of collection; refrigerate if determination not done immediately. Forward to outside laboratory frozen in plastic vial on dry ice.

REFERENCE RANGE: Negative

METHOD: Centrifuge specimen. Examine specimen microscopically for coarse brown granules in epithelial cells; if granules are seen, suspend the rest of the sediment in a fresh mixture of 5 ml of 2 percent potassium ferrocyanide and 5 ml of 1 percent HCl and allow to stand for 10 minutes. If iron is present, the ferric ion reacts with potassium ferrocyanide, $K_4Fe(CN)_6$, to form ferric ferrocyanide, $Fe_4(Fe(CN)_6)_3$ (Prussian blue). Centrifuge, and discard the supernatant. Examine the sediment microscopically. Coarse granules of hemosiderin appear blue.

INTERPRETATION: When excessive hemolysis occurs within the vascular space, hemoglobin is released. Up to 150 mg/dl may be bound to haptoglobin. Amounts in excess of this level are filtered through the renal tubules. Hemosiderin is formed in the tubular cells from hemoglobin; these cells slough into the urine. Urine hemosiderin is only found with intravascular hemolysis. The causes of intravascular and extravascular hemolysis are given in the next Table:

Causes of Intravascular and Extravascular Hemolysis	
Intravascular	Extravascular
Hemolytic Transfusion Reactions	Spherocytosis
Autoimmune Hemolytic Anemia	Sickle Cell Trait
Drug-Induced Hemolysis	Thalassemia Minor
Massive Burns	Other Hemoglobinopathies
Sickle Cell Anemia(Occ.)	
Thalassemia Major(Occ.)	
Paroxysmal Nocturnal Hemoglobinuria	
Intravascular Water (e.g., Prostatic Resection)	
Hemochromatosis	
Microangiopathic Hemolytic Anemia	

HEPATIC PANEL

HEPATIC PANEL
Serum Glutamic Oxalacetic Transaminase(SGOT)
Serum Glutamic Pyruvic Transaminase(SGPT)
Gamma Glutamyl Transpeptidase(GGTP)
Alkaline Phosphatase (Alk. Phos.)
Bilirubin, Total and Direct
Urine Bilirubin
Urine Urobilinogen
Prothrombin Time(PT)
Partial Thromboplastin Time(PTT)
Serum Total Protein, Albumin, Protein Electrophoresis

HEPATITIS A ANTIBODIES, IgM AND IgG

SPECIMEN: Red top tube, separate serum

REFERENCE RANGE: Negative

METHOD: Competitive binding of antibody in patient's serum or plasma with the
binding of a known amount of I-125 labeled antibody to hepatitis A virus (HAV)
coated on to a solid phase.

INTERPRETATION: IgM antibodies reflect recent acute infection with HAV, and
IgG antibodies reflect infection which occurred months to years before. The
changes in relative concentration of hepatitis A antibodies, IgG and IgM,
following exposure to hepatitis A, are shown in the Figure:

Changes in Relative Concentration of Hepatitis A Antibodies, IgG and IgM

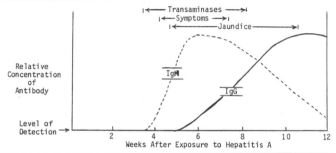

    IgM antibodies are detected in the serum at the same time as symptoms
develop and stay elevated for several months; IgG antibodies are detected for
years after the acute illness.
    Most people with hepatitis A, perhaps 90%, do not develop overt disease.
When it is overt, illness ranges from mild malaise, to prostration with nausea,
vomiting and diarrhea, fever, headaches and myalgia. Hepatomegaly is common but
not splenomegaly. Clinical disease resolves in 0.5 to 1.5 months. Hepatitis A
virus does not appear to cause chronic hepatitis.
    When hepatitis A occurs in epidemics, it is usually referable to a common
source such as contaminated water or the ingestion of shell-fish containing the
virus. The majority of U.S. Nationals returning from travel abroad with acute
viral hepatitis probably have type A hepatitis. There is a high incidence of
HAV disease in military personnel and Peace Corps volunteers living abroad.

HEPATITIS B SURFACE ANTIGEN (HB₈Ag, AUSTRALIAN ANTIGEN, HAA)

SPECIMEN: Red top tube, separate serum
REFERENCE RANGE: Negative
METHOD: RIA or ELISA. New ELISA methods (Auszyme II) are at least as sensitive as RIA methods.
INTERPRETATION: HB_SAg is the surface antigen of the hepatitis B virus. The HB_SAg is detected between one and 4 months following exposure to the hepatitis B virus. The presence of HB_SAg in blood indicates infection with the B virus. HB_SAg may persist in the blood in 3% to 5% of subjects; these patients often develop chronic liver disease. Some patients may develop chronic liver disease and be HB_SAg positive but have no history of acute hepatitis. HB_SAg may be present in some people who have no history of hepatitis or signs of chronic liver disease; these people are referred to as "carriers." The changes in the different antigens and antibodies in serum hepatitis are shown in the next Figure:

Changes in Antigens and Antibodies in Serum Hepatitis
Parenterally Transmitted

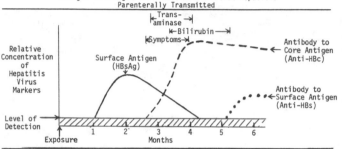

HEPATITIS B ANTIBODY (HB₈Ab: ANTIBODY TO HEPATITIS B SURFACE ANTIGEN: ANTI-HB₈)

SPECIMEN: Red top tube, separate serum
REFERENCE RANGE: Negative
METHOD: RIA or ELISA
INTERPRETATION: The presence of hepatitis B antibody indicates previous hepatitis B infection or exposure; it usually appears about 5 months after exposure to hepatitis.

HEPATITIS B CORE ANTIBODIES, IgM AND IgG

SPECIMEN: Red top tube, separate serum
REFERENCE RANGE: Negative
METHOD: RIA or ELISA
INTERPRETATION: Anti-HBc, IgM, is most useful as a marker for hepatitis B infection in the "window" between the time that the surface antigen (HB_SAg) disappears and the time that the antibody to the surface antigen (Anti-HB_S) appears. The presence of anti-HBc, IgM, indicates current hepatitis B infection.

The presence of anti-HBc, IgG, indicates previous hepatitis B infection. The anti-HBc, IgG antibody persists indefinitely.

HEPATITIS Be ANTIGEN (HBeAg)

SPECIMEN: Red top tube, separate serum
REFERENCE RANGE: Negative
METHOD: RIA
INTERPRETATION: The presence of the "e" antigen indicates the blood is
particularly infectious. The persistence of HBe from 8 to 10 weeks following
acute infection indicates the development of the chronic carrier state and
probable liver disease. Seroconversion of HBeAg to anti-HBe in the acute
stages of infection indicates reduced viral replication and pending resolution
of infection.

The "e" antigen is not found in healthy carriers but is often present in
patients with chronic liver disease particularly those with chronic aggressive
hepatitis. Seroconversion from HBeAg to antiHBe in the chronic carrier state is
prognostic for an improvement in the patient's liver disease status (Lofgren, B.
and Nordenfelt, E., J. Med. Virology 5, 323-330, 1980).

The changes in the different antigens and antibodies in serum hepatitis
are shown in the next Figure:

Changes in Antigens and Antibodies in Serum Hepatitis
Parenterally Transmitted

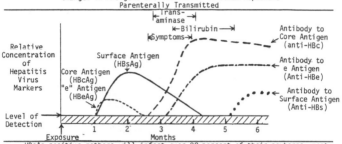

HBeAg positive mothers will infect over 90 percent of their newborns, most
of whom will become chronic carriers; these carriers tend to develop serious
chronic sequellae of hepatitis B infection, such as cirrhosis and may infect
others. Hepatitis B immunoglobulins should be administered to all newborns of
HBeAg positive mothers.

Hepatitis B,
Antibody to
Delta Agent

HEPATITIS B ANTIBODY TO DELTA AGENT, IgG AND IgM

SPECIMEN: Red top tube, separate serum
REFERENCE RANGE: Negative
METHOD: In development by Abbott Laboratories; probably available in 1985.
Currently, competitive inhibition radioimmunoassay is used (Rizzetto, M. et al.,
J. Immunol. 125, 318-324, 1980).
INTERPRETATION: The delta agent discovered in 1977 (Rizzetto, M. et al., Gut
18, 997-1003, 1977) is a virus-like particle consisting of delta antigen and a
ribonucleic acid core; this agent requires hepatitis B virus for replication.
The presence of the delta agent is associated with a higher incidence of
fulminant hepatitis and chronic hepatitis.

Delta agent has been found in groups at high risk for hepatitis B; these
include persons from southern Italy, parts of Africa, drug addicts and
hemophiliacs in western Europe and the United States.

Infection by the delta agent can occur as a coinfection with hepatitis B,
which usually causes acute hepatitis that resolves. When delta agent occurs as
a superinfection of a hepatitis B carrier, an episode of acute hepatitis and the
establishment of persistent delta infection may lead to chronic active hepatitis
and cirrhosis (Rizzetto, M. et al., Ann. Intern. Med. 98, 437-441, 1983).
Fulminant hepatitis is associated with coinfection or superinfection with delta
agent (Smedile, A. et al., Lancet 2, 945-947, 1982).

Delta agent may cause epidemic hepatitis in areas where hepatitis B
infection is endemic (Hadler, S.C. et al., Ann. Intern. Med. 100, 339-344,
1984).

HEPATITIS, CHRONIC, PANEL

The conditions to consider and the tests to perform in patients with chronic hepatitis are given in the next Table:

Chronic Hepatitis	
Condition	Tests
Hepatitis B	Hepatitis B Surface Antigen(HBsAg)
	Hepatitis B Surface Antibody
	Core Antibody
Cytomegalovirus	Cytomegalovirus Antibody Titer
Infectious Mononucleosis	Heterophil Antibody
"Autoimmune" Hepatitis	Anti-Smooth Muscle(ASM) Antibody
Metabolic Causes:	
Wilson's Disease	Ceruloplasmin; Urinary Copper
Alpha-I-Antitrypsin	Serum Electrophoresis
Deficiency	Quant. of Alpha-1-Antitrypsin
Hemochromotosis	Serum Iron and Iron Binding Capacity
	Serum Ferritin

Liver biopsy is the most definitive approach.

HEPATITIS PANEL (HB$_s$Ag: ANTI-HB$_s$: ANTI-HB$_c$: HEPATITIS A: IgM (ACUTE PHASE), IgG (CONVALESCENT)

SPECIMEN: Red top tube, separate serum
REFERENCE RANGE: Negative
METHOD: RIA or ELISA
INTERPRETATION: This profile is used for the diagnosis of hepatitis A and hepatitis B; the diagnosis of hepatitis C (non-A, non-B) is made by exclusion. The interpretation of results is given in the next Table:

Interpretation of Results of Tests for Hepatitis						
		Anti-HBc		Hepatitis A		
HBsAg	Anti-HBs	IgM	IgG	IgM	IgG	Interpretation
−	−	−	−	−	−	Compatible with hepatitis C
+	−	+	−	−	−	Early acute hepatitis B
−	+	−	+	−	−	Convalescent hepatitis B
−	−	−	+	−	−	Late convalescent hepatitis B
−	−	−	−	+	−	Early acute hepatitis A
−	−	−	−	−	+	Hepatitis A in the past

HEPATITIS, VIRAL, TYPES, NEONATE

Types of viral hepatitis and the tests to detect these different types are given in the next Table; (Zucker, G.M. and Clayman, C.B., JAMA 247, 2011, 1982):

Viral Hepatitis	
Virus	Assay
Hepatitis A	Hepatitis A, Antibodies,
	IgM: Early Acute Hepatitis A
	IgG: Hepatitis A in the Past
Hepatitis B	HB$_s$Ag; Anti-HB$_s$; Anti-HBc; HBeAg
Non-A, Non-B	No Specific Tests
Cytomegalic Inclusion-Virus	Cytomegalovirus(CMV) Antibodies
	IgM: Current Infection
	IgG: Infection in the Past
Rubella (Congenital or Acquired)	Rubella (German Measles) Antibodies
	IgM: Current Infection
	IgG: Infection in the Past
Herpes Simplex (Type 2 more than Type 1)	Herpes Simplex Antibodies
	IgM: Current Infection
	IgG: Infection in the Past
	These Tests Usually Do Not Distinguish Between Type 1 and 2.

In the neonate, hepatitis associated with viruses other than A, B or non-A, non-B may occur; these are listed in the previous Table. Other stigmata of these viral diseases are usually present.

When viral isolation and serologic findings are negative, and metabolic and hereditary causes are excluded, there remain up to 75 percent of cases in which the etiology remains unknown.

HEROIN

SPECIMEN: 50 ml random urine without preservatives; store at -20°C.

REFERENCE RANGE: None detected.

METHODS: Thin-layer chromatography; radioimmunoassay; gas liquid chromatography; latex agglutination-inhibition; hemagglutination-inhibition; spectrophotofluorometry; free radical assay.

INTERPRETATION: Heroin is diacetylmorphine and, in the body, is rapidly deacetylated to morphine; urine is the major route of excretion and morphine is excreted either free (5 to 20 percent) or conjugated, primarily to glucuronide. Following injection, 50 percent of the total dose is excreted in the urine within the first 8 hours and about 90 percent during the first 24 hours. Measurable urinary free morphine persists for 48 hours or longer. The time since the last dose at which morphine is detected in the urine varies with the method of detection as follows: hemagglutination-inhibition, >4 days; RIA, >4 days; thin-layer chromatography, 4 days (Catlin, D.H., Am. J. Clin. Path. 60, 719-728, 1973). The sensitivity of thin-layer chromatography is about 1000ng/ml of urine.

Morphine may appear in the urine within six minutes of an intravenous injection of heroin.

Quinine is a common diluent of heroin; it may be detected in urine for several days after detectable morphine has disappeared.

The medical complications of narcotic addiction are given in the next Table (Swartz, M.N., N. Engl. J. Med. 285, pg. 42, 1971):

Medical Complications of Narcotic Addiction
Hepatitis
Pulmonary Disease: Edema, Pneumonia, Lung Abscess, T.B., Angiothrombotic Pulmonary Hypertension
Vascular Lesions: Local Arterial Occlusion, Local Phlebitis, Mycotic Aneurysm, Necrotizing Angiitis
Cardiac Lesions: Endocarditis
Miscellaneous Infections: Local abscess, Septicemia, Malaria, Tetanus, Syringe-Transmitted Syphilis

Street heroin is generally 5-10% actual heroin; the usual euphoric dose taken by abusers is equivalent to 10-20 mg of morphine.

The classical triad of narcotic (heroin, morphine and codeine) overdose is coma, respiratory depression, and miosis (pinpoint pupils).

In the treatment of narcotic overdose, a specific antidote exists; naloxone (Narcan-Endo) can reverse the pharmacological effects of opioid agents. It reverses CNS depression caused by narcotics but has no effect on CNS depression of non-narcotic etiology. The most common complication of narcotic overdose is pulmonary edema (Cuddy, P.G., Crit. Care Quart. 4, 65-74, Mar. 1982).

HERPES SIMPLEX CULTURE

SPECIMEN: Specimens may be obtained from vesicular skin lesions and other body sites, specifically the throat, urine, leukocytes and cerebrospinal fluid. Use viral transport media (such as sterile one percent bovine albumin); this media is available in the hospital microbiology laboratory, in refrigerators in Pediatrics, Neonatal Intensive Care Units and the Emergency Department. Mucous membranes or other lesions should be swabbed; the swab is rinsed vigorously in the transport media and squeezed-out against the glass and discarded. Specimens should be brought to the laboratory as soon as possible; if delay is unavoidable, specimens must be stored at -70°C.

REFERENCE RANGE: Negative

METHOD: Herpes simplex virus grows readily on tissue culture cells; cell strains and lines of human origin (human embryo, Hela, etc.) support the growth of herpes simplex virus. Growth is rapid and the characteristic cytopathogenic effects, with ballooning and rounding of cells, are visible within 24 to 48 hours. With HSV-2 strains, syncytial formation usually occurs. Definitive typing of the virus, type 1, type 2, may be necessary.

INTERPRETATION: Virus isolation is the best method to confirm infection with herpes and should be used wherever possible in place of serology. Herpes simplex virus infection is probably the most common viral disease in man; the vast majority are not clinically apparent. (Herpes is derived from the Greek word, herpein, meaning "to creep"; herpes zoster is caused by the chicken pox virus and is also known as shingles.)

Two main strains of the virus exist; type 1 causes most nongenital infections and type 2 causes genital infections. Type 1 usually produces infections above the waist while type 2 primarily produces venereal lesions in males and females.

Genital Herpes: About 300,000 cases of genital herpes are diagnosed each year in the United States. Following sexual contact, the incubation period ranges from 3 to 14 days (average 5 to 7 days). In the male, the glans penis, the skin, the mucosal surfaces of the prepuce and the frenal area are commonly involved. In the female, lesions are common on the mucosal surfaces of the labia minora, clitoral hood, skin of the labia majora, buttocks, and thighs in severe cases; lesions heal in about 3 weeks.

Recurrent herpes lesions are less severe and are not due to reinfection; the mean time interval between initial and recurrent infection is about 120 days (25-360 days).

Congenital infections are rare. When infection, either primary or recurrent, occurs in the pregnant female, there is a high abortion rate. The neonate usually has a vesicular rash present at birth; 80 percent of these neonates suffer diffuse brain damage.

The most serious consequence of genital infections is the transmission of the virus to the newborn as the fetus passes through the birth canal, usually resulting in permanent neurologic sequelae or neonatal death. If the mother has herpes at the time of delivery, the risk to the neonate is high. However, up to 70 percent of babies with neonatal herpes simplex are born to mothers with no symptoms or signs of the disease at the time of delivery (Whitley, R.J. et. al., Pediatrics 66, 489-494, 1980). The risk of neonatal infection can be decreased by caesarean section.

Bryson and co-workers demonstrated the efficacy of oral acyclovir for first episodes of genital herpes (Bryson, Y.J. et al., N. Engl. J. Med. 308, 916, 1983).

References:
Adler, M.W. and Mindel, A., Brit. Med. J. 286, 1767-1768, June 4, 1983.
Oates, J.K., Brit. J. Hosp. Med. 29, 13-22, Jan., 1983.
Adler, M.W., Brit. Med. J. 287, 1864-1866, Dec. 17, 1983.

5-HIAA (HYDROXYINDOLE ACETIC ACID A SEROTONIN METABOLITE)

<u>SPECIMEN:</u> Random urine for screening test. <u>Quantitative</u> Test: Collect 24 hour urine in bottle containing 12g of boric acid, acetic acid or 25ml 6NHCl; remove a 100ml aliquot for analysis.

<u>REFERENCE RANGE:</u> Screening test positive if 5-HIAA present in large amount. <u>Quantitative:</u> 1.8-6.0mg/24 hour. To convert <u>conventional</u> units in mg/24 hour to <u>international</u> units in mcmol/day, <u>multiply conventional units by 5.230.</u>

<u>METHOD:</u> Reaction of 5-HIAA with nitrous acid and 1-nitroso-2-naphthol.

<u>INTERPRETATION:</u> The causes of <u>elevation</u> of urine 5-HIAA are given in the next Table:

Elevation of Urine 5-HIAA
Carcinoid Tumor
Food Products: Banana; Pineapple; Tomato; Avocado; Walnut; Glyceryl Guaiacolate

Carcinoid tumors are found throughout the gastrointestinal tract from the stomach to the rectum and are found in the biliary tree, pancreatic duct or gonads. They occur <u>most commonly</u> in the <u>terminal ileum</u> and <u>appendix</u>; when they occur in the appendix, they usually cause appendicitis or are incidental findings at operation or autopsy.

Carcinoid tumors (argentaffin cells) synthesize <u>5-hydroxytryptamine</u> (<u>serotonin</u>) which is metabolized to 5-HIAA and excreted in the urine.

The carcinoid <u>syndrome</u> is usually caused by <u>wide-spread hepatitic</u> <u>metastasis</u> of a carcinoid tumor. The hallmark of the carcinoid syndrome is the occurrence of attacks of flushing. These attacks do not occur or are minor in one-third of the patients. Other hallmarks of the carcinoid syndrome are abdominal pain, cramps, diarrhea, weight loss, facial or nasal cyanosis with telangiectasia; hypotension may occur.

High Density
Lipoprotein(HDL)

HIGH DENSITY LIPOPROTEIN (HDL)

<u>SPECIMEN:</u> Fasting overnight; red top tube, separate serum; or lavender(EDTA) top tube, separate plasma.

<u>REFERENCE RANGE:</u>

HDL-Cholesterol (mg/dl)		
Age	Male	Female
0-19	30-65	30-70
20-29	35-70	35-75
30-39	30-65	35-80
40-49	30-65	40-95
50-59	30-65	35-85

Manual of Laboratory Operations, Vol. I, NIH.
DHEW Publication No. (NIH 75-628, pg. 73)

<u>METHOD:</u> HDL is measured as cholesterol in the supernatant of serum after <u>precipitation</u> of other lipoproteins with a polyanion-divalent cation combination such as phosphotungstate-Mg^{++} or heparin-Mn^{++}.

<u>INTERPRETATION:</u> The incidence of coronary heart disease by HDL cholesterol level is shown in the next Table (Gordon et al, Am. J. Med. <u>62</u>, 707-714, 1977)

Incidence of Coronary Heart Disease by HDL Cholesterol Level			
HDL Cholesterol Level (mg/dl)	Incidence of Coronary Heart Disease		Interpretation
All Levels	Men	Women	
25	18%		CHD Risk at Dangerous Level
25 - 34	10%	17%	High CHD Risk
35 - 44	10%	5%	Moderate CHD Risk
45 - 54	5%	5%	Average CHD Risk
55 - 64	6%	4%	Below Average CHD Risk
65 - 74	2.5%	1.4%	Protection Probable
75	0%	2%	Associated with Longevity

CHD = Coronary Heart Disease

Decreased levels of high density lipoprotein (HDL) in plasma lead to an <u>increased risk in males</u> of coronary heart disease; among the various lipid risk factors, <u>HDL cholesterol</u> appears to have the <u>strongest relationship</u> to coronary <u>heart disease.</u> It may also be that raised levels exert a protective effect in

that premenopausal women have HDL concentrations 30 to 60 percent higher than
their male counterparts and subjects with familial hyper-alphalipoproteinemia
have an above-average life expectancy. It is not known whether people can
improve their chances of avoiding a heart attack by increasing their HDL
concentration.

The probable function of HDL is to act as an acceptor for tissue
cholesterol and to carry it back to the liver which can break it down and
excrete it.

HDL tends to be low in males, obese, sedentary, cigarette smokers,
diabetics and patients with uremia.

Estrogens increase concentrations of HDL cholesterol and decrease LDL
cholesterol; androgens have the opposite effect. Treatment of hypercholes-
terolemic post-menopausal females with estradiol decreases the LDL cholesterol
concentration and increases that of HDL cholesterol (Tikkanen, M.J. et al., Acta
Obstet. Gynecol. Scand. [Suppl.] 88, 83-88, 1979).

Ratio of Total Cholesterol: HDL Cholesterol: The total cholesterol: HDL
cholesterol ratio has been used as a risk factor for atherosclerosis. Using
this scheme, a ratio greater than 5 for males and greater than 4.5 for females
indicates an increase risk.

Tangier's disease is a rare familial disorder in which the levels of HDL
are extremely low. In these patients, there are marked accumulations of cholesterol in many tissues in the body, e.g., spleen, lymph nodes, intestinal
mucosa and blood vessels. It may be that HDL mobilizes or transports
cholesterol from peripheral tissues back to the liver where it can be excreted.

Appropriate diet and exercise may increase HDL.

HISTOPLASMOSIS ANTIBODY

SPECIMEN: Red top tube, separate serum.
REFERENCE RANGE: Negative; a positive histoplasmin skin test induces elevated
histoplasmin CF titers in 12 to 27 percent of persons (Buechner, N.A. et al.,
Chest 63, 259-270, 1973).
METHOD: Immunodiffusion by referral to Center for Disease Control (CDC),
Bureau of Laboratories, Atlanta, Georgia. Information required by physician:
patient's age, sex, occupation, address, date of onset of illness, brief
clinical history.

Latex agglutination test available from American Scientific Products.
Complement fixation tests are also done.
INTERPRETATION: Following infection with Histoplasma capsulatum, serum
antibody is detected with rising titer within six weeks of infection; the
antibody persists for a few weeks to more than one year. A rising titer is
associated with progression of the disease.

By immunodiffusion, two precipitin bands may be present, an H band and a M
band. The M band occurs before the H band. Both H and M bands indicate active
infection with histoplasmosis; M band alone indicates active infection, past
infection or recent skin test; the H antibody indicates active infection.

Histoplasmosis primarily causes pulmonary infections but may progress
systemically to affect other organs of the body.

HUMAN CHORIONIC GONADOTROPIN (HCG), BETA SUBUNIT, SERUM

<u>SPECIMEN:</u> Red top tube, separate serum
<u>REFERENCE RANGE:</u> Negative; Half-Life=5 days
<u>METHOD:</u> RIA; ELISA methods are being developed. The most specific antibody
for serum HCG will be raised against the carboxy-terminal beta-subunit segment.
<u>INTERPRETATION:</u> Serum HCG is increased in the conditions listed in the next
Table:

Increase of Serum HCG in Different Conditions
Normal Pregnancy
Ectopic Pregnancy
Abortion
Gestational Trophoblastic Tumors:
Hydatiform Mole and Choriocarcinoma
Testicular Tumors; Germinal Cell Origin:
Choriocarcinoma
Embryonal Carcinoma with Syncytiotrophoblastic
Giant Cells (STGC)
Seminoma with Syncytiotrophoblastic Giant Cells (STGC)
Other Tumors: Some Gastric Carcinomas; Some Hematomas
and Some Pancreatic Carcinomas and Other Tumors

<u>Normal Pregnancy:</u> The change of human chorionic gonadotropin (HCG) level in
maternal blood during pregnancy is shown in the next Figure; (Krieg, A.F. in
Clinical Diagnosis and Management by Laboratory Methods, Henry, J.B. (ed), W.B.
Saunders, Co., Phila. 1979, pg. 685):

Human Chorionic Gonadotropin (HCG) in Pregnancy

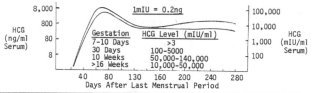

Gestation	HCG Level (mIU/ml)
7-10 Days	>3
30 Days	100-5000
10 Weeks	50,000-140,000
>16 Weeks	10,000-50,000

1mIU = 0.2ng

Serum HCG becomes detectable within 24 hours after implantation (5mIU/ml),
increases progressively and peaks about 10 weeks after the last menstrual
period. Elevated levels may be found in multiple pregnancies, polyhydramnios,
eclampsia and erythroblastosis fetalis.
Ectopic Pregnancy: The levels of serum HCG in ectopic pregnancy are in the range
of 10 to 800mIU/ml. Following surgical removal, the <u>disappearance</u> of serum HCG
in <u>ectopic pregnancy</u> may take <u>up to 24 days</u>, the time being determined by the
<u>initial HCG level</u> (Kamrava, M.M. et al., Obstet. Gynecol. <u>62</u>, 486-488, 1983).
    In a prospective study of patients with <u>suspected ectopic pregnancy</u>, <u>serum
HCG</u> and <u>ultrasound</u> provided for nearly <u>100 percent clinical accuracy</u> in
diagnosing ectopic pregnancy in suspected cases; the protocol is shown in the
next Figure; (Bryson, S.C.P., Am. J. Obstet. Gynecol. <u>146</u>, 163-165. 1983):

S. Bakerman

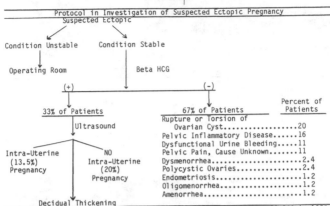

Protocol in Investigation of Suspected Ectopic Pregnancy

The number of ectopic pregnancies in the United States was 52,200 in 1980; the rate of ectopic pregnancies per 1000 pregnancies was 10.5. The death-to-case rate was 0.9 (Morbidity and Mortality Weekly Report, Centers for Disease Control, 33, No. 15, April 20, 1984).

Abortion: Following a complete first trimester abortion, HCG disappears in about 40 days.

Gestational Trophoblastic Tumors: Gestational trophoblastic tumors, hydatiform mole and choriocarcinoma, often present with markedly elevated levels of HCG. Following complete removal of hydatiform mole, serum HCG declines steadily and disappears in 100 days, on average. (Franke, H.R. et al., Obstet. Gynecol. 62, 467-473, 1983).

Testicular Tumors: HCG is present in the serum in patients with choriocarcinomas and in patients with tumors that may have syncytiotrophoblastic cells such as embryonal carcinomas or seminomas.

Tests for Malignant Insulinoma: In a relatively high percent of patients with malignant insulinoma, there is increased plasma levels of alpha-HCG (57%), beta-HCG (21%) or immunoreactive HCG (25%). Plasma levels of HCG were not elevated in any of 41 benign insulinomas. (Kahn et al, N. Engl. J. Med. 297, 565-569, 1977).

HUMAN CHORIONIC GONADOTROPIN(HCG) STIMULATION TEST

SPECIMEN: Blood specimens for testosterone, androstenedione and 17-hydroxyprogesterone are collected in red top tubes at 0900 hours on day 1 and days 4 and 5; separate serum.

REFERENCE RANGE: Normal response is an increase in steroids to above the upper limit of the normal range in at least one of the test samples.

METHOD: On day 1, collect blood specimen at 0900; then inject intramuscularly 3000 units of HCG per square meter body surface area. Repeat intramuscular injection on days 2, 3 and 4. On day 4, collect blood specimen prior to injection of HCG. On day 5, collect blood specimen.

INTERPRETATION: Human chorionic gonadotropin(HCG) stimulates interstitial cells and is used as a test of the ability of interstitial cells to secrete testosterone. A normal response is an increase in plasma testosterone to above the upper limit of the normal range in at least one of the test samples. In primary testicular disease, there is a reduced or absent response; in hypogonadotrophism, a normal response may be obtained.

HUMAN LEUKOCYTE ANTIGEN(HLA)

SPECIMEN: Three green(heparin) top tubes. Do not refrigerate or freeze.
Specimens must arrive in laboratory within 24 hours of drawing.
REFERENCE RANGE: Antigens present are reported.
METHOD: In the lymphocytotoxicity test, purified lymphocytes are mixed with
typing sera (known antibody) and incubated as illustrated in the next Figure:

HLA Testing

Rabbit Complement is Added
and the Preparation
Reincubated.

If Complement is Bound,
The Cell is <u>Killed</u>.

Indicator Dye is Added

Dye Binds to Altered
Cell Membrane of
Dead Cell

INTERPRETATION: HLA are antigens located on the surface of leukocytes as well
as on the surface of all nucleated cells in other tissues; they are <u>not</u> present
on the surfaces of erythrocytes. In HLA testing, the HLA antigens on
lymphocytes are determined; lymphocytes have a relatively high concentration of
HLA antigens and lack ABO antigens. HLA are glycoproteins and can be
demonstrated by serologic tests.

HLA typing is done for the reasons listed in the next Table:

Uses of HLA Typing
Matching Donor with Recipient for Organ Transplants, e.g., Kidney Transplants; Bone Marrow Transplants
Paternity Testing
Platelet and Leukocyte Transfusion of HLA-Matched Cancer Patients Receiving Extensive Chemotherapy
Anthropologic Studies: Caucasians, A-1, B-8; Orientals, A-9; Blacks, AW 30

Genetic coding for HLA is found on the short arm of chromosome 6. There
are loci of the HLA region; these loci are called A, B, C, D and DR (for
D-related). D-antigens are tested by a different method, that is, <u>mixed
lymphocyte</u> culture, because their immune function differs from that of other
antigens.

S. Bakerman

Some of the HLA antigens are associated with disease; these are listed in the next Table:

HLA and Disease	HLA
Ankylosing Spondylitis	B-27
Reiter's Syndrome	
Yersinia Enterocolitica Arthritis	
Multiple Sclerosis	DW-2
Chronic Active Hepatitis	B-8
Gluten-Sensitive Enteropathy	
Systemic Lupus Erythematosus	B-15

Juvenile- or early-onset insulin-dependent diabetes mellitus(IDDM) in Caucasians is associated with several HLA antigens, HLA-B8, HLA-B15, HLA-DR3 and HLA-DR4 (Ginsberg-Fellner, F. et al., Diabetes 31, 292-298, 1982; Pittman, W.B. et al., Diabetes, 31, 122-125, 1982; Patel, R. et al., Metabolism 26, 487-492, 1977). In adult- or late-onset IDDM, there is a significant increase in the frequency of HLA-DRW and a decreased frequency of HLA-DR2 (Pittman, W.B. et al., Diabetes 31, 122-125, 1982).

Antibodies to A, B and C are found in the following:

Antibodies to HLA Antigens
Polytransfused Patient Who Has Received Whole Blood
Multiparous Women
Recipients of Allotransplants
Kidney Dialysis Patients
Volunteers Immunized with Cells

Most of the antisera contain multispecific antibodies.

HUMAN TUMOR STEM-CELL ASSAY

SPECIMEN: Surgical specimen: Tumor Tissue; remove connective and adipose tissues; weigh specimen; mince into pieces less than 2mm in diameter in the presence of appropriate media.

REFERENCE RANGE: Growth or inhibition of growth in presence of particular drug.

METHOD: The cells of the tumor are enzymatically dissociated using enzymes such as DNase and collagenase. The tumor fragments are stirred and the free cells are decanted through gauze and centrifuged two times. The supernatant is removed and the free single cells are resuspended; drugs are added to the tumor cell suspension. The cells are grown on soft agar placed in wells. The plates are examined microscopically about twice a week; the number of colonies reach maximum in two to three weeks.

INTERPRETATION: The human tumor stem-cell assay has been used as an in-vitro test for determining the sensitivity or resistance of an individual patient's tumor to an anticancer drug. The objective is to allow the selection of patients who are more likely to respond to chemotherapy. The concept is similar to in-vitro testing of antibiotics to isolated bacteria. The human tumor stem-cell assay is commercially available and many tumor specimens have been tested.

There are data that have shown that in-vitro responses of individual tumors to selected drugs correspond well with the clinical responses of the patient to those drugs. Typically, using the results of human tumor stem-cell assay, the accuracy rate for prediction of drug sensitivity is correct in 40 to 90 percent of cases and the accuracy rate for prediction of drug resistance is correct in 90 to 95 percent of patients. The data indicate that this stem-cell assay may allow selection of patients who are more likely than others to respond to specific drug therapy.

There are problems with the use and interpretation of these tests; usually, less than 50 percent of the tumors cultured have sufficient in-vitro growth for drug testing. Other criticisms are discussed by Selby, et al. (N. Engl. J. Med. 308, 129-134, 1983) and Von Hoff (Von Hoff, D.D., N. Engl. J. Med. 308, 154-155, 1983). However, recent studies indicate highly significant associations between in-vitro chemosensitivities and the clinical course of the patient. (Kern, D.H. et al., Ann. Clin. Lab. Science 13, No. 1, 10-15, 1983)

HYDROGEN BREATH ANALYSIS IN GASTROENTEROLOGY

SPECIMEN: Breath hydrogen specimens for analysis following oral administration of carbohydrate

REFERENCE RANGE: Compare to normal curves; 2 to 20ppm.

METHOD: Gas chromatography using thermal-conductivity detector.

INTERPRETATION: Major clinical applications of the hydrogen breath test in gastrointestinal disease are given in the left hand column of the next Figure with test results in the right hand column; (Solomons, N.W., Comprehensive Therapy 7, No.8, 7-15, 1981; Brugge, W.R., Resident and Staff Physician, pgs. 29PC-37PC, Nov. 1983):

Major Clinical Applications of the Hydrogen Breath Test

Lactose Malabsorption
 (Lactase Deficiency)

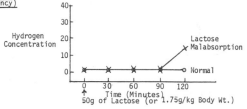

Patients with lactase deficiency do not absorb lactose in the small bowel; lactose passes unchanged into the colon where it is broken down to hydrogen and other products. The hydrogen is detected in the breath.

Bacterial Overgrowth, Small Bowel
 e.g. Fistulae
    Blind Loop
    Jejunal Diverticulosis
    Diabetic Gastroparesis

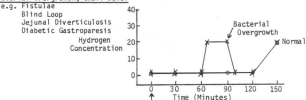

In patients with bacterial growth in the small bowel, the bacteria will will metabolize lactulose to hydrogen and hydrogen will appear in the breath at an earlier time as compared to normal.

Intestinal Transit
 Decreased Transit Time
    Irritable Bowel Syndrome
    Some patients with Vagotomy
    Some Patients with Gastro-
     jejunostomy or Fistula

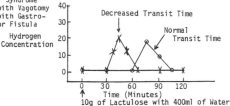

The most common indication for use of breath hydrogen analysis is in the patient who is suspected of having lactose malabsorption. Other major clinical applications of the hydrogen breath test are in the work-up of patients with possible sucrose malabsorption and D-xylose malabsorption.

Malabsorption of a carbohydrate is detected by an increase in hydrogen in the breath; values in excess of 20ppm are considered abnormal. Factors that influence interpretation of hydrogen breath tests are given in the next Table; (Solomons, N.W., Comprehensive Therapy 7, No.8, 7-15, 1981):

Influences on Hydrogen Breath Tests
Idiopathic Absence of the Appropriate Flora
Iatrogenically Induced Absence of the Appropriate Flora
Active Diarrhea
Acidic Colonic Milieu
Delayed Gastric Emptying
Uninterrupted Sleeping During Test
Tobacco Smoking
Inclusion of Fiber or Oligosaccharides in Test Meals

Hydrogen gas originates solely from the metabolism of carbohydrate from the bacteria in the gastrointestinal tract. Ninety percent of hydrogen production originates in the colon. In order for the hydrogen to be generated in the colon, the carbohydrate must pass unabsorbed by the small intestine to the colon. Thus, hydrogen in the breath reflects bacterial metabolism of carbohydrate in the colon. The hydrogen gas is absorbed into the venous blood; in the lung it diffuses into the alveoli and is finally expired.

17-HYDROXYCORTICOSTEROIDS (17-OHCS) IN URINE

<u>SPECIMEN:</u> 24 hour urine. Place 10gm boric acid in container prior to collection. Instruct the patient to void at 8:00 A.M. and discard the specimen. Then collect all urine including the 8:00 A.M. specimen at the end of the 24 hour collection period. Refrigerate urine as each specimen is collected. The specimen must be <u>refrigerated or frozen</u> after collection since 17-OHCS are destroyed at room temperature. The patient should be off all drugs with the exception of aspirin and barbiturates. <u>Record the 24 hour volume.</u> 50ml aliquot is required for the assay. Any test requiring a 24 hour urine collection may also be run on this specimen, e.g., 17-keto- steroids.

<u>REFERENCE RANGE:</u> Male: 6-16mg/24 hours; Female: 4-8mg/24 hours; Children: 2-10mg/24 hours; 0-2 years: 2-4mg/24 hours; 2-6 years: 6-10mg/24 hours; 6-10 years: 6-8mg/24 hours; 10-14 years: 8-10mg/24 hours

<u>METHOD:</u> Three steroids are assayed; these are <u>desoxycortisol</u> (Cpd S), <u>cortisol</u> (Cpd F) and <u>Cpd E</u>; all of these compounds have the dihydroxy acetone side chain configuration in ring structure D. The color-reaction involves the 17-hydroxy-steroid plus phenylhydrazine to yield the phenylhydrazone which absorbs at 410nm.

<u>INTERPRETATION:</u> Causes of increased urinary 17-OHCS are listed in the next Table:

Causes of Increased Urinary 17-OHCS
Cushing's Syndrome
Medical or Surgical Stress
Hyperthyroidism
Obesity (occasionally)
ACTH, Cortisone or Cortisol Therapy
11-Hydroxylase Deficiency

The causes of decreased urinary 17-OHCS are listed in the next Table:

Causes of Decreased Urinary 17-OHCS
Addison's Disease
Pituitary Deficiency of ACTH Secretion
Administration of Synthetic Potent Corticosteroids
21-Hydroxylase Deficiency; 17-Hydroxylase Deficiency
Inanition States, i.e., Anorexia Nervosa
Liver Disease; Hypothyroidism
Newborn Period (Due to Decreased Glucuronidation, Not Decreased Synthesis)

The steroids that are assayed as 17-hydroxycorticosteroids in the pathway for the synthesis of steroids by the adrenal are given in the next Figure:

Steroids Assayed as 17-Hydroxycorticosteroids

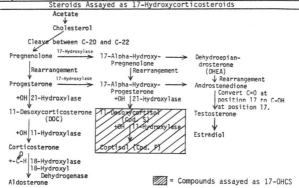

= Compounds assayed as 17-OHCS

17-HYDROXYPROGESTERONE (17-OH PROG.), SERUM

SPECIMEN: Two ml serum; freeze and keep frozen in dry ice.
REFERENCE RANGE: Male: 0.4-4.0ng/ml; Female: 0.1-3.3ng/ml; Pregnant Women:
2.3-7.6ng/ml; Postmenopausal: 0.3-0.9ng/ml; Children: up to 0.5ng/ml. To
convert conventional units in ng/ml to international units in nmol/liter,
multiply conventional units by 3.026.
METHOD: RIA
INTERPRETATION: Serum 17-OH progesterone is elevated in the adrenal enzyme
deficiency states, 21- and 11-hydroxylase deficiency; it is decreased in
17-hydroxylase deficiency. The adrenal enzyme deficiencies are illustrated in
the next Figure:

21-, 11- and 17-Hydroxylase Deficiencies in the Synthesis of Steroids

HYDROXYPROLINE, FREE, URINE

SPECIMEN: Add 20ml of toluene as preservative to urine container. The
patient discards the first voided AM specimen; then the patient collects all
voided urine specimens until the same time on the following day, 24 hours later.
Mix before obtaining 20ml sample. Indicate patient's age and the 24 hour urine
volume on request form.

REFERENCE RANGE: Undetectable to trace

METHOD: Amino acid analysis, quantitated by ninhydrin reaction

INTERPRETATION: Urinary hydroxyproline is a measure of resorption of bone
collagen. Hydroxyproline is an imino acid found only in collagen; during
periods of increased resorption of collagen, especially collagen of bone,
hydroxyproline increases in the urine. The causes of increased urinary
hydroproline are listed in the Table under HYDROXYPROLINE, TOTAL, URINE.

HYDROXYPROLINE, TOTAL, URINE

SPECIMEN: Add 20ml of toluene as preservative to urine container. The
patient discards the first voided AM specimen; then the patient collects all
voided urine specimens until the same time on the following day, 24 hours later.
Mix before obtaining 20ml sample. Indicate patient's age and the 24 hour urine
volume on request form.

REFERENCE RANGE: Adults: 14-45mg/24 hours; children: no normals (Mayo Medical
Laboratories). To convert conventional units in mg/24 hours to international
units in mcmol/day, multiply conventional units by 7.626.

METHOD: Amino acid analysis, quantitated by ninhydrin reaction

INTERPRETATION: Urinary hydroxyproline is a measure of resorption of bone
collagen. Hydroxyproline is an imino acid found only in collagen; during
periods of increased resorption of collagen, especially collagen of bone,
hydroxyproline increases in the urine. The causes of increased urinary
hydroproline are listed in the next Table:

Causes of Increased Urinary Hydroxyproline
Growth Period
Bone Diseases:
Primary and Secondary Hyperparathyroidism
Paget's Disease
Myeloma
Osteoporosis
Marfan's Syndrome

HYDROXYPROLINE, URINARY, 2 HOUR

SPECIMEN: 20ml from 2 hour collection. The patient should not eat or drink
after 6:00 p.m. Fasting will continue until the completion of the urine
collection at 10:00 a.m. An adequate urine specimen is insured by drinking
three or more eight ounce glasses of water. One half of this water should be
ingested between 7:30 and and 8:00 a.m.; the remaining water should be ingested
at 8:30 a.m. The bladder should be completely emptied at 8:00 a.m. and the
urine discarded and the subsequent urine collected in the container. Record the
date and time (the exact start and completion time of the two-hour collection)
on the container label. The collection ends at 10:00 a.m.

REFERENCE RANGE: Male: 0.4-5.0mg/2 hour specimen; Female: 0.4-2.9mg/2 hour
specimen. To convert conventional units in mg to international units in mcmol,
multiply conventional units by 7.626.

METHOD: Amino acid analysis, quantitated by ninhydrin reaction

INTERPRETATION: Urinary hydroxyproline is a measure of resorption of bone
collagen. Hydroxyproline is an imino acid found only in collagen; during
periods of increased resorption of collagen, especially collagen of bone,
hydroxyproline increases in the urine. The causes of increased urinary
hydroproline are listed in the Table under HYDROXYPROLINE, TOTAL, URINE.

25-HYDROXYVITAMIN D (see VITAMIN D)   25-Hydroxyvitamin D

S. Bakerman

HYPERTENSION PANEL

The causes of hypertension, as reflected by referrals to a community-based hypertension clinic, are given in the next Table; (Ferguson, R.K., Ann. Int. Med. 82, 761, 1975):

Causes of Hypertension - Patients with Hypertension Referred to a Community-Based Referral Hypertension Clinic		
Diagnosis	Percent	Suggestive of Diagnosis
Essential Hypertension	66	
Borderline Hypertension	15	
No Diagnostic Hypertension	10	
Secondary Hypertension		
(1) Oral Contraceptive-Induced Hypertension	3.3 (11/331 Patients)	History
(2) Renal Hypertension	3.9 (13/331 Patients)	
Renal Arterial Hypertension	2.1 (7/331 Patients)	
Chronic Renal Disease	0.9 (3/331 Patients)	Urinalysis
Polycystic Kidney Disease	0.9 (3/331 Patients)	Urinalysis
(3) Endocrine	0.6 (2/331 Patients)	
Primary Aldosteronism	0.3 (1/331 Patients)	Decreased Potassium, Increased Sodium
17-Hydroxylase Deficiency	0.3 (1/331 Patients)	Decreased Potassium, Increased Sodium

Diagnosis of Chronic Renal Disease (0.9%; 3/331 patients): All 3 patients who had chronic renal disease had at least one abnormal urinalysis containing cells, casts or protein.

Diagnosis of Polycystic Kidneys (0.9%, 3/331 patients): These 3 patients with polycystic kidneys had microscopic hematuria or casts or both in urinalysis. The diagnosis was confirmed radiologically.

Endocrine Hypertension; Hypertension (0.6%; 2/331 patients): Two patients had probable endocrine hypertension, one due to primary aldosteronism and the other was a suspected 17-hydroxylase deficiency. Diagnostic clues were provided by serum potassium which was low to borderline low. Additional support for the diagnosis of primary aldosteronism was obtained by decreased plasma renin and increased plasma aldosterone.

Most of the secondary causes of hypertension are suggested from history (oral contraceptives), urinalysis and electrolyte analysis.

Curable causes of hypertension are given in the next Table:

Curable Forms of Hypertension
Renovascular Disease (The Major Cause of Secondary Hypertension)
Primary Aldosteronism
Pheochromocytoma
Coarctation of the Aorta
Unilateral Renal Parenchymal Disease
Cushing's Syndrome
Adrenogenital Syndromes (17- and 11-Hydroxylase Deficiency)
Oral Contraceptives
Licorice Induced Hypertension
Renin-Secreting Neoplasms

The most common causes of hypertension are essential hypertension and malignant hypertension. As seen from the list in the Table, most of the potentially curable forms of hypertension are secondary to renal or adrenal diseases; adrenal diseases are not common causes of hypertension.

Essential tests in the routine evaluation of patients with hypertension are given in the next Table:

Routine Evaluation of Hypertension
Complete Blood Count(CBC)
Urinalysis
Serum Potassium
Blood Urea Nitrogen(BUN) or Serum Creatinine
Fasting Blood Sugar
Serum Cholesterol
Serum Uric Acid
Electrocardiogram(ECG)

These tests are done to evaluate possible complications associated with hypertension.

Screening tests for secondary hypertension are given in the next Table:

Screening Tests for Secondary Hypertension	
Suspected Condition	Test
Renovascular Hypertension	Intravenous Pyelogram Suppressed Plasma Renin Activity(PRA) (of Patient Supine for Greater Than 30 Min.) or Stimulated Plasma Renin Activity(PRA) (40-80 mg Furosemide P.O. Followed by 4 hour Upright or 40 mg I.V. followed by 30 Min. Upight).
Primary Aldosterism	Serum Potassium Urine Potassium Stimulated Plasma Renin Activity(PRA) (40-80 mg Furosemide P.O. Followed by 4 hour Upright or 40 mg I.V. followed by 30 Min. Upight).
Pheochromocytoma	Urinary Catecholamines or Urinary Metanephrines or Urinary Vanillylmandelic Acid(VMA)
Cushing's Syndrome	8 A.M. Plasma Cortisol After Suppression with Dexamethasone (1 mg) Taken at 11:00 P.M. the Night Before
Coarctation	Chest X-Ray

IMIPRAMINE (TOFANIL)(see TRICYCLIC ANTIDEPRESSANTS)   Imipramine

IMMUNE COMPLEX PROFILE   Immune Complex Profile
SPECIMEN: See Clq Binding Assay and Raji Cell Assay
REFERENCE RANGE: None detected
METHOD: See Clq Binding Assay and Raji Cell Assay; results are method
dependent since different methods detect complexes of different compositions.
INTERPRETATION: Some diseases associated with immune complexes are given in
the next Table; (Michael M. Frank, NIH 1980):

Diseases Associated with Immune Complexes

Rheumatologic Diseases:
    Systemic Lupus Erythematosus, Rheumatoid Arthritis,
    Sjögrens Syndrome, Mixed Connective Tissue Disease,
    Relapsing Polychondritis, Mixed Cryloglobulinemia,
    Necrotizing Vasculitis

Glomerulonephritis:
    Rapidly Progressive Glomerulonephritis,
    Systemic Lupus Erythematosus, Acute Glomerulonephritis

Infectious Diseases:
    Bacterial Endocarditis, Disseminated Gonorrhea,
    Acute and Chronic Schistosomiasis, Malaria, Leprosy,
    Viral Hepatitis, Dengue

Neoplastic Diseases:
    Hodgkin's Disease, Leukemias, Solid Tumors, Burkitt's Lymphoma,
    Malignant Melanoma

Miscellaneous:
    Primary Biliary Cirrhosis, Chronic Active Hepatitis,
    Idiopathic Pulmonary Fibrosis

Immune complex formation is illustrated in the next Figure:

Immune Complex Formation

Antigen   Antibody   Antigen-Antibody Complex

Antigen-Antibody Complex   Complement   Antigen-Antibody-Complement

Antigen-antibody soluble complexes may develop in the circulation.
Vasculitis and glomerulonephritis may occur.

IMMUNOGLOBULIN E (IgE)

SPECIMEN: Red top tube, separate serum

REFERENCE RANGE:

Age	Mean ± 2SD IgE (U/ml)
Newborn.......	(0.14-2)
1-11 months...	(0.11-56)
1 year........	(0.1 -83)
2 years.......	(0.3 -133)
3 years.......	(1.4 -101)
4 years.......	(0.4 -144)
5 years.......	(3.00-148)
6 years.......	(0.44-573)
7 years.......	(0.35-552)
8 years.......	(1.3 -270)
9 years.......	(0.6 -464)
10 years......	(1.9 -421)
11-14 years....	(1.6 -456)
15-19 years....	(1.5 -384)
20-30 years....	(0.86-239)
31-50 years....	(1.2 -324)
51-80 years....	(0.70-197)

Reference: Pharmacia Diagnostics

To convert conventional units in U/ml to international units in mcg/liter, multiply conventional units by 2.4.

METHOD: PRIST (paper-radioimmunosorbent test) Pharmacia Fine Chemicals, Piscataway, N.J.; methods include both RIA and EIA (enzyme-immunoassay).

INTERPRETATION: The causes of elevation of serum IgE are given in the next Table; (Ali, M., Nalebuff, D.J., Fadal, R.G. and Ramanarayanan, M.D., Diagnostic Medicine, May/June, 1982; CPC, Am. J. Med. 24, 887-897, 1983):

Causes of Elevation of Serum IgE
Parasitic Infestations:
Ascariasis; Visceral Larva Migrans, (Toxocara);
Capillariasis; Echinococcosis;
Hookworm (Necator); Amebiasis
Allergic Disorder:
Asthma
Allergic Rhinitis
Hayfever
Inhalant Allergy
Atopic Rhinitis and Sinusitis
Atopic Dermatitis and Urticaria
Bronchopulmonary Aspergillosis
Hypersensitivity Pneumonitis
Drug and Food Allergies
Immunological Disorders of Uncertain Pathogenesis:
Hyper-IgE and Recurrent Pyoderma (Job-Buckley Syndrome)
Thymic Dysplasias and Deficiencies
Wiskott-Aldrich Syndrome
Pemphigoid
Periarteritis Nodosa
Hypereosinophilic Syndrome
Neoplasms
IgE Myeloma

The most common causes of elevation of serum IgE are parasitic infestations (ascariasis, Toxocara canis, intestinal capilliariasis, bilharziasis, hookworm, Echinococcus, trichinosis) and atopic disorders [atopic allergy implies a familial tendency to manifest alone, or in combination, clinical conditions as bronchial asthma, rhinitis, urticaria and eczematous dermatitis (atopic dermatitis)].

IgE is frequently elevated in asthma, hay fever and other allergic disorders and tends to correlate with the degree of allergic hypersensitivity.

IgE participates in the immediate anaphylactic reaction by binding to the cell membrane of basophils and mast cells; these cells can be stimulated by antigen to release vasoactive amines as illustrated in the next Figure:

Anaphylaxis			
Antigen to Previously Sensitized Individual	I.V. Orally; Subcut. → Antigen - IgE Mast Cells	Release of Histamine and Other Agents from Mast Cells →	Difficulty in breathing due to bronchospasm and/or edema of upper respiratory tract; Hypotension due to dilatation of capillaries, small arterioles and venules.

The release of chemical mediators from mast cells may be visualized as follows:

Release of Chemical Mediators from Mast Cells

Sensitized Patient   Histamine and SRS-A → Smooth M. Contraction

IgE levels are <u>decreased</u> in the serum in conditions listed in the next Table:

Causes of Decrease in Serum IgE
Congenital Hypogammaglobulinemia
Acquired Hypogammaglobulinemia
Sex-linked Hypogammaglobulinemia
Ataxia-telangiectasia
IgE Deficiency

S. Bakerman

IMMUNOGLOBULIN PROFILE, (QUANTITATIVE SERUM IMMUNOGLOBULINS G, A, M)

SPECIMEN: Red top tube, separate serum and store at -20°C
REFERENCE RANGE:

	Normal Levels for IgG, IgM and IgA by Age		
Age	IgG (mg/dl)	IgM (mg/dl)	IgA (mg/dl)
Newborn	1,031 ± 200	11 ± 5	2 ± 3
1-3 mo	430 ± 119	30 ± 11	21 ± 13
4-6 mo	427 ± 186	43 ± 17	28 ± 18
7-12 mo	661 ± 219	54 ± 23	37 ± 18
13-24 mo	762 ± 209	54 ± 23	50 ± 24
25-36 mo	892 ± 183	61 ± 19	71 ± 37
3-5 yr	929 ± 228	56 ± 18	93 ± 27
6-8 yr	923 ± 256	65 ± 25	124 ± 45
9-11 yr	1,124 ± 235	79 ± 33	131 ± 60
12-16 yr	946 ± 124	59 ± 20	148 ± 63
Adults	1,158 ± 305	99 ± 27	200 ± 61

Ref. Lou, K., and Shanbrom, E., JAMA, 200:323, 1967.

IgG: To convert conventional units in mg/dl to international units in g/liter, multiply conventional units by 0.01.

IgM: To convert conventional units in mg/dl to international units in g/liter, multiply conventional units by 0.01.

IgA: To convert conventional units in mg/dl to international units in g/liter, multiply conventional units by 0.01.

METHOD: Single radial diffusion; turbidity
INTERPRETATION: The basic structure and properties of immunoglobulins are shown in the next Figure:

Properties of Immunoglobulins

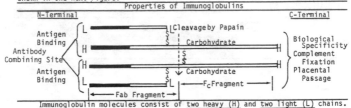

Immunoglobulin molecules consist of two heavy (H) and two light (L) chains. The classification of immunoglobulins is made on the basis of differences in the amino acids of the constant regions of the heavy (H) chains. Assay for gamma globulins are useful for determination of specific changes that are associated with diseases and for following the course of therapy. The diseases that are reflected in the alteration of the concentration of these components are shown in the next Table:

Decreased Values of Gamma Globulins		
Gamma G	Gamma A	Gamma M
Hypo-IgG	Hypo-IgA	Hypo-IgM
Protein-losing Enteropathies	Protein-losing Enteropathies	Protein-losing Enteropathies
	Heredita.y Ataxia Telangiectasia	
Nephrotic Syndrome	Nephrotic Syndrome	

A common clinical indication for immunoglobulin assay is in patients, particularly children, who have repeated severe infections. The infections may be related to impaired synthesis of one or more of the immunoglobulins. The most frequent site of infection is the respiratory tract. The age of onset of illness in hypoimmunoglobulinemia depends upon whether the deficiency is congenital or acquired later in life.

The genetics and serum levels of immunoglobulins are given in the next Table:

DISEASE	GENETICS	IgG	IgA	IgM
Genetics and Serum Levels of Immunoglobulins			Serum Levels of Immunoglobulins	
IgA Deficiency	Unknown	N	↓	N
Hereditary Agammaglobulinemia (Bruton)	Sex-linked	↓ to A	↓ to A	↓ to A
Combined Immunodeficiency	Autosomal Recessive or Sex-linked		Same as above	
Primary Acquired	Unknown	↓	↓	↓
Immunoglobulin M Deficiency	Sex-linked	N	N	↑
Hereditary Ataxia Telangiectasia	Autosomal Recessive	N	Def. in 60%	N
Wiskott-Aldrich Syndrome	Sex-linked	N	↑	N to ↓
Thymic Hypoplasia (DiGeorge)	Unknown	N	N	N

↓ = Decrease; ↑ = Increase; A = Absent

As seen in the above table, there may be a deficiency in one or more of the immunoglobulins in these diseases.

IgA deficiency is the most frequently recognized selective hypogamma-globulinemia; the incidence is estimated at 1 in 500 to 3500 persons. Both serum and secretory IgA are decreased or absent; the secretory piece is produced. Patients commonly have an associated autoimmune disease.

Gamma globulins may be elevated in the conditions listed in the next Table:

Elevated Levels of Serum Immunoglobulins		
Gamma-G	Gamma-A	Gamma M
Chronic Infections	Portal Cirrhosis	Parasitic Disorders
Autoimmune Diseases	Multiple-Myeloma	a. Trypanosomiasis
Multiple Myeloma	Hepatitis (Late)	b. Toxoplasmosis
	IgA Nephropathy	Nephrotic Syndrome
	(Berger's)	Waldenström's Macro-
	Henoch-Schölein	globulinemia
	Purpura	Hepatitis (Early)

Secretory immunoglobulin A is also present in circulating blood and concentrations are high in patients with carcinomas and chronic infectious diseases; a column enzyme immunoassay has been devised (Yamamoto, R. et al., Clin. Chem. 29, 151-153, 1983).

Waldenström's Macroglobulinemia: Laboratory abnormalities in patients with macroglobulinemia are as follows: anemia (88%), positive Sia test (76%), increased serum viscosity (41%), cryoglobulinemia (37%), Bence-Jones proteinuria (25%), thrombocytopenia (6%), leukocytosis (4%). (Deuel, T.F., Arch. Intern. Med. 143, 986-988, 1983)

INAPPROPRIATE ANTI-DIURETIC HORMONE SYNDROME(IADHS)

<u>Laboratory Findings in IADHS</u>: The laboratory findings in IADHS are given in the next Table:

Laboratory Findings in IADHS
Hyponatremia
Decreased Plasma Osmolality
Renal Sodium Loss (Normal Urine Sodium = 4 g/24 hours)
Increased Urine Osmolality
Normal or Expanded Extracellular Fluid Volume
Low Serum Urea Nitrogen (< 10 mg/dl)
Low Serum Uric Acid

The basis for the laboratory diagnosis is <u>decreased plasma osmolality</u> and increased urine osmolality.

<u>Clinical Features of IADHS</u>: The clinical features of the syndrome are: (1) those of the <u>underlying disease</u> and (2) those of <u>hyponatremia</u>.

The extracellular <u>hyponatremia</u> is clinically important because water moves into the body cells; overhydration occurs in brain cells giving rise to symptoms of increased intracranial pressure - confusion, convulsions and eventually coma.

Symptoms are usually G.I. and mild when serum sodium levels are above 120 mEq/L; the symptoms may include anorexia, nausea and vomiting. When the concentration of serum sodium falls below 115 mEq/L, symptoms are usually CNS and finally coma.

The treatment of excessive secretion of ADH in the ectopic ADH syndrome is radiotherapy or chemotherapy plus water restriction; or lithium or demeclocyline which inhibit arginine vasopressin sensitive adenylcylase in the collecting ducts.

<u>Causes of IADHS</u>: The IADHS is caused by the following conditions:

Conditions Causing IADHS
Ectopic ADH:
Lung: Oat Cell Carcinoma, Adeno- or Alveolar Carcinoma
Tumors of Pancreas and the Thymus
Infectious Disorders
Meningitis
Brain Abscess
Aneurysm
Cerebral Vascular Accidents
CNS Tumors
Drugs e.g. Vincristine
Post-Surgical on CNS

The conditions associated with IADHS are predominantly <u>pulmonary</u> and <u>central-nervous system abnormalities</u>.

The pathogenesis of the conditions causing IADHS may be obscure; some tumors produce ADH themselves; tuberculous lung tissue may produce ADH (Vorrher et al., Ann. Int. Med. 72, 383, 1970); drugs may cause IADHS.

In the IADHS, there is excess ADH; the excess ADH causes increased absorption of water from the renal tubular lumen.

INBORN ERRORS OF METABOLISM PANEL

Laboratory Clues:
   <u>Urine</u>: Odor, Ketones, Reducing Substances
   <u>Blood</u>: Anemia, Leukopenia, Thrombocytopenia, Acidosis, Hypoglycemia, Ketonemia

Complete Laboratory Evaluation: Complete laboratory testing for evaluation of patients suspected of having an inborn error of metabolism is given in the next Table; (Cederbaum, S.D. in Genetic Disease, Diagnosis and Treatment, Dietz, Albert A. ed., Am. Assoc. Clin. Chem., 1725 K. Street, N.W., Wash. D.C., pg. 149, 1983).

Laboratory Evaluation of An Inborn Error of Metabolism	
Blood Sugar	Urine Metabolic Screening Tests
Plasma and Urine Ketones	Urine Amino Acids
Blood Electrolytes (or pH and Bicarbonate)	Plasma Amino Acids
Blood Lactate and Pyruvate	Urine Organic Acids
Plasma Ammonia	Plasma Organic Acids(?)

INBORN ERRORS OF METABOLISM SCREEN

SPECIMEN: Obtain about 20ml of urine specimen, refrigerate
REFERENCE RANGE: Negative
METHOD: See below
INTERPRETATION: Urine is used to screen for metabolic diseases; in these diseases, there is an abnormal metabolite or a larger than normal amount of a normal metabolite. Reactions of $FeCl_3$, DNP-hydrazine, cyanide-nitroprusside and Benedict's with urine in different conditions are given in the next Table:

Reactions of Metabolic Screening Tests in Urine				
Disease	$FeCl_3$	DNPH	Nitroprusside	Benedict's
Phenylketonuria	Blue-Green	+++	-	-
Maple Syrup Urine	Blue	+++	-	-
Tyrosinosis (emia)	Green (fades quickly)	+++	-	+
TyrosyIuria	Green	+++	-	-
Histidinemia	Green	++	-	-
Hyperglycemia	-	+++	-	-
Hartnup's Disease	-	-	-	-
Fructose Intolerance	-	-	-	+
Galactosemia	-	-	-	+
Homocystinuria	-	-	+	-
Cystinuria	-	-	+	-
Wilson's Disease	-	-	-	-
Cystinosis	-	-	-	+
Lowe's Syndrome	-	-	-	+
Mucopolysaccharidoses	-	-	-	-
Salicylates	Purple	-	-	-
Phenothiazines	Purple	-	-	-
Acetone		+++		

The sensitivity of these screening tests for the amino acidurias is usually not as great as the quantitative column methods for amino acid analysis. Ferric chloride test is positive for several amino acids and other metabolic abnormalities. See FERRIC CHLORIDE TEST.

DNP-Hydrazine Test (2,4-Dinitrophenyl Hydrazine) (a yellow precipitate forms in the presence of keto acids): A positive test occurs in the following conditions:

Positive 2,4-Dinitrophenyl Hydrazine Test	
Condition	Metabolite in Urine
Maple Syrup Urine Disease	Keto Acids
Phenylketonuria (PKU)	Phenylpyruvic Acid
Histidinemia	Imidazole Pyruvic Acid
Oasthouse Syndrome (Methionine Malabsorption)	Alpha Ketopyruvic Acid
Hyperglycinemia	Acetone
Glycogen Storage Diseases Types 1,3,5 and 6	Acetone
Acetonemia (Ketonuria) from any cause	

The DNP-hydrazine test is positive in maple syrup urine disease, phenylketonuria (PKU) and histidinemia.

Cyanide-Nitroprusside Test: This is a test for cystine and homocystine (cystine is reduced by cyanide to cysteine and the sulhydryl groups then react with nitroprusside to produce a red-purple color).

Benedict's Test (cupric sulfate (blue) $\longrightarrow$ cuprous oxide (red)) for sugars and other reducing substances. The reactions with Benedict's test of substances found in the urine are given in the next Table:

Positive Benedict's Test on Urine
Glucose
Sugars Other than Glucose:
Fructose, Galactose, Lactose, Maltose, Pentose, Sucrose
Urine Constituents: Homogentisic Acid (Alkaptonuria); Creatinine and Uric Acid may cause False Positives
Many other reducing agents may cause False Positives

Glucose Oxidase Test (Glucose reacts with oxygen in the presence of glucose oxidase to cause the removal of two hydrogen ions and the formation of gluconic acid and hydrogen peroxide; the hydrogen peroxide oxidizes orthotolidine to a blue color): No substance, other than glucose, that is excreted in the urine has been found that will yield a positive reaction with the glucose oxidase test; false positive results occur as a result of contamination of the urine samples with H_2O_2 used for rinsing catheters or hypochlorite present in some detergents. False negative results occur in the presence of large quantities of ascorbic acid (vitamin C).

INDIRECT COOMBS (see COOMBS, INDIRECT)

INFANT, IRRITABLE, PANEL

Differential diagnosis of irritable infant is given in the next Table; (Mullen, N., Hosp. Pract. 19, No. 1, 209-213, Jan. 1984):

Irritable Infant, Differential Diagnosis
Otitis Media
Drugs, e.g., Aspirin, Anti-Histamine, Decongestant Anticholinergic; Maternal Alcohol or Drug Abuse in Neonate
Corneal Abrasion
Metabolic Abnormalities
Chalasia with Esophagitis
Incarcerated Hernias
DTP Reactions
CNS Infection
Thrush
Intussusception
Bone Pain
Colic

Laboratory studies for irritable infants are listed in the next Table:

Laboratory Studies for Irritable Infants
Complete Blood Count(CBC)
Urinalysis
Chemistries(Serum):
Glucose
Electrolytes(Na$^+$, K , Cl$^-$, CO$_2$ Content)
BUN or Creatinine
Calcium
Specific Drug Assays if Indicated
Lumbar Puncture
Abdominal Films
Fluoresein Stain in the Eyes(Corneal Abrasion)

INFECTIOUS DISEASE PANELS

The frequency of infectious diseases in ambulatory patients is shown in the next Table; (McHenry, M.C. and Weinstein, A.J., Med. Clin. of N. Am. 67, 3-16, 1983):

Frequency of Infectious Diseases in Ambulatory Patients

Ambulatory Patients	Frequency (Percent)
Family Practice	78
Pediatric Practice	73
Emergency Room	28

Infectious diseases are the most common illnesses in ambulatory patients, accounting for 87 percent of illnesses (Dingle, J.H. et al. Illness in the Home, A Study of a Group of Cleveland Families, Cleveland, Ohio, Case Western University Press, 1964, pp. 19-32); 78 percent of illnesses in family practice and 28 percent of visits to the emergency room of a large general hospital (Moffet, H.L., Ann. Intern. Med. 89, 264-277, 1978).

The frequency of various infections in ambulatory patients is shown in the next Table; (Moffet, H.L., Ann. Intern. Med. 89, 743-745, 1978):

Frequency of Various Infections in Ambulatory Patients

Infections	Frequency (Percent)
Upper and Lower Respiratory Infections and Pararespiratory Infections, such as, Pharyngitis, Otitis Media and Sinusitis	70
Gastrointestinal Infections	8
Sexually Transmitted Diseases	5
Dermatologic Infections	5
Urinary Tract Infections	4
Miscellaneous Infections	8

Frequently used antimicrobial drugs are given in the next Table; (McHenry, M.C. and Weinstein, A.J., Med. Clin. of N. Am. 67, 3-16, 1983):

Frequently Used Antimicrobial Drugs

Drugs	Indications
Ampicillin, amoxicillin, bacampocillin	Otitis media; sinusitis; bronchitis; urinary tract infections; gonorrheae
Benzathine penicillin G	Group A streptococcal pharyngitis; syphilis; prophylaxis for rheumatic fever
Phenoxymethy penicillin (penicillin V), penicillin G	Group A streptococcal pharyngitis; impetigo(children); otitis media(adults)
Cloxacillin or dicloxacillin	Cellulitis; impetigo (adults)
Erythromycin	Pneumonia caused by pneumococci, Mycoplasma pneumoniae; group A streptococcal pharyngitis in patients allergic to pencillin; prophylaxis of endocarditis in patients allergic to pencillin; nongonococcal urethritis in patients unable to tolerate alternative drugs
Erythromycin plus sulfisoxazole	Otitis media (children)
Metronidazole	Trichomoniasis; vaginitis associated with Gardnerella vaginalis
Sulfonamides	Acute uncomplicated urinary tract infections
Tetracyclines	Acute exacerbations of chronic bronchitis in patients with COPD; Mycoplasma pneumoniae; gonorrheae and nongonococcal urethritis; syphilis in patients allergic to penicillin
Trimethoprim-sulfamethoxazole	Urinary tract infection; bacterial prostatitis; shigellosis
Trimethoprim	Urinary tract infection in patients allergic to sulfonamides

Occasionally useful antimicrobial agents are given in the next Table;
(McHenry, M.C. and Weinstein, A.J., Med. Clin. of N. Am., 67, 3-16, 1983):

Occasionally Useful Antimicrobial Agents	
Drugs	Indications
Cephalosporins	
Cephalexin, cephradine cefadroxil	Urinary tract infections caused by susceptible bacteria, particularly in patients with minor allergy to penicillin; other mild infections in patients with mild allergy to penicillin, such as acute exacerbations of chronic bronchitis in patients with COPD
Cefaclor	Otitis media
Cefoxitin	Penicillinase-producing of Neisseria gonorrhoeae
Clindamycin	Streptococcal or staphylococcal infections in patients allergic to penicillin
Cinoxacin	Urinary tract infections
Ethambutal	Tuberculosis
Isoniazid	Tuberculosis
Nitrofurantoin	Urinary tract infections
Rifampin	Tuberculosis; meningococcal prophylazis or treatment of carriers
Spectinomycin	Gonorrheae in patients allergic to penicillin; penicillinase-producing strains of Neisseria gonorrhoeae

Examples of infections amenable to antimicrobial chemotherapy in ambulatory patients are given in the next Table; (McHenry, M.C. and Weinstein, A.J., Med. Clin. of N. Am., 67, 3-16, 1983):

Examples of Infections Amenable to Antimicrobial Chemotherapy in Ambulatory Patients	
Infection	Usual Causative Organism(s)
Streptococcal Pharyngitis and Tonsilitis	Group A Streptococcus
Acute Exacerbation of Chronic Bronchitis in Patients with Chronic Obstructive Pulmonary Disease (COPD)	Streptococcus Pneumoniae, Hemophilus Influenzae
Uncomplicated Community-Acquired Pneumonia	Mycoplasma Pneumoniae, Streptococcus Pneumoniae
Dental Abscess of Cellulitis	Penicillin-Susceptible Anaerobes
Otitis Media (Patients Over 2 Months of Age)	Streptococcus Pneumoniae, Hemophilus Influenzae
Sinusitis, Acute	Streptococcus Pneumoniae, Hemophilus Influenzae
Sinusitis, Chronic	Penicillin-Susceptible Anaerobes
Urethritis	Chlamydia Trachomatis, Neisseria Gonorrhoeae, Ureaplasma Urealyticum
Salpingitis (PID)	Neisseria Gonorrhoeae

Pharynx: Bacteria associated with pharyngitis are listed in the next Table:

Bacteria Associated with Pharyngitis
Group A Streptococci (Streptococcus Pyogenes)
Neisseria Gonorrhoeae
Corynebacterium Diphtheriae

The normal flora (bacteria) of the oropharynx are listed in the next Table:

Normal Flora of the Oropharynx
Alpha-Hemolytic Streptococci
Non-Hemolytic Streptococci
Streptococcus (Formerly Diplococcus) Pneumoniae
Neisseria Species
Staphylococcus Epidermidis
Diphtheroids

Throat cultures most often reveal acute <u>streptococcal infections</u> as the cause of <u>acute pharyngitis</u> and tonsillitis. <u>Most</u> acute <u>streptococcal diseases</u> result from <u>Group A</u> organisms; <u>sequalae</u> of Group A streptococcal infections are <u>acute rheumatic fever</u> and <u>acute glomerulonephritis</u>. Group D enterococci may cause a variety of diseases and Group B may cause <u>puerperal</u> and <u>neonatal</u> infections. The differentiation of Group A from non-group A beta hemolytic streptococci is of clinical importance primarily in terms of the <u>preventing</u> of rheumatic fever and <u>acute glomerulonephritis</u>.

Methods of diagnosis of agents that cause <u>exudative pharyngitis</u> are given in the next Table; (Levy, M.L.; et al., Med. Clin. of N. Am., <u>67</u>, 153-171, 1983):

Methods of Diagnosis of Agents that Cause Exudative Pharyngitis

Organism or Condition	Method of Diagnosis	
	Readily Available	Generally Available
Group A beta hemolytic streptococcus	Throat Culture	Serologic tests: ASO, AHT*
Epstein-Barr virus	Complete blood cell count with differential Heterophile	Specific Epstein-Barr antibodies titers
Paraquat poisoning	None	Urine and serum assays for paraquat
C. diphtheriae	Culture	Toxin assay
F. tularensis	Serologic test (agglutination)	Culture on appropriate media**
Vincent's angina	Gram's stain of mouth lesion	
Viruses (adenovirus, coxsackie-viruses, herpes simplex)	Usually none	Viral culture with confirmatory serologic test

*ASO = antistreptolysin-O; AHT = antihemolysin
**Poses risk of creating infectious aerosol in the laboratory.

Pneumonia: Causes of overwhelming <u>pneumonia</u> are given in the next Table; (Bradsher, R.W., Med. Clin. of N. Am. <u>67</u>, 1233-1250, November, 1983):

Causes of Overwhelming Pneumonia

	Normal Host	Abnormal Host
Usual Organisms	Pneumococcus	Pneumococcus
	Mycoplasma Pneumoniae	Gram-negative Bacilli
	Hemophilus Influenzae	Anaerobic Bacteria
	Influenza Viruses	Staphylococcus Aureus
	Fungi	Fungi
Unusual Organisms	Legionella	Aspergillus
	Chlamydia	Mucor. Absidia, Rhizopus
	Franciscella Tularensis	Candida
	Yersinia Pestis	Nocardia
	Coxiella Burnetii	Varicella-Zoster Virus
	Meningococcus	Cytomegalovirus
	Group A Streptococcus	Pneumocystis Carinii
	Actinomyces Israelii	Strongyloides Stercoralis
	Mycobacterium	Noninfectious

<u>Infectious Diarrhea</u>: Clues that are associated with specific infectious agents are given in the next Table; (Satterwhite, T.K. and DuPont, H.L., Med. Clin. of N. Am., <u>67</u>, 203-220, 1983):

Infectious Diarrhea - Clues Associated with Specific Infectious Agents

Clue	Suspect Agent
Recent administration of antibiotics	Clostridium Difficile
Contact with children attending day care center	Shigella, Campylobacter, Rotavirus, or Giardia
Foodborne illness without fever	Staphylococcus Aureus, Clostridium Perfringens, Bacillus Cereus, or Enterotoxigenic Escherichia Coli
Foodborne illness with fever	Salmonella, Shigella, Vibrio Parahaemolyticus, Campylobacter, or Yersinia
Specific food item incriminated	
poultry	Salmonella
eggs	Salmonella or Campylobacter
seafood	V. parahaemolyticus
raw milk	Salmonella, Campylobacter, or Yersinia
fried rice or bean sprouts	B. cereus
Prior travel from industrialized to developing region	Enterotoxigenic E. coli in one half of cases; other agents are Shigella (15%), Salmonella (10%), Giardia (4%), and Campylobacter (3%) (Hoffman, T., Hosp. Pract., pgs. 111-112, Jan. 1984)
Homosexual activity	Gonococcal and Chlamydial Proctitis, Amoeba, Shigella, or Campylobacter

Laboratory identification of pathogens is given in the next Table; (Satterwhite, T.K. and DuPont, H.L., Med. Clin. of N. Am., 67, 203-220, 1983):

Laboratory Identification of Pathogens Involved in Infectious Diarrhea

Agents detected by routine laboratory testing:
  Salmonella
  Shigella
  Campylobacter
  Entamoeba
  Giardia

Agents that can be detected by most laboratories by special request:
  Vibro parahaemolyticus
  Rotavirus
  Yersinia

Agents that are identified only by special laboratories:
  Enterotoxigenic Escherichia coli
  Invasive Escherichia coli
  Clostridium perfringens
  Clostridium difficile
  Norwalk agent

Agents that require quantitative food microbiology or toxin assay:
  Staphylococcus aureus
  Clostridium perfringens
  Clostridium difficile
  Bacillus cereus

Ref.: Wolf, J.L., Patient Care, pgs. 79-125, May 15, 1983.

Vaginitis: The most frequent causes and methods of diagnosis of vaginitis are shown in the next Table:

Causes and Methods of Diagnosis of Vaginitis

Causes	Methods of Diagnosis
Candida Albicans ("Yeast Vaginitis")	Potassium Hydroxide (Preferred Method) Gram Smears Saline Wet Mounts Routine Culture for Fungi Unnecessary
Trichomonas Vaginalis	Saline Wet Mount Culture (Sensitive but Expensive)
Gardnerella Vaginalis (Hemophilus vaginalis)	"Clue Cells" on Methylene Blue or Gram Stain; "Clue Cells" are Epithelial Cells whose surface is covered with Coccobacilli

Vaginitis is commonly due to yeast and Trichomonas; herpes can also cause symptoms. "Non-specific" vaginitis is related to the bacteria Gardnerella vaginitis, an organism which responds to sulfonamides, usually in the form of vaginal suppositories. (Amsel, R. et al., Amer. J. Med. 74, 14, Jan. 1983)
Cervicitis: The most frequent causes of cervicitis are N. Gonorrhoeae; Chlamydia Trachomatis; Herpes Simplex.
Sexually Transmitted Diseases: These are syphilis, gonorrhea, lymphogranuloma venereum, granuloma inguinale, chancroid and chylamydia trachomatis. Sexual promiscuity is a risk factor in the transmission of herpes and hepatitis B infection.
The prevalence, estimated annual incidence and sequelae of sexually transmitted diseases are given in the next Table; (Rudbach, J.A., Infectious Disease Forum, Abbott Laboratories, Dec. 1983):

Sexually Transmitted Disease		
Disease	Estimated Annual Incidence	Sequelae
Gonorrhea	2 Million	Females: Pelvic Inflammatory Disease (PID) including Fallopian Tube Damage; Pelvic Adhesions; Ectopic(Tubal) Pregnancy; and Sterility Males: Epididymitis which can Lead to Sterility. Disseminated Infection
Syphilis	More than 400,000	Brain and Heart Damage Birth Defects
Chlamydia Trachomatis	3 - 4 Million	Pelvic Inflammatory Disease (PID) Epididymitis
Genital Herpes	500,000	Transmission to the Newborn; May Cause Death or Severe Retardation

Chlamydia trachomatis is one of the most common sexually transmitted pathogens.
Perinatal Infections; (Bolande, R., Personal Communication): The causes of perinatal infections are given in the next Table:

Perinatal Infections
Transplacental Infections of Fetus
Congenital Syphilis
Toxoplasmosis
Rubella Syndrome
Cytomegalovirus Infection
Coxsackie B. Encephalo-myocarditis
Amniotic Infection Syndrome
Coliform organisms
Group B Streptococci
Listeriosis
Mycoplasma
Intrapartum Infections
Herpes Simplex
Chlamydia

INSULIN ANTIBODIES

SPECIMEN: Red top tube, separate serum
REFERENCE RANGE: Negative
METHOD: RIA
INTERPRETATION: Almost all patients receiving exogeneous (porcine (pig) or bovine (beef)) insulin develop serum antibodies after several months. The concentration of antibodies is usually moderate. The similarity in structure to human insulin is responsible for the usual low antigenicity of the commercial preparations. There are two types of antibodies, high affinity and low affinity. There is no apparent clinical significance except in the special case of the development of insulin resistance.

The presence of insulin antibodies makes it impossible to obtain accurate values for serum insulin; in the presence of insulin antibodies, C-peptide is the test of choice to reflect insulin secretion.

Recombinant insulin (insulin produced by bacteria into which the human insulin gene has been introduced) has recently been used therapeutically; theoretically, this type of insulin would not produce antibodies, but it may be due to impurities in the preparation.

INSULIN WITH GLUCOSE ASSAY

SPECIMEN: Patient should be fasting. Red top tube, separate serum and freeze immediately; hemolysis destroys insulin; heparin gives spuriously high values; if the patient has been treated with pork or beef insulin, antibodies usually develop and cause invalid test results.
REFERENCE RANGE: Oral glucose tolerance test values (glucose and insulin values) are given in the next Table;(Meites, S. ed. "Pediatric Clinical Chemistry," Am. Ass. Clin. Chem., 2nd. Edition, 1725 K Street, N.W., Wash. D.C. 20006, 1981):

Oral Glucose Tolerance Test Values (Glucose and Insulin Values)		
Fasting	Glucose (mg/dl)	Insulin (Micro-USP units/ml)
Fasting	56-96	5-40
30 min.	91-185	36-110
60 min.	66-164	22-124
90 min.	68-148	17-105
2 hours	66-122	6-84
3 hours	47-99	2-46
4 hours	61-93	3-32
5 hours	63-86	5-37

Fasting insulin values are lower in infants and children than in adults.

To convert conventional units in microunits/ml to international units in pmol/liter, multiply conventional units by 7.175.
METHOD: RIA using antibodies derived from guinea pigs.
INTERPRETATION: This test is used in the diagnosis of insulinoma. The laboratory features of insulinoma are as follows:
  (a) Hypoglycemia (glucose less than 60mg/dl) with simultaneous hyperinsulinemia (serum insulin >6 microI.U./ml).
  (b) Absence of insulin antibodies: Absence of insulin antibodies helps to rule out surreptitious self-administration of insulin.
  (c) "C"-peptide level; both insulin and "C" peptide are secreted in equimolar amounts in insulinoma.
Insulinoma: (Fajans, S.S. and Floyd, J.C., Ann. Rev. Med. 30, 313-329, 1979; Editorial, the Lancet, Jan. 5, 1980, pgs 22-23; Editorial, Brit. Med. J. 282, 927, March 21, 1981; Harrington, M.G. et al., The Lancet 1, 1094-1095, May 14, 1983): Insulinoma is a rare tumor. Eighty percent of patients have a single benign tumor, usually less than 2 cm in diameter, located about equally in head, body or tail of the pancreas; about 10 percent have multiple adenomas often associated with multiple endocrine neoplasia, Type I syndrome (see Adrenal Function Tests on Multiple Endocrine Adenomatosis). The remaining 10 percent of patients have metastatic malignant insulinoma.
Whipple's Triad: (1) Symptoms precipitated by fasting or exercise; (2) Association of symptoms of hypoglycemia with a low circulating glucose concentration; and (3) Relief of symptoms after administration of glucose.

Tests for Malignant Insulinoma: In a relatively high percent of patients with malignant insulinoma, there are increased plasma levels of alpha-HCG (57%), beta-HCG (21%) or immunoreactive HCG (25%). Plasma levels of HCG were not elevated in any of 41 benign insulinomas. (Kahn et al., N. Engl. J. Med. 297, 565-569, 1977).

Conditions in which plasma insulin secretion is increased or decreased in response to a glucose load are given in the next Table:

Conditions Associated with Increase or Decrease in Plasma Insulin
Decrease in Plasma Insulin: Type I or Insulin-Dependent Diabetes Mellitus (IDDM) (Formerly Juvenile; Diabetes and Ketosis-Prone Diabetes and Brittle Diabetes); Pancreatic Diabetes (Following Pancreatectomy)
Increase in Plasma Insulin: Type II or Non-Insulin-Dependent Diabetes Mellitus (NIDDM) (Formerly Maturity Onset Diabetes); Hormonogenic Diabetes, e.g., Cushing's Syndrome, Acromegaly, and Pheochromocytoma

The inadequate production and/or secretion of insulin is the primary cause of insulin-dependent diabetes (Type I). In insulin-dependent diabetes, there tends to be a decrease in islet cells; this could cause decrease of plasma insulin. The reason for the decrease in functional islet cells is not known.

In non-insulin-dependent diabetes (Type II), there may be increased plasma insulin; it may be that in this disease, the action of insulin is blocked.

Insulin is increased and glucose is decreased in about 8 percent of patients with falciparum malaria; quinine may induce insulin secretion (White, N.J. et al., N. Engl. J. Med. 309, 61-66, 1983).

INTRINSIC FACTOR (IF) BLOCKING ANTIBODY

SPECIMEN: Red top tube, separate serum; avoid injection of vitamin B_{12} for 1 week before drawing specimen; store at -20°C.

REFERENCE RANGE: None detected

METHOD: Blocking antibody inhibits binding of vitamin B_{12} to intrinsic factor(IF). The patient's serum is added to a solution containing purified IF; incubate; add [Co-57][vitamin B_{12}]; incubate; remove free [Co-57][vitamin B_{12}] by addition of IF bound to glass beads; count supernatant (Test kit; Corning, Inc., Medfield, MA).

INTERPRETATION: Intrinsic factor (IF) blocking antibody is detected in the sera of 50 to 60 percent of patients with proven pernicious anemia and rarely in any other condition. Test for IF blocking antibody in patients who have a low serum vitamin B_{12} concentration. It may be that the Schilling test is unnecessary for patients with positive IF-blocking antibody results and low serum vitamin B_{12} levels. The panel of tests in the work-up of patients with possible pernicious anemia are as follows: MCV, >100fl; serum vitamin B_{12}; serum and RBC folate; if serum vitamin B_{12} is less than 150pg/ml, consider performing IF blocking antibody test.

Pernicious anemia results from a failure of the gastric parietal cells to secrete IF; associated findings are atrophy of gastric mucosa and loss of gastric parietal cell function. Normally, IF combines with vitamin B_{12}, and this complex is absorbed by receptors in the mucosa of the ileum.

There are two types of IF antibodies; these are blocking and binding antibodies. As already mentioned, the blocking antibody inhibits the complexing of vitamin B_{12} with IF.

References: Bates, H.M., Laboratory Management, pgs. 19-22, Sept. 1982.
           Fairbanks, V.F. et al., Mayo Clin. Proc. 58, 203-204, 1983.

IRON (Fe), SERUM

SPECIMEN: Red top tube; separate serum from cells as soon as possible. The serum specimen is stable up to 4 days at room temp. and one week in the refrigerator at 4°C. Interferences: Gross hemolysis. The anticoagulants citrate, EDTA and fluoride-oxalate cause significant depression of serum iron values.

REFERENCE RANGE: Changes of serum iron with age are given in the next Table:

Changes of Serum Iron with Age	
Age	Total Serum Iron (mcg/dl)
Newborn	100-250
Infant	40-100
Child	50-120
Adult:	
Male	50-160
Female	40-150

Iron-Binding Capacity: Infant: 100-400mcg/dl; Thereafter: 250-400mcg/dl.

To convert conventional units in mcg/dl to international units in mcmol/liter, multiply conventional units by 0.1791.

METHOD: Iron (Fe^{+3}) is liberated from transferrin at pH 4.2; then,

$$Fe^{+3} \xrightarrow[\text{Reducing Agent}]{+ e} Fe^{+2}.$$ Iron is then reacted with a reagent such

as bathophenanthroline, or one of the newer chelating agents, to form a colored complex.

INTERPRETATION: The causes of decrease in serum iron and the laboratory findings are given in the next Table; (Cook, J.D., Seminars in Hematology, 19, 6-18, 1982):

Causes and Laboratory Findings in Conditions Associated with Decreased Serum Iron		
Decreased Serum Iron	Iron-Binding Capacity	Serum Ferritin
Iron Deficiency Anemia	Increased	Decreased
Anemia of Chronic Disease	Low or Normal	Normal or Elevated
Chronic Renal Failure	Low or Normal	Below 50mcg/liter

Iron Deficiency Anemia: Causes of iron deficiency anemia are given in the next Table:

Causes of Iron Deficiency Anemia
Nutritional Iron Deficiency
Chronic Blood Loss, Usually G.I., Uterine, Hookworm
Achlorhydria and Gastrectomy
Defective Absorption e.g., Sprue or Steatorrhea
Increased Demands, e.g., Pregnancy

Nutritional Iron Deficiency: This is by far the most common cause of anemia in infants, children and premenopausal women; this is a relatively mild iron deficiency anemia and is particularly relevant to outpatient practice. Iron deficiency anemia occurs in infants (milk anemia) who are fed only milk for the first six to nine months of life.

Chronic Blood Loss: Chronic blood loss may occur from gastrointestinal bleeding lesions of any type such as hemorrhoids, esophageal varices or colonic tumors. Excessive blood loss may occur during menstrual periods.

Achlorhydria: Iron in food is in both ferric (Fe^{+3}) and ferrous (Fe^{+2}) state; in the stomach, acid causes the conversion of ferric iron to ferrous iron. The ferrous form of iron is absorbed. In the absence of acid, achlorhydria, gastrectomy, there is an inadequate amount of ferrous iron available for absorption.

Defective Absorption: Iron is absorbed in the duodenum and jejunum; conditions that alter the mucosa, such as sprue or steatorrhea, may cause a decrease in iron absorption.

Increased Demand: During pregnancy, the iron requirements increase significantly for two basic reasons; increased iron requirements are necessary to meet the needs of the fetus and for the enlarging red cell mass.

Anemia of Chronic Disease (ACD): ACD is the most common cause of anemia in a hospital population. Conditions associated with ACD are rheumatoid arthritis, collagen vascular disease, tuberculosis, fungal infections, inflammatory bowel disease, lymphoma and metastatic carcinoma. Serum ferritin may be used to reliably distinguish iron deficiency anemia from ACD; in iron deficiency anemia, serum ferritin is low; in ACD, serum ferritin is normal or elevated.

Chronic Renal Disease: In patients on chronic hemodialysis, the typical picture is anemia of chronic renal failure and superimposed iron deficiency from blood loss associated with maintenance dialysis. Serum ferritin levels can be used to monitor iron balance in these patients. A serum ferritin below 50mcg/liter is highly suggestive if not diagnostic of iron deficiency in patients with chronic renal failure; serum ferritin measurements should be performed at 4-6 week intervals from the onset of maintenance hemodialysis (Cook, J.D., Seminars in Hematology 19, 6-16, 1982).

The causes of increase in serum iron are given in the next Table; (Halliday, J.W. and Powell, L.W., Seminars in Hematology 19, 42-53, 1982):

Increase in Serum Iron
Idiopathic (primary hereditary) Hemochromatosis
Secondary Hemochromatosis
(1) Anemia and Ineffective Erythropoiesis
(a) Thalassemia Major
(b) Sideroblastic Anemia
(c) Hemolytic Anemias
(2) Liver Disease
(a) Alcoholic Cirrhosis
(b) After Portacaval Anastomosis
(3) Excessively High Oral Intake
(a) Prolonged Ingestion of Medicinal Iron
(b) Intake of Iron with Brewed Beverages

The most common causes of hemochromatosis are idiopathic hemochromatosis and hemochromatosis secondary to ineffective erythropoiesis (e.g. thalassemia and sideroblastic anemias). The laboratory tests that are done in idiopathic hemochromatosis and secondary hemochromatosis are shown in the next Table:

Tests in Primary or Secondary Hemochromatosis	
Test	Hemochromatosis
Serum Iron	>175mcg/dl
Saturation of Transferrin	>50%
Serum Ferritin	>250ng/ml

Idiopathic Hemochromatosis: Liver biopsy is the definitive test for confirmation of increased iron stores. Since this disease is genetically determined (autosomal recessive inheritance; full expression in homozygote), it is possible to identify within families homozygous and heterozygous relatives by comparison of their HLA antigens with those of the affected who is presumed homozygote (Bassett, M.L. et al., Hepatology 1, 120-126, 1981).

Secondary Hemochromatosis: In hemochromatosis secondary to anemia, the diagnosis of the underlying anemia is usually apparent after routine hematologic studies. The diagnosis of the iron overload is made in the same manner as for primary hemochromatosis. In iron loading anemias, there are two groups: (a) hypoplastic bone marrow, (e.g., aplastic anemia) whose major source of excess iron is blood transfusion and (b) patients with hyperplastic bone marrow but ineffective erythropoiesis. In this latter group, the excess iron results from increased iron absorption secondary to the ineffective erythropoiesis and blood transfusions.

Alcoholic liver disease: (See Halliday, J.W. and Powell, L.W., Seminars in Hematology 19, 42-53, 1982).

Iron Poisoning: Iron is an important cause of accidental poisoning death in children. The average lethal dose is about 200 to 250mg/kg body weight. Iron causes a necrotizing gastro-enteritis with bleeding and accumulation of blood in the lumen and bloody diarrhea. After 16-24 hours, the patient may develop metabolic acidosis. Specific treatment is with the chelating agent for iron, deferoxamine (Desferal).

IRON-BINDING CAPACITY

SPECIMEN: Red top tube; separate serum from cells as soon as possible. The serum specimen is stable up to 4 days at room temperature and one week in the refrigerator at 4°C. Interferences: gross hemolysis. The anticoagulants, citrate, EDTA and fluoride-oxalate cause significant depression of iron-binding capacity.

REFERENCE RANGE: Iron-Binding Capacity: Infant: 100-400mcg/dl; Thereafter: 250-400mcg/dl. To convert conventional units in mcg/dl to international units in mcmol/liter, multiply conventional units by 0.1791.

METHOD: Add ferric ions to serum; adsorb excess iron with magnesium carbonate; assay for iron.

INTERPRETATION: The causes of decreased serum total iron-binding capacity are given in the next Table:

Decreased Serum Total Iron-Binding Capacity
Anemia of Chronic Disorders
Hemochromatosis
Sideroblastic Anemia
Protein Deficiency

The causes of increased serum total iron-binding capacity (TIBC) are given in the next Table:

Increased Serum Total Iron-Binding Capacity (TIBC)
Iron Deficiency Anemia
Pregnancy
Oral Contraceptives

Percent Saturation: The percent saturation is calculated from the serum iron (mcg/dl) and the total iron-binding capacity (TIBC) (mcg/dl) as follows:

$$\frac{\text{Serum Iron (mcg/dl)}}{\text{Total Iron-Binding Capacity (mcg/dl)}} \times 100 = \text{\% Saturation}$$

The normal value for percent saturation is 33%. The alterations in iron metabolism in several disorders are given in the next Table; (Cartwright, G.E., Diagnostic Laboratory Hematology, Grune and Stratton, New York, 4th ed., 1968, pg. 215):

Alterations in Iron Metabolism in Several Disorders			
Disorder	Serum Iron (mcg/dl)	Total Iron Binding Capacity (mcg/dl)	% Sat.
Iron Deficiency	D	I	D
Pregnancy	D	I	D
Anemia of Chronic Disorders	D	D	D
Hemochromatosis	I	D	I
Sideroblastic Anemia	I	D	I
Protein Deficiency	D	D	N

Serum transferrin levels may be calculated from serum total iron-binding capacity(TIBC) utilizing the conversion formula:

$$\text{Serum Transferrin} = 0.8 \times \text{TIBC} - 43$$

ISOPROPYL ALCOHOL

SPECIMEN: Red top tube

REFERENCE RANGE: Negative. To convert conventional units in mg/dl to international units in mmol/liter, multiply conventional units by 0.1664.

METHOD: Gas-liquid chromatography

INTERPRETATION: Rubbing alcohol contains 70 percent isopropyl alcohol (isopropanol) and is sometimes ingested by accident or in desperation by alcoholics.

The lethal dose of isopropyl alcohol is about 8 ounces or 240 ml. Isopropanol is metabolized to acetone; blood and urine acetone are very high. A patient survived with a blood level of isopropanol of 440mg/dl (King, L.H., et al., JAMA, 211, 1855, 1970), and another patient survived with a blood acetone level of 1,878mg/dl (Dua, S.L., JAMA, 230, 35, 1974); these patients were treated with hemodialysis and peritoneal dialysis respectively.

The clinical signs and symptoms are hypothermia, deep coma, areflexia, and hypotension.

JOINT PANEL [see ARTHRITIC (JOINT) PANEL]

17-KETOGENIC STEROIDS (17-KGS) URINE
SPECIMEN: 24 hour urine. Place 10gm boric acid in container prior to
collection. Instruct the patient to void at 8:00 A.M. and discard the specimen.
Then collect all urine including the 8:00 A.M. specimen at the end of the 24
hour collection period. Refrigerate urine as each specimen is collected. The
specimen must be refrigerated or frozen after collection since 17-OHCS are
destroyed at room temperature. The patient should be off all drugs with the
exception of aspirin and barbiturates. Record the 24 hour volume. 50ml aliquot
is required for the assay. Any test requiring a 24 hour urine collection may
also be run on this specimen e.g. 17-keto- steroids.
REFERENCE RANGE: Female: 3-15mg/24 hours; Male: 5-22mg/24 hours.
METHOD: The 17-KGS include the 17-OHCS (Cpd S and Cpd F) plus 17-OH
precursors of these compounds; 17-Keto compounds are not assayed.
    17-OH compounds are reduced using sodium borohydride, e.g.,

$-\overset{|}{\underset{|}{C}}=0 \longrightarrow -\overset{|}{\underset{|}{C}}-OH$; then, the reduced compounds are oxidized using
bismuthate or periodate to yield 17-keto compound, e.g.,

$\longrightarrow$. The 17-keto compounds are assayed using the same reaction as
for the 17-keto steroids (Zimmerman Reaction).
INTERPRETATION: 17-Keto steroids are not assayed by this method; 17-OHCS (Cpd
S and Cpd F) plus 17-OHCS precursors are assayed.

KETONES, SERUM (see ACETONE, SERUM)

17-KETOSTEROIDS (17-KS), URINE
SPECIMEN: 24 hour urine. Place 10gm boric acid in container prior to
collection. Instruct the patient to void at 8:00 A.M. and discard the specimen.
Then collect all urine including the 8:00 A.M. specimen at the end of the 24
hour collection period. Refrigerate urine as each specimen is collected. The
specimen must be refrigerated or frozen after collection since 17-KS are
destroyed at room temperature. The patient should be off all drugs with the
exception of aspirin and barbiturates. Record the 24 hour volume. 50ml aliquot
is required for the assay. Any test requiring a 24 hour urine collection may
also be run on this specimen e.g. 17-hydroxycorticosteroids.
REFERENCE RANGE: Male: 9-22mg/24 hour; Female: 4-14mg/24 hour; Neonate: up to
2mg/24 hour; 1 month-5 year: ≤0.5mg/24 hour; 6-8 year: 1-2mg/24 hour.
METHOD: Steroids having a ketone group at position 17 in the ring structure D
are assayed; testosterone has a hydroxyl group at position D. The color reaction
involves reaction of the 17-keto-group with meta- dinitrobenzene to yield a
purple compound that absorbs at 520nm.
INTERPRETATION: The causes of increased urinary 17-KS are listed in the next
Table:

Causes of Increased Urinary 17-KS
Cushing's Syndrome
11-Hydroxylase and 21-Hydroxylase Deficiency
Stressful Illness, e.g., burns, etc.
ACTH, Cortisone or Androgens (except Methyltestosterone)
Androgen Producing Gonadal Tumors (Rare)

The causes of decreased urinary 17-KS are listed in the next Table:

Causes of Decreased Urinary 17-KS
Addison's Disease
Anorexia Nervosa
Panhypopituitarism

The steroids that are assayed as 17-ketosteroids are given in the next Figure:

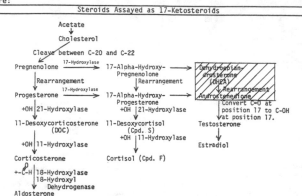

Steroids Assayed as 17-Ketosteroids

▨ = Compounds Assayed as 17-KS

KIDNEY STONE ANALYSIS (see RENAL STONE ANALYSIS)

KLEIHAUER-BETKE STAIN (see FETAL HEMOGLOBIN STAIN)

KOH EXAMINATION

SPECIMEN: Obtain fluid, such as sputum, rather than a swab of fluid.
REFERENCE RANGE: Negative
METHOD: Suspend a drop of exudate such as vaginal discharge in a drop of 10 or 20 percent KOH on a clean glass slide. Cover the drop with a clean glass cover slip and press gently to make a thin mount. Allow specimen to sit at room temperature for about 20 to 30 min. Scan under low power with reduced lighting; change to high power to inspect for presence of suspected fungi.
INTERPRETATION: The polysaccharides of fungal cell walls are relatively resistant to alkali that dissolves mammalian cells or renders them translucent.

Fungal diseases, known as mycoses, are classified as systemic (deep), subcutaneous or superficial (cutaneous) mycoses. Fungi capable of causing mycoses are given in the next Table:

Fungi Causing Mycoses		
Systemic	Subcutaneous	Cutaneous
Cryptococcus Neoformans	Sporotricum Schenckii	Trichophyton
Coccidioides Immitis	Candida Albicans	Epidermophyton
Histoplasma Capsulatum		Keratinomyces
Blastomyces Dermatidides		Candida Albicans
Candida Albicans		

The KOH preparation is used to examine vaginal discharge for Candida Albicans ("yeast vaginitis"); typical yeast cells and pseudomycelia may be observed.

Cytologic preparations have greater sensitivity than KOH preparations for rapid diagnosis of pulmonary coccidiodomycoses. (Warlick, M.A. et al., Arch. Intern. Med. 143, 723-725, 1983.)

LABORATORY PANEL - GENERAL PRACTICE

Routine laboratory studies are listed in the next Table:

Routine Laboratory Studies, General Practice
Complete Blood Count(CBC)
Urinalysis
Erythrocyte Sedimentation Rate(ESR)
Chemistries:
Blood Sugar
Electrolytes(Na^+, K^+, Cl^-, CO_2 Content)
Blood Urea or Creatinine
Chest X-Ray
Electrocardiogram(ECG)

The classification of cases and the importance of diagnostic laboratory investigations are given in the next Table; (Sandler, G., Brit. Med. J. 2, 21-24, July 7, 1979):

Classification of Cases and Importance of Diagnostic Laboratory Investigations					
	Percent of	Diagnostic Laboratory Investigations		Laboratory Investigations Management	
Classification	Patients(%)	Routine(%)	Special(%)	Routine(%)	Special(%)
Cardiovascular:	44	3	6	4	13
Chest Pain,					
Hypertension,					
Arrhythmias,					
Heart Failure,					
Murmurs, Other					
Neurological:	20	3	14	3	20
Loss of					
Consciousness,					
Headache,					
Paresthesia and					
Weakness, Other					
Endocrine:	10	11	42	32	43
Diabetes,					
Thyroid Disorder,					
Other					
Alimentary:	8	0	58	0	63
Indigestion,					
Bowel Change,					
Other					
Respiratory:	4	17	14	20	17
Dyspnea,					
Bronchitis,					
Other					
Urinary:	2	5	26	20	40
Miscellaneous	10	8	20	10	20

LACTASE (LACTOSE TOLERANCE TEST)

SPECIMEN: 4 Red top tubes; specimens obtained at time 0, 30 min., 1 hour and 2 hours following oral dose of lactose 50g/m^2 body surface area.

REFERENCE RANGE: Rise of serum glucose greater than 20mg/dl between 30 minutes and 1 hour following lactose load.

METHOD: Glucose assay.

INTERPRETATION: A significant number of patients with the irritable bowel syndrome (gas and bloating, abdominal pain and diarrhea) have lactase deficiency and relief may be obtained by restricting the intake of milk. High prevalence of lactase deficiency occurs in Blacks, Orientals, Jews and American Indians.

Lactase deficiency is normally defined for subjects who have a blood glucose rise of less than 20mg/dl after 50g/m^2 of a oral lactose dose; a tolerance test with 25g of glucose and 25g of galactose should be normal. The results of a oral lactose dose to normal subjects and to subjects with lactase deficiency are shown in the next Figure:

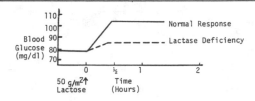

Oral Lactose Test

Hypolactasia is a common cause of gastrointestinal symptoms. Lactase is found in the brush border of the enterocyte; it catalyses the reaction:

$$\text{Lactose} \xrightarrow{\text{Lactase}} \text{Glucose} + \text{Galactose}$$

Glucose and galactose are then actively transported into the epithelial cell; only trace amounts of lactose pass through the epithelial cell.

In patients with lactase deficiency, lactose remains within the intestinal lumen; it produces an osmotic diarrhea and may be associated with secondary malabsorption of other nutrients.

Lactase deficiency may be detected by analysis of respiratory hydrogen. The test is based on the principle that lactose (or other carbohydrate) that is not absorbed in the small intestine is metabolized by colonic bacteria to various products, including H_2. The presence of H_2 in the breath is indicative of bacterial degradation. The diagnosis of lactase deficiency is determined by measurement of respiratory H_2 excretion before and 1, 2 3 and 4 hours after a lactose load of 50g dissolved in 500ml of water. Respiratory H_2 values of more than 0.30ml/minute above the fasting value are considered indicative of lactase deficiency.

A test on urine, using a simple strip impregnated with galactose oxidase, has been developed for possible use in mass screening to detect subjects with lactose malabsorption (Arola, H. et al., The Lancet, pgs. 524-525, Sept. 4, 1982).

Lactase deficiency has been found to be infrequently responsible for symptoms among patients with the irritable bowel syndrome who are white, of northern-western European extraction, and without a milk intolerance. (Newcomer, A.D. and McGill, D.B., Mayo Clin. Proc. 58, 339-341, 1983).

Individuals who cannot tolerate lactose in milk may be able to tolerate the same sugar in yogurt (Kolars, J.C. et al, N. Engl. J. Med. 310, 1-3, Jan. 5, 1984).

LACTATE (see LACTIC ACID)

LACTIC ACID

SPECIMEN: Arterial blood preferred because contraction of muscles can cause increase in lactate in venous blood. Grey (sodium fluoride) top tube; keep tube on ice until delivered to laboratory; separate plasma and refrigerate plasma.

REFERENCE RANGE: Arterial blood: 0.5-1.66 mmol/liter (4.5-14.4mg/dl). Venous blood: 0.5-2.2 mmol/liter (4.5-19.8mg/dl).

METHOD: Lactate + NAD$^+$ $\xrightarrow{\text{LDH}}$ Pyruvate + NADH + H$^+$
(340nm)

INTERPRETATION: Lactic acidosis may be divided into two types; in one type, there is poor tissue oxygenation and in another type, there is no clear evidence of tissue hypoxia. The two types and the conditions associated with these types are given in the next Table:

Types of Lactic Acidosis	
Poor Tissue Oxygenation	Non-Hypoxic Conditions
Cardiopulmonary Failure:	Common Disorders:
Congestive Heart Failure(CHF)	Diabetes Mellitus
Severe Anemia	Renal Failure
Hemorrhage	Liver Disease
Hypertension	Infection
	Malignancies
	Convulsions
	Alkalosis
	Drugs/Toxins
	Hereditary Enzyme Defects

The normal arterial concentration of lactate is less than 1.6 mmol/liter (approximately 15mg/dl). A concentration greater than 2 mmol/liter is abnormal. When lactic acidosis is present, the arterial lactate concentration is usually higher than 7 mmol/liter. The normal arterial concentration of pyruvate is less than 0.15 mmol/liter. The arterial lactate-pyruvate ratio is normally about 10:1 (1.5 mmol/liter lactate: 0.15 mmol/liter pyruvate). When lactic acidosis is present, the ratio may rise to 60/1 (Narins, R.G. et al., Hosp. Pract. pgs 91-98, June, 1980). Lactic acidosis has been successfully treated with dichloroacetate (Stacpoole, P.W. et al., N. Engl. J. Med. 309, 390-396, 1983); dichloroacetate reduces circulating lactate concentration by stimulating the activity of pyruvate dehydrogenase, the enzyme that catalyzes pyruvate to acetyl CoA.

S. Bakerman

LACTIC DEHYDROGENASE(LDH)

SPECIMEN: Special care must be taken in obtaining samples since minimal hemolysis will cause elevated LDH. Do not freeze or refrigerate specimens.

REFERENCE RANGE: 30-200IU/liter

METHOD: Lactate + NAD$^+$ $\xrightarrow{\text{LDH}}$ Pyruvate + NADH + H$^+$
(340nm)

INTERPRETATION: Serum lactate dehydrogenase is elevated in a wide variety of conditions reflecting widespread tissue distribution; it is elevated in the conditions listed in the next Table:

Causes of Increased Lactic Dehydrogenase
Myocardial Infarction (Acute)
Myocarditis - from any source
Liver Disease
Congestive Heart Failure
Infectious Mononucleosis
Malignancy
Hemolytic Anemia
Reabsorbing Extravasated Blood
Megaloblastic Anemia
Skeletal Muscle Disease
Trauma
Tissue Necrosis
Shock
Acute Pancreatitis (Occ.)
Multisystem Diseases
Collagen-Vascular Diseases
Cerebral Vascular Accidents
Acute Renal Infarction
Pulmonary Infarction
Pulmonary Disease (Active)
Transfusions
Hemolysis
Failure to Separate Clot
Heparin Therapy

This enzyme is elevated in <u>acute myocardial infarct</u>, <u>liver necrosis</u>, <u>acute pulmonary infarct</u>, <u>primary muscular dystrophy</u>, <u>pernicious anemia</u>, <u>hemolytic anemia</u>, <u>cerebral infarcts</u> and <u>malignancy</u>. These diseases may be differentiated by LDH isoenzyme patterns.

Lactic dehydrogenase has been recommended as a prognostic factor in following patients with advanced <u>colorectal carcinoma</u>; patients with an initially normal level of LDH versus those with an abnormal level of LDH had median survivals of 16 and 7.0 months respectively. (Kemeny, N. and Braun, D.W., Am. J. Med. 74, 786-794, 1983).

Sixty percent of hypertensive patients with <u>pheochromocytoma</u> had elevated serum LDH (O'Connor, D.T. and Gochman, N., JAMA 249, 383-385, 1983).

Lactic dehydrogenase(LDH) was elevated in patients with ovarian dysgerminoma (Awais, G.M., Obstet. Gynecol. 61, 99-101, 1983).

LDH is elevated in about one-third of patients on heparin therapy; LDH-5, the liver band, is elevated (Dukes, G.E. et al., Ann. Int. Med. 100, 645-650, 1984).

LACTIC DEHYDROGENASE ISOENZYMES(LDH ISOENZYMES) SERUM

<u>SPECIMEN:</u> Red top tube; separate serum; do not freeze because freezing destroys LDH-4 and LDH-5; do not use hemolyzed specimen - hemolysis causes elevation of LDH-1 and LDH-2. Isoenzymes should <u>not</u> be done if total LDH is less than 130IU/L.

<u>REFERENCE RANGE:</u>

Isoenzyme	%	IU/L
Adult Values:		
LDH-1 (Heart)	22 to 37	22 to 85
LDH-2	30 to 45	30 to 100
LDH-3	15 to 30	15 to 65
LDH-4	5 to 11	5 to 25
LDH-5 (Liver)	2 to 11	2 to 25
		Total: 100 to 225

Neonates: As compared to adults, the percentage of LDH-5 is increased in neonates; the percentage of LDH-1 is decreased while LDH-2, LDH-3 and LDH-4 are at the same percentage in neonates as in adults.

<u>METHOD:</u> Electrophoresis; LDH isoenzymes should be reported in <u>both</u> percent and international units.

<u>INTERPRETATION:</u> The normal LDH isoenzyme pattern of serum is given in the next Figure. The tissues having high concentration of LDH-1, LDH-3, and LDH-5 are given just below the corresponding isoenzyme. Those that have clinical significance are underlined.

Serum Isoenzyme Pattern and Tissue Distribution of LDH Isoenzymes

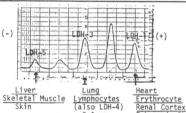

Liver	Lung	Heart
Skeletal Muscle	Lymphocytes	Erythrocyte
Skin	(also LDH-4)	Renal Cortex
	Spleen	Embryonic Skeletal Muscle
	Pancreas	Brain
	Placenta	

Serum LDH isoenzyme patterns found in various diseases are given in the next six Figures:

Elevation of LDH-1

Pattern	Causes
	Acute Myocardial Infarction
	Hemolytic Anemia
	Megaloblastic Anemia
	Acute Renal Infarction
	Hemolyzed Specimen
	Testicular Cancer

Elevation of LDH-1: As expected from the tissue distribution (see previous Figure), elevation of serum LDH-1 is seen in <u>acute myocardial infarction</u>; <u>hemolytic anemia</u>; <u>megaloblastic anemia</u>, <u>both</u> folic acid deficiency and vitamin B_{12} deficiency and <u>acute renal infarction</u>. LDH-1 occurs in the brain but this isoenzyme probably does not normally pass the blood-brain barrier. The LDH-1, which may be elevated in patients with strokes, probably originates from the red blood cells in the thrombus, or in the area of hemorrhage in the brain. <u>Hemolyzed serum specimen will cause an elevated LDH-1.</u> Serum LDH-1 elevation may occur in patients with testicular carcinoma (Liu, F. et al., Am. J. Clin. Path. <u>78</u>, 178-183, 1983), and in patients with ovarian carcinoma.

Elevation of LDH-3

Pattern	Causes
	Acute Pulmonary Infarction Extensive Pulmonary Pneumonia Advanced Cancer Acute Pancreatitis Lymphocytosis

Elevation of LDH-3: Elevation of LDH-3 occurs occasionally in acute pulmonary infarction and extensive pulmonary pneumonia. Elevation of LDH-3 may occur in patients with advanced cancer and may be useful in following the effectiveness of therapy for cancer. An example of the clinical usefulness of LDH-3 was recently seen in a patient who initially was thought to have a uterine leiomyoma (fibroid). The clinician was informed that elevation of LDH-3 is observed in malignancy; at operation a uterine malignancy was removed.

Elevation of LDH-5

Pattern	Causes
	Hepatic Congestion Hepatitis or other liver injury or inflammation Skeletal Muscle Injury

Elevation of LDH-5: Elevation of LDH-5 may occur in congestive heart failure, and in acute, subacute, and chronic hepatitis; it is not elevated in intra- or extrahepatic bile duct obstruction. LDH-5 is elevated in skeletal muscle injury whether caused by trauma or by surgery.

Elevation of All Isoenzymes

Pattern	Causes
	Systemic Diseases: Carcinomatosis Collagen Vascular Disease Overwhelming Sepsis Disseminated Intravascular Coagulopathy

Elevation of all isoenzymes of LDH occur in systemic diseases, such as carcinomatosis, collagen vascular disease, overwhelming sepsis, disseminated intravascular coagulopathy.

Elevation of Both LDH-1 and LDH-5

Pattern	Causes
	Following Acute Myocardial Infarction Chronic Alcoholics with Hepatitis and Megaloblastic Anemia

Clinically, this pattern is observed most often several days following an acute myocardial infarction; elevation of LDH-1 is due to the acute myocardial infarction; elevation of LDH-5 is due to hepatic congestion which occurs in 30 percent of patients 1.5 days following infarction (West, et al., AM. J. Med. Sci., 241, 350, 1961). This pattern may also be seen in chronic alcoholics who develop both hepatitis and megaloblastic anemia due to folic acid deficiency.

Elevation of Both LDH-3 and LDH-5	
Pattern	Causes
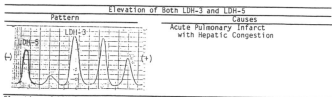	Acute Pulmonary Infarct with Hepatic Congestion

Elevation of LDH-3 and LDH-5: This pattern may be seen in patients with acute pulmonary infarcts and right-sided heart failure resulting in hepatic congestion.

LATEX AGGLUTINATION(LA)(see BACTERIAL ANTIGENS)

LEAD, BLOOD

SPECIMEN: Lavender (EDTA) top tube; or green (heparin) top tube.
REFERENCE RANGE: Children: <30mcg/dl; adults: <40mcg/dl; industrial exposure: <60mcg/dl. To convert conventional units in mcg/dl to international units in mcmol/liter, multiply conventional units by 0.04826.
METHOD: Atomic absorption; anodic stripping voltammetry (ESA Trace Metals Analyzer)
INTERPRETATION: The concentration of lead in whole blood is about 75 times that of serum or plasma.

The laboratory tests that have been used in evaluating lead exposure are listed in the next Table:

Laboratory Tests in Evaluating Lead Poisoning
Increased lead levels in blood, urine and/or hair
Increased delta amino levulinic acid(DALA) and coproporphyrin III in urine
Decreased delta amino levulinic acid(ALA) dehydrase
Increased erythrocyte zinc protoporphyrin(ZPP)

The inhibition of the activity of delta ALA dehydrase is a sensitive indicator of acute and chronic lead poisoning. The activity of this enzyme decreases with increasing lead levels.

Toxic effects of lead are observed in the systems listed in the next Table:

Systems Involved in Toxic Effects of Lead
Hematopoietic System
Nervous System
Gastrointestinal System
Kidneys

Hematopoietic System: The hematologic manifestations of lead poisoning are listed in the next Table:

Hematologic Manifestations of Lead Poisoning
Basophilic stippling of red blood cells and precursors
Anemia
Reticulocytosis
Erythroid hyperplasia with dyserythropoiesis
Autofluorescence of red blood cells and precursors

Lead interferes with the biosynthesis of heme; the location of metabolic blocks in the biosynthesis of heme is shown in the next Figure:

Pathway for Synthesis of Heme and Enzymes Inhibited by Lead

Lead inhibits the activity of <u>three enzymes</u> in the synthesis of heme; these are indicated in the previous Figure and are listed in the next Table (Cambell, B.C., Brodie, M.J., Thompson, G.G., Meredith, P.A., Moore, M.R., and Goldberg, A., Clinical Science and Molecular Medicine, 53, 335-340, 1971):

Enzymes Inhibited by Lead

Delta Amino Levulinic Acid (ALA) Dehydrase
Coproporphyrinogen Oxidase
Ferrochelatase (Heme Synthetase)

Lead is an electropositive metal and has a high affinity for the negatively charged sulfhydryl group; ALA dehydrase, coproporphyrinogen oxidase and ferro-chelatase are sulfhydryl-dependent enzymes.

There is increased activity of the enzyme, ALA synthetase, probably secondary to decreased production of heme.

<u>Nervous System:</u> Both the central and peripheral nervous system may be affected. The characteristic neurologic abnormalities are motor paralysis (wrist drop) and acute encephalopathy. There may be nonspecific symptoms such as headache, dizziness, sleep disturbances, memory deficit and increased irritability, and in severe cases, profound central nervous system disturbances with convulsions.

<u>Gastrointestinal System:</u> Gastrointestinal symptoms may be related to disturbed innervation of the intestines; there may be loss of appetite, weight loss, abdominal discomfort and crampy abdominal pains, e.g., "lead colic."

<u>Kidney:</u> Lead exposure may cause interstitial nephritis.

LEAD EXPOSURE PANEL

The laboratory tests for evaluation of lead exposure are listed in the next Table;

Laboratory Tests for Evaluation of Lead Exposure

Screening Test:
  Erythrocyte Zinc Protoporphyrin(ZPP)
    Increased in Lead Poisoning
Enzyme Test:
  Erythrocyte Delta Aminolevulinic Acid(DALA) Dehydrase
    Decreased in Lead Poisoning
Definitive Test:
  Lead Levels in Blood

Erythrocyte zinc protoporphyrin(ZPP) is an excellent <u>screening</u> test to detect lead exposure (see Zinc Protoporphyrin, Blood).

The inhibition of <u>DALA dehydrase activity</u> is a <u>sensitive indicator</u> of acute and chronic lead poisoning; this enzyme's activity <u>decreases</u> with <u>increasing</u> <u>lead levels</u>.

The blood lead level is a definitive indicator of the extent of current or recent lead absorption. Following exposure to the metal, the whole-blood concentration of lead (which localizes primarily in the erythrocytes) is about 75 times that in the serum or plasma. The reference levels for lead in the blood are 30 mcg/dl for children and 40 mcg/dl for adults; the industrial exposure level is 60 mcg/dl. Toxicity is probable when the blood lead level exceeds these limits. Lead can be quantitated with atomic absorption(AA) spectrometry or anodic striping voltammetry(ASV).

LEAD, SERUM

SPECIMEN: Red top tube, separate serum.
REFERENCE RANGE: 0.80-2.46ng/ml
METHOD: Atomic absorption; anodic stripping voltammetry
INTERPRETATION: The concentration of lead in whole blood is about 75 times that of serum or plasma. Thus, whole blood is the preferred sample for determination of toxic lead levels in blood rather than serum. See LEAD, BLOOD.

LEAD, URINE

SPECIMEN: Collect 24 hour urine in acid washed-containers. Use plastic containers (polyethylene or polypropylene); add 10% HCl solution to the container and allow to "soak" for 10 minutes; rinse with five volumes of tap water and then five volumes of deionized or distilled water. The patient should urinate at 8:00 A.M. and the urine is discarded. Then, urine is collected for 24 hours including the next day 8:00 A.M. specimen. Indicate 24 hour volume. A 50ml aliquot is used for analysis.
REFERENCE RANGE: <125mcg/24 hours. To convert conventional units in mcg/24 hours to international units in mcmol/day, multiply conventional units by 0.004826.
METHOD: Atomic absorption; anodic stripping voltammetry
INTERPRETATION: See LEAD, BLOOD. Urine lead levels may not correlate with clinical symptoms of lead toxicity or blood lead levels; "normal" urinary lead excretions have been observed in the presence of significantly elevated lead levels and clinical evidence of lead poisoning (Berman, E., Laboratory Medicine 12, 677-684, 1981).

The CaNa2EDTA lead mobilization test has been used to identify lead nephropathy. This test is done as follows: Urine is collected for 24 hours for lead analysis as control. Then, 1 gram of CaNa2EDTA is given I.V. in 250ml of 5% dextrose/water over 1 hour. Urine is collected for lead assay for 3 days. Normal subjects excrete less than 600mcg during the 3 days following parenteral EDTA; subjects with lead nephropathy excrete more than 600mcg. (Wedeen,R.P., Clin. Exper. Dialysis and Apheresis, 6, 113-146, 1982; Wedeen,R.P. et al., Am. J. Med. 59, 630, 1975; Emmerson, B.T., Kidney Int. 4, 1, 1973). The EDTA lead-mobilization test may also be performed by intramuscular injection of a total of 2 g of EDTA with 1.0 ml of 2 percent xylocaine, given in two divided, one gram, doses 8 to 12 hours apart. Twenty-four-hour urinary excretion of lead is measured over three consecutive days, starting with the first injection. Subjects without unusual lead exposure excrete less than 600 mcg (2.9 mcmol) of lead per three days during this test (Batuman, V. et al., N. Engl. J. Med. 309, 17-21, 1983).

LEAD MOBILIZATION TEST (see LEAD, URINE)

LECITHIN/SPHINGOMYELIN (L/S) RATIO AND PHOSPHATIDYLGLYCEROL (PG)

SPECIMEN: Amniotic fluid. If amniotic fluid is contaminated by <u>maternal blood</u>, it may falsely lower the L/S ratio of amniotic fluid if the ratio is greater than 1.5/1. Amniotic fluid contaminated by <u>meconium</u> gives uninterpretable L/S ratio.

REFERENCE RANGE: When L/S ratio is >2/1 and phosphatidylglycerol (PG) is 3% or greater, respiratory distress syndrome (RDS) has not been observed.

METHOD: Two dimensional thin-layer chromatography (Kulovich, M.V., Hallman, M.B., Gluck, L., Am. J. Obstet. Gynecol. <u>135</u>, 57-63, 1979; Kulovich, M.V. and Gluck, L., Am. J. Obstet. Gynecol. <u>135</u>, 64-70, 1979).

INTERPRETATION: The surface-active lipid, lecithin, originates from Type II alveolar epithelial cells of the fetal lung and during gestation finds its way into amniotic fluid; its presence and chemical character (palmitic acid content) reflect fetal pulmonary maturation. The changes in <u>amniotic fluid lecithin</u> and <u>sphingomyelin</u> during gestation are shown in the next Figure:

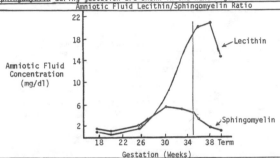

Amniotic Fluid Lecithin/Sphingomyelin Ratio

The phospholipid, <u>lecithin</u>, increases with gestational age; another phospholipid, <u>sphingomyelin</u>, remains at a relatively low and <u>constant</u> concentration throughout gestation and can be used as a <u>baseline</u> for comparison.

The L/S ratio is close to 1:1 until the 35th week of gestation. At that time, the lecithin content increases <u>dramatically</u>, attaining a ratio of 2:1, indicating <u>pulmonary maturity</u>.

This test is very accurate in determining the <u>absence</u> of RDS; but, subjects with a ratio <2:1 may not have RDS.

Phosphatidylglycerol (PG) in Amniotic Fluid: The content of phosphatidylglycerol in amniotic fluid during gestation is shown in the next Figure; (Saunders, B.S. et al., Clin. Perinatology, <u>5</u>, 231-242, 1978):

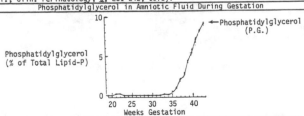

Phosphatidylglycerol in Amniotic Fluid During Gestation

Phosphatidylglycerol is first detected at 35 to 36 weeks' gestation. The concentration increases and it is the second most common phospholipid at term. <u>The presence of phosphatidylglycerol in amniotic fluid is indicative of pulmonary maturity</u>.

Diabetes Mellitus: In diabetes mellitus, a L/S ratio of 2.0 or greater does not insure fetal lung maturity. However, the presence of 3% or more PG in surfactant in amniotic fluid is used as the guideline for delivery in diabetics; RDS has not been observed when PG is 3% or greater.

Hypertension: In mild hypertension, the development of surfactant, as reflected by L/S ratio and PG, is similar to the normal. In severe hypertension, there may be accelerated maturation, sometimes with L/S ratios >2.0 and early PG.

Prolonged Rupture of Membranes(PROM): In PROM, there is acceleration of fetal lung maturation with early elevation of L/S ratios and early appearance of PG.

Lee-White
Clotting Time

LEE-WHITE CLOTTING TIME(SEE CLOTTING TIME, LEE-WHITE)

Legionnaires'
Antibody

LEGIONNAIRES' DISEASE, ANTIBODY SERUM
SPECIMEN: Obtain two blood specimens: an acute phase specimen plus a convalescent phase specimen. The convalescent phase specimen is obtained 21 or more days after onset of fever.
      Red top tube, separate and refrigerate serum.
REFERENCE RANGE: Acute infection is indicated by a fourfold rise in titer between acute and convalescent samples. A single titer greater or equal to 1:256 is evidence of previous infections.
METHOD: Indirect immunofluorescence
INTERPRETATION: In Legionella infections, antibody titers tend to be low during the first week, rise during the second and third weeks and reach maximum levels at about five weeks. After recovery, titers drop slowly and tend to remain elevated for many years.

Legionnaires'
Culture

LEGIONNARIES' DISEASE, CULTURE
SPECIMEN: Fresh lung tissue, such as needle biopsy, pleural fluid, bronchial washings or brushings, transtracheal aspirates; expectorated sputum is not acceptable. Collect specimen in sterile container. Transport to laboratory immediately. Specimen may be forwarded to reference laboratory in sterile, screw-capped container on ice. Do not freeze.
REFERENCE RANGE: Negative
METHOD: Semi-selective media (Edelstein, P.H., J. Clin. Microbiol. 14, 298-303, 1981).
INTERPRETATION: The etiologic agent of Legionnaires' disease is Legionella pneumophilia and related species; it is a gram-negative, non-acid fast bacillus. The pathologic findings are largely limited to the lungs, in which pneumonia and acute diffuse alveolar damage are seen (Chandler, F.W., et al., N. Engl. J. Med. 297, 1218-1220, 1977). The incubation period is two to ten days; patients with underlying pulmonary disease and smokers are particularly susceptible. Patients develop fever, headache, malaise and pneumonia.
      Epidemics are associated with public buildings, such as hotels and hospitals; it has been found in the water supply (Editorial, Lancet., pgs. 381-383, Aug. 13, 1983).
      Legionella causes two types of illness: a severe form associated with pneumonia (Legionnaire's Disease) and a mild self-limiting form characterized by general symptoms of fever and aching (Pontiac fever). Erythromycin is the drug of choice; rifampin is used if there is not a satisfactory response to erythromycin alone.

Legionnaires',
Direct Immunofluorescence

LEGIONNARIES' DISEASE, DIRECT IMMUNOFLUORESCENCE
SPECIMEN: Lung tissue, respiratory tract fluids and cultures.
REFERENCE RANGE: Negative
METHOD: Direct immunofluorescence
INTERPRETATION: This test is specific for Legionella.

LEUKEMIA, ACUTE, PANEL

Acute leukemia is divided into two divisions: <u>acute lymphocytic leukemia</u> (ALL) and acute nonlymphocytic leukemia(ANLL); this division is illustrated in the next Figure; (modified by Kiely, J.M., Mayo Clin. Proc. <u>58</u>, 774-775, Nov. 1983):

Differentiation of Acute Leukemia

Morphology
Cytochemical Stains if Necessary

Acute Lymphocytic Leukemia(ALL)

Acute Nonlymphocytic Leukemia(ANLL)

Surface Immunoglobulins (Immunofluorescence) (Third Component of Complement; and Fc Portion of Immunoglobulins) Rosettes Newer Immunocytochemical Techniques and Monoclonal Antibody Reagents

Cytochemical Stains
See Below

T-ALL   B-ALL   Common   Non-T, Non-B
                              ALL   (Undifferentiated ALL)

B-cells predominate in the follicles and medulla of the lymph nodes and in the follicles of the spleen; they make up 10 to 15% of circulating lymphocytes.

Differentiation of <u>acute leukemias</u> using <u>cytochemical stains</u> is given in the next Table:

Differentiation of Acute Leukemias Using Cytochemical Stains

Acute Leukemia	Sudan Black B	Chloroacetate Esterase	Naphthyl Acetate Esterase (Non-Specific)	Periodic Acid-Schiff (PAS)
Acute Lymphocytic Leukemia(ALL)	Neg.	Neg.	+ to -	Pos.
Acute Myelogenous Leukemia(AML)	Pos.	Pos.	Neg.	+ to-
Acute Monoblastic Leukemia	+ to -	Neg.	Pos.	Neg.
Acute Myelomonocytic Leukemia	Pos.	Pos. Myel.	Pos. Mono.	Neg.

LEUKEMIA, LYMPHOCYTIC, CHRONIC(CLL) PANEL

Chronic lymphocytic leukemia(CLL) constitutes a spectrum of disorders, subclassified with the aid of immunocytochemical techniques; a classification is given in the next Figure:

Differentiation of Chronic Lymphocytic Leukemia(CLL)

Absent   Present

T-Cell Type (Uncommon)   B- Cell Type (Most Cases)
(Phyliky,R.L., Li, C-Y and Yam, L.T.
Mayo Clin. Proc. <u>58</u>, 709-720, Nov. 1983)

                                          Tartrate-Resistant
                                          Isoenzyme of Acid
                          Present   Phosphatase

                    Hairy-Cell
                    Leukemia

The spectrum of chronic lymphocytic leukemia(CLL) includes malignant lymphomas in the leukemic phase and Waldenström's macroglobulinemia.

S. Bakerman

LEUKOCYTE ALKALINE PHOSPHATASE (LAP) STAIN

Leukocyte
Alkaline Phosphatase
(LAP) Stain

SPECIMEN: Six fresh finger stick smears; air dry.

REFERENCE RANGE: Score of 50-150; each laboratory should determine its own normal range.

METHOD: Positive and negative controls must be run with each unknown slide; a positive control is obtained from women in their third trimester of pregnancy or within a few days following delivery. Negative controls are prepared by immersing an air dried fixed smear into boiling water for 1 minute. Slides are fixed in formalin:methanol (1:9) for 30 sec. at 0-5°C; wash in running water and air dry; incubate with substrate.

$$\text{Naphthol Phosphate} + H_2O \xrightarrow{\text{Leukocyte Alkaline Phosphatase}} \text{Naphthol} + H_3PO_4$$

$$\text{Naphthol} + \text{Diazo Cpd.} \longrightarrow \underset{\text{(colored compound)}}{\text{Diazotized Naphthol}}$$

Alkaline phosphatase activity is reflected by colored granules (usually red, blue or purple depending on diazo cpd) in the cytoplasm of mature leukocytes; cells other than reticulum cells, osteoblasts and endothelial cells show no activity. A kit for determining LAP activity is available from Sigma Corp. St. Louis, Mo.

Using oil immersion, 100 consecutive segmented and band form neutrophils are rated 0 to 4+ on the basis of granule size, staining and background of cytoplasm. The score is the sum of these grades; the maximum score is 400.

INTERPRETATION: LAP has no relationship to serum alkaline phosphatase. The causes of increased LAP are given in the next Table:

Causes of Increased Leukocyte Alkaline Phosphatase
Neutrophilia Secondary to Infections
Lymphomas: Acute Lymphatic Leukemia; Hodgkin's Disease
Myeloproliferative Disorders: Polycythemia Vera; Myelofibrosis, Agnogenic Myeloid Metaplasia
Stress, Trauma and Burns, Immediate Postoperative Period, Tissue Necrosis
Oral Contraceptives
Pregnancy and During Lactation
Newborns to 2 weeks of age

The causes of decreased LAP are given in the next Table:

Causes of Decreased Leukocyte Alkaline Phosphatase (LAP)
Acute (50%) and Chronic Granulocytic Leukemia
Erythroleukemia
Infectious Mononucleosis (Early)
Collagen Diseases
Paroxysmal Nocturnal Hemoglobinuria
Hypophosphatasia

LAP is normal in secondary polycythemia, chronic lymphocytic leukemia and viral infections.

LAP may be useful in differentiating the conditions listed in the next Table:

Differentiation Using Leukocyte Alkaline Phosphatase (LAP)		
Chronic Granulocytic Leukemia (Low LAP)	versus	Polycythemia Vera Granulocytic Leukemoid Reactions Myelosclerosis Myeloid Metaplasia (Elevated LAP)
Acute Myelogenous Leukemia (Normal or Low LAP)	versus	Acute Lymphatic Leukemia; Acute Myelomonocytic Leukemia (Elevated LAP)
Secondary Polycythemia (Normal LAP)	versus	Polycythemia Vera (High LAP)

In paroxysmal nocturnal hemoglobinemia, the LAP score is very low.

LIDOCAINE (XYLOCAINE)

SPECIMEN: Red top tube, separate serum and refrigerate; <u>or</u> green (heparin) top tube.

Spuriously low plasma lidocaine concentrations may be obtained when blood is collected in certain commercial rubber-stoppered collection tubes. Use all-glass tubes or test any commercial collecting tube thoroughly before use.

REFERENCE RANGE: Therapeutic: 2-5mcg/ml. Time to Steady State (Sampling Time): 12 hours after beginning lidocaine. <u>Half-Life</u>: 74-140 minutes. <u>Cardiac Failure</u>: 115 minutes. <u>Uremia</u>: 77 minutes. <u>Cirrhosis</u>: 296 minutes. <u>Toxic</u>: >5mcg/ml.

METHOD: EMIT; GLC; HPLC; Fluorescence Polarization (Abbott).

INTERPRETATION: Lidocaine is used for the prevention and treatment of ventricular arrhythmias. It had been used as a local anesthetic. Guidelines for <u>therapy</u> and serum specimen <u>monitoring</u> are given in the next Table:

Guidelines for Therapy and Monitoring	
Therapy	Monitoring
Loading Dose: Infusion <50mg/min	12 hours after beginning lidocaine
Normal: 200 to 300mg	infusion; at least every 12 hours
Cardiac Failure: 110 to 160mg	in patients with signs or symptoms
Cirrhosis: 200 to 300mg	of cardiac or hepatic insufficiency
Maintenance Infusion:	Obtain specimen whenever toxicity is
Normal: 1.5 to 3mg/min	suspected.
Cardiac Failure: 0.9 to 1.8mg/min	Monitor whenever ventricular arrhythmias
Cirrhosis: 0.9 to 1.8mg/min	occur despite lidocaine administration.

The majority of lidocaine's adverse effects involve the central nervous or circulatory systems. Milder adverse reactions, such as drowsiness, dizziness, transient paresthesias, muscle twitching, and nausea, are common after loading doses or can occur at plasma levels greater than 5mcg/ml. Severe adverse reactions, such as central nervous system, cardiovascular and respiratory depression, convulsions, coma, may appear at concentrations greater than 9mcg/ml.

Plasma levels and signs or symptoms of toxicity need to be closely monitored in patients with altered drug elimination such as cardiac failure, acute myocardial infarction, liver disease and prolonged intravenous infusion (Dr. Ginette Lapierre, personal communication).

LIPASE, SERUM

SPECIMEN: Red top tube, separate serum; store refrigerated or frozen.

REFERENCE RANGE: <200U/liter

METHOD: Triolein $\xrightarrow{\text{Lipase}}$ Glycerides + Fatty Acids(Turbidity at 340nm)

INTERPRETATION: Serum lipase is <u>elevated</u> in patients with <u>acute pancreatitis</u>. The change in serum lipase, serum amylase and urine amylase in acute pancreatitis is illustrated in the next Figure:

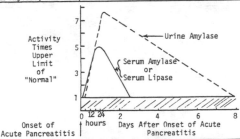

Serum Amylase, Serum Lipase and Urine Amylase Following Acute Pancreatitis

Serum lipase changes in a manner similar to that of serum amylase following onset of acute pancreatitis. It begins to rise in 2 to 6 hours, reaches a maximum in 12 to 30 hours and remains elevated for 2 to 4 days.

LIPOPROTEIN PROFILE/LIPOPROTEIN ELECTROPHORESIS, CHOLESTEROL, TRIGLYCERIDES, HIGH DENSITY LIPOPROTEIN (HDL)

SPECIMEN: Fasting, 12-14 hours; red top tube, separate serum.

REFERENCE RANGE: The normal distributions for plasma lipids and lipoprotein cholesterol are given in the next Table; (Manual of Laboratory Operations, Vol I, N.I.H., Bethesda, Maryland 20014, DHEW Publication No. (NIH. 75-628, pg. 73); Consensus Conference, JAMA 253, 2080-2086, 1985):

Normal Distributions for Plasma Lipids and Lipoprotein Cholesterol

Age	Total Plasma Cholesterol mg/dl	Plasma Triglyceride mg/dl	VLDL Cholesterol mg/dl	LDL Cholesterol mg/dl	HDL Cholesterol mg/dl Male	Female
0-19	<180	10-140	5-25	50-170	30-65	30-70
20-29	<200	10-140	5-25	60-170	35-70	35-75
30-39	<220	10-150	5-35	70-190	30-65	35-80
40-49	<240	10-160	5-35	80-190	30-65	40-95
50-59	<240	10-190	10-40	80-210	30-65	35-85

Cholesterol: To convert conventional units in mg/dl to international units in mmol/liter, multiply conventional units by 0.02586.

Triglycerides: To convert conventional units in mg/dl to international units in mmol/liter, multiply conventional units by 0.01129.

METHOD: Cholesterol is assayed by a cholesterol oxidase method; triglycerides are assayed by an enzymatic method. HDL is measured as cholesterol in the supernatant of serum after precipitation of other lipoproteins with a polyanion-divalent cation combination such as phosphotungstate-Mg^{++} or heparin-Mn^{++}. The lipoproteins are separated electrophoretically; the chylomicrons are found at the line of application; LDL migrates to the beta-globulin region; VLDL migrates to the pre-beta region and HDL migrates to the alpha region.

INTERPRETATION:

High Density Lipoprotein (HDL): It has been found that decreased levels of high density lipoprotein (HDL) in plasma lead to an increased risk in males of coronary heart disease; among the various lipid risk factors, HDL cholesterol appears to have the strongest relationship to coronary heart disease. It may also be that raised levels exert a protective effect in that premenopausal women have HDL concentrations 30 to 60 percent higher than their male counterparts and subjects with familial hyper-alphalipoproteinemia have an above-average life expectancy. It is not known whether people can improve their chances of avoiding a heart attack by increasing their HDL concentration. (Editorial by Jean Marx, Science 205, 677-679, Aug. 17, 1979; Grundy, Scott M., Resident and Staff Physician, February 1980, 3s-17s; Coulter Currents, Issue No.5, May, 1980).

The incidence of coronary heart disease by HDL cholesterol level is shown in the next Table (Gordon et al., Am. J. Med. 62, 707-714, 1977):

Incidence of Coronary Heart Disease by HDL Cholesterol Level

HDL Cholesterol Level (mg/dl) All Levels	Incidence of Coronary Heart Disease Men	Women
25	18%	
25 - 34	10%	17%
35 - 44	10%	5%
45 - 54	5%	5%
55 - 64	6%	4%
65 - 74	2.5%	1.4%
75	0%	2%

Lipoprotein Electrophoresis: Lipoprotein electrophoresis is done to place disorders of lipoprotein into convenient groups for evaluation and treatment. It is not necessary to perform lipoprotein electrophoresis on the secondary hyperlipoproteinemias.

S. Bakerman

Hyperlipidemias are classified using Fredrickson's Classification, or using the classification by Goldstein et al (Goldstein et al., J. Clin. Invest. 52, 1544-1568, 1973).

Fredrickson's Classification: Fredrickson classified the hyperlipidemias into five distinctive clinical syndromes based on elevation in the plasma level of one or more of the lipoproteins, beta, pre-beta, or chylomicrons.

Classification of Hyperlipidemia

Type of Hyperlipidemia	Blood Lipid and Lipoprotein Increase		
	Cholesterol	Triglyceride	Lipoprotein
I	Usually Normal	Elevated	Chylomicrons
IIa	Elevated	Normal	Beta lipoprotein
IIb	Elevated	Moderately Elevated	Beta and pre-beta lipoprotein
III	Elevated	Elevated	Abnormal lipoprotein with beta-electro-phoretic mobility and pre-beta density
IV	Normal or Slightly Elevated	Elevated	Pre-beta lipoprotein
V	Elevated	Elevated	Chylomicrons and pre-beta lipoprotein

Classification of hyperlipidemias may also be done on the basis of fasting serum levels of cholesterol and triglycerides alone. The lipoprotein pattern is normal if plasma cholesterol and triglycerides are both normal. If fasting plasma cholesterol is elevated and plasma triglycerides are both normal, then the patient has a Type IIa hyperlipoproteinemia. Thus, it is not necessary to perform lipoprotein electrophoresis if fasting plasma cholesterol and plasma triglycerides are both normal; nor is it necessary to perform lipoprotein electrophoresis on a specimen if fasting plasma cholesterol is elevated and plasma triglycerides are normal.

If the fasting plasma triglycerides are elevated and plasma cholesterol is normal, then VLD lipoprotein or chylomicrons are elevated; fasting elevated triglyceride is usually due to elevated VLD lipoprotein and Type IV hyperlipoproteinemia; if chylomicrons are also elevated, then the patient may have Type V; if only chylomicrons are elevated, then the patient has Type I hyperlipidemia which is a very rare disorder.

The alterations of lipids and possible interpretation are summarized in the next Table:

Alterations of Plasma Lipids and Fredrickson's Classification

Lipid	Interpretation
Cholesterol Increased; Triglyceride Normal	Type IIa only
Triglyceride Increased; Cholesterol Normal	Type IV, V, I
Both Cholesterol and Triglyceride Increased	Type IIb, III, IV

Differentiation of Type IIb from Type IV: It is often necessary to differentiate whether a patient with elevated triglyceride with accompanying elevated cholesterol be classified as Type IIb or Type IV. Is hypercholesterolemia due to abnormally high concentration of VLDL or of LDL? The following formula may be used if the triglyceride concentration is less than 400mg/dl.

$$\text{LDL·Cholesterol} = \text{Serum Cholesterol} - \left(\frac{T.G.}{5} + \text{HDL·Cholesterol}\right)$$

The upper limit of normal for LDL is 170.

Secondary Hyperlipoproteinemia: The causes of secondary hyperlipoproteinemia and the change in plasma cholesterol and triglycerides are given in the next Table:

Changes in Plasma Lipids in Secondary Hyperlipoproteinemias		
Condition	Cholesterol	Triglycerides
Pregnancy	↑	
Hypothyroidism	↑	
Cholestasis; Intrahepatic or Extrahepatic	↑	
Acute Intermittent Porphyria	↑	
Nephrotic Syndrome	↑	↑
Chronic Renal Failure	↑	↑
Corticosteroid Immuno-Suppressive Therapy, e.g., Renal Transplant Patients	↑	↑
Oral Contraceptives		↑
Diabetes Mellitus		↑
Acute Alcoholism		↑
Acute Pancreatitis		↑
Gout		↑
Gram-Negative Septicemia		↑
Type I Glycogen Storage Disease		↑

The conditions above the first dashed line cause elevation of cholesterol; those between the dashed lines cause elevation of both cholesterol and triglycerides while those below the second dashed line cause elevation of triglycerides.

LITHIUM

SPECIMEN: Red top tube, separate serum; refrigerate at 4°C; stable for 24 hours at room temperature. <u>Reject</u> specimen if collected in lithium heparin.

REFERENCE RANGE: Serum lithium range: 0.8mmol/liter to 1.3mmol/liter for the treatment of acute mania; 0.5 to 1.0mmol/liter for sustained prophylactic use. The levels in saliva are two times greater than those in serum.

METHOD: Flame emission photometry; atomic absorption.

INTERPRETATION: Lithium (more precisely lithium ion) is used for the treatment of patients with <u>acute mania</u> and for prophylaxis against <u>recurrent manicdepressive illness</u>. Lithium may <u>prevent depression</u> but is not effective in lifting a patient <u>out of the depressive phase</u>. The serum <u>half-life</u> varies from about <u>18 hours</u> in the young to about <u>36 hours</u> in the elderly. The usual therapeutic range of serum lithium in psychiatric patients is 0.8mmol/liter to 1.3mmol/liter for the treatment of <u>acute mania</u> and levels of <u>0.5 to 1.0mmol/liter</u> are recommended for sustained <u>prophylactic use</u>. For the average adult with normal creatinine clearance, these serum levels can be achieved with a dose of lithium carbonate of approximately <u>300mg to 600mg orally three times daily</u>. However, the dosage must be individualized because there is variation in the rate at which lithium is eliminated by the kidney.

Lithium is rapidly absorbed following oral doses with <u>peak plasma levels</u> occurring in <u>one to three hours</u>. Therapeutic and toxic drug levels are obtained on specimens <u>drawn 8 to 12 hours after administration of the last oral dose</u>. A specimen drawn earlier than this period of time will give the false conclusion of an abnormally elevated serum level whereas a specimen drawn after the 12 hour interval will lead to the conclusion of a drug level lower than it really is.

Side Effects of Lithium: The common side-effects of lithium therapy are given in the next Table:

Common Side Effects of Lithium Therapy
Thyroid: (Four forms of thyroid dysfunction)
Transitory Biochemical Changes (Elevated Thyroid Stimulating Hormone (TSH) Levels in Serum; Decreased Levels of Free Thyroxine (T-4) and Triiodothyronine (T-3) in Serum without Clinical Change)
Goiter with Euthyroidism (4%)
Hypothyroidism (4% to 15%)
(a) Thyroiditis with Goiter; High Thyroid Antibody Titer; Low RAIU
(b) No Goiter; No Antibodies
Hyperthyroidism (rare)
Renal Function:
Polyuria (40% of patients) Plasma Osmolality, Normal; Urine Osmolality, Decreased
Granulocytes:
Benign, reversible leukocytosis (14,000-24,000/cu mm) or total count normal, increase in granulocytes and platelets

Thyroid: Patients who have a personal or family history of thyroid disorder may be especially prone to develop <u>hypothyroidism</u>; the incidence in women is about ten times that in men. Hypothyroidism may occur <u>any time</u> during the course of therapy and is <u>treated by thyroid hormone</u>. The usual course of events is that thyroxine levels decrease initially and the thyrotropin (TSH) levels become elevated in the first few weeks and finally stabilize. Thyrotropin (TSH) levels have been found to be abnormal in up to 30% of lithium treated patients at some time during the course of therapy. They return to normal in about half of these patients over a period of time (Emerson, C.H., Dysin, W.L., Utiger, R.D., J. Clin. Endocrinol. Metab. 36, 338-346, 1973). (Calabrese, J.R. and Gulledge, A.D. Cleve. Clin. Q. 50, 32-33, 1983). <u>Withdrawal of lithium usually results in the return of normal thyroid function.</u>

Renal Function: Toxic renal effects including tubular lesions, interstitial fibrosis and decreased creatinine clearance have been reported in patients treated with lithium; these are <u>uncommon</u> (Hansen, H.E., Drugs, 22, 461, 1981; Ramsey, T.A. and Cox, M., Am. J. Psychiatry 139, 443, 1982).

Polyuria has been reported to occur in over 40% of patients on lithium; in some patients the urinary output has exceeded 11 liters/day. Lithium is excreted almost entirely by the kidneys; in states of sodium depletion, lithium will be reabsorbed by the tubules, leading to increase in serum and tissue lithium levels and development of clinical lithium toxicity. Conditions associated with sodium loss, such as renal loss due to diuretics and renal failure, gastrointestinal losses due to vomiting or diarrhea and excessive perspiration, may be associated with lithium toxicity.

S. Bakerman   Lithium

The thiazide diuretics which act on the cells of the distal convoluted tubules cause a significant rise in the serum lithium. These diuretics cause a net sodium loss, which may lead to a compensatory increase in proximal sodium (and hence lithium) reabsorption. Furosemide, which acts on the cells of the ascending limb of the loop of Henle proximal to the thiazide diuretics, does not cause a significant increase in serum lithium levels (Jefferson, T.W., Kalin, N.H., JAMA 241, 1134-1136, 1979). Conditions that are associated with reduced renal function may result in an abnormal retention of lithium.
Granulocytes: A benign, reversible leukocytosis is frequently seen during lithium therapy (14,000-24,000/cu mm) or the total white count may remain normal with an absolute increase in mature granulocytes and platelets. The leukocytosis begins within a week of the start of administration, persists during therapy, and resolves within a week of discontinuance of the drug. This lithium effect is of no clinical significance other than to avoid confusion with an occult infection or blood dyscrasia (Editorial, The Lancet, pgs. 626-627, Sept.20,1980).

The side effects of lithium, that is, hypothyroidism, polyuria and leukocytosis have been utilized in treatment of patients with hyperthyroidism, inappropriate secretion of antidiuretic hormone and leukopenia, respectively.
Laboratory Studies Performed Before Lithium Administration: Before administration of lithium the laboratory studies listed in the next Table should be performed:

Laboratory Studies Performed Before Lithium Administration	
Test	Possible Change During Lithium Therapy
Urinalysis	Lithium excreted by kidneys; retention in renal disease; for baseline specific gravity.
Serum Creatinine or BUN	Lithium excreted by kidneys; retention in renal disease.
Serum Sodium	Low sodium associated with lithium retention; high sodium associated with lithium elimination.
Thyroxine (T-4) TSH Thyroid Microsomal Antibody Titers	Hypothyroidism - High incidence (4% to 15%)
White Cell Count	Leukocytosis - See above
Electrocardiogram	Flattened or Inverted T wave - Benign and reversible Arrhythmias, Ventricular Tachycardia, Sinoatrial and Atrioventricular block (Jaffe, C.M., Am. J. Psy. 134, 88-89, 1977). Frequent ECG monitoring recommended.

Toxic Manifestations of Lithium: Toxic manifestations of lithium generally do not appear unless plasma levels exceed 1.5mmol/l. The toxic manifestations are listed in the next Table; (Coyle, J.T., Med. Clin. N. Am. 61(4), 891-905, 1977):

Toxic Manifestations of Lithium Therapy	
Plasma Level (mmol/l)	Symptoms
<1.5 (Therapeutic)	Nausea (Especially on Initiation of Treatment) Fine Tremor: Mild Polyuria
>1.5 <2.5	Diarrhea and Vomiting; Polyuria; Coarse.Tremor, Ataxia; Muscle Weakness and Fasciculation; Sedation, Langour
>2.5 <4.0	Muscle Hypertonia; Choreiform Movements; Increased Deep Deep Tendon Reflexes; Impairment of Consciousness with Somnolence, Confusion, Stupor; Transient Focal Neurologic; Signs; Seizures
>4.0	Coma: Death

Most of the effects are due to the action of lithium on the central nervous system. Neurological side effects are many and vary from tremor, usually seen at the start of treatment and occurring at therapeutic drug levels, to marked alterations in consciousness, neuromuscular irritability, seizures, incoordination, delirium, irreversible brain damage and death.
The degree of toxic symptoms does not always correlate well with serum lithium levels and in fact, individuals have been seen with documented lithium toxicity in spite of "normal" lithium serum values (Strayhorn J.M., Nash, J.L., Diseases of the Nervous System 38, 17-111, 1977).

LIVER ANTIBODY PANEL

Smooth muscle antibodies, antimitochondrial antibodies and antinuclear antibodies in hepatic conditions are given in the next Table; (Mayo Medical Laboratories Test Catalogue, page 319, 1983).

Antibodies in Hepatic Conditions			
Condition	Anti-Mitochondrial	Anti-Smooth Muscle	Anti-Nuclear
Biliary Cirrhosis, Primary	75-95	0-50	25-50
Chronic Active Hepatitis (Autoimmune Type; Lupoid Hepatitis)	0-30	50-80	20-30
Extra-Hepatic Biliary Obstruction	0-5	0	0
Cryptogenic Cirrhosis	0-25	0-1	0-1
Viral (Infectious) Hepatitis	0	1-2	---
Drug-Induced Hepatitis	50-80	---	---

LIVER DISEASES: CONGENITAL, GENETIC AND CHILDHOOD DISEASE PANEL

The conditions and tests in childhood liver diseases are given in the next Table:

Congenital, Genetic and Childhood Liver Diseases	
Condition	Tests
Defects in Bilirubin Metabolism: (Berk, P.D. and Javitt, N.B., Am. J. Med. 64:311-326, 1978)	Liver Function Tests are Normal Except for Serum Bilirubin
Gilbert's Syndrome	Increased Indirect Bilirubin; Bile Acids are Normal; Caloric Restriction: 200 Calorie Diet for 24-48 Hours; Increased Serum Bilirubin by Two-Fold or More.
Crigler-Najjar Type I	Indirect Bilirubin, >20 mg/dl. Direct Bilirubin, Absent Response to Phenobarbital: None
Type II	Indirect Bilirubin, <20 mg/dl Direct Bilirubin, Absent Response to Phenobarbital: Lowers Bilirubin Concentration
Dubin-Johnson Syndrome	Increased Direct Bilirubin
Hereditary Hemochromatosis	Serum Iron Increased; Decreased Iron Binding Capacity; >70 Percent Saturation Increased Serum Ferritin Biopsy of Liver: Increased Iron
Hepatolenticular Degeneration (Wilson's Disease)	Decreased Ceruloplasmin Serum Decreased Total Copper in Serum: Decreased Indirect Reacting Copper Increased Direct Reacting Copper Increased Copper in Urine Increased Copper Deposited in Liver
Alpha-1-Antitrypsin Deficiency	Serum Protein Electrophoresis; Carefully examine Alpha-1-Band Assay Alpha-1-Antitrypsin when Alpha-1-Band is Decreased or Absent (Usual Technique is Radial Immunodiffusion) Perform Phenotyping if Indicated
Neonatal Hepatitis	Liver Function Tests
Biliary Atresia	Liver Function Tests

LUTEINIZING HORMONE (LH)

SPECIMEN: Hormone levels undergo rapid and large oscillations; obtain three equally-spaced samples at about 10 min. intervals and pool the three specimens. Red top tube, separate serum

REFERENCE RANGE: Prepubertal children: less than 5mIU/ml; LH values are high in the first several months of life. Adult female: follicular: 3-14 IU/ml; luteal: 2-14 IU/ml. Post Menopausal or Castrate: 15-70 IU/ml. Adult Male: 3-12 IU/ml.

METHOD: RIA

INTERPRETATION:

(1) Male: The gonadotropins, LH and FSH, are under the control of a single hypothalamic hormone, the decapeptide gonadotropin releasing hormone (GnRH); secretion of GnRH must be pulsatile in order to induce normal adult sexual function. Pulsatile release of GnRH begins early in adolescence; the gonadotropin ratios (FSH,LH) shift from the prepubertal pattern of FSH dominance to the postpubertal pattern of LH dominance. Without GnRH pulses, there is no puberty. FSH stimulates the seminiferous tubules and spermatogenesis; LH stimulates the Leydig cells which produce testosterone (Spark, R.F., Ann. Int. Med. 98, 103-105, 1983).

Knowledge of the characteristics of the pathway from the hypothalamus to the Leydig cells and secretion of testosterone is utilized in the successful treatment of patients with advanced prostatic cancer; these patients were treated with repeated administration of a long acting analogue of GnRH; suppression of the gonadotropins and testosterone and objective and subjective signs of regression of disease occurred in 75% of the patients (Waxman, J.H. et al., Brit. Med. J. 286, 1309-1312, April 23, 1983).

The integrity of the adult male reproductive system may be tested by measuring parameters of ejaculation. If the following values are achieved or exceeded, then the entire hormonal system is normal.

Minimal Values of Ejaculation
Reflecting Normal Male Hormonal Reproductive System

Parameter	Minimum Value
Volume	2ml
Sperm Count	70,000,000/ml
Motility	80%
Normal Morphology	80%

The American Society for the Study of Sterility has proposed a volume of 2-5 ml., a minimum count of 60 million/ml., and at least 60% active forms as adequate for fertility. Others in this field feel that these figures are too high and that a minimum sperm count of 30-40 million/ml. with 40% or more active forms is adequate. Morphology is classified as poor when less than 60% are normal.

In hypogonadism, serum testosterone is decreased.

(a) Testicular Failure (Hypergonadotropic Hypogonadism): In these conditions there is testicular failure: LH and FSH are elevated and testosterone is decreased; parenteral testosterone is the preferred treatment. Conditions associated with hypergonadotropic hypogonadism are shown in the next Table:

Conditions Associated with Testicular Failure
(Hypergonadotropic Hypogonadism)

Condition	Prepubertal	Postpubertal
Damaged seminiferous tubules with variable Leydig cell failure	Klinefelter's syndrome Germinal aplasia Reifenstein syndrome (and other related syndromes)	Cryptorchidism Mumps and other orchitis Irradiation Alkylating agents Idiopathic Myotonia dystrophica
Total gonadal failure	Functional prepubertal castrate (anorchia, etc.)	Adult castrate (mumps, trauma, surgery)

Klinefelter's Syndrome (XXY): Klinefelter's syndrome presents in one of the following ways:

(1) Adolescent boy with gynecomastia
(2) Married male with infertility
(3) Adult male with a clinical history of testosterone deficiency

Reifenstein Syndrome: This is a hereditary condition characterized by hypospadia, postpubertal atrophy of the seminiferous tubules, and varying degrees of eunuchoidism and gynecomastia.

(b) Pituitary Failure (Hypogonadotropic Hypogonadism): When gonadotropin levels are low, or low-normal, hypothalamic pituitary disease must be suspected; further evaluation must include determination of serum prolactin values for hyperprolactinemia (Carter, J.N. et al., N. Engl. J. Med. 299, 847-852, 1978). Conditions associated with pituitary failure are shown in the next Table:

Conditions Associated with Pituitary Failure (Hypogonadotropic Hypogonadism)		
Condition	Prepubertal	Postpubertal
LH failure alone	"Fertile eunuch"	
FSH and LH failure	Delayed puberty Hypogonadotrophic eunuchoidism: (a) with anosmia (Kallman) (b) without anosmia (c) Laurence-Moon-Biedl	Adult isolated gonadotropin failure (rare)
Failure of all pituitary hormones	Prepubertal panhypopituitarism	Adult panhypopituitarism

Kallman's Syndrome: In this syndrome, hypogonadism is associated with anosmia or hyposmia and often with other neurologic defects (anosmia - loss or impairment of the sense of smell). The presumed cause of Kallman's syndrome is lack of pulsatile release of GnRH; sexual function can be restored by administering GnRH via an infusion pump programmed to mimic normal pulsatile secretion (Crowley, W.F. and McArthur, J., Proc. 62nd Ann. Meeting of Endocrine Society, Bethesda, Maryland, Endocrine Soc., 1980 (Abstract 743).
Lawrence-Moon-Biedl Syndrome: This syndrome is characterized by retinitis pigmentosa, obesity, mental deficiency, polydactylia and hypogonadotropic hypogonadism.
(c) Precocious Puberty: Premature gonadotropin releasing hormone (GnRH) pulses cause precocious puberty (Crowley, W.F. et al., J. Clin. Endocrinol. Metab. 52, 370-372, 1981).
(2) Female: Serum LH levels are low in pituitary insufficiency and sustained elevations are seen in ovarian failure or postmenopausally. The tests that may be used to differentiate pituitary, ovarian and end-organ failure, as a cause of amenorrhea, are given in the next Table:

Evaluation of Amenorrhea				
Failure	Plasma FSH and LH	Urinary Estrogens	Urinary Estrogens Following HCG	Remarks
Pituitary	↓	↓	↑	Probable Prolactinoma
Ovarian	↑	↓	No Change	Patients with poly-cystic ovarian disease have ↑ FSH
End-Organ	N	N	↑	

In isolated pituitary gonadotropin (FSH and LH) deficiency, there are no clinical signs before the age of puberty. Secondary sexual development fails to occur during adolescence. If isolated pituitary gonadotropin deficiency occurs in the female adult, amenorrhea develops.

LUTEINIZING HORMONE(LH)-RELEASING HORMONE(RH) TEST (LH-RH TEST)

SPECIMEN: Red top tubes for each blood specimen collected at -20 minutes and at 0, 20, 40, 60, 90, 120 and 180 minutes for LH and FSH assay; allow blood to clot in tubes, centrifuge and separate serum.

REFERENCE RANGE: Normal responses for prepubertal subjects, adult males, and adult females, early follicular, preovulatory and luteal phase are given in the next Table:

Gonadotropin-Releasing Hormone(LH-RH) Test Results		
	Results Following LH-RH	
Subject	LH	FSH
Prepubertal	Little Change	Increase: 1/2 to 2-Fold
Adult Male	Increase: 4 to 10-Fold	Increase: 1/2 to 2-Fold
Adult Female, Early Follicular Phase	Increase: 3 to 4-Fold	Increase: 1/2 to 2-Fold
Adult Female Preovulatory	Increase: 3 to 5-Fold	Increase: 1/2 to 2-Fold
Adult Female, Luteal	Increase: 8 to 10-Fold	Increase: 1/2 to 2-Fold

METHOD: NPO from evening snack or midnight. In A.M., start slow I.V. drip of normal saline solution. Obtain blood specimen at -20 and at 0 minutes for LH and FSH assay. After 0 specimen is drawn, give luteinizing releasing hormone(LRH), dose 100 mcg in children, 100 to 150 mcg in adults, I.V., push. Then, obtain blood specimens. After specimen obtained at 180 minutes, I.V. may be discontinued, and the patient may eat.

INTREPRETATION: The gonadotropin-releasing hormone or luteinizing hormone-releasing hormone(LH-RH) stimulates LH and, to a lesser degree, FSH secretion by the pituitary. This test is a measure of the functional integrity of the hypothalamic-pituitary-gonadal axis.

If the gonadotropin values(LH and FSH) are unequivocally elevated, there is no need to perform a LH-RH test.

Females: The LH-RH test is most often used to evaluate females presenting with menstrual disturbances or infertility. The conditions in which the LH-RH test may be useful are given in the next Table; (Wills, M.R. and Harvard, B., "Laboratory Investigation of Endocrine Disorders" Butterworths, London, 2nd edition, pgs. 20-23, 1983):

LH-RH Test
Hypothalamic Lesions
Isolated Hypogonadotrophic Hypogonadism
Post-Operative Assessment Following Hypophysectomy
Polycystic Ovaries (Stein-Leventhal Syndrome)
Primary Amenorrhea
Secondary Amenorrhea
Without Galactorrhea
With Galactorrhea

Males: The LH-RH test is used in the investigation of males with hypogonadotropic hypogonadism in conditions listed in the next Table:

LH-RH Test in Males
Primary Hypogonadotropic Hypogonadism:
Adult
Pubertal
Cryptorchism
Secondary Hypogonadotropic Hypogonadism:
Systemic Diseases
Traumatic
Tumors
Infections
Psychogenic

Disease State Responses: LH and FSH are low in panhypopituitarism and response to stimulation is absent or blunted. A blunted response also occurs in the majority of patients with hypothalamic lesions. An exaggerated response with high basal values indicates primary gonadal failure.

LYSERGIC ACID DIETHYLAMIDE (LSD)

SPECIMEN: Urine or plasma
REFERENCE RANGE: Negative
METHOD: RIA; fluorometric
INTERPRETATION: Lysergic acid diethylamide (LSD) is a potent hallucinogen derived from ergot, the fungus that spoils rye grain. It may cause a delirium, accompanied by hallucinations and delusions.

LYSOZYME, (MURAMIDASE) SERUM

SPECIMEN: Red top tube, separate serum and freeze
REFERENCE RANGE: 3.0-12.8 mg/liter
METHOD: Worthington lysozyme assay
INTERPRETATION: Lysozyme is a lysosomal enzyme which is useful in diagnosing and following the course of certain leukemias and sarcoidosis. Relatively high concentrations occur in monocytes and polymorphonuclear leukocytes. Conditions associated with an increase in serum lysozyme are given in the next Table:

Conditions Associated with Increased Lysozyme

Leukemias:
    Acute Monocytic Leukemia
    Acute Myelomonocytic Leukemia
    Chronic Myelocytic Leukemia (Philadelphia Chromosome Positive)
    Acute Granulocytic
Histiocytic Medullary Reticulosis
Sarcoidosis
Crohn's Disease

Renal Disease; Acute Bacterial Infections; Tuberculosis; Leukemoid Reactions; Megaloblastic Anemias; Neutropenic Disorders Associated with Increased Granulocytic Turnover

Serum lysozyme assay has been valuable in diagnosis or following patients in conditions listed above the dashed line.

Leukemia: (Perillie and Finch, Med. Clin. N. Am. 57, 395, 1973): Assay of serum and urinary lysozyme is useful to distinguish acute myeloid and acute monocytic leukemia from acute lymphocytic leukemia. In the acute leukemias, marked lysozymuria is virtually pathognomonic of acute myelomonocytic leukemia or acute monocytic leukemia; this is so except for an occasional patient with acute granulocytic leukemia and in the absence of renal dysfunction. Serial measurements of serum lysozyme in those acute leukemias associated with marked elevation may be useful in following the effect of therapy. Serum enzyme measurements are not diagnostically useful in chronic leukemia.

Histiocytic Medullary Reticulosis: (Duffy et al, N. Engl. J. Med. 294, 167, 1976): Histiocytic medullary reticulosis is a monocyte-macrophage disorder, categorized as a lymphoproliferative disease and characterized by erythrophagocytosis, fever, lymphadenopathy and hepatosplenomegaly. Considering the peripheral monocytosis, urinary muramidase was measured. Urinary muramidase measured in a patient with this disease was 140 mcg/dl (normal, 2 mcg/dl).

Crohn's Disease: Falchuk et al (N. Engl. J. Med. 292, 395, 1975) found serum lysozyme levels to be useful in the diagnosis and follow-up of patients with Crohn's disease. Mean lysozyme concentrations are greater in patients with active Crohn's disease than in those with uncomplicated ulcerative colitis; elevated lysozyme levels are found in patients with complicated ulcerative colitis. (Falchuk and Perrotto, Gastroenterology 68, 890, 1975).

Sarcoidosis: (Pascual et al, N. Engl. J. Med. 289, 1074, 1973): Mononuclear phagocytes contain high lysozyme activity. The serum lysozyme tends to be elevated in granulomatous diseases such as sarcoidosis and tuberculosis which are associated with increased monocytes.

LYSOZYME, (MURAMIDASE) URINE

SPECIMEN: 50ml random urine, freeze
REFERENCE RANGE: 0-2.9 mg/liter
METHOD: Worthington lysozyme assay
INTERPRETATION: See LYSOZYME, SERUM. Renal Disease; (Daniels et al, Texas, Rep. Biol. Med. 30, 1, 1972): Lysozyme has a molecular weight between 14,000 and 15,000; the glomerular membrane is about 0.8 as permeable to lysozyme as it is to creatinine. The proximal tubule is the site of catabolism and reabsorption of lysozyme (Ottonsen and Naunsback, Kidney International 3, 315-326, 1973).

Lysozyme is found in the urine in a wide variety of renal diseases and urine lysozyme is increased in uremics. It is used to differentiate glomerular disease from renal tubular disease in that lysozyme is excreted in renal tubular disease and minimally in glomerular disease.

MAGNESIUM, SERUM

SPECIMEN: Red top tube, separate serum. Hemolysis may yield elevated results; the red cell magnesium level concentration is almost three times that of serum magnesium level.
REFERENCE RANGE:

Change of Serum Magnesium with Age			
Age	mg/dl	mEq/liter	mmol/liter
Newborn	1.8-2.8	1.5-2.3	0.75-1.15
Children	1.7-2.3	1.4-1.9	0.7-0.95
Adults	1.7-2.4	1.4-2.0	0.7-1.0

METHOD: Fluorometric, 8-hydroxyquinoline; colorimetric, titan yellow
INTERPRETATION: The causes of hypomagnesemia are given in the next Table; (Massry, S.G., "Magnesium Homeostasis and Its Clinical Pathophysiology", Resident and Staff Physician, pgs. 105-109, June, 1981; Rude, B.K. and Singer, F.R., Annu. Rev. Med. 32, 245-259, 1981):

Causes of Hypomagnesemia	

Excessive Urinary Losses
  Chronic alcoholism
  Diuretic therapy
  Diuretic phase of acute
    renal failure
  Primary aldosteronism
  Hypercalcemic states:
    malignancy, hyper-
    parathyroidism, and
    vitamin D excess
  Renal tubular acidosis
  Diabetes, especially during
    and following treatment
    of acidosis
  Hyperthyroidism
  Hypoparathyoidism
  Idiopathic renal magnesium
    wasting
  Chronic renal failure with
    renal magnesium wasting
  Gentamicin toxicity
  Cis-platinum nephrotoxicity
 	
Decreased Intake
  Protein-calorie malnutrition
  Starvation
  Prolonged intravenous therapy
	Decreased Intestinal Absorption
	  Malabsorption syndromes, including
	    nontropical sprue
	  Massive surgical resection of small
	    intestine
	  Neonatal hypomagnesemia with selective
	    malabsorption of magnesium
	
	Excessive Loss of Body Fluids
	  Prolonged nasogastric suction
	  Excessive use of purgatives
	  Intestinal and biliary fistulas
	  Severe diarrhea, as in ulcerative
	    colitis and infantile gastroenteritis
	  Rarely, prolonged lactation
	
	Miscellaneous
	  Idiopathic hypomagnesemia
	  Acute pancreatitis
	  Porphyria with inappropriate secretion
	    of antidiuretic hormone
	  Multiple transfusions or exchange
	    transfusions with citrated blood
	  Primary hypomagnesemia (Hennekam, R.C.M.,
	    and Donckerwolcke, R.A., Lancet, 1, 927,
	    April 23, 1983)

Probably the most common cause of hypomagnesemia is acute or chronic alcoholism; the excretion of magnesium by the kidney is increased during the time when there is a significant concentration of alcohol in the blood.

Magnesium deficiency is treated as follows: Assess renal function before initiating magnesium therapy; reduce replacement dose if glomerular filtration is reduced. One-quarter of the normal dose has been suggested for renal impairment; obtain daily serum magnesium levels in these patients. (Note that

S. Bakerman

in the presence of magnesium deficiency, hypocalcemia and hypokalemia cannot be corrected without magnesium replacement.)

Give 4 to 6 grams of magnesium sulfate per day for 5 days, intravenously in 250-1000 ml of I.V. fluid, administered continuously. Monitor serum magnesium daily.

The laboratory findings in magnesium depletion are given in the next Table:

Laboratory Findings in Magnesium Deficiency
Hypomagnesemia, Hypocalcemia
Hypokalemia
Hypophosphatemia, occasionally
Hyperphosphatemia
Low urinary Mg and calcium

The conditions associated with hypermagnesemia are listed in the next Table:

Conditions Associated with Hypermagnesemia
Acute renal failure
Chronic renal failure
Infants of mothers treated with Mg for eclampsia
Adrenal insufficiency
Administration of pharmacologic doses of Mg and use of oral purgatives or rectal enemas containing Mg, especially in patients with impaired renal function

The most common cause of hypermagnesemia is acute or chronic renal failure; the elevation of serum magnesium tends to occur when creatinine clearance is less than 30 ml/min.

MALABSORPTION PANEL (see FAT MALABSORPTION PANEL

MALIGNANT HYPERPYREXIA PANEL

In malignant hyperpyrexia, rhabdomyolysis occurs late in the syndrome; laboratory tests in malignant hyperpyrexia are listed in the next Table; (Nelson, T.E. and Flewellen, E.H., N. Engl. J. Med. 309:416-418, 1983).

Laboratory Tests in Malignant Hyperpyrexia
Serum Creatine Phosphokinase(CPK)
Serum and Urine Myoglobin
Electrolyte Abnormalities with Hyperkalemia
Disseminated Intravascular Coagulation(DIC)

MARIHUANA, HASHISH (HASH)

SPECIMEN: Random urine
REFERENCE RANGE: None
METHOD: Gas-liquid chromatography; mass spectroscopy; EMIT (detection limit; greater than 50 ng/ml).
INTERPRETATION: Serum levels of THC peak within 10 to 30 minutes after inhalation and within 3 hours after ingestion; THC declines rapidly. Urine levels peak from 2 to 6 hours and may remain detectable for more than 24 hours and longer than 8 days using sensitive methods.

There are great variations in absorption and distribution of the drug, manner of ingestion and urinary volume of THC; thus, assay is useful only as an indicator of recent use and not as a measure of intoxication. Psychological effects do not correlate with urinary metabolite levels.

Marihuana is an Asiatic herb; the principal active ingredient is tetrahydrocannabinol (THC). The parts with the highest THC content are the flowering tops of the plant. Hashish (hash) is the dark brown resin that is obtained from the tops of the plants. Marihuana or hashish is generally smoked in self-rolled cigarettes called "joints" or in ordinary pipes; they may be added to foods or drinks.

The effects of marihuana vary widely; it may act as a stimulant or a depressant, or as a hallucinogen with sedative properties.

Chronic heavy marihuana use has been associated with depression of plasma testosterone levels, gynecomastia, severe debilitation of bronchial tract and lungs, and neural and possibly immunologic effects.

S. Bakerman

MEAN CORPUSCULAR HEMOGLOBIN (MCH)
(see RED CELL INDICES)

MEAN CORPUSCULAR HEMOGLOBIN CONCENTRATION (MCHC)
(see RED CELL INDICES)

MEAN CORPUSCULAR VOLUME (MCV) (see RED CELL INDICES)

MELANIN, URINE
SPECIMEN: Freshly voided random urine
REFERENCE RANGE: Negative
METHOD: Thormahlen's Test: Nitroprusside $\xrightarrow{\text{Melanin}}$ Prussian Blue;
Ferric chloride test.
INTERPRETATION: This test is used to evaluate urine melanin in patients with
known or suspected metastatic melanoma.

MERCURY, URINE
SPECIMEN: Collect 24 hour urine in acid washed-containers. Use plastic
containers (borosilicate, polyethylene or polypropylene); add 10% HCl solution
to the container and allow to "soak" for 10 minutes; rinse with five volumes of
tap water and then five volumes of deionized or distilled water. The patient
should urinate at 8:00 A.M. and the urine is discarded. Then, urine is collected
for 24 hours including the next day 8:00 A.M. specimen. Indicate 24 hour
volume. A 50ml aliquot is used for analysis.
REFERENCE RANGE: 0-20mcg/liter; symptoms: >200mcg/liter; remove from
occupational exposure: >300mcg/liter; toxicity in chronic exposure can be seen
at levels: 50-100mcg/liter. To convert conventional units in mcg to
international units in nmol, multiply conventional units by 4.985.
METHOD: Qualitative: Reinsch test sensitive to 1000mcg/liter; Quantitative:
Neutron activation and atomic absorption.
INTERPRETATION: The organ systems involved in acute mercury poisoning are
gastrointestinal tract, the kidneys and the central nervous system; in chronic
mercury poisoning, the same organ systems are involved. The three cardinal
signs of mercury poisoning are dysarthria, ataxia and constricted visual fields.
In addition, pulmonary injury occurs when mercury vapor is inhaled.
  Both inorganic (elemental) mercury and organic mercury poisoning occur.
Chronic elemental mercury poisoning is insidious in onset, and may take years to
develop. It is an occupational disease of miners, mirror makers, gilders,
hatters ("mad as a hatter", Lewis Carrol's "Alice-in- Wonderland"), factory and
laboratory workers. Organic mercury poisoning is usually a more serious disease
and may develop precipitously.
  One death and several instances of elevated serum mercury levels have been
reported from the practice of instilling merthiolate into external ears (FDA
Drug Bulletin, 13, No. 1, April 1983).

METABOLIC SCREEN (see INBORN ERRORS OF METABOLISM SCREEN)

METANEPHRINES, TOTAL URINE

SPECIMEN: 24 Hour urine. The specimen may be used for assay of catecholamines and for vanillylmandelic acid (VMA). Instruct the patient to void at 8:00 A.M. and discard the specimen. Add 25ml of 6N HCL to container prior to collection. Then, collect all urine including the 8:00 A.M. specimen at the end of the 24 hour collection period. Refrigerate container as each specimen is collected. Following collection, add 6N HCL to pH 1 to 2 (Do not use boric acid). Metanephrines are unstable in urine at alkaline pH. Record 24 hour urine volume. In urine collected at a pH of less than 3 and stored at 4°C, the metanephrines are stable for at least one week. Forward 100ml aliquot of 24 hour urine for analysis.

High doses of guanethidine, hydrocortisone, unipramine, isoetharine, levodopa, phenobarbital and phenylephrine cause false positive results in the spectrophotometric assays for urinary metanephrines. Propanolol and theophylline cause false negative results (Spilker, B. et al., Ann. Clin. Lab. Sci. 13:16-19, 1983). In order to ensure reliability of test results, discontinue all drugs for at least 3 days before urine is collected.

REFERENCE RANGE: The reference range for urinary metanephrines is given in the next Table; (Gitlow, S.E. et al., J. Lab. Clin. Med. 72:612-620, 1968).

Reference Range for Metanephrines	
Age (years)	Upper Limit (mcg/mg Creatinine)
<1	4.3
1-2	3.9
2-10	2.8
10-15	1.6
15-Adults	1.0

METHOD: Conjugates of metanephrines are hydrolyzed and the metanephrines are absorbed onto an ion-exchange resin; after elution, metanephrines are converted to vanillin which is measured spectrophotometrically.

INTERPRETATION: Urinary metanephrines are often measured with urinary catecholamines and vanillylmandelic acid (VMA) in the work-up of patients with possible pheochromocytoma, neuroblastomas, ganglioneuromas and ganglioneuroblastomas.

METHANOL (METHYL ALCOHOL)

SPECIMEN: Red top tube or random urine or gastric contents; store at 4°C.

REFERENCE RANGE: Normally not present; toxicity: >250mg/liter (25mg/dl) in serum. To convert conventional units in mg/dl to international units in mmol/liter, multiply conventional units by 0.3121.

METHOD: Gas-liquid Chromatography; Sunshine, Methodology for Analytical Toxicology, CRC Press, 1975.

INTERPRETATION: Most laboratories do not offer gas-liquid chromatography on a "stat" basis. However, there are two tests that may be done to screen for methanol ingestion; these are: osmolality and acid-base imbalance with increased anion gap metabolic acidosis. The increase in serum osmolality for each 100mg/dl of methanol is 31 mosm/liter. Only slightly elevated or decreased serum osmolality suggests other causes of coma such as trauma or drugs other than methanol. Most patients with methanol toxicity will have elevated serum and urine amylase.

The minimum lethal dose of methanol is approximately 30g. Methanol itself is non-toxic; its metabolic products, formaldehyde, and formic acid, are toxic. Following ingestion of methanol, it takes 12 to 24 hours for the toxic products to form and take effect. The mortality rate is related to both duration and severity of the metabolic acidosis (Pappas, S.C. and Silverman, M., Can. Med. Assoc. J. 126, 1391-1394, 1982) which develop in 8 to 12 hours. The clinical effects of methanol are as follows: drunkenness without alcohol on the breath, convulsions, coma, retinal edema, blindness due to optic nerve damage and atrophy.

Treatment is directed toward correction of metabolic acidosis (usually bicarbonate) and removal of methanol and toxic metabolites (hemodialysis). Dialysis is indicated when a patient has metabolic acidosis, mental, visual, or funduscopic changes, blood methanol greater than 50mg/dl or has ingested more than 30g of methanol. Ethanol serves as an antidote for methanol by competitive inhibition since both are metabolized by alcohol dehydrogenase.

Ref. Editorial, The Lancet 1, 910-912, April 23, 1983.

METHAQUALONE (QUAALUDES, "QUADS", "LUDES", "SOPERS", "SOAPERS", "MANDIES", "LOVE DRUG")

SPECIMEN: Quantitation: Red top tube, separate serum. Qualitative: 50ml random urine.

REFERENCE RANGE: Toxic: 1.0-3.0mg/dl; Lethal: >3.0mg/dl. Half-life: 10-42 hours.

METHOD: Gas-liquid chromatography; mass spectroscopy; color

INTERPRETATION: Methaqualone is a prescription sedative-hypnotic drug which is used as an illegal "street drug" obtained by production in illegal laboratories or by illegal importation. When used recreationally, methaqualone produces a feeling of both sedation and mild euphoria.

Methaqualone and alcohol act synergistically; most methaqualone-related deaths result from overdose plus ethanol.

The clinical symptoms include hypertonia, vomiting, salivation, delirium, muscle twitching, tonic-clonic convulsions and coma.

Methemoglobin

METHEMOGLOBIN (FERRI-HEMOGLOBIN)

SPECIMEN: Green (heparin) or lavender (EDTA) or blue (citrate) top tube. Transport specimen to laboratory on ice immediately. The specimen is rejected if specimen takes longer than 2 hours in transit or specimen clotted.

REFERENCE RANGE: Up to 3% of total hemoglobin.

METHOD: Spectrophotometric

INTERPRETATION: Methemoglobin is hemoglobin with iron in the ferric form; methemoglobin is not capable of carrying oxygen. Assay of methemoglobin is done to evaluate cyanosis; concentration of methemoglobin greater than 10% of the total hemoglobin or 1.5 to 2.0g/dl of methemoglobin may cause cyanosis.

The diagnosis can be suspected by the characteristic chocolate brown color of a freshly obtained blood sample. The blood sample does not become bright red when oxygen is bubbled though it. A screening test is as follows: Place one drop of patient's blood and one drop of your own blood or other appropriate blood specimen as control side by side, on a piece of filter paper. Chocolate color indicates presence of methemoglobin.

The causes of methemoglobinemia are given in the next Table; (Cartwright, G.E., Diagnostic Laboratory Hematology, Grune and Stratton, New York, 4th Edition, 1968, pg. 234.):

Causes of Methemoglobinemia
I. Acquired: Nitrites, Nitrates, Chlorates, Quinones, Aminobenzenes, Nitrobenzenes, Nitrotoluenes, Ferrous sulfate (large doses), Some Sulfonamides. Aniline Dye Derivatives, Phenacetin, Acetanilid Pyridium, Benzocaine
II. Hereditary
A. Enzymatic or metabolic
DPNH-Methemoglobin reductase (diaphorase) deficiency (recessive)
TPNH-Methemoglobin reductase deficiency (recessive)
Glutathione deficiency (dominant)
B. Hemoglobin M (dominant)

The causes of methemoglobinemia are acquired or hereditary; most causes are acquired and are due to drugs and chemicals especially those containing nitro- and amino-groups such as aniline and derivatives, nitrites and some sulfonamides, large doses of ferrous sulfate and some bacteria.

Patients with methemoglobinemia may develop polycythemia as a compensatory mechanism and may have dyspnea and headache. Acute methemoglobinemia may be treated with the reducing agent methylene blue; treatment is not necessary unless 35% or more of the hemoglobin is in the methemoglobin form. The definitive treatment of significant methemoglobinemia includes intravenous 1% methylene blue; the dose is 1-2mg/kg given slowly over a 5-minute period; give oxygen. The lethal concentration of methemoglobin is probably 70% and above.

References: Chilcote, R., et al., Pediatrics, 59, 280-282, 1977; Mack, R.B., North Carolina Medical Journal, 43, 292-293, April, 1982.

METHOTREXATE, HIGH DOSE

SPECIMEN: Red top tube, separate serum as soon as possible. Stable at 4°C for at least 24 hours and at -20°C for one month. Collect specimens at 24, 48 and 72 hours past high-dose infusion. Keep specimen in the dark.

REFERENCE RANGE: More than 90 percent of methotrexate is excreted by the kidney and renal toxicity is common; determine serum creatinine and creatinine clearance before starting therapy.

High dose methotrexate is usually accompanied by citrovorum rescue depending on the concentration of methotrexate at 24, 48 and 72 hours.

Sampling Times: Sampling times are usually 24, 48 and 72 hours after starting methotrexate infusion, and ideally everyday during citrovorum rescue.

Toxicity: Without citrovorum rescue: $>8 \times 10^{-8}$M for > 42 hours. With citrovorum rescue: 10^{-7}M when citrovorum is discontinued; 10^{-5}M at 24 hours; 10^{-6}M at 48 hours; 10^{-7}M at 72 hours.

Half-Life: After intravenous infusion, methotrexate is distributed within the total body water and the initial half-life is about one hour. After this initial distribution, the half-life is about 3 hours.

METHOD: EMIT; RIA

INTERPRETATION: Methotrexate in high-dose is used to treat cancer and amounts up to 30 g per square meter of body surface area are given by intravenous infusion. Massive doses of methotrexate must be coupled with citrovorum factor because of the toxicity induced by methotrexate. Methotrexate blocks the action of the enzyme, dihydrofolate reductase, as illustrated in the next Figure:

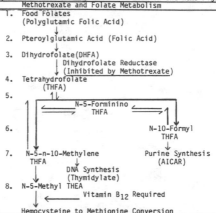

Methotrexate and Folate Metabolism

1. Food Folates
   (Polyglutamic Folic Acid)

2. Pteroylglutamic Acid (Folic Acid)

3. Dihydrofolate(DHFA)
   ↓ Dihydrofolate Reductase
   ↓ (Inhibited by Methotrexate)

4. Tetrahydrofolate
   (THFA)

5.   N-5-Forminino
          THFA

6.   N-10-Formyl
                      THFA

7. N-5-n-10-Methylene   Purine Synthesis
   THFA   (AICAR)
        DNA Synthesis
        (Thymidylate)

8. N-5-Methyl THFA
        ← Vitamin B_{12} Required

Hemocysteine to Methionine Conversion

As illustrated in this Figure, methotrexate inhibits the synthesis of thymidine, which is required for DNA synthesis and cell replication.

Toxicity: With high-dose methotrexate, nephrotoxicity is a common problem. More than 90 percent of methotrexate is excreted by the kidney within 24 to 48 hours. At acid pH, methotrexate may deposit in the renal cells; one of the major metabolites, 7-OH methotrexate, is relatively soluble at acid pH. Nephrotoxicity is reversible.

Toxicities, other than nephrotoxicity that occurs with high-dose methotrexate include B-lymphocyte dysfunction and acute dermatitis, bone marrow depression, leukopenia, thrombocytopenia, anemia, stomatitis, vomiting and diarrhea.

METOPIRONE TEST

PROCEDURE: The metopirone test is performed as follows: Collect four complete 24-hour urine specimens for urinary 17-OHCS assay; collect two days of control specimens, day of metopirone administration specimen and the day after metopirone administration.

Days 1 and 2: Collect 24 hour urine for 17-OHCS
Day 3: Give metopirone 300mg/sq.meter (up to 750mg) every 4 hours by mouth; collect 24 hour urine for 17-OHCS.
Day 4: Collect 24 hour urine for 17-OHCS.

INTERPRETATION: Metopirone is a drug that is used to differentiate adrenal hyperplasia from adrenal tumor and is used to measure pituitary reserve. Metopirone selectively inhibits the synthesis of cortisol from 11-deoxy-cortisol, that is the 11-hydroxylation of deoxycortisol to cortisol as shown in the next reaction:

Desoxycortisol(Cpd S) $\xrightarrow{\text{11-Beta Hydroxylase}}$ Cortisol(Cpd F)

The metabolic pathway, which is blocked by metopirone, is shown in the next Figure:

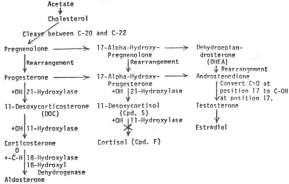

Inhibition of Pathway for Synthesis of Cortisol by Metopirone

In normal patients, the resultant fall in serum cortisol causes an increase in ACTH, which causes increased steroid synthesis to the point of inhibition; this is illustrated in the next Figure:

Secretion of ACTH in Subjects Treated with Metopirone

Normal: Urinary 17-OHCS increases 2.5 to 6 times control by day 3 or 4; 17-OHCS should reach 9mg/m^2/24 hour.

Abnormal: Little or no increase in urinary 17-OHCS is seen with ACTH deficiency, hypothalamic tumors and adrenal suppression secondary to exogenous pharmacologic doses of steroids.

Metopirone test may be used to differentiate adrenal hyperplasia from adrenal adenoma in that there is an increase in urinary steroids in hyperplasia but the urinary levels are not changed in patients with adenoma.

MINIMUM INHIBITORY CONCENTRATION (MIC)

SPECIMEN: Isolated culture of organisms to be tested, prepared by the microbiology laboratory. The test cannot be done if bacterium isolated from patient is not available or fails to grow. The physician must request that the laboratory save the patient's infecting organism within 48 hours of submission of the specimen for initial culture in order to perform an MIC test. If the isolate has not been saved, the test cannot be performed.

REFERENCE RANGE: The reference range is the least concentration (minimum inhibitory concentration-MIC) of antibiotic which will inhibit growth of that organism.

METHOD: MIC Test Panels are molded plastic plates containing 80 wells in which are presented eleven antimicrobic agents at multiple dilutions each, one or more singles at breakpoint concentrations plus sterility and growth controls. Different drugs are presented in panels for testing gram-positive or gram-negative organisms. The wells are inoculated, incubated for 15-18 hours, and read for the presence of visible turbidity. The lowest concentration of each antimicrobic which demonstrates no visible turbidity (no visible growth) is recorded as the minimum inhibitory concentration for the antimicrobic in mcg/ml against the test organism.

INTERPRETATION: MIC is done in order to determine the susceptibility of a given organism to an antimicrobic. Information on susceptibility is important in order to select intelligently the choice of antibiotic to be used in therapy. Antimicrobic panels are available for both gram-positive and gram-negative organisms. Antimicrobics and the range of concentration used in test panels are given in the next Table:

Antimicrobics and Their Range of Concentration Used in Test Panels			
Gram Positive Panel		Gram Negative Panel	
Antimicrobic	mcg/ml range	Antimicrobic	mcg/ml range
Clindamycin	16-0.25	Ampicillin	16-0.25
Erythromycin	16-0.25	Cephalothin	64-1
Nafcillin	16-0.25	Gentamicin	16-0.5
Penicillin G	4-0.06	Tetracycline	16-0.25
Ampicillin	16-0.25	Carbenicillin	512-8
Cephalothin	64-1	Chloramphenicol	32-0.5
Gentamicin	16-0.5	Tobramycin	16-0.5
Tetracycline	16-0.25	Amikacin	32-1
Chloramphenicol	32-0.5	Cefamandole	64-1
Trimeth/Sulfa	32/608-0.5/9.5	Cefoxitin	64-1
Vancomycin	32-0.5	Cefotaxime	32-2
		Kanamycin	8 (only)
		Trimeth/Sulfa	2/38 (only)

MIC susceptibility can be requested for drugs not in the routine panels.

Ref.: Ellner, P. and Neu, H.C., JAMA 246, 1575-1578, 1981.
  Jones, R.L. et al., Am. J. Clin. Pathol. 78, 651-658, 1982.

MITOCHONDRIAL ANTIBODIES

SPECIMEN: Red top tube, separate serum

REFERENCE RANGE: Negative; positive titer at 1:10

METHOD: Indirect immunofluorescence. The usual substrate is rat kidney and a granular pattern is seen in epithelial cells.

INTERPRETATION: Antibodies against mitochondria are found in the sera of 93 percent of patients with primary biliary cirrhosis (Sherlock, S. and Scheuer, P.J., N. Engl. J. Med. 289:674-678, 1973).

Mitochondrial antibodies may also be found in the serum of patients with other liver diseases: chronic hepatitis and cryptogenic (obscure origin) cirrhosis - 25%; viral hepatitis - less than 2%, and drug hypersensitivity, especially halothane jaundice. A raised titer of mitochondrial antibodies may be found in all types of non-hepatic autoimmune disease.

The antibodies are rarely found in extrahepatic biliary tract obstruction or normal patients; thus, this test is very valuable in the differential diagnosis of primary biliary cirrhosis versus extrahepatic biliary tract obstruction.

Primary biliary cirrhosis is a disease of unknown etiology characterized by severe derangement of hepatic excretory function and progressive cirrhosis. Clinical features of the disease are listed in the next Table:

Clinical Features of Primary Biliary Cirrhosis
Middle-aged Females
Insidious Onset
Pruritis, Skin Pigmentation, Jaundice
Skin Xanthomata, Xanthelasmata
Hepatomegaly, Splenomegaly
Steatorrhea (Fat in Stool)
Signs and Symptoms of Portal Hypertension
Associated Diseases (Sicca Syndrome, Thyroiditis, etc.)
Death in Liver Failure

About 90 percent of the patients are women. The disease starts insidiously. Pruritis may be the presenting complaint, and usually precedes jaundice. Pruritis is usually attributed to raised bile acid concentrations. Hepatosplenomegaly is not usually a presenting finding. Steatorrhea is caused by defective bile acid excretion and thus poor fat and fat-soluble vitamin absorption leading to vitamin K deficiency and vitamin D deficiency. As the disease progresses, portal hypertension may develop.

Primary biliary cirrhosis often occurs in association with other autoimmune conditions such as Hashimoto's thyroiditis or Sjögrens secca (dry) syndrome. Diagnosis of Primary Biliary Cirrhosis: Laboratory findings in primary biliary cirrhosis are those of biliary obstruction, e.g., serum alkaline phosphatase increased, serum bilirubin increased, serum cholesterol increased, serum IgM values usually increased and positive serum mitochondrial antibody test.

MONOSPOT SCREEN

SPECIMEN: Red top tube, separate serum and refrigerate.
REFERENCE RANGE: Negative
METHOD: The test is performed on a slide as illustrated in the next Figure:

Monospot Screen

```
  /  Serum   \        /  Serum   \
 /     +      \      /     +      \
(  Guinea Pig  )    (  Beef Red    )
 \   Kidney    /      \ Cell Stroma/
  \    +      /        \    +      /
   \Horse Red/          \Horse Red/
    \ Cells /            \ Cells /
```

Absence of Antibody to
Infectious Mononucleosis: No Agglutination   No Agglutination
Presence of Antibody to
Infectious Mononucleosis: Agglutination   No Agglutination

Antibodies for horse red cells may occur in normal sera and in conditions other than infectious mononucleosis such as serum sickness and other infections; these antibodies are absorbed by guinea pig kidney but not by beef red cells. Beef red cells remove antibodies developed in infectious mononucleosis. The antibodies for horse red cells, present in infectious mononucleosis are not absorbed by guinea pig kidney but are absorbed by beef red cells.

INTERPRETATION: Infectious mononucleosis is apparently caused by the Epstein-Barr virus. The virus preferentially infects B lymphocytes; the immune response includes the activation of T cells which, when present in the blood, appear as atypical lymphocytes (altered T cells).

The hallmark of infectious mononucleosis is the heterophil antibodies, IgM immunoglobulins. These antibodies are present in over 50 percent of patients during the first two weeks of illness and up to 95 percent at the end of one month; these disappear within four to six weeks.

Positive results may be obtained in patients who have high titers of serum sickness antibodies and the infectious mononucleosis antibodies have been found in patients with leukemia, Epstein-Barr virus, cytomegalovirus, Burkitt's lymphoma, rheumatoid arthritis and viral hepatitis.

MONOSPOT SCREEN PLUS HETEROPHILE ABSORPTION

SPECIMEN: Red top tube, separate serum.
REFERENCE RANGE: Monospot: Negative. Heterophile Absorption: Negative titer <1:56.
METHOD: When the monospot test is positive and antibody titers are desired then the heterophil absorption test may be requested. In this test, titration is done using sheep cells, guinea pig kidney antigen and beef red cell stroma.

There are three possible sources of heterophile antibody; these are infectious mononucleosis, serum sickness and Forssman antibodies. The differentiation of these heterophile antibodies by the Davidsohn differential text is given in the next Table:

Differentiation of Antibodies Formed in Infectious Mononucleosis, Serum Sickness and Normal Forssman Antibodies		
	Observation after Absorption	
	with Guinea Pig Kidney	Beef RBC
Disease:		
Infectious Mononucleosis	Antibody Remains	Antibody Absorbed
Serum Sickness	Antibody Absorbed	Antibody Absorbed
No Disease:		
Forssman Antibodies	Antibody Absorbed	Antibody Remains

INTERPRETATION: The heterophile absorption test is done only when the monospot test is positive. The sensitivity and specificity of the heterophile agglutination test are very high: the sensitivity of the heterophile test is 80% to 90% in infectious mononucleosis (see EPSTEIN-BARR VIRUS(EBV) ANTIBODIES, SERUM).

MORPHINE

SPECIMEN: 50ml random urine without preservatives, store at -20°C
REFERENCE RANGE: None detected
METHOD: Thin-layer chromatography; radioimmunoassay; gas-liquid chromatography; latex agglutination-inhibition; hemagglutination-inhibition, spectrophotofluorometry; free radical assay.
INTERPRETATION: Morphine is used for the relief of pain (analgesia); it relieves tension and anxiety and produces sedation.

The finding of morphine in specimens often implicates heroin use.

Heroin is diacetylmorphine and, in the body, is rapidly deactelylated to morphine; urine is the major route of excretion and morphine is excreted either free (5 to 20 percent) or conjugated, primarily to glucuronide.

MUCIN COAGULATION, JOINT FLUID (ROPE'S TEST)

SPECIMEN: 2ml of joint fluid; no anticoagulant
REFERENCE RANGE: Normal: A tight, ropy clump or soft mass with a clear or slightly cloudy solution. Abnormal: A small, friable mass in a cloudy solution or a few flecks in a cloudy solution.
METHOD: A few drops of synovial fluid are added to 20ml of 5% acetic acid.
INTERPRETAION: The mucin coagulation test reflects polymerization of hyaluronate; the normal concentration of hyaluronate in synovial fluid is 0.3 to 0.4g/l. Normally, a clot forms within 1 minute; a "poor" clot indicates inflammation. Mucin clot test results in different conditions are shown in the next Table:

Mucin Clot Results in Different Conditions
Good to Fair Mucin Clot Formation: Osteoarthritis, Traumatic Arthritis, Systemic Lupus Erythematosus, Rheumatic Fever
Fair to Poor Mucin Clot Formation: Rheumatoid Arthritis, Gout, Pseudogout
Poor Mucin Clot Formation: Acute Bacterial Arthritis, Tuberculous Arthritis

Both viscosity and clot formation reflect polymerization of hyaluronate; synovial fluids with poor viscosity also form poor clots.

MUCOPOLYSACCHARIDES (MPS), QUALITATIVE, URINE

SPECIMEN: Random urine; no preservative. Stable for 1 week at 4°C. Stable
indefinitely at -20°C.

REFERENCE RANGE: Negative

METHOD: Acid turbidity test: The acid albumin turbidity test is based on the
reaction of acid MPS with albumin, at acid pH, to form a precipitate.

A spot test may be done by spotting urine on filter paper and then adding
toluidine blue O; Alcian blue may be used instead of toluidine blue.

Another spot test is based on the reaction of cetyltrimethylammoniumoramide
with mucopolysaccharide (Dorman, A. and Matalon, R. in "The Metabolic Basis of
Inherited Disease," pgs. 1241-1248, 3rd ed., Stanbury, J.B., Wyngaarden, J.B.
and Fredrickson, D.S., eds., N.Y., McGraw-Hill Book Co., 1972).

INTERPRETATION: The mucopolysacchariddoses (MPS) are a group of heritable
diseases characterized by intracellular storage and abnormal excretion of
urinary MPS secondary to enzyme deficiency. All are inherited as autosomal
recessives, except Hunter's syndrome, which is sex-linked.

The mucopolysacchariddoses (MPS) are listed in the next Table:

Mucopolysacchariddoses		
Disease	Excessive Urinary MPS	Enzyme Deficiency
Hurler's Syndrome	Dermatan Sulfate Heparin Sulfate	Alpha-L-Iduronidase
Scheie Syndrome	Dermatan Sulfate Heparin Sulfate	Alpha-L-Iduronidase
Hurler-Scheie	Dermatan Sulfate Heparin Sulfate	Alpha-L-Iduronidase
Hunter Syndrome, Severe	Dermatan Sulfate Heparin Sulfate	Sulfoiduronate Sulfatase
Hunter Syndrome, Mild	Dermatan Sulfate Heparin Sulfate	Sulfoiduronate Sulfatase
Sanfilippo Syndrome A	Heparin Sulfate	Heparin Sulfate Sulfatase
Sanfilippo Syndrome B	Heparin Sulfate	N-Acetyl-Alpha-D Glucosaminidase
Morquio's Syndrome	Keratin Sulfate	Chondroitin Sulfate N-Acetylhexosamine Sulfate Sulfatase
Maroteaux-Lamy Syndrome, Classic Form	Dermatan Sulfate	Arylsulfatase B
Maroteaux-Lamy Syndrome, Mild Form	Dermatan Sulfate	Arylsulfatase B
Beta-Glucuronidase Deficiency	Dermatan Sulfate	Beta-Glucuronidase

Heparin sulfate and/or dermatan sulfate are excreted in all of the MPS
except for Morquio's syndrome in which keratin sulfate is excreted.

Specific diagnosis is done by demonstrating deficiency of a specific
enzyme; this is usually done on cultured fibroblasts obtained following skin
biopsy.

MUCOPOLYSACCHARIDES (MPS), QUANTITATIVE, URINE

SPECIMEN: 24 hour urine collection. The patient voids at 8:00 A.M and
discards the specimen. Then, all urine is collected over the next 24 hours
including the 8:00 A.M. specimen collected the next morning. The urine should
be kept cold during the collection. The specimen is stable for one week at room
temperature.

REFERENCE RANGE: Method dependent; age-dependent.
See Dorfman, A. and Matalon, R. in "The Metabolic Basis of Inherited Diseases,"
pgs. 1248-1249, 3rd. ed., Stanbury, J.B., Wyngaarden, J.B. and Fredrickson,
D.S., eds. N.Y., McGraw-Hill Book Co., 1972.

Rapid identification of MPS excreted in urine may be made by electrophoresis.
In one method, concentrated urine is placed on a Sephadex G-25 column; the
MPS fraction is collected. The acid MPS are precipitated with cetylpyridinium
chloride; the polysaccharide-cetylpyridinium complex is dissolved and uronic
acid is determined by the carbazole method. The acid MPS keratin sulfate, which
is excreted in Morquio's syndrome, is not detected using this method.

INTERPRETATION: See MUCOPOLYSACCHARIDES, QUALITATIVE, URINE

MULTIPLE SCLEROSIS, PANEL

Multiple sclerosis panel is given in the next Table.

Multiple Sclerosis - Panel
Cerebrospinal Fluid:
CSF Protein Electrophoresis
Myelin Basic Protein

Typical CSF findings in multiple sclerosis are given in the next Table:

Typical CSF Findings in Multiple Sclerosis				
Condition	Pressure mm CSF	Gross Appearance	Cells (cu. mm.)	Protein (mg/dl)	Glucose (mg/dl)
Multiple Sclerosis	Normal	Clear, Colorless	Normal or 10-50 Lymphs	Normal or 45-100	Normal

CSF Protein Electrophoresis: The CSF protein electrophoretic pattern of patients with active multiple sclerosis and other diseases associated with increased CSF gamma globulin fraction (normal CSF gamma globulins, <13%; increased gamma globulins, >15%) is shown in the next Figure:

Cerebrospinal Fluid Protein Electrophoretic Pattern in Active Multiple Sclerosis

Diseases Associated With Increased CSF Gamma Globulin
Active Multiple Sclerosis
Inflammatory Diseases of the CNS
  Encephalitis
  Meningitis (particularly T.B.)
  Neurosyphilis
  Arachnoiditis
Benign and Malignant Intracranial Tumors

In multiple sclerosis, CSF gamma globulin is found to be elevated in about 75% of patients with an established clinical diagnosis of multiple sclerosis.

Oligoclonal Banding: Normal CSF immunoglobulin migrates as a diffuse band; abnormal immunoglobulins migrate as discrete sharp bands or oligoclonal bands. This is the pattern observed in multiple sclerosis; this pattern of discrete bands within the gamma globulins supports the hypothesis that these globulins arise from a few clones of cells. However, ologoclonal bands are not unique to multiple sclerosis and are observed in the conditions listed in the next Table:

Oligoclonal Bands on Electrophoresis
Multiple Sclerosis
Inflammatory Polyneuropathy
Neurosyphilis
Cryptococcal Meningitis
Chronic Rubella Panencephalitis
Subacute Sclerosing Panencephalitis
(Persistent Rubella Viral Infection)

Oligoclonal bands are observed using agarose electrophoresis; 3ml of CSF is concentrated to 0.05ml. A specimen of serum from the patient is electrophoresed at the same time. The results of electrophoresis are illustrated in the next Figure;(Johnson, K.P. and Hosein, Zobeeda, Lab. Man., pgs. 36-40, May, 1981):

Agarose Electrophoresis of Serum and CSF Specimens

S. Bakerman

An abnormal result is the finding of two or more bands in the CSF that are not present in the serum specimen. The sensitivity of CSF oligoclonal bands in the diagnosis of multiple sclerosis is 79-94% (Markowitz, H. and Kokmen, E., Mayo Clin. Proc. 58, 273-274, Apr. 1983). A scan of cerebrospinal fluid, obtained from a patient with multiple sclerosis, using high resolution, is shown in the next Figure:

High Resolution Scan of Cerebrospinal Fluid

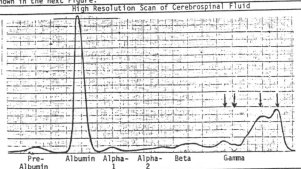

| Pre-
Albumin | Albumin | Alpha-
1 | Alpha-
2 | Beta | Gamma |

Myelin Basic Protein: The conditions associated with elevated levels of <u>myelin basic protein</u> in cerebrospinal fluid(CSF) are given in the next Table:

Conditions Associated with Elevated Levels of Myelin Basic Protein in Cerebrospinal Fluid (CSF)

Multiple Sclerosis
Myelinopathies Other Than Multiple Sclerosis
    Transverse Myelitis
    Hereditary Leukodystrophy
    Metachromatic Leukodystrophy
    Central-Pontine Myelinolysis
Neurologic Disease:
    Strokes Near the Surface of the Brain
    Brain Damage After Head Injury
    Subarachnoid Hemorrhage

Myelin basic protein may also be detected in the serum of patients with <u>brain damage after head injury</u>; in these patients, levels of myelin basic protein correlate well with the <u>type and severity</u> of brain damage and also with final outcome (Thomas et al., Lancet 1, 113-115, 1978; Palfreyman et al., Clin. Chim. Acta 92, 403-409, 1979).

Multiple Sclerosis: Multiple sclerosis has a characteristic geographic distribution. Patients with acute multiple sclerosis have very high levels of myelin basic protein in their <u>cerebrospinal fluid (CSF)</u>; those patients whose disease is in remission and clinically inactive have no detectable myelin basic protein. The levels of myelin basic protein are <u>low</u> in those patients whose disease is chronically active (Cohen, S.R. et al., N. Engl. J. Med. 295, 1455-1457, 1976).

Muramidase, Serum

MURAMIDASE, SERUM (SEE LYSOZYME, SERUM)

Muramidase, Urine

MURAMIDASE, URINE (SEE LYSOZYME, URINE)

MYELIN BASIC PROTEIN
SPECIMEN: Cerebrospinal fluid (CSF)
REFERENCE RANGE: MetPath: Reference Range of Myelin Basic Protein

Result	Interpretation
Less than 4ng/ml	No demyelination
4-8ng/ml	Weakly positive result
	Slowly progressive form of demyelination or the recovery phase from acute exacerbation
Greater than 9ng/ml	Positive result. Consistent with active demyelinating process

METHOD: RIA(Day, Eugene D., Clin. Immunology Newsletter, 3, No.9, 53-59, 1982)
INTERPRETATION: The conditions associated with elevated levels of myelin basic protein in cerebrospinal fluid (CSF) are given in the next Table:

Conditions Associated with Elevated Levels of Myelin Basic Protein in Cerebrospinal Fluid (CSF)
Multiple Sclerosis
Myelinopathies Other Than Multiple Sclerosis
Transverse Myelitis
Hereditary Leukodystrophy
Metachromatic Leukodystrophy
Central-Pontine Myelinolysis
Neurologic Disease:
Strokes Near the Surface of the Brain
Brain Damage After Head Injury
Subarachnoid Hemorrhage

Myelin basic protein may also be detected in the serum of patients with brain damage after head injury; in these patients, levels of myelin basic protein correlate well with the type and severity of brain damage and also with final outcome (Thomas et al., Lancet 1, 113-115, 1978; Palfreyman et al., Clin. Chim. Acta 92, 403-409, 1979).
Multiple Sclerosis: Multiple sclerosis has a characteristic geographic distribution. Patients with acute multiple sclerosis have very high levels of myelin basic protein in their cerebrospinal fluid (CSF); those patients whose disease is in remission and clinically inactive have no detectable basic protein. The levels of myelin basic protein are low in those patients whose disease is chronically active (Cohen, S.R. et al., N. Engl. J. Med. 295, 1455-1457, 1976).

MYOGLOBIN, SERUM
SPECIMEN: Red top tube, separate serum
REFERENCE RANGE: 30-90ng/ml
METHOD: RIA
INTERPRETATION: Myoglobin is the main protein of striated muscle (skeletal and cardiac). It resembles hemoglobin; however, it is unable to release oxygen except at extremely low oxygen tension. Myoglobin is increased in the serum in the conditions listed in the next Table:

Increase in Serum Myoglobin
Acute Myocardial Infarction
Skeletal Muscle Conditions: Trauma (Crush Injury); Surgical Procedures; Polymyositis; Acute Alcoholism with Delirium Tremors

Serum myoglobin is elevated following onset of acute myocardial infarction, as follows: 4 hours, 89% of patients; 6 hours, 92% of patients. It reached its peak within 12 hours. The 4 hour specimen had a sensitivity of 92% and specificity of 95%; 6 hour specimens have a sensitivity of 93% and specificity of 89%. It is concluded that serum myoglobin determination can complement serial serum enzyme determinations and can be of considerable value in the early confirmation of the diagnosis of acute myocardial infarction (McComb, J.M. et al., Br. Heart J. 51, 189-194, 1984).

MYOGLOBIN, URINE
SPECIMEN: Random urine, refrigerate
REFERENCE RANGE: Qualitative: Negative; Quantitative: 0-2mcg/ml
METHOD: Qualitative: Dipstick (Ames or BMC) will detect both hemoglobin and
myoglobin; if test is positive use ammonium sulfate to precipitate hemoglobin,
filter and test supernatant for myoglobin with dipstick. Quantitative:
Antigen-antibody reaction measured by nephelometry.
INTERPRETATION: Myoglobin is increased in the urine in the conditions listed
in the next Table:

Increase in Urine Myoglobin
Acute Myocardial Infarction
Viral Infection (Cunningham, E. et al., JAMA 24, 2428-2429, 1979)
Skeletal Muscle Conditions
Trauma (Crush Injury)
Surgical Procedures
Polymyositis
Acute Alcoholism with Delirium Tremors
Exertional Myoglobinuria in Untrained Individuals
Metabolic Conditions Affecting Striated Muscle

Following acute myocardial infarction, myoglobin appears in the urine
within 48 hours.

Mysoline

MYSOLINE (see PRIMIDONE)

Nasopharyngeal Culture
For Bordetella Pertussis

NASOPHARYNGEAL CULTURE FOR BORDETELLA PERTUSSIS
SPECIMEN: The patient must not be on antimicrobial therapy. A tiny
nasopharyngeal (not throat) swab, in Culturette, is used. With the patient's
head immobilized, pass the swab gently into the nostril until it reaches the
posterior nares and is left in place for 15 to 30 seconds. If resistance is
encountered during insertion of the swab, the other nostril should be tried.
Throat swabs may be collected at the same time as the nasal swab.
Direct plating on special media, such as Bordet-Gengou agar, should be done
at the bedside. Therefore, notify the laboratory beforehand so that the special
isolation media can be warmed.
REFERENCE RANGE: No B. pertussis or B. parapertussis.
METHOD: The genus Bordetella consists of aerobic, gram-negative, coccoid
bacilli; Bordetella pertussis is strictly a human parasite. Growth of B.
pertussis requires the presence of blood; charcoal or ion-exchange resin are
incorporated in the medium to absorb substances, such as peroxides, fatty acids,
etc., that inhibit the growth of B. pertussis. The media should contain
cephalexin to prevent the overgrowth of indigenous organisms. The organism
grows on charcoal agar after 3 to 4 days of incubation at 35° with added CO_2.
Microscopically, the Bordetella are non-spore-forming, encapsulated, bipolar,
pale-obtaining, small, gram-negative bacilli.
INTERPRETATION: Bordetella pertussis is the causative agent of pertussis, a
childhood disease characterized by an inspiratory whoop followed by a long
paroxysm of coughing. B. parapertussis causes a pertussis-like illness. The
organism multiplies in the nasopharynx, releases toxins, causing inflammation.
In the classical pertussis syndrome, there is an incubation period of 7 to 10
days, followed by coughing which progresses over 1 to 2 weeks. The "whoop"
represents a forced inspiration over a partially closed glottis. Pertussis is
prevented by vaccination. Lymphocytosis occurs during this period.
The very young infant may develop choking, apnea and cyanosis rather than a
"whoop."

NEURON-SPECIFIC ENOLASE(NSE)

SPECIMEN: Red top tube, separate serum; freeze serum.

REFERENCE RANGE Normal: 5.4-12.9ng/ml; Children with Metastatic Neuroblastoma: 10-1240ng/ml. (Zeltzer, P.M. et al., The Lancet, II, pgs. 361-363, Aug. 13, 1983).

METHOD: Double-antibody radioimmunoassay.

INTERPRETATION: The serum neuron-specific enolase(NSE) level is a very sensitive marker and a prognostic indicator in children with metastatic neuroblastoma. Neuroblastoma is the second most common solid tumor of childhood.

Neuron-specific enolase(NSE) is an isoenzyme of the glycolytic enzyme, 2-phospho-D-glycerate-hydrolase; the enzyme is found in neuronal and neuroendocrine cells of the central and peripheral nervous systems.

Serum NSE levels has prognostic value for patients with neuroblastoma. NSE levels below 100ng/ml correlates with longer survival; this relation is highly significant in the subgroup of infants less than 1 year old at diagnosis. (Zeltzer, P.M. et al., The Lancet II, pgs. 361-363, Aug. 13, 1983).

NITROGEN BALANCE

SPECIMEN: Twenty-four hour urine urea nitrogen determination is required. Collect urine as follows: Refrigerate specimen as it is collected. Instruct the patient to void at 8:00 A.M. and discard the specimen. Then collect all urine including the final specimen voided at the end of the 24 hour collection; i.e., 8:00 A.M. the next morning. Transport to lab and refrigerate specimen at 4°C.

REFERENCE RANGE: Normally, the 24 hour urinary output of nitrogen is 12 to 24 grams. The range of values for nitrogen balance in adults is positive 4 to negative 20 grams of nitrogen per day.

METHOD: Urine urea nitrogen, see Urea Nitrogen.

INTERPRETATION: Nitrogen balance is used as a measure of the effectiveness of nutritional therapy; it is calculated from the formula:

$$\text{Nitrogen Balance} = \frac{\text{24-Hr Protein Intake(g)}}{6.25} - [\text{24 Hr Urine Urea Nitrogen(g)} + 4]$$

With this equation, the net gains (positive nitrogen balance) or losses (negative nitrogen balance) by the body are assessed. A gram of nitrogen excreted represents the degradation of 6.25 grams of protein. The 4 in the above equation is an estimate of the non-urinary nitrogen loss, i.e., skin and feces and non-urea urinary losses; this approximation is reasonable in the absence of severe diarrhea, hemmorhage or fistula losses.

Normally, the 24 hour urinary output of urea nitrogen is 12 to 20 grams. The range of values for nitrogen balance seen clinically in adults is from positive 4 to negative 20 grams of nitrogen per day. One gram of nitrogen represents the protein content of one ounce of lean tissue. A high-protein diet will not necessarily result in a positive nitrogen balance because ingestion of large amounts of protein without an adequate intake of calories results in utilization of much of the protein as energy (Calloway, D.H., Spector, H., Am. J. Clin. Nutr. 2, 405-412, 1954).

NORPACE (see DISOPYRAMIDE)

NORTRIPTYLINE (see TRICYCLIC ANTIDEPRESSANTS)

NUTRITION, PARENTAL, PEDIATRIC MONITORING PANEL

Some of the indications for parenteral nutrition in infants are given in the next Table:

Indications for Parenteral Nutrition in Infants
Premature Infants:
Severe Respiratory Disease
Congenital Anomalies of the Gastrointestinal Tract
Inflammatory Disease of the Intestinal Mucosa
(Necrotizing Enterocolitis)
Older Infants:
Intractable Diarrhea
Short Bowel Syndrome
Severe Malnutrition
Inflammatory Bowel Disease
Extensive Body Surface Burns
Malignancy
Cardiac Failure
Renal Failure

A suggested schedule for monitoring parenteral nutrition in pediatric patients patients is given in the next Table; (Committee on Parenteral Nutrition, Pediatrics 71, 547-552, 1983; Heird, W.C. in Pediatric Nurtition Handbook; Evanston, Il., Am. Acad. Pediatr. pgs. 392-408, 1979):

Suggested Monitoring Schedule During Total Parenteral Nutrition		
	Suggested Frequency	
Variable Monitored	Initial Period	Later Period
Serum Electrolytes	3-4 times/wk	2-3 times/wk
Serum Urea Nitrogen	3 times/wk	2 times/wk
Serum Calcium, Magnesium, Phosphorus	3 times/wk	2 times/wk
Serum Glucose	See Below	See Below
Serum Acid-Base Status	3-4 times/wk	2-3 times/wk
Serum Ammonia	2 times/wk	Weekly
Serum Protein (Electrophoresis or Albumin/Globulin)	Weekly	Weekly
Liver Function Studies	Weekly	Weekly
Hemoglobin	2 times/wk	2 times/wk
Urine Glucose	Daily	Daily
Clinical Observations (Activity, Temperature, etc.)	Daily	Daily
WBC Count and Differential Count	As Indicated	As Indicated
Cultures	As Indicated	As Indicated
Serum Triglyceride	As Indicated	As Indicated

Initial period is period before glucose intake is achieved, or any period of metabolic instability.
Later period is period during which patient is in a metabolic steady state.
Blood glucose should be monitored closely during period of glucosuria (to determine degree of hyperglycemia) and for two to three days after cessation of parenteral nutrition (to detect hypoglycemia). In latter instance, frequent Dextrostix determination constituents adequate screening. Total parenteral intravenous nutrition is given to those patients that are unable to ingest, digest or absorb sufficient nutrients. Total parenteral nutrition (TPN) can preserve body weight in almost all groups of patients studied. Conditions that have been treated with TPN are given in the Table of conditions associated with malnutrition; a critical assessment of the indications for TPN has been reviewed (Goodgame, J.T., "A critical assessment of the indications for total parenteral nutrition," Surg., Gynecol. & Obstet. 151, 433-552, 1980).

NUTRITIONAL PROFILE FOR PROTEIN-
   CALORIE MALNUTRITION

The objective evaluation of protein-calorie malnutrition states is based on clinical laboratory testing, evaluation of immunocompetence and anthropometric measurements. Malnutrition is classified into three general categories; these are called kwashiorkor-like syndrome, marasmus and combinations of these two states.

Marasmus (derived from the Greek marasmos meaning a dying away) is a variant of protein-calorie malnutrition, characterized by a depression of the anthropometric measurements, weight and creatinine-height index in the presence of normal serum albumin. There is a loss of lean body mass and fat stores and in the more advanced cases, depressed immune function. This condition is brought about by prolonged inadequate intake of both protein and calories (prolonged negative nitrogen balance) leading to wasting of muscle mass and subcutaneous fat. This condition is often readily recognized clinically; edema is absent.

Kwashiorkor-like syndrome occurs as a result of inadequate protein intake with a low or normal caloric intake; this syndrome is characterized by depletion of visceral protein stores, that is, serum albumin and serum transferrin and normal anthropometric measurements. In addition, there is decreased immunological resistance reflected by low total lymphocyte count and delayed hypersensitivity as measured by skin tests. Clinically, patients have pitting edema, ascites, growth failure, liver enlargement and diarrhea.

Both marasmus and kwashiorkor commonly coexist in malnourished patients. Characteristics of marasmus and kwashiorkor are shown in the next Table; (Blackburn, G.L. et al., J. Parenter. Nutr. 1, 11-22, 1977):

Classification of Malnutrition

Classification Marasmus:	Percent Ideal Weight	Creatinine-Height Index (Percent)	All Three Skin Tests (mm)
Moderate	60-80	60-80	
Severe	<60	<60	<5

Kwashiorkor-Like:	Serum Albumin (g/dl)	Serum Transferrin (mg/dl)	Total Lymphocyte Count	All Three Skin Tests (mm)
Moderate	2.1-3.0	100-150	800-1200	<5
Severe	<2.1	<100	<800	<5

Patients who are at particularly high risk for complications of malnutrition are surgical patients and alcoholics.

Clinical Laboratory Testing : The clinical laboratory assays used to evaluate the malnutrition, reflecting the kwashiorkor-like syndrome, are given in the next Table:

Nutritional Assessment Profile

Measurement of Serum Proteins (Hepatic Secretory Proteins)
      Albumin
      Transferrin or Total Iron-Binding Capacity(TIBC)
        (Transferrin = (0.8 x TIBC)-43)
      Thyroxine-Binding Prealbumin(TBPA)
      Retinol-Binding Protein(RBP)
Immunocompetence:
      Total Lymphocyte Count
      Delayed Cutaneous Hypersensitivity

Protein Malnutrition: Measuring the proteins, serum albumin and serum transferrin, has been the mainstay of nutritional screening in clinical practice. In a comparative study of the four serum proteins, albumin, transferrin, thyroxine-binding prealbumin(TBPA) and retinol binding protein (RBP), it was shown that albumin has low sensitivity and transferrin has intermediate while TBPA and RBP have the greatest sensitivity to an alteration in the nutritional state (Ingenbleek, Y. et al., Clin. Chim. Acta. 63, 61-67, 1975).

Serum albumin is not sensitive to short-term low protein diet because it has a relatively long half-life (15 to 19 days), large total albumin mass, capacity to maintain hepatic synthesis of albumin and reduced albumin catabolism;(James, W.P.T. et al., in Bianchi, R., Mariani, G., McFarlane, A.S., eds., London, MacMillan, pgs. 251-263, 1976) in addition, intravascular albumin mass is also maintained by mobilization of albumin from the extravascular pool on dietary restriction (Hoffenberg, R. et al., J. Clin. Invest. 45, 143-151,

1966). Thus, given all these factors and others, it is concluded that albumin is a poor index of the short-term adequacy of protein and energy intake (Shetty, P.S. et al., Lancet, pgs. 230-232, Aug. 4, 1979).

Serum transferrin is reduced in severe protein-energy malnutrition; however, serum transferrin is often not reliable in mild cases of malnutrition and the response to treatment is unpredictable (Ingenbleek, Y. et al., Clin. Chim. Acta. 63, 61-67, 1975). Serum transferrin is not sensitive to short-term restriction of protein or energy: transferrin has a relatively long half-life (8 days); it is increased in iron deficiency; it is an "acute phase" reactant and is increased in patients with infections and as a response to stress.

The serum proteins, thyroxine-binding prealbumin (TBPA) and retinol-binding protein (RBP) are more sensitive indices of protein-energy malnutrition than either serum albumin or serum transferrin. Both TBPA and RBP are synthesized in the liver; the half-life of TBPA is 2 days while that of RBP is only 12 hours, TBPA and RBP could be used to detect subclinical malnutrition and monitor the effectiveness of dietary treatment (Shetty, P.S. et al., Lancet, pgs. 230-232, Aug. 4, 1979). However, measurement of TBPA and RBP are not ordinarily available in most clinical laboratories and assay of these proteins is not usually included in a nutritional profile.

Evaluation of Immunocompetence: The lymphocyte count and skin sensitivity reactions are used to reflect immunocompetence; immune response is modified by inherited and environmental factors; dietary intake is the most important environmental factor and malnutrition is the most frequent cause of secondary immunodeficiency. Lymphoid tissue, such as the thymus and tonsils, are small in subjects with malnutrition; this is reflected in a decreased number of lymphocytes in the peripheral blood.

Skin Testing and Delayed Hypersensitivity Reactions: Intradermal skin tests are used employing recall antigens, e.g., mumps skin test antigen, diptheria toxoid, streptokinase/streptodornase, Candida albicans skin test antigen, trichophytin, and tuberculin purified protein derivative (PPD). The diameter of induration is measured and recorded at 24 and 48 hours. The skin test is regarded as positive or reactive if the induration at either reading is 5mm or greater for any of the three antigens (Mullen, J.L. et al., Arch. Surg. 114, 121-125, 1979). The utility of skin testing in nutritional assessment has been critically reviewed (Twomey, P. et al., J. Parenter. Nutr. 6, pgs. 50-58, 1982).

Anthropometric Measurments: Anthropometric measurements, used to evaluate the malnutrition found in marasmus, are given in the next Table:

Anthropometric Measurements
Height
Calculate Creatinine-Height Index
Weight
Calculate Percent Ideal Weight
Arm Measurements:
Triceps Skinfold Thickness(mm)
Arm Circumference(cm)
Calculate Arm Muscle Circumference(cm)

Creatinine-Height Index (CHI): The creatinine-height index(CHI) is a measure of lean muscle mass; it is given by the equation:

$$\text{Creatinine Height Index(CHI)} = \frac{\text{Actual Urinary Creatinine}}{\text{Ideal Urinary Creatinine}} \times 100$$

Two measurements are required, height(cm) and 24 hour urinary creatinine. The ideal urinary creatinine is obtained from standard Tables (Blackburn, G.L. et al., J. Parenter. Nutr., 1, 11-22, 1977).

Percent Ideal Weight: Percent ideal weight is calculated from the formula:

$$\text{Percent Ideal Body Weight} = \frac{\text{Actual Weight}}{\text{Ideal Body Weight}} \times 100$$

The ideal body weight is obtained from standard Tables (Blackburn, G.L. et al., J. Parenter. Nutr., 1, 11-22, 1977).

Arm Measurements: Arm measurements are made to estimate muscle mass and fat stores; arm muscle circumference is calculated from the equation:

$$\text{Arm Muscle Circumference(cm)} = \text{Arm Circumference (cm)} - [0.314 \times \text{Triceps Skinfold(mm)}]$$

Triceps Skinfold Thickness: There is a good correlation between triceps skinfold and body fat (Forse, R.A. et al., Surgery, 88, 17-24, 1980). The triceps skinfold is measured on the nondominant arm with a caliper; tables of average values for triceps skinfold are given in the article by Butterworth and Blackburn (Butterworth, C.E. and Blackburn, G.L., Nutrition Today, pgs. 8-18, March/April 1975). The percentage of body fat may be predicted from the skinfold thickness measurements of the triceps, subscapular and suprailiac folds (Shephard, R.J.. Diagnosis, pgs. 157-172, January, 1984).

Mid-Upper Arm Circumference: The mid-arm circumference reflects both caloric adequacy and muscle mass. It is measured using a tape measure.

Therapy : During the course of nutritional therapy, the following measurements are obtained; (Blackburn, G.L. et al., J. Parenter. Nutr. 1, 11-22, 1977):

Protocol for Monitoring Efficacy of Nutritional Therapy	
Suggested Frequency	Variable Monitored
Daily	Body Weight
Twice Weekly	Nitrogen Balance
Weekly	Total Lymphocyte Count
Every Three Weeks	Anthropometrics
	Serum Transferrin
	Skin Tests

OSMOLALITY, SERUM

SPECIMEN: Red top tube, separate serum
REFERENCE RANGE: 285-295milliosmols/liter
METHOD: Freezing point lowering or vapor pressure; volatiles are not detected using vapor pressure method.
INTERPRETATION: The serum osmolality is increased in the conditions listed in the next Table:

Causes of Increased Serum Osmolality
Increase in normal constituents of serum:
Increased sodium ion (hypernatremia)
Increased glucose (hyperglycemia)
Increased BUN (uremia)
Increased toxic substances:
Ethanol; Methanol; Ethylene Glycol
Dehydration
Diabetes insipidus

Serum osmolality measurements are done for two reasons: Determine whether serum water content deviates widely from normal; or screen for the presence of foreign low molecular-weight substances in the blood.

The steps in the evaluation of serum osmolality are given in the next Table; (Gennari, F.J., N. Engl. J. Med. 310, 102-105, 1984).

Steps in the Evaluation of Serum Osmolality

1. Measure Serum Osmolality, Sodium, Glucose and Urea.
2. Calculate Osmolality:
$$mOsm/kg\ H_2O = 2Na^+ + \frac{Glu}{18} + \frac{BUN}{2.8}$$
3. If Measured minus Calculated Osmolality>10 mOsm/kg H_2O, then consider the following Differential:
   Decreased Serum Water Content
   Hyperlipidemia
   Hyperproteinemia (If Total Protein > 10 g/dl)
   Low-Molecular-Weight Substances in Serum: Ethanol; Methanol;
   Ethylene Glycol; Mannitol; Other Low Molecular Substances, e.g.,
   Acetone, Isopropanol, Paraldehyde, etc.
   Laboratory Error
4. If Measured minus Calculated Osmolality < 10 mOsm/kg H_2O, then the Above Possibilities are Ruled Out.

The osmolality of the serum is approximately 285 to 295; all but 25 milliomols are contributed by sodium ion and the anions, chloride and bicarbonate. An osmolality above 350mOsm per liter or below 240mOm per liter is necessary to produce clinical signs and symptoms; however, symptoms may occur at values closer to the normal range if the change in osmolality has developed rapidly.

The contribution to the serum osmolality of the following substances is given in the next Table:

Contribution to Serum Osmolality	
Substance	Contribution
Glucose	5.5 (Each 100mg/dl)
BUN	7.1 (Each 20mg/dl)
Ethanol	21.7 (Each 100mg/dl)
Methanol	31.0 (Each 100mg/dl)
Ethylene Glycol	16.3 (Each 100mg/dl)

There is little evidence that hyperosmolality per se is harmful; hyperosmolality may reflect severe fluid shifts. Fluid shifts depend upon whether or not the substance has free access to intracellular water; substances with ready access, such as ethanol and urea, cause little fluid shift; substances with limited access, such as glucose and sodium ion, cause significant fluid shift.

Serum osmolality is decreased in the conditions listed in the next Table:

Causes of Decreased Serum Osmolality
Overhydration
Inappropriate antidiuretic hormone syndrome(IADHS)
Loss of sodium ion (hyponatremia)

See SODIUM, SERUM for causes of hyponatremia.

OSMOTIC FRAGILITY

<u>SPECIMEN</u>: Green (heparin) top tube; fresh blood must be used.
<u>REFERENCE RANGE</u>: Normal Values for Osmotic Fragility

Sodium Chloride (Conc. in %)	Percent Hemolysis	
	Prior to Incubation	Following Incubation
0.85	0	0
0.75	0	0-2
0.65	0	0-20
0.60	0	10-40
0.55	0	15-70
0.50	0-5	40-85
0.45	5-45	55-95
0.40	50-90	65-100
0.35	90-99	75-100
0.30	97-100	80-100
0.20	100	91-100
0.10	100	100

<u>Pre-Incubation</u>: Normal red blood cells will begin lysis at 0.5% and lysis is complete at about 0.3%.
<u>Incubation</u>: Following incubation, the normal red blood cells are more <u>sensitive</u> and will begin lysis at a higher concentration of saline.
<u>METHOD</u>: Red blood cells are placed in decreasing concentrations of NaCl phosphate buffered solutions (1.0 to 0.1%). The solutions are mixed and allowed to stand at room temperature for 30 minutes. Then, the solutions are remixed, centrifuged for 5 minutes at 2000RPM and the supernatant is pipetted off. Measure the absorbance of the supernatant in each tube at 540nm.

The values for percent hemolysis are recorded and plotted on ordinary graph paper with percent hemolysis on the ordinate and the concentrations of sodium chloride on the abscissa.
<u>Incubation</u>: Blood is defibrinated by shaking with glass beads; duplicate samples of the defibrinated blood are incubated at 37°C for 24 hours in sterile screw-capped containers; then, osmotic fragility is determined as described.
<u>INTERPRETATION</u>: Red cells obtained from patients with <u>hereditary spherocytosis</u> are especially susceptible to rupture when placed in <u>hypotonic</u> solution; the causes of increased and decreased susceptibility to osmotic fragility are given in the next Table:

Osmotic Fragility	
Increased Susceptibility to Osmotic Fragility	Decreased Susceptibility to Osmotic Fragility
Hereditary Spherocytosis	Glucose-6-Phosphate Dehydrogenase (G6PD)
Acquired Immune Hemolytic Anemias	Iron Deficiency
ABO Hemolytic Disease	Liver Diseases (Obstructive Jaundice,
Antihuman Globulin-Positive Anemias	Cirrhosis, Gilbert's Disease)
Zieve's Syndrome	Hereditary Anemias: Sickle Cell Anemia,
Pregnancy (32 percent in last	Hgb C Disease, Hgb E Disease,
trimester)	Thalassemia Major
	Plumbism

Incubation increases the sensitivity of the test; mild cases of hereditary spherocytosis can be more readily detected following incubation.
<u>Hereditary Spherocytosis</u>: Hereditary spherocytosis is inherited as an <u>autosomal dominant</u>. The circulating red cells are rounder (spherocytes) and sometimes smaller than normal. The life span of the spherocyte is <u>shorter</u> than normal. The disease usually manifests itself in <u>children</u> but may occur at <u>any age</u>. The triad in the young is usually <u>anemia</u>, <u>jaundice</u> and <u>splenomegaly</u>. Treatment is splenectomy.

The reason these cells are relatively more susceptible to rupture than normal cells is because of their <u>shape</u>; the spherocyte has a <u>smaller surface area to cell size</u> as compared to the normal cell which has the shape of a biconcave disc.

OSTEOMALACIA PANEL

The tests for osteomalacia are given in the next Table:

Screening Tests for Osteomalacia	
Test	Expected Value in Osteomalacia
Screening:	
Serum Calcium (Corrected for Serum Albumin)	Decreased
Serum Phosphate	Decreased
Serum Alkaline Phosphatase	Increased
25-Hydroxyvitamin D	Decreased
Definitive:	
Bone Biopsy	Histology

If the corrected serum calcium or serum phosphate level is below or the serum alkaline phosphatase level is above the normal range, then bone biopsy may be considered. 25-Hydroxvitamin D is expensive (about $50).

Serum alkaline phosphatase is the best single <u>routine</u> biochemical screening test for osteomalacia; however, the false-negative rate is 10% and the false-positive rate is 32% (Peach, H. et al., J. Clin. Pathol. <u>35</u>, 625-630, 1982). <u>Bone Biopsy</u>: Osteomalacia is characterized histologically by <u>defective bone mineralization</u>; there is an increase in osteoid volume and seam thickness.

Transiliac biopsy specimens are obtained. The antibiotic, tetracycline, taken orally, will <u>deposit</u> in <u>sites</u> of <u>bone formation</u> and subsequently can be studied in uncalcified sections by fluorescence microscopy (Frost, H.M., Calc. Tiss. Res. <u>3</u>, 211-237, 1969). Tetracycline may be given as follows: Give demethylchlortetracycline (i.e., Declomycin) 10-15 mg/kg daily for four days, orally; three weeks later, repeat tetracycline schedule. Obtain biopsy 2 to 8 days after the last dose; examine biopsy by fluorescence microscopy. Bone histology showed a false-negative rate of 10% and a false-positive rate of 11%; the predictive value of the negative result was 97% (Peach, H. et al., The Lancet, pgs. 1347-1349, Dec. 10, 1983).

P-50 (OXYGEN DISSOCIATION CURVE)

<u>SPECIMEN:</u> Green (heparin) top tube; also submit a normal control.
<u>REFERENCE RANGE:</u> See curve below.
<u>METHOD:</u> Sample of blood is exposed to varying oxygen partial pressure (PO_2) and the fraction of oxyhemoglobin (HbO_2) is measured. The PO_2 is measured by an oxygen electrode. The HbO_2 is measured by dual wavelength spectrometry at 560 and 576nm. The data are plotted (PO_2 versus HbO_2 fraction); <u>the P-50, which is defined as the partial pressure of oxygen at which the given hemoglobin sample is 50 percent saturated, is read from the curve.</u>

The equipment for analyzing and plotting hemoglobin-oxygen dissociation curves (Hem-O-Scan) is available from American Instrument Company, a division of Travenol Laboratories, Inc.

<u>INTERPRETATION:</u> A wide variety of conditions will cause a shift of the dissociation curve of hemoglobin. These conditions may cause a shift of the dissociation curve of oxyhemoglobin to the left or to the right.

The principal significance of the shift in the oxyhemoglobin dissociation curve in different conditions lies in the delivery of oxygen to the tissue. When the <u>curve is shifted to the left</u>, less oxygen is delivered to the tissue for a given percent saturation of hemoglobin. When the <u>curve is shifted to the right</u>, more oxygen is delivered to the tissue for a given percent saturation of hemoglobin. This is illustrated in the next Figure:

Conditions Influencing Oxyhemoglobin Dissociation Curve	
Shift to the Left	Shift to the right
High Affinity of Hgb for Oxygen Delivers Less Oxygen to Tissue	Low Affinity of Hgb for Oxygen Delivers More Oxygen to Tissue

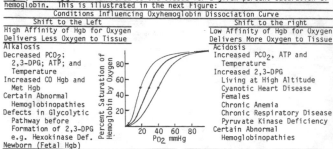

Alkalosis	Acidosis
Decreased PCO_2; 2,3-DPG; ATP; and Temperature	Increased PCO_2, ATP and Temperature
Increased CO Hgb and Met Hgb	Increased 2,3-DPG
Certain Abnormal Hemoglobinopathies	Living at High Altitude
	Cyanotic Heart Disease
Defects in Glycolytic Pathway before Formation of 2,3-DPG e.g. Hexokinase Def.	Females
	Chronic Anemia
	Chronic Respiratory Disease
	Pyruvate Kinase Deficiency
Newborn (Fetal Hgb)	Certain Abnormal Hemoglobinopathies

At 50 percent saturation of hemoglobin by oxygen, PO_2 at the tissue level is 20mmHg is instead of 27mmHg when the curve is shifted to the left and PO_2 is 40mmHg instead of 27mmHg when the curve is shifted to the right.

PANCREATIC INSUFFICIENCY, PANEL

Tests for pancreatic insufficiency are listed in the next Table:

Tests for Pancreatic Insufficiency
Tests for Steatorrhea: (see FAT MALABSORPTION PANEL)
Fecal Fat (72 Hours)
Microscopic Examination of the Stool Fat
Tests for Pancreatic Insufficiency:
Bentiromide
Trypsin, Immunoreactive Serum
Abdominal Radiograph (Pancreatic Calcification)
Secretin Test
Lundh Test
Schilling Test
[^{14}C] Triolein Breath Test

The bentiromide test and the assay of immunoreactive trypsin in serum are two new simple outpatient tests for diagnosis of pancreatic exocrine insufficiency. See Trypsin, Immunoreactive, Serum. The bentiromide test is described below.

Bentiromide Test: (see The Medical Letter on Drugs and Therapeutics 26, pg. 50, May 11, 1984) Bentiromide (Chymex-Adria) has been approved by the Food and Drug Administration for clinical use as a diagnostic procedure for evaluation of pancreatic insufficiency in both children and adults. This test may be done on outpatients. The patient fasts overnight; the patient should be observed for about 30 minutes after taking bentiromide as a precautionary measure in case of a rare adverse reaction.

Bentiromide is taken orally; it is cleaved by pancreatic chymotrypsin as follows:

$$\text{N-Benzoyl-L-Tyroxyl-p-Aminobenzoic Acid} \xrightarrow[\text{Chymotrypsin}]{\text{Pancreatic}} \text{p-Aminobenzoic Acid(PABA)} + \text{N-Benzoyl-L-Tyrosine}$$

PABA is absorbed in the intestine, conjugated in the liver and excreted in the urine; urinary PABA is measured in a 6 hour urine sample.

The sensitivity (positivity in disease) of the bentiromide test for pancreatic insufficiency is listed in the next Table:

Sensitivity of the Bentiromide Test	
Sensitivity(%)	Reference
86	
97	Lankisch, P.G. et al., Dig. Dis. Sci. 28, 490(1983)
100	Toskes, P.P., Gastroenterology 85, 565 (1983)

The sensitivity is decreased in patients with mild to moderate pancreatic disease (Ventrucci, M. et al., Am. J. Gastroenterol. 78, 806, 1983).

False positive results occur in the following conditions: vomiting, gastric retention, impaired gut mucosal function, renal insufficiency and severe liver disease. Thus, false-positive results occur in patients with gastrointestinal diseases (Mitchell, C.J., Pharmacotherapy 4, 79, Mar.-Apr. 1984).

Adverse effects are as follows: nausea, vomiting, diarrhea, headache and transient elevations of liver function tests; one patient developed acute respiratory symptoms.

Interferences that occur with the analytical method are as follows; Drugs: acetaminophen, e.g., Tylenol, phenacetin, benzocaine, lidocaine, procaine, chloramphenicol, procainamide, sulfanamides and thiazide diuretics. These drugs are metabolized to products that absorb at the same wavelength as PABA and should be discontinued three days before the test. Pancreatic enzyme supplements, tanning and sunscreen lotions and multivitamin preparations that contain PABA should be discontinued for seven days before the test.

The patient charge for the bentiromide test is about $40-$50.

PANCREATITIS, ACUTE, PANEL

SPECIMEN: Red top tube, separate serum for serum amylase, serum lipase, serum trypsinogen, serum calcium; random urine specimen for amylase.

Changes in enzymes in acute pancreatitis are summarized in the next Table:

Enzyme	Enzyme Changes in Acute Pancreatitis		
	Beginning of Increase (hrs.)	Maximum (hrs.)	Return to Normal (Days)
Serum Amylase	2-6	12-30	2-4
Urine Amylase	4-8	18-36	7-10
Serum Lipase	2-6	12-30	2-4

It is important to note that urine amylase is elevated when serum amylase is normal. This occurs because renal glomerular filtration for amylase is increased in acute pancreatitis; thus, amylase appears in the urine several days after the serum amylase returns to normal.

The change in serum amylase, serum lipase and urine amylase in acute pancreatitis is illustrated in the next Figure:

Serum Amylase, Serum Lipase and Urine Amylase Following Acute Pancreatitis

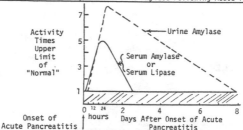

Following onset of acute pancreatitis, serum amylase begins to rise in 2 to 6 hours, reaches a maximum in 12 to 30 hours and remains elevated for 2 to 4 days. Urine amylase begins to rise 4 to 8 hours following onset of acute pancreatitis; this is several hours after the initial increase of serum amylase. Urine amylase remains elevated for 7 to 10 days, about five days after the serum amylase returns to normal. There is an increased renal clearance of amylase. Serum lipase changes in a manner similar to that of serum amylase following onset of acute pancreatitis. Apparently, serum lipase is elevated as often as serum amylase in patients with acute pancreatitis.

Prolonged elevation of serum and urine amylase and serum lipase suggests either continued inflammation, pseudocyst or renal disease. However, in renal diseases, serum amylase activity rarely exceeds twice the upper limit of normal.

Serum calcium is often depressed during the first 24 hours following onset of acute pancreatitis; fatty acids released from triglycerides, combine with calcium to form insoluble calcium salts.

There are two types of acute pancreatitis, edematous pancreatitis and hemorrhagic pancreatitis. Hemorrhagic pancreatitis has a high mortality. If methemalbumin is detected in the serum of a patient with acute pancreatitis, then the diagnosis of hemorrhagic pancreatitis should be strongly considered. (Geokas, M.C., et al "Acute Pancreatitis" Ann. Int. Med. 76, 105-117 (1972)).

Increased amylase in pleural fluid is practically pathognomonic of acute pancreatitis; this is not that far-fetched considering the fact that pleural effusion occurs in approximately 10 percent of patients with acute pancreatitis. One incidence, other than acute pancreatitis whereby increased amylase was found in pleural fluid occurred in a patient who had a ruptured lower esophagus.

Serum trypsinogen is considered to be the most sensitive and specific test for acute pancreatitis (Hughes, G.S., personal communication).

There are no diagnostic chemical tests with high sensitivity and specificity for chronic pancreatitis. Calcium precipitates may be detected by X-ray. If there is extensive destruction of acinar tissue, malabsorption will occur. If there is extensive destruction of islets of Langerhans, which tend to be preserved, diabetes mellitus will occur.

In macroamylasemia, amylase is combined in a macromolecular complex, too large to be excreted in the urine; this condition may be diagnosed by finding a very low amylase/creatinine clearance.

PARATHYROID HORMONE (PTH)

SPECIMEN: Fasting; red top tube, separate serum and freeze. Fasting specimen is recommended because lipids interfere with the test.

REFERENCE RANGE: It is essential to relate the plasma PTH level to the plasma calcium level; levels of plasma PTH and plasma calcium in different conditions are given in the next Table; (MetPath Lab Report):

Levels of Plasma PTH and Plasma Calcium in Different Conditions			
Calcium (mg/dl)	PTH (pg/ml)	Result Consistent with	Most Common Causes
8.8-10.8	100-600	Normal Parathyroid Function	
11.0-16.2	>310	Primary Hyperparathyroidism	Benign Adenoma Hyperplasia Carcinoma
11.2-21.0	<280	Nonparathyroid Hypercalcemia	See list; Causes of hypercalcemia
<9.4	>800	Secondary Hyperparathyroidism	Chronic Renal Disease
<7.5	<300	Hypoparathyroidism	
<7.0	350-700	Pseudohypoparathyroidism	

METHOD: RIA. Progress had been slow in the development of radioimmunoassays of human PTH in plasma because of the lack of sufficient quantities of human PTH to make antibodies. Furthermore, there is immunochemical heterogeneity of circulating PTH; PTH in the peripheral circulation consists of several forms: intact PTH (the biologically active 84 amino acid form); the N-terminal PTH fragment which has physiological activity but a short half-life; and the inactive C-terminal PTH fragment which has a longer half-life than the N-terminal fragment. There is controversy as to whether the antibody directed toward the N-terminal fragment or the antibody directed toward the C-terminal fragment is most effective in differentiating the various causes of hypercalcemia. Martin et al (N. Engl. J. Med. 301, 1092-1098, 1979) present evidence to indicate that the antibody directed toward the C-terminal fragment correlates best with the activity of the parathyroid glands. Raisz et al (Raisz, L.G., Yajnik, C.H., Bockman, R.S. and Bower, B.F., Ann. Int. Med. 91, 739-740, 1979) present data that suggest that the antibody used in the determination of the N-terminal fragment might be more useful to differentiate between the hypercalcemia of primary hyperparathyroidism and the hypercalcemia associated with malignancy. Most often, clinicians order the assay of intact PTH and C-terminal fragment for work-up of patients for possible hyperparathyroidism.

INTERPRETATION: The parathyroid hormone is increased in two conditions: primary and secondary hyperparthyroidism; it is decreased in hypoparathyroidism. Conditions associated with increased parathyroid hormone are listed in the next Table:

Conditions Associated with Increased Parathyroid Hormone
Primary Hyperparathyroidism
Secondary Hyperparathyroidism (Associated with Hypocalcemia)
Chronic Renal Failure
Pseudohypoparathyroidism, Types I and II
Pseudoidiopathic Hypoparathyroidism
Magnesium Deficiency (Some Patients)
Other Causes of Hypocalcemia
Drug-Induced: Anticonvulsives
Rickets
Osteomalacia

Primary Hyperparathyroidism: PTH is increased in primary hyperparathyroidism because the parathyroid gland secretes PTH autonomously since feedback is not intact.

Secondary Hyperparathyroidism: Conditions associated with hypocalcemia are associated with increased PTH; the exception is hypoparathyroidism. In hypocalcemia, suboptimal levels of serum calcium stimulate PTH.

Conditions associated with decreased PTH are listed in the next Table:

Conditions Associated with Decreased PTH
Hypercalcemic states other than primary hyperparathyroidism
Primary Hypoparathyroidism
Wilson's Disease (Carpenter, T.O. et al., N. Engl. Med. 309, 873-877, 1983)

PARTIAL THROMBOPLASTIN TIME, ACTIVATED (APTT)

<u>SPECIMEN</u>: Blue (citrate) top tube filled to capacity; or pediatric blue top tube; avoid contamination with tissue thromboplastin as follows: If multiple tests are being drawn, draw coagulation studies last; if only an APTT is being drawn, draw 1 to 2ml into another vacutainer, discard, and then collect blood for APTT. If the test cannot be assayed immediately, centrifuge, separate plasma from cells and freeze plasma.

The specimen is <u>rejected</u> if the tube is <u>not full</u>, specimen <u>hemolyzed</u>, specimen <u>clotted</u> or specimen received more than <u>2 hours</u> after collection.
<u>REFERENCE RANGE</u>: 20 to 34 seconds.
<u>METHOD</u>: Reagents: Lipid from brain or soybean as a substitute for platelet factor 3, i.e., partial thromboplastin; calcium chloride; activator is celite kaolin.

<u>INTERPRETATION</u>: Triplett, D.A., Anticoagulant Therapy: Monitoring Techniques, Laboratory Management <u>20</u>, 31-42, 1982. The APTT test is a measure of all of the blood factors except factor VII. This test is most often used to monitor therapy with heparin. The causes of prolonged APTT are given in the next Table:

Causes of Prolonged Activated Partial Thromboplastin Time
Heparin Therapy
Vitamin K Deficiency
Liver Disease
Disseminated Intravascular Coagulation(DIC)
Factor Deficiency:
Factor VIII (Classical Hemophilia; Hemophilia A)
Factor IX (Christmas Disease; Hemophilia B)
Factor XII (Hageman Factor)
Factor XI (Partial Thromboplastin Antecedent)
Factor X (Stuart-Prower Factor)
Factor V (Labile Factor)
Factor II (Prothrombin)
Factor I (Fibrinogen)
Nephrotic Syndrome
Deficiencies of Prekallikrein and Kininogens
Dysproteinemias, i.e., Waldenström's
Some Dysfibrinogenemias
Gaucher's Disease

The PTT is an excellent <u>presurgical coagulation screening test</u> and is used to monitor <u>heparin therapy.</u>

<u>Heparin Therapy</u>: Heparin is used in the prophylaxis and management of thromboembolic disease on the venous side of the circulation, that is, <u>venous thrombosis</u> and <u>pulmonary embolism</u>. In heparin therapy, the APTT is usually maintained at from 1.5 to 2.0 times the upper limits of normal (50 to 65 seconds). In another approach, the APTT is maintained at from 1.5 to 2.5 times the patient's admission APTT.

Heparin mediates its actions by first interacting with <u>antithrombin III</u>, converting antithrombin to its active form as illustrated in the next Figure:

Interaction of Heparin with Antithrombin III

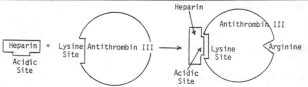

Heparin is a negatively charged mucopolysaccharide that reacts with the basic amino acid lysine of the antithrombin molecule; following this interaction, another basic amino acid, arginine is exposed on antithrombin III. Then, this complex inhibits the active site of the activated blood coagulation factors XIIa, XIa, IXa, Xa, and thrombin. The coagulation cascade is shown in the next Figure:

Coagulation Cascade

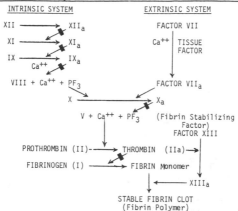

INTRINSIC SYSTEM   EXTRINSIC SYSTEM

= Block by Heparin   PF$_3$ = Platelet Factor

Heparin also inhibits plasmin, the active serine protease of the fibrinolytic system.

Thrombocytopenia occurs in association with heparin therapy; a platelet count every three days allows this complication to be detected early (Stein, P.D. and Willis, P.W., Arch. Intern. Med. 143, 991-994, 1983).

When cephalosporins are given with heparin, there is possible increased bleeding risk with moxalactam (additive); avoid concurrent use of over 20,000 U/day of heparin with moxalactam (The Medical Letter 26, 11-14, Feb. 3, 1984).

Blood coagulation may be initiated by (1) contact activation triggered in-vivo by skin and collagen (intrinsic system) or (2) release of tissue thromboplastin (tissue activated extrinsic system).

Intrinsic System: The following factors are involved in the intrinsic system: XII, XI, IX, VIII, X, V, II, I; the intrinsic system is measured by the PTT. This pathway does not require tissue thromboplastin or Factor VII.

Extrinsic System: This pathway involves tissue thromboplastin and Factors X, V, VII.

Warfarin (4-Hydroxycoumarin Derivative): Vitamin K Antagonist: The laboratory control of oral anticoagulation is usually done using the prothrombin-time (PT); Factor VII is the procoagulant first depressed with coumadin therapy and PTT does not measure Factor VII.

Liver Disease: In liver disease, levels of factors XII, XI, IX, X, VII, V, II and I may be deficient. Factor VIII activity remains unchanged. Deficiency of clotting factors may be due to defective synthesis such as occurs in parenchymal liver disease; or deficiency may be due to lack of vitamin K as in biliary obstruction; vitamin K is required for the synthesis of factors IX, X, VII, and II.

Disseminated Intravascular Coagulation (DIC): The partial thromboplastin time(PTT) is usually prolonged (see Disseminated Intravascular Coagulopathy (DIC) Panel).

Factor Deficiency: Deficiency of individual factors are measured by specific assays; the most common disorders are classic hemophilia, A or B, von Willebrand's disease and anticoagulants used in therapy. Factors VIII and IX deficiencies are sex-linked; women with congenital defects usually have either factor XI deficiency or von Willebrand's disease.

von Willebrand's Disease: In von Willebrand's disease, factor VIII levels are usually low (pseudohemophilia), and there is a prolongation of bleeding time (vascular hemophilia).

The best single test for von Willebrand's disease is inhibition of platelet aggregation in the presence of the antibiotic, ristocetin.

Factor VIII complex consists of a glycoprotein complex, consisting of at least two parts. There is the portion of the complex (factor VIII) which, when deficient, causes classic hemophilia and is controlled by an X-chromosomal gene. There is another portion of the molecule (carrier protein) which is missing or reduced in concentration in von Willebrand's disease and is called von Willebrand's factor; this portion is controlled by an autosomal gene. This factor is necessary for normal platelet adherence to surfaces.

Nephrotic Syndrome: About 10% of patients with the nephrotic syndrome in relapse have a factor IX deficiency secondary to loss of factor IX in the urine. Factor IX deficiency is accompanied by a prolonged PTT and normal PT. However, most patients with nephrotic syndrome have hypercoagulability syndrome secondary to loss of antithrombin-III (Paul Bakerman, Roberta Gray, Personal Communication).

PAROXYSMAL NOCTURNAL HEMOGLOBINURIA (PNH) SCREEN (HAM'S TEST)

SPECIMEN: Purple(EDTA) top tube, refrigerate.
REFERENCE RANGE: Negative
METHOD: Red cells are incubated at 37°C with complement containing serum at pH 6.5 to 7.0 (Ham acid hemolysis test) or placed in a medium of low ionic strength (sugar or sugar water hemolysis). Only PNH cells are lysed.
INTERPRETATION: This test is used to diagnose paroxysmal nocturnal hemoglobinuria(PNH). The pathognomonic feature of PNH is that one population of the red cells is markedly sensitive to lysis by normal complement; another population is less sensitive to lysis.

Paroxysmal nocturnal hemoglobinuria(PNH) is a rare acquired chronic hemolytic anemia; episodic urinary excretion of hemoglobin after a night's sleep occurs in 25 percent of cases. Half of PNH cases develop between ages 20 and 40; other conditions that may be associated with PNH are aplastic anemia, leukemia, and myelofibrosis. Also, PNH cells have deficient acetylocholinesterase activity.

PERICARDIAL EFFUSION ANALYSIS

SPECIMEN: Collect 4 tubes; purple (EDTA) for cell count; special Bactec-vials for microbiological studies, blue (aerobic), yellow (anaerobic); red or green (heparin) top tube for chemistries.
REFERENCE RANGE: Normal volume, 10 to 50 ml, clear and pale yellow.
METHOD: See individual methods.
INTERPRETATION: An algorithm for differentiating causes of pericardial effusion is given in the next Figure: (Kindig, J.R. and Goodman, M.R., Am. J. Med. 75, 1077-1079, 1983):

Algorithm for Pericardial Effusion		
WBC's>15,000 per cubic mm		
Glucose<55 mg/dl	Glucose, 45-130 mg/dl	Glucose, 95-120 mg/dl
pH <7.1	pH 7.2-7.4	pH >7.4
Connective Tissue Disorders Bacterial	Neoplasm Idiopathic Tuberculosis Uremia	Hypothyroidism Post-Pericardiotomy

Additional data should be obtained in order to confirm the results of Kindig and Goodman.

PERITONEAL FLUID ANALYSIS (see ASCITIC FLUID ANALYSIS)

PHENCYCLIDINE ("PCP", "ANGEL DUST", "MIST", "HOG", "ROCKET FUEL","PEACE PILL", "MONKEY TRANQUILIZER")

SPECIMEN: 50ml random urine
REFERENCE RANGE: None detected
METHOD: Thin-layer chromatography; mass spectrometry
INTERPRETATION: Phencyclidine (PCP) was used as an intravenous anaesthetic in the late 1950s; however, 10 to 20 percent of the patients developed postanaesthetic delirium with unmanageable behavior and, thus, human use was discontinued. Now, PCP is a commonly used animal tranquilizer and popular in the drug culture. PCP is taken either orally or inhaled by smoking ("angel dust"), mixed with marihuana or other smokable substances.

The plasma half-life of PCP ranges from 11 hours to 4 days. There is a poor correlation between degree of toxicity and urine levels. PCP is rapidly metabolized to hydroxylated derivatives; there is little physiological activity of PCP metabolites.

With oral intake, effects begin within 15 minutes and lasts from several hours to as long as several days.

In low doses (5 to 10mg; serum level 5 to 10mcg/dl), PCP may produce effects resembling alcohol intoxication; at higher doses (>25mg), acute dystonic reactions, vertical and horizontal nystagmus, and miosis are observed. The mental stage ranges from euphoria to psychosis. People with prior psychiatric illness appear to be highly vulnerable to PCP-induced psychosis. (Aniline, O. and Pitts, F.N., "Phencyclidine(PCP): A Review and Perspectives," CRC Critical Reviews in Toxicology, April, 1982); Krenzelok, E.P., Crit. Care Quart. 4, 55-63, Mar. 1982).

The laboratory tests to monitor phencyclidine toxicity are as follows: blood gases(pH, PCO_2, PO_2); electrolytes(Na$^+$, K$^+$, Cl$^-$, CO_2 content); BUN; ammonia; pH of the urine and urinary output.

                                                                    Phenobarbital

PHENOBARBITAL

SPECIMEN: Red top tube, separate serum; or green (heparin) top tube, separate plasma.
REFERENCE RANGE: Therapeutic: 15-40mcg/ml (Adults); 15-30mcg/ml (Infants and Children). Toxic:>60mcg/ml. Time to Obtain Specimens (Steady State): 11-25 days (Adults); 8-15 days (Children); Half-Life: 50-120 hours (Adults); 40-70 hours (Children).
METHOD: EMIT; RIA; GLC; HPLC; TLC (qualitative); Nephelometry(ICS); Fluorescence Polarization(Abbott)
INTERPRETATION: Phenobarbital is given by mouth and is used as an anticonvulsant; it is effective in all seizure disorders including status epilepticus but not petit mal seizures. Dosage guidelines and blood specimen monitoring schedule are given in the next Table:

The Dosage Guidelines and the Blood Specimen Monitoring for Phenobarbital	
Dosage Guidelines	Blood Monitoring
Newborns:	Frequent because the liver of the
Loading Dose: 20mg/kg I.V.Stat;	neonate metabolizes phenobarbital
Repeat if necessary. 8-10mg/kg/	slowly during the first 4 weeks of
day for two days	life.
Maintenance Dose: 5-6mg/kg/day	
Children:	Monitor in 10-15 days after starting
Loading Dose: 16-20mg/kg/day	therapy; then monitor every 3 to 4
for two days	months
Maintenance Dose:	
Age 1-5: 8-10mg/kg/day	
Age 6-Puberty: 3-8mg/kg/day	
Adults:	Monitor in 15-25 days
Loading Dose: 3-6mg/kg for 4 days	
Maintenance Dose: 1.5-3.0mg/kg/day	

Therapeutic monitoring should be done just prior to the next dose. Monitoring should be for the following patients: poorly controlled, toxic symptoms, change of medication or dosage allowing 2 to 3 weeks for new steady state and when drugs, such as primidone and mephobarbital, are given. These latter drugs are metabolized to phenobarbital.

S. Bakerman

PHENOTHIAZINES, URINE

SPECIMEN: Random urine
REFERENCE RANGE: None detected
METHOD: Ferric chloride (Phenistix) urine test, liquid chromatography
INTERPRETATION: Diagnosis of phenothiazine toxicity can be aided by history
of ingestion, suspicious physical findings, ferric chloride (Phenistix) urine
test and a plain abdominal film - phenothiazines are radio-opaque. Blood levels
do not correlate well with symptoms or prognosis (Mack, R.B., N. Carolina Med.
J., 43, 222-223, 1982).
  The ferric chloride (Phenistix) test is non-specific; a list of selected
substances in urine that react with ferric chloride are given in the next Table;
(Bradley, M. and Schumann, G.B., in Clinical Diagnosis and Management, 17 Ed.,
Editor, Henry, J.B.; W.B. Saunders Co., Phila., pg. 451, 1984):

Ferric Chloride Test in Urine

Substance	Color Change
Acetoacetic Acid	Red or Red-Brown
Pyruvic Acid	Deep Gold-Yellow or Green
Drugs:	
Aminosalicylic Acid	Red-Brown
Antipyrines and	Red
Acetophenetidines	
Cyanates	Red
Phenol Derivatives	Violet
Phenothiazine Derivatives	Purple-Pink
Salicylates	Stable Purple

  A more complete list of substances that react with ferric chloride is given
under the FERRIC CHLORIDE TEST.
  Phenothiazines are used for the following conditions: antipsychotic,
antinausea, antiemetic and antihistaminic and are used to potentiate analgesics,
sedatives and general anesthetics.
  Toxic reactions associated with phenothiazines include: peculiar posturing,
weakness and muscular fatigue, motor restlessness and jerking movements. Other
toxic reactions include hypotension, miosis, hypothermia and myocardial
depression.
  The classes of phenothiazines are shown in the next Table; (Mack, R.B., N.
Carolina Med. J. 43, 22-223, 1982):

Classes of Phenothiazines

Chemical		Trade Name	Dosage (Daily Oral) (ng/ml)	Toxic
	Aliphatic			
Chlorpromazine		Thorazine	10 mg, 3 or 4x; or 25 mg, 2 or 3x	>500
Promethazine		Phenergan		
Promazine		Sparine		
Triflupromazine		Vesprin		
	Piperdine			
Mesoridazine		Serentil		
Thioridazine		Mellaril		>500
Promethazine		Phenergan		
Piperacetazine		Quide		
	Piperazine			
Perphenazine		Triliafon		
Fluphenazine		Prolixin		
Prochlorperazine		Compazine	5 or 10 mg, 3 or 4x	>500
Trifluoperazine		Stelazine	1 or 2 mg, 2x; or 2 to 5 mg, 2x	
Acetophenazine		Tindal		

The underlined are the most commonly used drugs.

PHENYLALANINE BLOOD OR PLASMA

SPECIMEN: Collect specimen immediately prior to discharge from newborn
nursery. If an infant has been tested prior to 24 hours of age, rescreen;
collect a second specimen no sooner than 96 hours and no later than three weeks
of age (American Academy of Pediatrics, Committee on Genetics, Pediatrics 69,
104, 1982).
Blood: Special filter paper is provided by the laboratory as illustrated in the
next Figure:

Filter Paper for Blood Specimens for Phenylalanine Testing in the Newborn

All three circles are soaked with blood; be sure blood soaks through. Use
several drops of blood applied to each circle.
   These specimens on filter paper may be used to test for phenylalanine,
thyroid stimulating hormone(TSH), galactose and cystine.
Plasma: The filter paper method is used for neonatal screening; the method is
semi-quantitative only; quantitative values are obtained from plasma specimens
and are used to evaluate subjects with suspected phenylketonuria(PKU), to
monitor therapy of PKU patients who are on phenylalanine restricted diet or to
test siblings of patients known to have PKU.
   Green (heparin) top tube, separate plasma promptly from cells from fasting
patient; a fast of 4 hours or more in infants. Send specimen frozen in plastic
vial on dry ice.
REFERENCE RANGE: <2 mg/dl for samples collected between 12 and 48 hours of
age; <4 mg/dl on the third day of life or later, (McCabe, E.R.B. et al.,
Pediatrics 72, 390-398, 1983).
METHOD: Fluormetric; ion-exchange, bacterial inhibiton assay or paper
chromatography.
INTERPRETATION: Phenyketonuria(PKU) is an autosomal recessive disease; each
sibling of an identified patient has a 25 percent chance of having PKU. The
carrier rate in the United States is 2%. Phenylketonuria(PKU) is caused by a
deficiency of the enzyme phenylalanine hydroxylase and is characterized by an
increased level of the amino acid phenylalanine in the blood; the enzyme,
phenylalanine hydroxylase catalyzes the reaction:

Phenylalanine+O$_2$ + Tetrahydro- $\xrightarrow{\text{Phenylalanine}}$ Tyrosine + H$_2$O +Quinonoid
                  biopterin Hydroxylase   Dihydrobiopterin

   Tetrahydrobiopterin is an essential, nonprotein coenzyme.
   Clinically, the main signs and symptoms relate to neurologic findings and
to the finding of mental retardation due to the accumulation of the amino acid
phenylalanine. Genetically, the disease is transmitted as autosomal recessive
by the mating of parents who are heterozygotes; the disease tends to be fully
expressed clinically only in patients who are homozygotes. Prevention of mental
retardation associated with the disease depends on diet low in the amino acid
phenylalanine.
   The diagnosis of phenylketonuria(PKU) is made by testing the level of
phenylalanine in the plasma of the newborn 24 hours following feeding of
protein; misdiagnosis is associated with early discharge from the hospital,
e.g., before milk feeding has been established. In patients with PKU, tested
within the first two or three days following delivery, the plasma level of
phenylalanine is usually above normal (2 mg/dl), but not in the classic PKU
range, e.g., above 25 mg/dl. The plasma phenylalanine level tends to rise
markedly following adequate protein (milk) intake; this tends to occur after the
third day of life.
   Detection of Carriers of Phenylalanine(PKU): Heterozygotes for PKU are
clinically normal and can be detected by biochemical tests, the most widely used
being loading tests with phenylalanine given either orally or intravenously.
Loading tests indirectly assess the subject's ability to convert phenylalanine
to tyrosine. Compared with controls, carriers have higher phenylalanine levels
and lower tyrosine levels.
   The human gene for the enzyme, phenylalanine hydroxylase has been cloned.
Using the techniques developed, prenatal diagnosis is possible by analysis of
fetal DNA, obtained through amniocentesis, for matings between 2 heterozygotes
or a homozygote with a heterozygote (Woo, S.L. et al., Nature 306, 151, 1983).

PHENYTOIN, (DILANTIN)

SPECIMEN: The prerequisites for therapeutic monitoring are dosage must be stable for at least one week and dosage cannot be changed or missed. Specimens for monitoring therapeutic response should be drawn just before next dose. Specimens for suspected toxicity are obtained at least six hours after last dose. Blood: Red top tube, separate serum and freeze; or green (heparin) top tube. Urine: 50ml random urine.

Specimens should be obtained as follow:

Oral: Draw trough concentrations one week after initiating therapy and then again 3-5 weeks later.

I.V.: Obtain specimen within 2-4 hours after I.V. loading dose.

REFERENCE RANGE: Adults, Children, Neonates, Pre-term to 12 Weeks: 10-20 mcg/ml.

METHOD: RIA; EMIT; GLC; HPLC: SLFIA(Ames); Nephelometry(ICS); Fluorescence Polarization(Abbott).

INTERPRETATION: Phenytoin is given by mouth or intravenously; it should not be given intramuscularly. Dilantin is one of the most widely used anticonvulsants; it is effective against both grand mal and the generalized spread of seizures with focal origin. It is also used as an antiarrhythmic, in particular for the treatment of digoxin toxicity. The characteristics of Dilantin are as follows: half-life, adults: 18-30 hours; children, 12-22 hours; neonates 30-60 hours; time to steady state, adults, 4-6 days; children, 2-5 days; time to peak plasma level, 4-8 hours. Phenytoin exhibits nonlinear pharmocokinetics, resulting in highly variable time to steady state. The effectiveness of phenytoin is increased at higher plasma levels within the "therapeutic range."

Therapeutic monitoring is done for the reasons listed in the next Table; (Baer, M., "Interpretation of Drug Concentration", Am. Soc. Clin. Path., 1981):

Monitoring Dilantin Levels for Patients
Seizures Poorly Controlled
Toxic Symptoms
Change of Medication or Dosage; Allow One Week to Reach Steady State.
Children 10 to 13 years old every Three to Four Months until Dose-Serum Relationship has Stabilized.

Phenytoin may be given I.V. for acute treatment of seizures at a concentration of 6.7mg or less in normal saline via piggyback at a rate of 40mg/min or less (using an infusion pump). (Ernest, M.P. et al. JAMA 249, No. 6, 762-765, Feb. 11, 1983). Toxicity: Correlation of serum level of Dilantin and symptoms are given in the next Table:

Serum Level of Dilantin and Symptoms	
Serum Level (mcg/ml)	Symptoms
20mcg/ml	Nystagmus
30mcg/ml	Ataxia
40mcg/ml	Disorientation and Somnolence

The most common side effect of phenytoin therapy is gingival hyperplasia (overgrowth of the gums over the teeth;) this occurs in about 20 percent of all patients receiving the drug.

Patients with low serum albumin have a higher percentage of free to bound drug ratio (protein binding, 90 percent) and thus therapeutic or toxic reactions occur at a lower serum level. Hypoalbuminemia in liver disease and renal disease is associated with increased percentage of free phenytoin. In addition, elevated metabolic waste products, bilirubin and urea, can displace phenytoin from albumin binding sites. Accumulation of phenytoin metabolites in patients with end-stage renal disease is associated with artifactually elevated serum concentrations of phenytoin; phenytoin concentrations in uremic serum were 20% (fluorescence polarization immunoassay); 60% (enzyme immunoassay) and 80% (rate nephelometric inhibition immunoassay) higher than corresponding values determined by high performance liquid chromatography (Haughey, D.B. et al., J.Anal. Toxicol. 8, 106-111, May/June 1984).

Serum free testosterone is decreased in patients on phenytoin although total testosterone is elevated; these patients have decreased libido. Phenytoin induces the synthesis of sex hormone binding protein; the increase in the circulating level of sex hormone binding globulin elevates total serum testosterone but depresses free testosterone (Toone, B.K. et al., J. Neurol. Neurosurg. and Psych. 46, 824-826, 1983).

Phenytoin's metabolite, HPPH (see <u>metabolism</u> below), which inhibits the binding of phenytoin to albumin, may accumulate in renal failure. Since the free phenytoin increases in these conditions, optimal anticonvulsive activity is attained at lower total serum phenytoin concentration. Thus, direct measurement of <u>free phenytoin</u> would be the <u>best index</u> of anticonvulsive activity (Finn, A.L. and Olanow; C.W. in Individualizing Drug Therapy, Vol. 2, Taylor, W.J. and Finn, A. eds., Gross, Townsend, Frank, Inc., N.Y., N.Y., 1981, pgs. 64-85). <u>Serum phenytoin is unaltered by dialysis.</u>

Dilantin may cause <u>hepatotoxicity</u>, <u>non-dose related</u>, in susceptible individuals; this reaction is characterized by fever, skin rash, lymphadenopathy, eosinophilia, leukocytosis and hemolytic anemia (Spielberg, S.P. et al., N. Engl. J. Med. <u>305</u>, 722-727, 1981).

One of the <u>less common</u> complications of this drug is the induction of the movement disorder, <u>choreoathetosis</u> (Krishnamoorthy, K.S. et al., Pediatrics <u>72</u>, 831-834, Dec. 1983).

<u>Metabolism:</u> Dilantin is oxidized in the liver by cytochrome P-450; about 60 to 70 percent of phenytoin is metabolized to an inactive, hydroxylated derivative (5-p-hydroxyphenyl-5-phenylhydantoin, HPPH) which is conjugated and excreted in the urine. Dilantin potentiates the metabolism of other drugs, e.g., phenobarbital, by this same system (P-450). Five percent of dilantin appears in the urine unchanged.

The enzymatically mediated reaction utilized for the metabolism of phenytoin is nearly saturated at therapeutic plasma concentrations of the drug. The metabolism of phenytoin is thus said to be saturable, or capacity-limited. When the steady-state serum phenytoin level is within the therapeutic range of 10 to 20 mcg/ml, a <u>very small increase</u> in dose may result in clinical evidence of <u>toxicity</u> and serum levels well above the 20 mcg/ml (Raebel, M.A., N. Engl. J. Med. <u>309</u>, 925, 1983).

PHOSPHORUS, SERUM

<u>SPECIMEN:</u> Red top tube, separate serum; refrigerate at 4°C.
<u>REFERENCE RANGE:</u> Adults: 2.3-4.3mg/dl; Cord: 3.7-8.1mg/dl; <u>Premature:</u> 5.4-10.9mg/dl; <u>Newborn:</u> 3.5-8.6mg/dl; <u>Infant:</u> 4.5-6.7mg/dl; <u>Child:</u> 4.5-5.5mg/dl. To convert <u>conventional</u> units in mg/dl to <u>international</u> units in mmol/liter, multiply <u>conventional units by 0.3229.</u>
<u>METHOD:</u> Phosphomolydate
<u>INTERPRETATION:</u> The causes of <u>increased serum phosphorus</u> are listed in the next Table:

Causes of Increased Serum Phosphorus
Renal Disease
Malignancy Involving Bone
Immobilization
Magnesium Deficiency
Drug Induced:
Ca^{++} Containing Antacids plus milk (Milk-Alkali Syndrome)
Vitamin D Intoxication
Sarcoidosis
Hypoparathyroidism
Pseudohypoparathyroidism
Cushing's Disease
Acromegaly (growth hormone excess)
Transfusions
High Phosphate Diet
Hemolysis
Failure to Separate Clot from Serum Promptly

Elevation of serum phosphorus in patients with ischemic bowel disease indicates <u>extensive bowel injury</u>, acute renal insufficiency, and acidosis; mortality is significantly increased in the patients (May, L.D. and Berenson, M.M., Am. J. Surg. <u>146</u>, 266-268, 1983).

S. Bakerman

The causes of hypophosphatemia are given in the next Table; percentages
are given in parentheses (Kreisberg, R.A., "Phosphorus Deficiency and
Hypophosphatemia", Hospital Practice, pgs. 121-128, March, 1977):

Causes of Hypophosphatemia
Glucose Administration (40%)
Nasogastric Suction or Vomiting (14%)
Liver Disease (7%)
Insulin Administration (5%)
Antacid Abuse (4%) (Phosphorus may bind to antacids)
Various Disorders (4%)
No Identifiable Condition (26%)

The most common cause of hypophosphatemia is physician-induced
administration of glucose or use of insulin; phosphorus follows glucose into the
cell.

Hypophosphatemia is commonly present in diabetes. DeFronzo and Lang (N.
Engl. J. Med. 303, 1259-1263, 1980) found that hypophosphatemia can cause
insulin resistance and glucose intolerance.

Other causes of hypophosphatemia are listed in the next Table:

Other Causes of Hypophosphatemia
Primary Hyperparathyroidism
Malnutrition
Malabsorption
Renal Tubular Acidosis
Tubular Insufficiency
Fanconi's Syndrome
Familial Hypophosphatasia
Gram Negative Septicemia

PITUITARY PANEL

The tests on serum that are done in the work-up of patients with possible
pituitary tumors are listed in the next Table; (Tucker, H.St.G. et al., Ann.
Int. Med. 94, 302-307, 1981):

Serum Tests in Work-Up of Patients with Possible Pituitary Tumors
Serum Prolactin
Serum Growth Hormone(GH)
Serum Adrenocorticotropin(ACTH)
Follicle Stimulating Hormone(FSH)
Luteinizing Hormone(LH)
Thyroid Stimulating Hormone(TSH)
Total Thyroxine(T-4)
Triiodothyronine Resin Uptake(T-3 Resin Uptake)
Cortisol
Estradiol

Pituitary reserve is determined by measuring urinary 17-hydroxycortico-
steroids or serum desoxycortisol before and after administration of oral
metyrapone, 750mg every 4 hours for six doses.

Intravenous injection of insulin is given and blood is drawn every 30
minutes for 2 hours and analyzed for glucose, prolactin, growth hormone and
cortisol. The test is considered invalid if the blood sugar does not fall below
45mg/dl.

Intravenous thyrotropin releasing hormone(TRH) is given; serum specimens
for prolactin, growth hormone and TSH are obtained before and every 15 minutes
for 1 hour.

Other stimulation tests are done (Tucker, H.St.G., et al., Ann. Int. Med.
94, 302-307, 1981).

Stimulation tests (response to insulin, response to thyrotropin-releasing
hormone, and response to luteinizing hormone-releasing hormone) may be performed
simultaneously on a single day (Lufkin, E.D. et al., Am. J. Med. 75, 471-475,
1983). This approach would save a patient's time and money.

Pituitary adenoma cell type in patients with pituitary tumors is given in the next table; (Kovacs et al., Pathology Annual 12, 341, 1977):

Adenoma Type in Pituitary Tumors	
	Incidence (%)
Prolactin Cell Adenoma	32
Undifferentiated Cell Adenoma	23
Growth Hormone Cell Adenoma	21
Corticotroph Cell Adenoma	13
Mixed Growth Hormone and Prolactin Cell Adenoma	6
Acidophil Stem Cell Adenoma	3.5
Gonadotroph Cell Adenoma	1
Thyrotroph Cell Adenoma	0.5

Of the secreting neoplasms, prolactin-secreting tumors are the most common. In the study by Randall et al (Mayo Clin. Proc. 58, 108-121, 1983) prolactin-secreting pituitary adenomas constituted 40 percent of patients with pituitary adenomas. In women, this is usually manifested by amenorrhea, with or without galactorrhea; in men, this is usually manifested by decreased libido and sexual potency. The mechanism as to how prolactin causes amenorrhea and impotence is not known; hyperprolactinemia may induce a central defect, possibly dampening normal secretion of pulsatile GnRH, thus blunting pituitary gonadotropin release and testosterone falls. Supplemental testosterone therapy alone is ineffective in the presence of hyperprolactinemia; this indicates that hyperprolactinemia may cause peripheral androgen resistance (Magrini, G. et al., J. Clin. Endocrinol. Metab. 43, 944-947, 1976).

Jordan et al (Jordan, R.M. et al., Ann. Intern. Med. 85, 49-55, 1976) measured ACTH, GH, TSH, prolactin, LH and FSH in CSF in patients with pituitary tumors; their data suggest that an elevated CSF pituitary hormone is a sensitive indicator of suprasellar extension of a pituitary tumor, and post-treatment measurements are useful in determining efficacy of treatment.

For evaluation of pituitary anatomy, contrast enhanced computed tomography is the preferred radiologic procedure (Editorial, Lancet 1, 430-431, 1982).
Diabetes Insipidus: The causes of diabetes insipidus are given in the next Table; (Coggins, C.H. in Case Records of the Massachusetts General Hospital, N.Engl. J. Med. 309, 420, 1983):

Causes of Diabetes Insipidus
Idiopathic - 30 Percent
Familial - Rare
Traumatic - 30 Percent
Head Injury; Neurosurgical Operation
Neoplastic - 30 Percent
Primary Brain Tumor
Pituitary Adenoma; Craniopharyngioma; Meningioma; Optic Glioma
Metastic Tumor
Lung; Breast; Colon and Others
Vascular
Hemorrhage; Aneurysm; Hemangioma; Postpartum Necrosis; Postanoxic
Infection
Bacterial or Fungal Abscess; Meningitis/Encephalitis; Tuberculoma
Systemic
Sarcoidosis; Langerhans-Cell Granulomatosis; Wegener's Granulomatosis

Suprasellar Tumors: The most common primary tumors in the suprasellar region are listed in the next Table; (Kjellberg, R.N., Craniopharyngiomas in: Tindall, G.T. and Collins, W.F. eds., Clinical Management of Pituitary Disorders, N.Y., Raven Press, 373-388, 1979):

Suprasellar Tumors	
Tumor	Percent
Pituitary Adenomas	68
Craniopharyngiomas	14
Suprasellar Meningiomas	11
Gliomas	12

PLASMA EXCHANGE (SEE THERAPEUTIC APHERESIS)

PLASMINOGEN

SPECIMEN: Blue (citrate) top tube, separate plasma. Do <u>not</u> use plasma
collected in the presence of fluoride, EDTA or heparin.
REFERENCE RANGE: 73-122% (DuPont ACA Methodology).
METHOD: DuPont ACA Methodology: The plasminogen method is based on the
reaction of streptokinase, which has been added in excess, to form an
enzymatically active complex with plasminogen in the sample as follows:

$$\text{Plasminogen} + \text{Streptokinase} \longrightarrow \text{Plasminogen-Streptokinase}$$
$$(\text{Excess}) \qquad\qquad (\text{Active Complex})$$

The <u>active complex</u> hydrolyzes a synthetic substrate; the product reacts
with a chromogen to form a complex that absorbs at 405 nm. The <u>increase in</u>
<u>absorbence</u> at 405 nm is <u>directly proportional</u> to the amount of <u>functional</u>
<u>plasminogen</u>.
INTERPRETATION: The fibrinolytic system is illustrated in the next Figure:

Fibrinolytic System

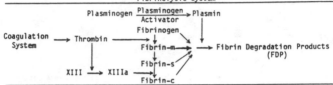

Plasminogen is proteolytically activated to plasmin; plasmin attacks
fibrinogen and fibrin to produce fibrin degradation products(FDP).
    Depressed values are seen in <u>active fibrinolysis</u>, such as <u>disseminated</u>
<u>intravascular coagulopathy</u>(DIC).

PLATELET AGGREGATION

SPECIMEN: Blue (citrate) top tube; do not refrigerate. Patient should not
receive aspirin, phenylbutazone, phenothiazines or antihistamines for 10 days
prior to the test. Platelet count should be less than 100,000/cu mm.
REFERENCE RANGE: Interpreted by laboratory
METHOD: Platelet aggregometer; this test is based on increase in the
transmission of light as platelets aggregate.
INTERPRETATION: Platelet aggregation studies are done to evaluate platelet
function. The following conditions cause alteration of platelet aggregation:

Alterations of Platelet Aggregation

Glanzmann's Thrombasthenia
Abnormality of Platelet Release:
    Aspirin, Uremia, Myeloproliferative Disorders,
    Severe Liver Disease, Dysproteinemia
Impaired Aggregation to Ristocetin with Normal Aggregation
    to ADP, Epinephrine and Collagen: von Willebrand's disease
Other Platelet Abnormalities

Glanzmann's Thrombasthenia: Glanzmann's thrombasthenia is a congenital disorder
characterized by prolonged <u>bleeding</u> time.

PLATELET ANTIBODIES

SPECIMEN: Red top tube, separate serum
REFERENCE RANGE: Negative
METHOD: Indirect immunofluorescence
INTERPRETATION: The conditions that are associated with platelet antibodies
are given in the next Table; (Mayo Medical Laboratories Communique 8, No. 12,
Dec. 1983):

Conditions Associated with Platelet Antibodies
Idiopathic Thrombocytopenia Purpura(ITP)
Post-Transfusion Purpura
Platelet Refractoriness
Neonatal Isoimmune Purpura
Drug-Induced Thrombocytopenia (Quinidine, Quinine,
Lasix, Sulfonamides)

Antibodies that develop to platelet antigens are of two types,
autoantibodies and alloantibodies. Autoantibodies are antibodies that develop
in response to one's own platelets; these antibodies are found in idiopathic
thrombocytopenia purpura(ITP). Alloantibodies develop following exposure to
antigens from foreign platelets.

Idiopathic Thrombocytopenia Purpura(ITP): About 90 percent of children with
acute ITP have autoantibodies in their serum against platelets. In children,
ITP is usually an acute process while in adults, it is usually chronic.

Post-Transfusion Purpura: Following whole blood transfusion, antibody
(anti-PLA^1) against a platelet specific antigen found in 98% of the
population, may develop. Profound thrombocytopenia, manifested by purpura and
mucosal bleeding, occurs in about one week after the transfusion; the antibody
tends to disappear in about six weeks. It may recur with subsequent
transfusions.

Platelet Refractoriness: On repeated infusions of platelet concentrates, the
increment in platelet counts after each infusion may become steadily smaller;
this is due to development of antibodies in the recipient to HLA antigens on the
infused platelets. HLA-matched platelets will usually increase the platelet
count.

Neonatal Isoimmune Purpura: The phenomena, observed in neonatal isoimmune
purpura, is similar to that of hemolytic disese of the newborn(HDN) (see COOMBS,
DIRECT) in that, the mother, on exposure to fetal platelets antigens of the
father, develops antibodies to these antigens. These antibodies pass from
maternal blood into the fetus. In the fetus, the antibodies combine with the
antigens on the fetal platelets.

If a pregnant mother has ITP, antibodies can cross the placenta and cause a
similar reaction.

Drug-Induced Thrombocytopenia: Drugs, such as quinidine, quinine, lasix and
sulfonamides, may be associated with thrombocytopenia. The fall in platelets
may occur acutely following administration of the drug or it may develop weeks
after the drug is discontinued. The thrombocytopenia may be due to antiplatelet
antibodies or progressive bone marrow damage.

PLATELET COUNT

SPECIMEN: Lavender (EDTA) top tube; specimen rejected if clotted.

REFERENCE RANGE: Adults: 150,000-400,000/mm^3; cord: 100,000-300,000/mm^3; premature: 100,000-300,000/mm^3; newborn: 140,000-300,000/mm^3; neonate: 150,000-300,000/mm^3; infant: 200,000-475,000/mm^3; child: 150,000-450,000/mm^3

METHOD: Automated methodology; rapid estimation of the platelet count can be made from the blood smear; there should be about one platelet for every 20 red cells; an estimation may be obtained by multiplying the number of platelets per oil immersion field by 20,000.

INTERPRETATION: The causes of decreased platelet count are given in the next Table (Bauer, J.D. in Gradwohl's Clinical Laboratory Methods and Diagnosis, Sonnenwirth, A.C. and Jarett, L., eds., 8th Ed., C.V. Mosby Co., St. Louis, 1980, pg. 990):

Causes of Decreased Platelet Count
Decreased Production:
Marrow Depression: Aplastic Anemia, Radiation
Marrow Infiltration: Acute Leukemia, Carcinoma,
Myelofibrosis, Multiple Myeloma
Megaloblastic Anemia
Congenital: Wiskott-Aldrich Syndrome, Immune Deficiency States,
Bernard-Sonlier Syndrome, Thrombocytopenia with absent Radius,
May-Hegglin Anomaly, Hereditary Thrombocytopenia resembling ITP
Increased Destruction:
Immunologic
Isoimmune
Autoimmune: Idiopathic Thrombocytopenic
Purpura (Auto-Thrombocytopenia), Evans' Syndrome,
Infectious Mononucleosis
Antigen-Antibody Complexes: Systemic Lupus Erythematosus,
Lymphoma, Chronic Lymphocytic Leukemia
Drugs (Cimo, P.L., Arch. Intern. Med. 143, 1117-1118, June, 1983)
Dilution: Exchange Transfusion
Coagulopathies: Diffuse Intravascular Coagulation,
Septicemia, Hemolytic-Uremic Syndrome, Thrombotic
Thrombocytopenic Purpura and Large or Multiple Hemangiomas
Severe Hemorrhage
Hypersplenism

There is little tendency to bleed until the platelet count falls below 20,000-50,000.

Drug-induced thrombocytopenia is rare; the number of drugs that have been invoked as possible cause of thrombocytopenia is extensive.

The causes of increased platelet count are given in the next Table:

Causes of Increased Platelet Count
Reactive Thrombocytosis: Infection, Acute Blood Loss,
Disseminated Carcinoma, Splenectomy (e.g. in Hereditary
Spherocytosis) and Surgery (Stress)
Thrombocythemia:
Myeloproliferative Disorders, Polycythemia Vera,
Chronic Granulocytic Leukemia, and Myelosclerosis
Essential Thrombocythemia

PLEURAL FLUID ANALYSIS

SPECIMEN: Collect 3 tubes; purple (EDTA) for cell count; special Bactec-vials for microbiological studies, blue (aerobic), yellow (anaerobic); red or green (heparin) top tube for chemistries. Culture, gram stain and Ziehl-Neelsen staining should be done on a centrifuged specimen.

Red top tube, separate serum (for serum protein and LDH analysis). Obtain within 30 minutes of pleural specimen.

REFERENCE RANGE: Appearance: Clear and colorless to pale yellow; less than 1000 WBC/cu mm; less that 25% polys; 0 RBC's; glucose level approximates serum glucose level.

METHOD: The tests are given in the next Table:

Tests of Pleural Fluid
Specific Gravity
Total Protein (Serum and Pleural Fluid)
Glucose
Lactate Dehydrogenase (LDH) (Serum and Pleural Fluid)
White Blood Cell Count and Differential
Red Cell Count
Culture, Gram Stain, Zeehl-Nielsen Stain for Mycobacteria
Cytologic Examination
CEA for Tumor
Amylase for Diagnosis of Acute Pancreatitis

INTERPRETATION: The causes of pleural fluid effusions are given in the next Table; (Krieg, A.F. in Henry, J.B., Clinical Diagnosis and Management, W.B. Saunders, Philadelphia, 17 Edition, pg. 484, 1984):

Pleural Fluid Effusions	
Transudates	Exudates
Congestive Heart Failure	Neoplasms
Hepatic Cirrhosis	Bronchogenic Carcinoma
Hypoproteinemia (e.g.	Metastatic Carcinoma
Nephrotic Syndrome)	Lymphoma
	Mesothelioma (Increased Hyaluronate in Effusion)
	Infections
	Tuberculosis
	Bacterial Pneumonia
	Viral or Mycoplasma Pneumonia
	Trauma
	Pulmonary Infarction
	Rheumatoid Disease (Usually Low Glucose)
	Systemic Lupus (Occ. L.E. Cells)
	Pancreatitis (Elev. Amylase)
	Ruptured Esophagus (Elevated Amylase, Low pH)
	Chylous Effusion: Damage or Obstructed Thoracic Duct

Differentiation of transudate from exudate is given in the next Table; (Light, R.W. et al., Ann. Intern. Med. 77, 507, 1972):

Differentiation of Transudate from Exudate		
Characteristic	Transudate	Exudate
Specific Gravity	<1.016	>1.016
Protein	<3 gms/dl	>3 gms/dl
$\frac{Pleural\ Fluid\ Protein}{Serum\ Protein}$ =	<0.5	>0.5
$\frac{Pleural\ Fluid\ LDH}{Serum\ LDH}$ =	<0.6	>0.6

Pleural effusions have been reported to occur in 3 to 18 percent of patients with acute pancreatitis or pseudocyst. The effusions are more commonly left-sided.

A level of amylase in pleural fluid, which is significantly raised above normal and higher than simultaneous serum values, is practically pathognomonic of pancreatitis. Increased amylase in pleural fluid may occur by one of the following mechanisms: transdiaphragmatic from peritoneal cavity to pleural cavity by lymphatic transfer; intrapleural rupture of mediastinal extensions of pseudocysts; or diaphragmatic perforation.

PORPHOBILINOGEN, QUALITATIVE AND QUANTITATIVE, URINE

SPECIMEN: <u>Qualitative</u>: 50ml of fresh random urine; keep specimen protected from light during transit to lab by wrapping in foil; freeze until ready for assay. <u>Quantitative</u>: The specimen should be protected from light at all times. Wrap container in foil. The patient should discard the urine specimen on arising in the morning; then collect urine specimens including the specimen obtained just prior to completion of 24 hour collection. Record the total urine volume and freeze a 100ml aliquot.

REFERENCE RANGE: Qualitative, negative; quantitative, up to 4mg/day. To convert <u>conventional</u> units in mg/day to <u>international</u> units in mcmol/day, multiply conventional units by 4.420.

METHOD: Watson-Schwartz Test:

Porphobilinogen + p-Dimethylaminobenzaldehyde $\xrightarrow{H}$ Red Color

INTERPRETATION: Porphobilinogen is increased in the conditions listed in the next Table:

Increase in Porphobilinogen
Hereditary Hepatic Porphyrias:
Acute Intermittent Porphyria (AIP)
Variegate Porphyria (VP)
Hereditary Coproporphyria (HCP)

Porphobilinogen is <u>not</u> increased in the urine in porphyria cutanea tarda (PCT) nor lead poisoning. The metabolic defect in AIP is shown in the next Figure:

Metabolic Defect in Acute Intermittent Porphyria (AIP)

"Activated Glycine" + Succinyl CoA

$\downarrow$ ALA Synthetase (+Activity)

2 Delta-Aminolevulinic Acid

$\downarrow$ ALA Dehydrase

Porphobilinogen

Uroporphyrinogen Synthase

$\downarrow$ Uroporphyrinogen Isomerase

Uroporphyrinogen III

$\downarrow$ Uroporphyrinogen Decarboxylase

Coproporphyrinogen III

$\downarrow$ Coproporphyrinogen Oxidase

Protoporphyrinogen III

$\downarrow$

Protoporphyrin III

$+Fe^{+2} \downarrow$ Ferrochelatase

Heme

PNEUMOCYSTIS CARINII ANTIGEN(S)

SPECIMEN: Red top tube, separate serum; forward on dry ice to reference laboratory.

REFERENCE RANGE: Negative

METHOD: Counterimmunoelectrophoresis(CIE); latex agglutination tests are being developed. Currently, (July, 1983), this test is done only at some medical centers. Dr. Linda Pifer, University of Tennessee Center for the Health Sciences and LeBonheur Children's Medical Center, Memphis, TN, has served as the "unofficial reference laboratory for P. carinii antigen detection;" she will accept serum specimens for analysis at no charge. A brief summary of the patient's history and antibiotic therapy should accompany the specimen. The specimens should be forwarded by overnight air express; "turn-around" time is usually 24 hours from time of specimen receipt to the reporting of results by telephone.

The antibody developed in Dr. Pifer's laboratory is prepared against cell culture-grown P. carinii. (Pifer, Linda L., "Penumocystis Carinii: A Diagnostic Dilemma," Pediatric Infectious Disease, 2:177-183, 1983.)

INTERPRETATION: Pneumocystis carinii is an opportunistic parasite which can cause an interstitial pneumonia in the immunocompromised patient.

Pneumocystis carinii pneumonia(PCP) is the most common life threatening opportunistic infection in AIDS patients. About fifty percent of patients with acquired immunodeficiency syndrome(AIDS) present with pneumocystis carinii pneumonia; another 80% present with Kaposi sarcoma plus PCP.

PCP is treated with trimethoprim-sulfamethoxazole(TMP=SMX). However, many of the fatal AIDS-associated cases of PCP who were unresponsive to TMP-SMX were switched to pentamidine isethionate.

POTASSIUM, SERUM

SPECIMEN Red top tube or green top (heparin) tube; the serum or plasma must be separated from the red cells promptly; otherwise spurious elevation of serum potassium occurs.

REFERENCE RANGE: Adult: 3.5-5.3 mmol/liter; premature (cord): 5.0-10.2 mmol/liter; premature (48 hours): 3.0-6.0 mmol/liter; newborn (cord): 5.6-12.0 mmol/liter; newborn: 3.7-5.0 mmol/liter; infant: 4.1-5.3 mmol/liter; child: 3.4-4.7 mmol/liter. The following are life-threatening values and, after confirmation, preferably by repeat determination on a different specimen, should be telephoned to the responsible nursing staff or, if unavailable, to the responsible physician so that corrective therapy may be immediately undertaken: serum potassium: <2.5 mmol/liters; serum potassium: >6.5 mmol/liter.

METHOD: Ion-selective electrode or flame emission.

INTERPRETATION: The concentration of intracellular K^+ is 130 mmol/liter; the serum concentration of K^+ is 3.5 to 5.0 mmol/liter. Thus, only two to three percent of the K^+ is extracellular. Generally, serum K^+ reflects K^+ stores.

There are only three mechanisms to consider in hypokalemia (1) urinary loss, (2) gastrointestinal loss and (3) possible movement of potassium from the extracellular to the intracellular fluid. These mechanisms are listed in the next Table:

Causes of Hypokalemia		
Urine Loss	G.I. Loss	Movement into the Intracellular Space
Diuretics (Thiazides (Chlormetric Sulfonamides): Lasix (Furosemide); Edecrin (Ethacrynic Acid))	Vomiting Nasogastric suction Pyloric obstruction Diarrhea Malabsorption Villous adenoma Enema and laxative abuse	Diabetic ketoacidosis (treated) Familial periodic hypokalemic paralysis
Magnesium Depletion Antibiotics Carbenicillin Amphotericin B		
Increased Mineralocorticoid Licorice Abuse		Other
Renal Tubular Acidosis, Type 1 (distal) and type 2 (proximal) Bartter's Syndrome		Decreased K^+ intake Acute myeloid leukemia

Diuretic therapy and gastrointestinal loss are major causes of hypokalemia.
Drug-induced hypokalemia is usually due to diuretics, adrenal
corticosteroids, or carbenoxolone; carbenoxolone has a mineralocorticoid-like
action on the distal tuble.

Small potassium supplements do not reliably prevent hypokalemia in patients
taking thiazides or loop diuretics. Elderly individuals and patients taking
both thiazides or loop diuretics plus digoxin should receive potassium
supplements sufficient to maintain a normal plasma potassium concentration. The
blood potassium concentration should be checked at least once every 3 months.

Generally, serum K^+ reflects K^+ stores; two important exceptions are
alkalosis and insulin hypersecretion which promote entry of K^+ into cells
resulting in hypokalemia despite adequate or increased intracellular K^+.
Excessive cell breakdown, e.g., hemolysis, results in a relative excess of
extracellular K^+ despite normal stores.

More than 90 percent of potassium is excreted in the urine; it is filtered
and totally reabsorbed proximally and is excreted by the distal tubules. The
factors that determine the degree of distal tubular secretion are dietary K^+
intake, mineralocorticoid secretion, distal tubular flow and the anion
accompanying Na^+ to the distal tubular site.

Both increased dietary K^+ intake and aldosterone promote movement of K^+
into cells and favor secretion into the tubular lumen and subsequent excretion.

Normally, less than 10 percent of potassium that is excreted appears in the
stool; however, gastrointestinal abnormalities can cause loss of K^+.

The increase of serum potassium in different conditions is shown in the
next Table:

Increase of Serum Potassium in Different Conditions
(1) Acidosis:
Metabolic or Respiratory Acidosis
(2) Renal Failure:
Acute Renal Failure
Chronic Renal Failure with Oliguria
(3) Aldosterone Antagonists, i.e., Spironolactone
(4) Adrenogenital Syndrome; 21-Hydroxylase Deficiency
(5) Adrenal Insufficiency, e.g., Patient on Long Term
Corticosteroids and Sudden Discontinuence
(6) Massive Muscle Necrosis
(7) I.V. Therapy, i.e., Especially with K^+ supplements
or to patients with renal disease
(8) Blood Transfusions of Aged Blood
(9) Significant Thrombocytosis
(10) Hyperkalemic Periodic Paralysis
(11) Artefactual Increase:
1) Hemolyzed Serum
2) Repeated Fist Clenching during Venipunture
3) Delayed Separation of Serum from Cells

Acute acidemia usually results in hyperkalemia. Hyperkalemia occurs most
often during the course of acute renal failure.

Hyperkalemia may be drug-induced, such as, distal tubular diuretic, e.g.,
aldosterone antagonists, to a patient with renal failure; potassium supplement
together with an aldosterone antagonist as diuretic.

Acute hyperkalemia may occur in recipients of massive blood transfusions;
it can be avoided by use of blood less than one week old.

The causes of hyperkalemia in a hospital population are given in the next
Table; (Paice, B. et al., Brit. Med. J. 286, 1189-1192, Apr. 9, 1983):

Hyperkalemia in a Hospital Population
Renal Disease, Acute or Chronic
Drug Treatment:
Oral Potassium Supplements
I.V. Potassium (Overenthusiastic)
Aldosterone Antagonists
Redistribution of Potassium
Catabolic States
Severe Acidosis e.g., Diabetic Ketoacidosis

Potassium supplementation in active diabetic ketoacidosis and renal
impairment may lead to prolonged severe hyperkalemia.

PREGNANCY TEST, URINE, ROUTINE
SPECIMEN: First morning voided urine; refrigerate of freeze.
REFERENCE RANGE: Negative; positive in normal pregnant female.
METHOD: Newer agglutination methods use antibody to beta chain of human
chorionic gonadotropin.
INTERPRETATION: The most sensitive urine tests are positive at the first
missed period and most kits will yield positive results two to three weeks after
the first missed period.
      False positive results are obtained with patients who have hematuria,
proteinuria, or opiates.
      The serum pregnancy test, instead of the urine pregnancy test, should be
used for early detection of pregnancy, detection of ectopic pregnancy and when a
false-positive or negative test result is suspected.

PREGNANCY TEST, SERUM
   (see HUMAN CHORIONIC GONADOTROPIN, HCG)

PRENATAL SCREENING PANEL
      A prenatal screening panel is given in the next Table:

Prenatal Screening Panel	
Test	Specimen
Complete Blood Count(CBC)	Lavender(EDTA) Top Tube; Two Unfixed Blood Smears
Blood Group-ABO	Red Top Tube
Rh(D) Type	Red Top Tube
Atypical Antibody Screen, (Coombs, Indirect)	Red Top Tube
VDRL	Red Top Tube

Specimens: Red top tube; lavender(EDTA) top tube; blood smears.
Atypical Antibody Screen: On routine prenatal screening for antibodies, 3
percent of Rh-negative patients and 1.8 percent of Rh-positive patients had
irregular antibodies. However, of the Rh-positive patients who had atypical
antibodies, only 6 percent had irregular antibodies that were considered
potentially clinically significant (anti-M, anti-Kell and anti-E) (Solola, A.
et al., Obstet. Gynecol. 61:25-29, 1983).

PREOPERATIVE PANEL

A preoperative panel is given in the next Table:

Preoperative Panel
Urinalysis
Complete Blood Count(CBC)
Prothrombin Time(PT) and Partial Thromboplastin Time(PTT)
Bleeding Time
Chemistry Tests:
Electrolytes(Na$^+$, K$^-$, Cl$^-$, CO_2 Content)
Glucose
BUN or Creatinine
Serum Glutamic Oxalacetic Transaminase(SGOT)
Lactate Dehydrogenase(LDH)
Bilirubin, Total and Direct
Chest X-Ray
Electrocardiogram(ECG)
Blood Type and Screen

A proposal for <u>preoperative testing</u>, generated to help <u>eliminate unnecessary tests</u>, is given in the next Table; (Blery, C. et al., Effective Health Care 1, 111-114, 1983):

Pre-Operative Testing - A Proposal							
	CBC	Type & Screen	ECG	Chest X-Ray	Blood Glucose	Electrolytes, Creat.& BUN	PT PTT	Platelets Bleeding Time
Surgical Procedures:								
Minor								
Major	X	X						
Age:								
<40								
40-70				X				
>70				X		X		
Associated Condition:								
Cardiovascular Disorder			X	X				
Pulmonary Disorder			X	X				
Malignant Disorder							X	
Hepato-Biliary Disorder							X	
Renal Disorder						X		
Bleeding Disorder							X	X
Diabetes					X	X		
Medications:								
Diuretics						X		
Digitalis						X		
Corticoids					X	X		
Anticoagulants							X	

Chest X-Ray: It was found that routine screening chest x-rays, obtained solely because of hospital admission for schedule surgery, were not <u>warranted in patients under 20 years</u> (Sagel, S.S. et al., N. Engl. J. Med. 291, 1001-1004, 1974). For patients over 40 years, the incidence of <u>unexpected x-ray findings</u> with therapeutic consequences was 0.2% (Thomsen, H.A. et al., Ugerskrift for Laeger 140, 765-768, 1979). The English Royal College of Radiologists concluded that routine preoperative chest x-ray was not clinically useful and recommended a temporary norm of utilization in no more than 12% of patients scheduled for <u>non-acute non-cardiopulmonary surgery</u> (Lancet 2, 83-86, 1979). Routine x-ray is recommended in populations where the prevalence of undiagnosed chest disease is likely to be high, e.g., immigrants.

PRIMIDONE (MYSOLINE) PLUS PHENOBARBITAL

SPECIMEN: Red top tube, separate serum and refrigerate; green (heparin) top tube, separate plasma. Reject if specimen hemolyzed.

REFERENCE RANGE: Therapeutic: 5-12mcg/ml (adults); 7-10mcg/ml (children under 5 years). Toxic: 12 to 15 mcg/ml. Time to Obtain Blood Specimen (Steady State): After 4 to 7 days; Half-Life: Primidone: 3-12 hours, Phenobarbital: 50 to 120 hours (adults), 40 to 70 hours (children); Peak Time: 0.5-9.0 hours (variable).

METHOD: EMIT; GLC; HPLC; SLFIA(Ames); Nephelometry(ICS); Fluorescence Polarization(Abbott)

INTERPRETATION: Primidone is used as an anticonvulsant including grand-mal and psychomotor seizures. Primidone is metabolized in the liver to phenobarbital and phenylethylmalonamide (PEMA). Both have anticonvulsant activity. Dosage guidelines and time to obtain blood specimens are given in the next Table:

Dosage Guidelines and Blood Specimens	
Dosage Guidelines	Blood Specimens
Adults: 500 to 1000mg (5 to 10mg/kg/day)	After 4 to 7 days and just before next dose
Pediatrics: 10 to 15mg/kg/day	

Toxicity: If toxicity is suspected, monitor at least six hours after last dose; primidone serum levels above 12mcg/ml are likely to be associated with serious toxicity.

Phenobarbital: Within 5 to 7 days, phenobarbital, from the metabolism of primidone, is detectable. Monitor blood phenobarbital levels; the optimal range of phenobarbital is 15-30mcg/ml; when both primidone and phenobarbital levels are measured together the phenobarbital level is two to three times the primidone level.

PRIST (TOTAL IgE) (see IgE)

PROCAINAMIDE (PRONESTYL)

SPECIMEN: Red top tube, separate serum and refrigerate; reject if serum left standing on cells for several days or serum frozen on cells.

REFERENCE RANGE: Therapeutic: 4-8mcg/ml for procainamide; 10-30mcg/ml for procainamide plus N-acetyl procainamide (NAPA). Toxic: >30mcg/ml for both procainamide plus N-acetyl procainamide (NAPA); up to 30mcg/ml may be tolerated by some individuals. Half-Life: Procainamide: 3 hours in normal subjects; 9 hours with renal impairment. NAPA: 7 hours in normal subjects; 10 hours to 40 hours with renal impairment. Determine creatinine and BUN before starting therapy. Steady State (Time to obtain serum specimens): Oral: 48 hours; Intravenous: 24 hours after loading dose.

METHOD: EMIT (Two separate assays, one for procainamide, other for NAPA); HPLC and GLC (Differentiates procainamide and NAPA).

INTERPRETATION: Procainamide is used in the treatment of cardiac arrhythmias; it is most often used to treat life-threatening ventricular arrhythmias resistant to lidocaine and as prophylaxis against ventricular arrhythmias following acute myocardial infarction.

The major metabolite of procainamide is N-acetylprocainamide (NAPA) which also possesses anti-arrhythmic activity. The serum concentration of NAPA is genetically determined by the activity of the hepatic enzyme, N-acetyltransferase; different individuals acetylate procainamide at different rates. The marked variations in absorption, metabolism and elimination are reasons to monitor patients receiving procainamide.

Procainamide (Pronestyl)   S. Bakerman

Therapy and Blood Specimens:

Intravenous Therapy	Blood Specimens

Loading Dose:
  17mg/kg in 100ml D5W
  given over one hour
	First blood specimen obtained at end of Loading Dose

Second Infusion:
(Maintenance Dose)
  2.8mg/kg/hour (Add
  total daily dose to
  500ml D5W and infuse
  at 20ml/hour for 24
  hours.
	Second: Two hours after maintenance infusion begins. Note: Loading infusion level should be higher than maintenance infusion level; neither sample should exceed 15ug/ml.
	Third: Six to 12 hours after maintenance infusion starts.
	Fourth: At 24 hours. (This is the time that steady-state is achieved)

Oral Therapy	Blood Specimens
Dosing interval should not exceed 4 hours; a sustained release preparation is given every 6 hours. Oral Dose = maintenance infusion rate (see above) x dosing interval in hours.	Blood specimens should be obtained as follows: First Blood Specimen: 48 hours after starting oral therapy and when dose is administered. Second Blood Specimen: Middle of same dosing interval. Third Blood Specimen: End of same dosing interval. Obtain second and third blood specimens at equally spaced time intervals.

Example:

$$\frac{2.8mg}{kg \times hr} \times 4 \text{ hrs} = \frac{11.2mg}{kg}$$

Renal Disease, Pulmonary Edema and Cardiogenic Shock: About half of the available dose of procainamide appears in the urine. In patients with renal impairment, the dose of procainamide must be adjusted downward.
Dose as follows in patients with renal failure:

Dosage of Procainamide in Renal Failure

Intravenous:
  Loading Dose (Normal): 17mg/kg
    Moderate Renal Failure (BUN 25 to 40mg/dl): 17mg/kg
    Severe Renal Failure (BUN >40mg/dl) or pulmonary edema or cardiogenic
      shock: 14mg/kg
  Maintenance Dose: Normal: 2.8mg/kg/hour
    Moderate Renal Failure (BUN 25 to 40mg/dl): 2mg/kg/hour
    Severe Renal Failure (BUN >40mg/dl) or pulmonary edema or cardiogenic
      shock: 1mg/kg/hour
Oral: Normal: 11.2mg/kg/4 hours
  Moderate Renal Failure: Dosing interval, 4 to 6 hours.
  Severe Renal Failure: Dosing interval, 6 to 12 hours.
Toxicity: Nausea, anorexia, lupus-like reaction, fever, urticaria, cardiac conduction disturbances; agranulocytosis may occur in association with the sustained release preparation.

    Severe neutropenia (granulocytopenia) is associated with sustained-release procaiamide; granulocytopenia generally occurs in the first 3 months and it usually develops rapidly. Therefore, leukocyte determinations should be done at 2-week intervals. Discontinuation of therapy has generally resulted in a prompt return to normal leukocyte counts (Ellrodt, A.G. et al., Ann. Intern. Med. 100, 197-201, 1984; Gabrielson, R.M., ibid 100, 766, 1984).

PROGESTERONE, SERUM

SPECIMEN: Red top tube, separate serum and freeze.
REFERENCE RANGE:

Female:	Follicular Phase:	<150 ng/dl
	Luteal Phase:	>300 ng/dl
	Midluteal:	may exceed 2000 ng/dl

Pregnancy:	First Trimester: 1500-5000 ng/dl
	Third Trimester: 8000-20,000 ng/dl

	Postmenopausal:	<50 ng/dl
Males:	_____:	<50 ng/dl

To convert conventional units in ng/dl to international units in nmol/liter, multiply conventional units by 0.03180.

METHOD: RIA

INTERPRETATION: Assay of serum progesterone is most often done to answer the question, "Has the patient ovulated?" It is the best single test for detecting infertility in females. The change in serum progesterone during the menstrual cycle is given in the next Figure:

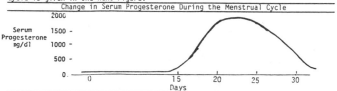

Change in Serum Progesterone During the Menstrual Cycle

In the normal menstrual cycle, low levels of progesterone (of adrenal origin) are present in the blood during the follicular phase.

Ovulation occurs as a result of the midcycle surge of luteinizing hormone (LH). Follicle cells undergo luteinization to form a corpus luteum which secretes progesterone; progesterone prepares the endometrium for implantation and development of the embryo and prepares the breast for lactation. Progesterone peaks in the mid-luteal phase.

To assess the formation and functional state of the corpus luteum, obtain blood specimens as follows: Mid-follicular phase: day 5 to 10; luteal phase: day 19 to 23 for normal menstrual cycle.

To pinpoint day of ovulation, obtain blood specimens on alternate days as follows: day 10, 12, 14, 16 of the menstrual cycle; for the most reliable results, all these samples should be assayed at the same time.

To diagnose premenstrual syndrome, obtain blood specimens for progesterone and estradiol measurements; progesterone is significantly decreased between 9 and 5 days prior to menstruation; the mid-luteal estradiol/progesterol ratio is increased.

Progesterone assays are useful in evaluating the corpus luteum during the first trimester of pregnancy; some cases of spontaneous abortion have been associated with low progesterone reflecting luteal insufficiency (Hensleigh, P.A. and Fainstat, T., Fertility and Sterility 32, 396, 1979). Serial samples of progesterone should be obtained early in pregnancy and continuing through the twelfth week; at that time, the placenta produces progesterone.

PROLACTIN

SPECIMEN: Red top tube, separate serum and freeze; keep frozen during delivery to reference laboratory.

REFERENCE RANGE: (Nichols Institute, San Pedro, Cal.)

Age	Value; Mean $\pm$S.D. (ng/ml)
4 Hours after Birth***	190$\pm$18
1 Day**	278$\pm$118
3 Days**	204$\pm$66
4 Days**	114$\pm$61
5 Days**	82$\pm$36
4 Weeks**	87$\pm$58
6 Weeks**	17$\pm$6
1 Year**	10
2-12 Years*	7$\pm$1.2
13-16 Years**	5
Adults	20 or less

*Foley, T.P. et al., J. Clin. Invest. 51, 2143-2150 (1972).
**Gayda, H. and Frilsen, H.G., Pediat. Res. 7, 534-540 (1973)
***Sack, J. et al., J. Pediat. 89, 298-300 (1976).

Pregnancy: 1st trimester: <80ng/ml
          2nd trimester: <160ng/ml
          3rd trimester: <400ng/ml.

To convert conventional units in ng/ml to international units in mcg/liter, multiply conventional units by 1.00.

METHOD: RIA

INTERPRETATION: The causes of inappropriate galactorrhea are given in the next Table; (Kleinberg, D.L. et al., N. Engl. J. Med. 296, 589-600, 1977):

Causes of Inappropriate Galactorrhea	
Cause	Number
Pituitary Tumors	48/235
Idiopathic:	
With Menses	76/235
With Amenorrhea	20/235
Chiari-Frommel	18/235
Tranquilizing Drugs	13/235
Other Drugs	2/235
Post-Oral Contraceptives	12/235
Hypothyroidism	10/235
Empty Sella	5/235
Miscellaneous	30/235

Drugs that may cause galactorrhea are as follows:
    Various drugs (e.g. chlorpromazine, phenothiazine, reserpine, cimetidine)
    Oral contraceptives
    Estrogens
    Thyrotropin-releasing hormone (TRH)
    Insulin-induced hypoglycemia
    Arginine

Chiari-Frommel Syndrome: Persistent galactorrhea and amenorrhea beginning after childbirth.

Empty Sella: Patients with empty sella syndrome have an enlarged sella turcica. The empty sella syndrome arises as a result of a defect in the dura which allows for prolapse of subarachnoid membranes into the sella fossa. Empty sella syndrome is infrequently associated with hormonal loss.

A serum level of prolactin greater than 200ng/ml strongly suggests pituitary tumor; a serum level above 300ng/ml is practically pathognomonic of pituitary tumor.

Reference: Edwards, C.R.W. and Fink, C.M., Brit. Med. J. 283, 1561-1562, Dec. 12, 1981.

Prolactin Response to Thyrotropin-Releasing Hormone(TRH):

In order to distinguish causes of galactorrhea, stimulation is done using thyrotropin-releasing hormone(TRH) and is injected I.V. (500 mcg) and serum specimens for prolactin are obtained at times 0, 15 and 30 minutes; the prolactin levels in patients with pituitary tumors show no increment, while prolactin levels in patients with other disorders increase 6-to 20-fold (Kleinberg, D.L. et al., N. Engl. J. Med. 296, 589-600, 1977).

The most common tumor of the pituitary is prolactin secreting adenoma; pituitary adenoma cell type in patients with pituitary tumors is given in the next Table; (Kovacs et al., Pathology Annual 12, 341, 1977):

Adenoma Type in Pituitary Tumors	
Tumor	Incidence(%)
Prolactin Cell	32
Undifferentiated Cell	23
Growth Hormone Cell	21
Corticotroph Cell	13
Mixed Growth Hormone and Prolactin Cell	6
Acidophil Stem Cell	3.5
Gonadotroph Cell	1
Thyrotroph Cell	0.5

Pituitary tumors most often secrete prolactin and secondary amenorrhea with associated infertility is often caused by hyperprolactinemia; a serum prolactin should be measured in unexplained amenorrhea.

Pergolide is a potent dopamine agonist; it has a long duration of action and is effective for treatment for hyperprolactinemia (Perryman, R.L. et al., J. Clin. Endocrinol. Metab. 53, 772-778, 1981).

Protein Electrophoresis,
CSF

PROTEIN ELECTROPHORESIS, CSF

SPECIMEN: 2ml CSF

REFERENCE RANGE: Total CSF protein is 15 to 45mg/dl; the normal albumin/globulin ratio is 2:1.

METHOD: Normal CSF contains only 0.5% to 1% of the protein concentration of serum (15mg/dl to 45mg/dl). Therefore, the CSF must be concentrated one hundred times to two hundred times prior to electrophoresis.

INTERPRETATION: The electrophoretic pattern of CSF resembles that of serum except for the pre-albumin fraction and lower proportion of gamma globulin. Interpretation of the pattern is dependent on the relative proportions of the protein fractions rather than their actual concentrations.

The most important clinical application of CSF protein electrophoresis is in the work-up of patients with possible multiple sclerosis. In multiple sclerosis, there is an increase in total protein concentration, primarily due to IgG; in addition, there is an increased cell count and unusual patterns on electrophoresis of the CSF proteins.

Increases in gamma globulin have been associated with disease. The CSF protein electrophoretic pattern of patients with active multiple sclerosis and other diseases associated with increased CSF gamma globulin fraction (normal CSF gamma globulins, <13%; increased gamma globulins, >15%) is shown in the next Figure:

Cerebrospinal Fluid Protein Electrophoretic Pattern in Active Multiple Sclerosis and other Conditions

Diseases Associated With Increased CSF Gamma Globulin-
Active Multiple Sclerosis
Inflammatory Diseases of the CNS
  Encephalitis
  Meningitis (particularly T.B.)
  Neurosyphilis
  Arachnoiditis
Benign and Malignant Intracranial Tumors

In multiple sclerosis, CSF gamma globulin is found to be elevated in about 75% of patients with an established clinical diagnosis of multiple sclerosis.

Elevation of cerebrospinal fluid gamma globulin may also occur when the serum gamma globulin is elevated as in multiple myeloma, collagen disease, chronic liver disease and lymphomas. It is, therefore, recommended that a serum protein electrophoresis be done at about the same time as the cerebrospinal fluid protein.

Oligoclonal Banding: Normal CSF immunoglobulin migrates as a diffuse band; abnormal immunoglobulins migrate as discrete sharp bands or oligoclonal bands. This is the pattern observed in multiple sclerosis; this pattern of discrete bands within the gamma globulins supports the hypothesis that these globulins arise from a few clones of cells. However, oligoclonal bands are not unique to multiple sclerosis and are observed in the conditions listed in the next Table:

Oligoclonal Bands on Electrophoresis

Multiple Sclerosis
Inflammatory Polyneuropathy
Neurosyphilis
Cryptococcal Meningitis
Chronic Rubella Panencephalitis
Subacute Sclerosing Panencephalitis
    (Persistent Rubella Viral Infection)
Burkitt's Lymphoma

Oligoclonal CSF IgG was observed in 12 of 13 patients with Burkitt's lymphoma (Wallen, W.C. et al., Arch. Neurol. 40, 11-13, 1983).

Oligoclonal bands are observed using agarose electrophoresis; 3ml of CSF is concentrated to 0.05ml. A specimen of serum from the patient is electrophoresed at the same time. The results of electrophoresis are illustrated in the next Figure; (Johnson, K.P. and Hosein, Zobeeda, Laboratory Management, pgs. 36-40, May, 1981):

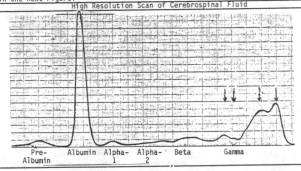

Agarose Electrophoresis of Serum and CSF Specimens

An abnormal result is the finding of two or more bands in the CSF that are not present in the serum specimen. The sensitivity of CSF oligoclonal bands in the diagnosis of multiple sclerosis is 79-94% (Markowitz, H. and Kokmen, E., Mayo Clin. Proc. 58, 273-274, Apr. 1983). A scan of cerebrospinal fluid, obtained from a patient with multiple sclerosis, using high resolution, is shown in the next Figure:

High Resolution Scan of Cerebrospinal Fluid

PROTEIN ELECTROPHORESIS, SERUM

SPECIMEN: Red top tube; separate serum; <u>plasma</u> should <u>not</u> be used because
fibrinogen separates as an extra band; hemolyzed specimens should not be used
since hemoglobin separates as an extra band.

REFERENCE RANGE:

	gm/dl
Total Protein	6.2 to 8.3
Albumin	3.6 to 5.2
Alpha-1-Globulin	0.15 to 0.40
Alpha-2-Globulin	0.50 to 1.00
Beta Globulin	0.60 to 1.20
Gamma Globulin	0.60 to 1.60

METHOD: Electrophoresis at pH 8.6

INTERPRETATION: Although in many instances the analysis of albumin and total
protein and the calculation of the globulins is sufficient, it is sometimes
important to have a quantitative analysis of the globulins. This is done by
protein electrophoresis. The patterns are diagnostic of some diseases and
suggestive of others. Specific alterations in patterns that are diagnostic
include analbuminemia and bisalbuminemia. The diseases that cause depression of
albumin levels have been given (see Albumin). Protein electrophoresis is helpful in the work-up of patients for liver disease, myeloma, macroglobulinemia,
hypoglobulinemia, collagen diseases, nephrotic syndrome and nutritional status.
Electrophoresis separates globulins into alpha-1, alpha-2, beta and gamma
components. Certain disease processes cause change in the concentration of
components as shown in the next Table:

Elevated Values of Globulins			
alpha-1	alpha-2	beta	gamma
Acute and chronic infections	Biliary cirrhosis	Biliary cirrhosis	Chronic infections
	Obstructive jaundice	Obstructive jaundice	Hepatic diseases
Febrile reactions	Nephrosis	Multiple Myeloma (Occasionally)	Autoimmune diseases
	Multiple Myeloma (rarely)		Collagen diseases
	Ulcerative colitis		Multiple myeloma
			Waldenström's macroglobulin
			Leukemias and some cancers

Depressed Values of Globulins			
alpha-1	alpha-2	beta	gamma
Nephrosis	Acute Hemolytic Anemia	Nephrosis	Agammaglobulinemia
Alpha-1 Anti-Trypsin Deficiency			Hypogammaglobulinemia
			Nephrotic Syndrome

The major components of the alpha-1, alpha-2 and beta globulins are given
in the next Figure:

Protein Electrophoretic Pattern-Major Components of Bands

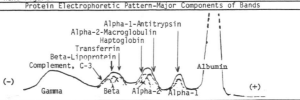

A split band in the beta region occurs with very fresh serum specimens,
excessively long electrophoresis runs, increase in beta lipoproteins or the
presence of a paraprotein.

A summary of protein electropohoretic patterns is shown in the next Figure:

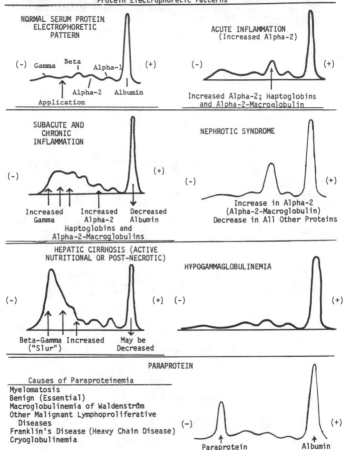

Protein Electrophoretic Patterns

NORMAL SERUM PROTEIN
ELECTROPHORETIC
PATTERN

(-) Gamma Beta   Alpha-1   (+)

Alpha-2   Albumin

Application

ACUTE INFLAMMATION
(Increased Alpha-2)

(-)   (+)

Increased Alpha-2; Haptoglobins
and Alpha-2-Macroglobulin

SUBACUTE AND
CHRONIC
INFLAMMATION

(-)   (+)

Increased   Increased   Decreased
Gamma   Alpha-2   Albumin
            Haptoglobins and
            Alpha-2-Macroglobulins

NEPHROTIC SYNDROME

(-)   (+)

Increase in Alpha-2
(Alpha-2-Macroglobulin)
Decrease in All Other Proteins

HEPATIC CIRRHOSIS (ACTIVE
NUTRITIONAL OR POST-NECROTIC)

(-)   (+)

Beta-Gamma   Increased   May be
("Slur")   Decreased

HYPOGAMMAGLOBULINEMIA

(-)   (+)

PARAPROTEIN

Causes of Paraproteinemia

Myelomatosis
Benign (Essential)
Macroglobulinemia of Waldenström
Other Malignant Lymphoproliferative
  Diseases
Franklin's Disease (Heavy Chain Disease)
Cryoglobulinemia

(-)   (+)

Paraprotein   Albumin

S. Bakerman

Protein Electrophoresis,
Serum

High Resolution Pattern: A serum protein electrophoretic pattern obtained using agarose gel is shown in the next Figure:

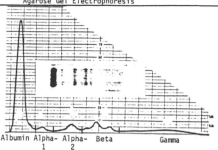

Agarose Gel Electrophoresis

Albumin Alpha- Alpha- Beta Gamma
1 2

The advantages of using high resolution are as follows: The bands are better separated so that they are quantitated more accurately; this is especially important for quantitation of the alpha-1-band. About 90 percent of this band is alpha-1-antitrypsin and deficiency is reflected in decrease in this band. Note that the beta band is resolved into three bands, transferrin, beta-lipoprotein and complement. "Spikes" can be detected more readily using high resolution.

Protein, Total,
Serum

PROTEIN , TOTAL, SERUM
SPECIMEN: Red top tube, separate serum; refrigerate at 4°C.
REFERENCE RANGE: Adult: 6.0-8.5g/dl; Premature: 4.3-7.6g/dl; Newborn: 4.6-7.6g/dl; Child: 6.2-8.0g/dl. To convert conventional units in g/dl to international units in g/liter, multiply conventional units by 10.0.
METHOD: Biuret: Complex of cupric ions with protein or peptide bond H_2O; Reaction takes place at alkaline pH; Coomassie Blue; refractometry -N-C-
INTERPRETATION: The total serum protein is the sum of the circulating proteins in the serum. The significance of the level of total protein is difficult to interpret without knowledge of the level of the individual fractions, e.g., albumin and globulins. The quantitative distribution of these is obtained by protein electrophoresis. Total protein is increased in the serum in conditions listed in the next Table:

Causes of Increased Total Serum Protein

Chronic Infection
Liver Diseases
Collagen - Autoimmune Diseases; e.g., Vascular-Systemic Lupus
        Erythematosus; Scleroderma; Rheumatoid Arthritis
Hypersensitivity States
Sarcoidosis
Cryoglobulinemia
Monoclonal Gammopathies: Multiple Myeloma; Macroglobinemia
Franklin's Disease
Dehydration
Hemolysis

Total protein is decreased in the serum in conditions listed in the next Table:

Causes of Decreased Total Serum Protein

Decreased Albumin
    Subacute and Chronic Debilitating Diseases; Liver Disease; Renal
        Disease; Nephrotic Syndrome; Malabsorption and Malnutrition
        Gastrointestinal Loss; Third Degree Burns; Exfoliative Dermatitis;
        Exfoliative Dermatitis; Dilution by I.V. Fluids
Hypo- and Agammaglobulinemia
Pregnancy

PROTEIN, TOTAL URINE

SPECIMEN: Twenty-four hour urine is unnecessary. Random urine; this single specimen is used to obtain quantitative protein value.

REFERENCE RANGE: <0.2 mg urinary protein/mg urinary creatinine. 25-150 mg/24 hours.

METHOD: Urinary Protein: Trichloroacetic acid precipitation; Biuret method; turbidimetric method; nephelometric method.

INTERPRETATION: The usual protein for quantitating urinary protein is to obtain a 24 hour urine; however, one of the most difficult tasks, especially in a hospital setting, is accurate collection of a 24 hour urine.

An excellent correlation has been established between the protein content of a 24-hour urine collection and the protein/creatinine ratio in a single urine sample (Ginsberg, J.M. et al, N. Engl. J. Med. 309,1543-1546, 1983). A ratio of urinary protein(mg)/urinary creatinine(mg), less than 0.2, is normal; a urinary protein(mg)/urinary creatinine(mg), greater than 3.5, is compatible with levels seen in nephrotic range. To convert the observed ratio to grams per 24 hours use the following formulas:

Males: $\dfrac{(140\text{-Age in Yrs})(\text{Wt in Kg})}{5000} \times \dfrac{(\text{Urinary Prot.(mg)})}{(\text{Urinary Creat.(mg)})} = \text{approx.} \dfrac{\text{grams}}{24 \text{ Hours}}$

Females: $\dfrac{(140\text{-Age in Yrs})(\text{Wt in Kg})}{5000} \times 0.85 \times \dfrac{(\text{Urinary Prot.(mg)})}{(\text{Urinary Creat.(mg)})} = \text{approx.} \dfrac{\text{grams}}{24 \text{ Hours}}$

The causes of increased protein in urine are given in the next Table:

Causes of Increased Protein in Urine	
Level of Protein	Causes
Heavy (> 4g/Day)	Nephrotic Syndrome
	Acute and Chronic Glomerulonephritis
	Lupus Nephritis
	Amyloid Disease
	Severe Venous Congestion of Kidney
Moderate (0.5 to 4 g/Day)	Diseases Listed Above plus
	Nephrosclerosis
	Pyelonephritis with Hypertension
	Multiple Myeloma
	Diabetic Nephropathy
	Pre-Eclampsia of Pregnancy
	Toxic Nephropathies Including Radiation
Minimal (< 0.5 g/Day)	Diseases Listed Above plus
	Chronic Pyelonephritis
	Polycystic Kidney Disease
	Renal Tubular Diseases
	"Benign" Postural Proteinurias

Differentiation; Glomerular Proteinuria versus Renal Tubular Proteinuria: The proteinuria of patients with renal tubular disease is qualitatively different from the proteinuria of glomerular disease. Both lysozyme (muramidase) and beta-2-microglobulin are increased in the urine in renal tubular disease and in renal parenchymal disease (Sherman, R.L. et al., Arch. Intern. Med. 143,1183-1185, June, 1983). Measurement of beta-2-microglobulin is a more sensitive marker for renal tubular disease than lysozyme (Barratt, M., Brit. Med. J. 287,1489-1490, Nov. 19, 1983). Lysozyme is increased 100-fold in the Fanconi syndrome (Barratt, T.M. and Crawford, R. Clin. Sci. 39,457-465, 1970).

Trace Proteinuria: Trace proteinuria is associated with significant increases in both cardiovascular and overall mortality; trace proteinuria is much more common in persons with diabetes and those with hypertension (Framingham Population Study, Hosp. Pract. 17(2), 32, 1982). Microalbuminuria in patients with maturity-onset (Type II) diabetes is predictive of clinical proteinuria and increased mortality (Mogensen, C.E., N. Engl. Med. 310, 356-360, 1984).

PROTHROMBIN TIME (PT)

SPECIMEN: Blue (citrate) top tube filled to 4.5ml with blood; pediatric blue top tube filled to 2.7ml with blood; tubes must be filled to capacity. Avoid contamination with tissue thromboplastin as follows: If multiple tests are being drawn, draw coagulation studies last; if only a PT is being drawn, draw 1 to 2ml of blood into another vacutainer, discard and then collect blood for PT.

Assay of PT is usually done on plasma. If test cannot be assayed immediately, centrifuge and separate plasma from cells and freeze plasma.

The specimen is rejected if the tube is not full, specimen hemolyzed, specimen clotted or specimen received more than 2 hours after collection.

REFERENCE RANGE: 10 to 13 seconds

METHOD: Tissue thromboplastin (complete thromboplastin) from brain; calcium chloride

INTERPRETATION: (Triplett, D.A., Anticoagulant Therapy: Monitoring Techniques, Laboratory Management 20, 31-42, 1982) The PT test is a measure of the extrinsic coagulation system and measures factors X, VII, V, II and I; this test is often used to monitor coumarin or warfarin effect. The causes of prolonged PT are given in the next Table:

Causes of Prolonged Prothrombin Time(PT)
Coumarin or Warfarin Therapy
Heparin Therapy
Vitamin K Deficiency
Liver Disease
Disseminated Intravascular Coagulation (DIC)
Factor Deficiency:
Factor X (Stuart Factor)
Factor VII (Proconvertin)
Factor V (Proaccelerin)
Factor II (Prothrombin)
Factor I (Fibrinogen)

Warfarin: The optimum prothrombin range when warfarin is used is 1.5 to 1.8 times the normal prothrombin time. Warfarin acts by decreasing the synthesis of the vitamin K-dependent clotting factors, prothrombin, factor VII, factor IX, and factor X. After administration of an oral anticoagulant, the level of factor VII is the first to decrease followed by the level of factor IX, followed by factor X and then factor II. The following conditions cause an increased sensitivity to warfarin: diarrhea, abnormalities of the small intestine, such as sprue.

Therapy with warfarin is begun with 10 to 15mg given once daily for 2 consecutive days, with a prothrombin time determination made before the third daily dose. Daily determinations of the PT are then made until the PT is less than two-times normal, and the maintenance dose is determined. The dose of warfarin given for daily maintenance is generally in the range of 2 to 15mg.

Drug interactions with oral anticoagulants are given in the next Table; (The Medical Letter, 26, 11-14, 1984):

Drug Interactions with Oral Anticoagulants		
Interacting Drugs	Adverse Effect (Probable Mechanism)	Comments and Recommendations
Aminoglutethamide	Decreased Anticoagulant Effect (Increased Metabolism)	Monitor Prothrombin Time(PT)
Erythromycins	Increased Anticoagulant Effect (Possible Decreased Metabolism)	Monitor Prothrombin Time(PT)
Moxalactam	Possible Increased Anticoagulant Effect (Mechanism not Established)	Avoid Concurent Use
Trimethoprim-Sulfamethoxazole	Increased Anticoagulant Effect (Decreased Metabolism)	Monitor Prothrombin (PT)

Heparin Therapy: See PTT

Abnormal Prothrombin: There is an abnormal prothrombin which circulates in the blood of some patients with primary hepato-cellular carcinoma; this abnormal prothrombin may be useful in the laboratory diagnosis of primary hepato-cellular carcinoma (Liebman, H.A. et al., N. Engl. J. Med. 310, 1427-1434, May 31, 1984).

Protoporphyrin, Free

PROTOPORPHYRIN, FREE (see ZINC PROTOPORPHYRIN, BLOOD)

PSITTACOSIS ANTIBODIES (see CHLAMYDIA ANTIBODIES)

PULMONARY CAPILLARY WEDGE PRESSURE
(SWAN-GANZ CATHETER)

REFERENCE RANGE: <12 mm Hg

METHOD: A Swan-Ganz catheter is introduced usually in a vein in the antecubital area and then the tip is guided into a sufficiently large vein, e.g., the subclavian vein. Then, a balloon is inflated in the large vein; blood flow propels the balloon and catheter into the right atrium, right ventricle, pulmonary artery and pulmonary artery branch. Once the balloon is in position, the balloon is deflated and subsequently reinflated when pulmonary capillary wedge pressure is to be recorded. The catheter has three or four lumens: a lumen that terminates at the catheter tip and is used to measure pulmonary artery and pulmonary capillary wedge pressure and to sample blood; a lumen terminates 30 cm from the tip of the catheter and is used to measure right atrial or central venous pressure. Cardiac output can also be reliable determined by the thermodilution method. For more details about the procedure see McIntyre, K.M. and Lewis, A.J., eds., Textbook of Advanced Cardiac Life Support, American Heart Association, pgs. 183-194, 1983.

INTERPRETATION: Pulmonary capillary wedge pressure is a useful and reliable indicator of left ventricular dynamics and pulmonary congestion. It is especially important to assess left ventricular dynamics following acute myocardial infarction. Pulmonary capillary wedge pressure, chest X-ray findings and pulmonary congestion are closely correlated, while the older method of measuring central venous pressure is not well correlated with pulmonary congestion.

Hemodynamics in acute myocardial infarction are given in the next Table; (Rose, John, Personal Communication):

Pulmonary Capillary Wedge Pressure in Acute Myocardinal Infarction			
Clinical Classification	Mean PC Wedge Pressure (mmHg)	Cardiac Index	Mortality
No Pulmonary Congestion or Peripheral Hypoperfusion	<18	>2.2	3%
Pulmonary Congestion without Hypoperfusion	>18	>2.2	9%
Peripheral Hypoperfusion Without Pulmonary Congestion	<18	<2.2	23%
Both Peripheral Hypoperfusion and Pulmonary Congestion	>18	<2.2	51%

(Forrester, J.S. et al, N. Engl. J. Med. 295, 1356, 1976)

Cardiac index is the cardiac output per square meter of body surface. Under basal conditions, it is equal to 2.5-4.2 liter/min/meter^2.

In the majority of patients who recover from acute myocardial infarction, the initially elevated pulmonary capillary wedge pressure returned toward normal over the first few days after the infarction.

The types of pulmonary edema are discriminated on the basis of pulmonary capillary wedge pressure as illustrated in the next Figure; (McIntyre, K.M. and Lewis, A.J., eds. Textbook of Advanced Cardiac Life Support, American Heart Association, pg. 237, 1983):

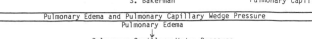

Pulmonary Edema and Pulmonary Capillary Wedge Pressure

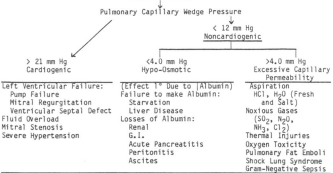

> 21 mm Hg Cardiogenic	<4.0 mm Hg Hypo-Osmotic	>4.0 mm Hg Excessive Capillary Permeability

Left Ventricular Failure:
  Pump Failure
  Mitral Regurgitation
  Ventricular Septal Defect
Fluid Overload
Mitral Stenosis
Severe Hypertension
	(Effect 1° Due to |Albumin)
	Failure to make Albumin:
	  Starvation
	  Liver Disease
	Losses of Albumin:
	  Renal
	  G.I.
	  Acute Pancreatitis
	  Peritonitis
	  Ascites
		Aspiration
		  HCl, H_2O (Fresh
		   and Salt)
		Noxious Gases
		  (SO_2, N_2O,
		  NH_3, Cl_2)
		Thermal Injuries
		Oxygen Toxicity
		Pulmonary Fat Emboli
		Shock Lung Syndrome
		Gram-Negative Sepsis

Cardiogenic pulmonary edema can occur with pulmonary capillary wedge pressure < 21 mmHg if the colloid osmotic wedge pressure is less than 25 mmHg.

Osmotic pressure is reduced when plasma proteins, particularly albumin levels are reduced.

PULMONARY EMBOLI PANEL

Laboratory tests in patients with pulmonary emboli are given in the next Table; (Hayes, S.P. and Bone, R.C., Med. Clin. N. Am. <u>67</u>:, No. 6, 1179-1191, Nov. 1983):

Laboratory Tests in Patients with Pulmonary Emboli
Definitive Diagnosis: Pulmonary Angiograms; Lung Scan(not truly definitive)
Required: Arterial Blood Gases; Chest X-Ray; Electrocardiogram

Rarely Helpful: Serum Bilirubin; Lactate Dehydrogenase(LDH); Serum Glutamic
    Oxalacetic Transaminase(SGOT); Leukocyte Count (Leukocytosis);
    Fibrin Split Products (Elevated)

Definitive Diagnosis:

Pulmonary Angiography: A bolus of radiopaque contrast medium is injected into a peripheral vessel. Serial films are obtained in an attempt to detect a filling defect. This is the "gold standard"!

Lung Scan: Technetium-99 albumin microspheres or technetium-99 macroaggregated albumin are injected intravenously. Particles become trapped in those areas of the pulmonary bed that are adequately perfused; regions of obstructed blood flow due to emboli are seen as nonradioactive defects. False positive results may occur, even when ventilation scans are added. A negative lung scan is usually reliable in excluding embolism (Rose, John, Personal Communication).

The mortality rate associated with pulmonary angiography is 0.2 percent; deaths are associated with severe pulmonary hypertension.

Required Studies:

Arterial Blood Gases: PO_2 almost always below 80 mm Hg; patients usually develop a respiratory alkalosis.

Chest X-Ray: May be normal even with massive emboli; pleural effusions, etc.

Electrocardiogram: Abnormal in 80 percent of patients with massive emboli.

Treatment of Pulmonary Emboli: Medical therapy for pulmonary emboli consists of the anticoagulants, heparin and warfarin, and thrombolytic agents, streptokinase or urokinase.

A heparin infusion is used during the acute phase and warfarin compounds for as long as 4 to 6 months. Prior to initiation of thrombolytic therapy, heparin should be discontinued and the thrombin time or activated partial thromboplastin time should be less than twice the normal control.

PULMONARY FUNCTION TESTS

Tests that may be used to evaluate pulmonary function airway resistance using a simple office spirometer are given in the next Table:

Evaluation of Pulmonary Function
Forced Expiratory Volume in 1 Second(FEV-1)
Total Forced Expiratory Volume (Vital Capacity)(FVC)
Ratio: (FEV-1) x 100/FVC

Lung Volume Subdivisions: Lung volume subdivisions are illustrated in the next Figure:

Lung Volume Subdivisions

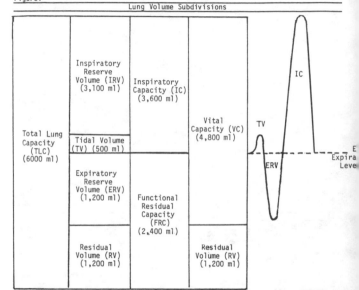

Spirometry: A hand-held peak flow meter, convenient for office use, is available, (Hospital Medicine, pg. 178, February, 1984).
Total Lung Capacity(TLC): The maximum amount of air the lungs can hold; it is divided into four primary volumes:
(1) Tidal Volume(TV): The amount of air moved into or out of the lungs; resting tidal volume is the normal volume of air inhaled or exhaled in one breath.
(2) Inspiratory Reserve Volume(IRV): The amount of air that can be forcefully inspired after a normal tidal volume inhalation.
(3) Expiratory Reserve Volume(ERV): The amount of air that can be forcefully exhaled after a normal tidal volume exhalation.
(4) Residual Volume(RV): The amount of air left in the lungs after a forced exhalation.

FORCED EXPIRATORY VOLUME IN 1 SECOND(FEV-1)
SPECIMEN: After maximum inspiration, FEV-1 is the volume of a single, maximally fast expiration in the first second.
REFERENCE RANGE: Normally between 3 and 5 liters. FEV-1 is expressed in one of two ways: (1) as a percentage of the expected FEV-1 for a normal individual of the same age, height and sex; or (2) as a percentage of the individual's own total forced expiratory volume (vital capacity); that is, (FEV-1) x 100 /FVC; this value is usually greater than 75 percent.
METHOD: Expiration into a spirometer.

INTERPRETATION: FEV-1 is the most important clinical tool for assessing the
severity of airway obstructive disease. The FEV-1 is a measure of flow early in
expiration and is somewhat dependent on effort. Forced expiratory volume in 1
second(FEV-1) is decreased in the presence of obstructive disease and
restrictive disease. Typical patterns are shown in the next Figure:

Patterns of FEV-1 in Normal, Obstructive and Restrictive Disease

Normal   Obstructive   Restrictive

Maximum
Inspiration

Liters
of
Expired
Air

FEV = 4.0	FEV = 1.3	FEV = 2.8
FVC = 5.0	FVC = 3.1	FVC = 3.1
% = 80	% = 42	% = 90

In obstructive airway disease, FEV-1 and FVC are diminished and the ratio
of FEV-1 to FVC is diminished. In restrictive airway disease, FEV-1 is somewhat
diminished, FVC is diminished but the ratio of FEV-1 to FVC is normal or
supranormal.

RATIO: FORCED EXPIRATORY VOLUME IN 1 SECOND(FEV-1): TOTAL FORCED VITAL
CAPACITY(FVC)
SPECIMEN: The patient takes the deepest inspiration possible; this is done
with the respirometer attached. The inhaled air will not be measured. At the
peak of inspiration, the patient is told to exhale as fast and completely as
possible.
REFERENCE RANGE: Normally, FEV-1 is between 3 and 5 liters and FVC is between
4 and 6 liters; the ratio, (FEV-1) x 100/FVC is usually greater than 75 percent.
METHOD: Expiration into a spirometer.
INTERPRETATION: Forced vital capacity reflects ventilatory reserve. See
previous Figures.
    The degree of risk in obstructive lung disease, as calculated from the
ratio, (FEV-1) x 100/FVC, is given in the next Table:

Degree of Risk in Obstructive Lung Disease	
Degree of Risk	(FEV-1) x 100/FVC
Normal..................	>75
Mild....................	60-75
Moderate...............	45-60
Severe.................	35-45
Extreme................	<35

Syndromes of Obstructive Lung Disease: Syndromes of obstructive lung disease are
listed in the next Table:

Syndromes of Obstructive Lung Disease
Asthma
Simple Chronic Bronchitis
Chronic Obstructive Bronchitis
Emphysema

Asthma: Asthma is characterized by recurrent episodes of generalized airway
obstruction that abate spontaneously or with treatment. The hallmark is airway
obstruction; symptoms result from the impedance of the movement of air into or
out of the lungs. The condition is reversible.
    The obstruction is due to smooth muscle contriction, mucosal edema, and the
secretion and/or lack of clearance of mucus.
    Classic childhood asthma tends to remit during adolescence; about 25
percent of childhood asthmatics continue to have symptoms as adults. If the
disease begins after 35, only a minority will show complete remission.

Syndromes of Restrictive Lung Disease: Syndromes of restrictive lung disease are listed in the next Table:

Syndromes of Restrictive Lung Disease
Chest Wall or Space Occupying Processes:
Kyphoscoliosis
Post-Surgery
Paralysis
Obesity
Ascites
Pleuritis
Pleural Effusion
Interstitial and Infiltrative Disorders of the Lung:
Sarcoidosis
Environmental Diseases, e.g., Asbestosis, Silicosis, Coal, Graphite, Berylliosis, etc.
Interstitial Pneumonias
Connective Tissue Disorders, e.g., Scleroderma, Lupus, Rheumatoid Lung
Pulmonary Vascular Disease, e.g., Multiple Emboli and Idiopathic Hypertension

QUINIDINE

SPECIMEN: Red top tube, separate serum; or lavender (EDTA) top tube, separate plasma.

REFERENCE RANGE: Therapeutic: 2.0-5.0mcg/ml; Toxic: >5.0mcg/ml
Half Life: 6 hours; Steady State: 24 to 42 hours (4 to 7 half-lives)
METHOD: EMIT, RIA, Fluorometric, GLC

INTERPRETATION: Quinidine depresses the excitability of cardiac muscle, slows the heart rate and is used as a prophylaxis and treatment of atrial and ventricular arrhythmias.

Specimens: The first serum specimen should be obtained when first initiating quinidine therapy after 24 to 42 hours (4 to 7 half-lives) following initial dose and just prior to the next dose (trough). Trough concentration should be above the minimum effective concentration to maintain the desired therapeutic effect (2.0mcg/ml).

Different preparations peak at different times; quinidine sulfate peaks in 1.5 hours; quinidine gluconate peaks in 4-5 hours.

If it is necessary to determine the half-life of quinidine, collect at least 3 timed specimens 30-60 minutes apart, and plot log concentration versus time.

Conditions which alter the half-life of quinidine are listed in the next Table:

Conditions Which Alter Half-Life of Quinidine	
Decrease	Increase
Phenobarbital	End Stage Liver Disease
Phenytoin	Severe Heart Failure

Toxicity: Toxicity may occur if serum levels exceed 5mcg/ml. The common toxic effects of quinidine are listed in the next Table:

Common Toxic Effects of Quinidine
Gastrointestinal Disturbances:
Nausea, Vomiting, Anorexia, Diarrhea
Cardiovascular:
Slow Intraventricular Conduction
Bradycardia, Hypotension, Syncope
Cinchonism:
"Tinnitus, Light-Headedness, Giddiness" etc.
Eight Cranial Nerve Damage

The major route of elimination is hepatic; renal failure does not generally lead to toxicity on the usual doses.

RAJI CELL ASSAY

SPECIMEN: Red top vacutainer; allow the blood to clot at 4°C; separate the
serum and freeze it immediately. Ship the frozen specimen in dry ice.

REFERENCE RANGE:
   Normal = 0 to 12 microgram Aggregated Human Gammaglobulin Equivalents
   Borderline = 12 to 25 microgram Aggregated Human Gammaglobulin Equivalents
   Abnormal = >25 microgram Aggregated Human Gammaglobulin Equivalents

METHOD: The Raji cell is a human lymphoblastoid derived cell line; it has B
cell characteristics but no surface immunoglobulins. Raji cells have C-3
receptors and can bind immune complexes that have fixed C-3. This is
illustrated in the next Figure:

Binding of Immune Complexes by Raji Cells

Antigen-Antibody-Complement   Raji Cell   Antigen-Antibody-Complement-
                                                      Raji Cell

Patient serum is first reacted with Raji cells, and I-125 anti-human IgG is
then added to label bound immune complexes.

INTERPRETATION: Some diseases reported to be associated with circulating
immune complexes are given in the Table in IMMUNE COMPLEX PROFILE.

RAST (SPECIFIC IgE)

Rast, IgE

SPECIMEN: Red top tube, separate serum

REFERENCE RANGE: Negative

METHOD: Rast=Radioallergosorbent Test. Specific allergens conjugated to
cellulose are used to bind specific IgE antibodies from the patient's sera;
after incubation, radiolabeled anti-IgE antibody is added to detect the specific
IgE-allergen complexes.

INTERPRETATION: Laboratory allergy tests (RAST) complement skin (or prick)
tests in the work-up of patients with possible allergy. The allergens assayed
by Dr. Donald Hoffman in our Department are shown in the next Table:

Allergens Assayed by RAST Test (Dr. Donald Hoffman)		
Rye Grass	Bee Venom	Milk Casein
Timothy grass	Yellow Jacket Venom	Milk Lactalbumin
Bermuda Grass	Yellow Hornet Venom	Milk Lactoglobulin
Johnson Grass	White Face Hornet Venom	Egg White
	Paper Wasp Venom	Codfish
	Fire Ant Venom	Shrimp
Alder or Birch Tree	Fire Ant or Honeybee WBE	Abalone
Box Elder Tree	Kissing Bug	Crab or Clam
Elm Tree		Lobster
Oak Tree	Alternaria	Oyster
Olive Tree	Aspergillus	Scallop
Poplar Tree	Chaetomium	Corn
Walnut Tree	Cladosporium	Oat
Mulberry Tree	Dreschlera	Rice
	Monilia	Wheat
	Penicillium	Kidney bean
English Plantain	Many Other Molds	Navy Bean
Pigweed	Available	Pea
Short Ragweed		Peanut
Russian Thistle	Penicilloyi	Soy
Sagebrush	KLH	Almond
Scale	Beef and Pork Insulin	Brazil Nut
Lambsquarter	Proquesterone BSA & BSA	Cashew Nut
		Hazel Nut
	Chocolate	Pecan or Walnut
House Dust and Mite	Coconut	Sesame or Orange
Cat, Dog, Horse Dander	Yeast or Buckwheat	Tomato or Cantaloupe

Rast (IgE)   S. Bakerman

Specific antibodies are measured in patients with respiratory allergy, (asthma, chronic bronchitis, persistent rhinitis, or sinusitis) food allergy, insect-venom allergy and drug hypersensitivity. (Hoffman, D.R., Ann. Allergy 42, 224-230, 1979; Hoffman, D.R. and Haddad, Z.H., J. Allergy Clin. Immunol. 54, 165-173, 1974; Hoffman, D.R. and Haddad, Z.H., Pediatrics 54, 151-6, 1974).
When choice of allergens is a problem, then the common allergens are assayed. The most common environmental allergens are a grass, pollens, house dust mite, cat or dog dander, and a mold; in children, food allergens such as egg white or cow's milk are also important. House dust is the most common environmental allergen; it is not a single substance but is a group of substances; the most important are the house dust mite.
Total serum IgE levels should not be used as a screening test but should be used to complement specific IgE results obtained by RAST. (Thompson and Bird, The Lancet, pg 169, Jan. 22, 1983; Atkinson, P. et al., The Lancet, pgs. 706-707, March 26, 1983).
In the skin test, allergen is applied to the skin by prick or scratch; a positive reaction occurs when allergen binds to specific IgE molecules which are bound to mast cells; this reaction causes release of vasoactive substances, histamine, from granules of mast cells (David, J. and Rocklin, R.E., Scientific American Medicine 6, IX:1-2, 1983).

RED BLOOD CELL COUNT
SPECIMEN: Lavender(EDTA) top tube or microtube containing EDTA. The specimen is stable at room temperature for up to 6 hours and in the refrigerator for up to 24 hours.
REFERENCE RANGE: The normal ranges at different ages are given in the next Table:

Red Blood Cell Count at Different Ages		
	Red Blood Cell Count(millions/mm^3)	
Age	Male	Female
Birth	5.0-6.3	5.0-6.3
1 Month	4.7-5.9	4.7-6.0
3 Months	3.8-5.2	3.8-5.2
6 Months	3.8-5.1	3.5-4.9
9 Months	3.7-5.2	3.5-4.9
1 Year	3.5-4.9	3.4-5.0
2 Years	3.5-4.9	3.5-5.0
4 Years	3.7-5.0	3.8-5.2
8 Years	4.0-5.1	3.9-5.1
14 Years	3.9-5.3	3.8-5.2
Adult	4.3-5.9	3.5-5.5

METHOD: Coulter Counter; Ortho; Technicon; and manual techniques.
Spurious Red-Cell Values with the Coulter Counter: Spurious red blood count, hematocrit and MCV are obtained using the Coulter counter when a patient has "cold" hemolytic anemia. Red cells are intermittently agglutinating; the particles being counted are larger than normal single red cells so the MCV's are elevated and the red cell count is decreased. The spurious lowering of the red-cell count is disproportionately greater than the false elevation of MCV, so the hematocrit is falsely depressed (Lawrence, C. and Zozicky, O., N. Engl. J. Med. 309, 925-926, 1983).
INTERPRETATION: The causes of decreased red blood count are the anemias which are classified as follows: microcytic hypochromic anemia; normocytic, normochromic anemia, e.g., hemolytic anemia; and macrocytic anemia (see Anemia Panel).
The causes of increased red cell count are polycythemia vera and secondary polycythemia, vigorous exercise and high altitude.

RED CELL INDICES (MCHC, MCV, MCH)

MCHC = Mean Cell Hemoglobin Concentration;
MCV = Mean Cell Volume;
MCH = Mean Cell Hemoglobin

SPECIMEN: Lavender (EDTA) top tube or EDTA microtainer.

REFERENCE RANGE: The value of the indices at different ages are given in the next Table; (Guest, G.M. and Brown, E.W., Am. J. Dis. Child. 93, 486, 1957):

		Indices at Different Ages	
Age	MCV	MCH	MCHC
Newborn	113 ± 0.8	36.9 ± 0.28	32.6 ± 0.18
1 day	110 ± 0.7	36.6 ± 0.25	33.5 ± 0.17
7 days	106 ± 0.6	36.2 ± 0.30	34.3 ± 0.18
20 days	100 ± 0.8	34.4 ± 0.32	34.5 ± 0.25
45 days	94 ± 0.8	32.8 ± 0.29	34.9 ± 0.16
76 days	88 ± 0.6	30.4 ± 0.21	34.6 ± 0.16
122 days	82 ± 0.7	27.7 ± 0.26	33.7 ± 0.17
6 months	78 ± 0.7	26.1 ± 0.28	33.5 ± 0.20
10 months	73 ± 1.0	23.7 ± 0.47	32.4 ± 0.28
12 months	73 ± 1.1	23.5 ± 0.51	32.3 ± 0.32
18 months	72 ± 1.2	23.2 ± 0.60	31.9 ± 0.39
2 years	75 ± 0.7	24.7 ± 0.35	32.6 ± 0.21
3 years	78 ± 0.7	26.6 ± 0.28	34.0 ± 0.20
5 years	80 ± 0.4	27.5 ± 0.15	34.2 ± 0.12
10 years	81 ± 0.6	27.6 ± 0.26	34.3 ± 0.14
14 years	81 ± 1.1	27.9 ± 0.48	34.3 ± 0.27
Adult	76-96	27-32	30-35

MCV decreases from birth to 18 months, and then increases slowly to adult level at ages 4 to 10. MCH changes in a similar fashion while MCHC remains relatively constant throughout life.

METHOD: Using Coulter, MCV red blood count and hemoglobin are measured directly; other indices are calculated using formulas below; the hematocrit is calculated.

INTERPRETATION: MCHC (Mean Cell Hemoglobin Concentration) is calculated from the formula:

$$MCHC = \frac{Hemoglobin\ (g/dl)}{Hematocrit\ (\%)} \quad (grams/dl\ of\ RBCs)$$

The causes of decreased MCHC are given in the next Table:

Causes of Decreased MCHC
Iron Deficiency Anemia
Defects in Porphyrin Synthesis
Hereditary Sideroblastic Anemia
Acquired Sideroblastic Anemia
Pyridoxine-Responsive Anemia
Lead Poisoning
Hemolytic Anemia, especially Thalassemia Minor

Current automated hematology instruments (Coulter Counter) measure red cell MCV, red cell count and hemoglobin and calculate hematocrit, MCH and MCHC. The originally derived low MCHC in iron deficiency anemia was caused by over estimation of the hematocrit due to increased plasma trapping during centrifugation. Low MCHC does not occur in iron deficiency anemia when the determination is done on the automated hematology instruments (Fischer, S.L. et al., Arch. Intern. Med. 143, 282-283, 1983).

Hereditary spherocytosis is one of the few conditions associated with an increase in MCHC, that is, greater than 36.

MCV (Mean Cell Volume): The MCV is given by the formula:

$$MCV = \frac{Hematocrit\ (\%)}{Red\ Cell\ Count\ (millions/mm^3)} \quad (femtoliters)$$

The causes of increased MCV are given in the next Table:

Causes of Increased MCV
Megaloblastic Anemia
Spurious Macrocytic Anemia
Aplasia
Myelofibrosis
Reticulocytosis

The causes of decreased MCV are given in the next Table:

Causes of Decreased MCV
Iron Deficiency Anemia
Defects in Porphyrin Synthesis
Hereditary Sideroblastic Anemia
Acquired Sideroblastic Anemia
Pyridoxine-Responsive Anemia
Lead Poisoning
Hemolytic Anemia, especially Thalassemia Minor

Thalassemia minor and iron deficiency are the only hematologic abnormalities that are associated with a marked decrease in the MCV. Patients with iron deficiency and beta-thalassemia trait may be differentiated using serum ferritin and MCV (Hershko, C., Acta Haematol. 62, 236-239, 1979).

MCH (Mean Cell Hemoglobin): The MCH is calculated from the formula:

$$MCH = \frac{Hemoglobin\ (g/liter)}{Red\ cell\ Count\ (millions/mm^3)}\ (picograms)$$

RED CELL SURVIVAL

SPECIMEN: The first blood specimen is used to tag the hemoglobin of the red blood cells with ^{51}Cr. Then, following incubation of the red blood cells with ^{51}Cr, the tagged hemoglobin of the red blood cells is injected and blood specimens are obtained at 24 hours and every two to three days for 30 days and counted. Specific instructions are available from the laboratory.

REFERENCE RANGE: 25-35 days. In this technique, cells are "randomly" labeled; thus the cells are not of a uniform age group but are erythrocytes of all ages representing the mixture present in the circulation. Thus, the red cell survival is an "average" survival and does not measure the actual longevity of the red cell.

METHOD: ^{51}Cr has a half-life of 27.8 days; the mode of decay is EC, gamma. The gamma ray energy is 322,000 electron volts. In comparison, the energies of other gamma emitters are as follows: ^{125}I, 35,000 e.v.; ^{131}I, 364,000 e.v.

INTERPRETATION: The red cell survival test is indicated when there is an obscure cause for anemia and in the evaluation of therapy for hemolytic anemia. The causes of shortened red cell survival are given in the next Table:

Causes of Shortened Red Cell Survival
Hemolysis
Sequestration of Red Cells by Spleen
Blood Loss

The causes of prolonged survival are prior splenectomy or an abnormality of red cell production.

The ^{51}Cr-labeled red blood cell survival test has been used to evaluate in-vivo survival of a small sample of donor red blood cells in patients with blood compatibility problems (Pineda, A.A. et al., Mayo Clin. Proc. 59, 25-30, 1984).

REMINDERS

Reminder "rules" were generated and clinical action was indicated for certain clinical conditions; these are listed in the next Table; (McDonald, C.J. et al., Ann. Intern. Med. 100, 130-138, 1984):

Reminder "Rules" and Clinical Action	
Action	Indication
Stool Occult Blood	Yearly Screening if Age Between 45 and 85
Serum Potassium	Yearly in Patients on Diuretic Agents or Potassium Supplement
Liver Enzyme Levels	Yearly if Patients on Hepatotoxic Drugs and Baseline for New Patients
Serum Amylase	Abdominal Pain
T-4 Index	Suspicion of Hypo- or Hyper- Thyroidism
Reticulocyte Count	Once for Undiagnosed Anemia
Iron and Total Iron Binding	Once for Undiagnosed Anemia and Median Cell Volume under Lower Limit of Normal
Hematocrit and Hemoglobin	Initial Baseline Study
Cervical Smear	Three Year Intervals
Urine Culture	Pyuria
Serum Fluorescent Treponemal Antibody Tests	Positive VDRL
Tuberculosis Skin Test	If Age is Under 35 and if No Past History of Positive Test or Tuberculosis (Inner City Population)
Prothrombin Time(PT)	After Coumadin Treatment
Vitamin K	Unexplained Prothrombin Time Elevations
Chest X-Ray	Initial Screening of Patients and Follow-up of Initial Infiltrates and Congestive Heart Failure
Mammogram	Every 18 Months in Women Between 50 and 70
Colon X-Rays	Hemoglobin-Positive Stools
Pneumococcal Vaccine	U.S. Public Health Service Criteria Except Over 60 instead of Over 50
Influenza Vaccine	U.S. Public Health Service Criteria If Patient Visits Clinic in Fall

RENAL FAILURE, ACUTE, PANEL

In acute renal failure, there is an <u>abrupt decline</u> in <u>renal function</u> with retention of creatinine and urea.

The tests that are useful to <u>differentiate pre-renal</u>, <u>renal</u> and <u>post-renal</u> are listed in the next Table:

Tests to Differentiate Pre-Renal, Renal and Post-Renal Oliguria
Examination of Urinary Sediment
Serum BUN/Serum Creatinine
Urine Specific Gravity
Urine Osmolality
Urine to Plasma Ratio for
Osmolality
Creatinine
Urea
Urine Sodium
Fractional Excretion of Sodium

The test <u>results</u> that may be useful for differentiating of pre-renal, renal parenchymal and post-renal states are given in the next Table; (Nissenson, A.R., Hosp. Med., pgs. 22-43, Sept., 1979; Schrier, R.W., Hosp. Pract., pgs. 93-112, March, 1981; Goldstein, M., Med. Clinics of N. Am. <u>67</u>, 1325-1344, Nov., 1983):

Differentiation of Pre-Renal, Renal Parenchymal and Post-Renal States			
Test	Pre-Renal	Renal	Post-Renal
Examination of Urinary Sediment	Bland	Renal Tubular Cells; Granular, Hyaline and Cellular Casts; Occ. R.B.C.'s and R.B.C Casts	Similar to Pre-Renal
Serum BUN/Serum Creatinine	>10/1	10/1	5/1
Urine Specific Gravity	High	Low	Low
Urine Osmolality	>500 mOsm/kg	<350 mOsm/kg(ATN)	Non-Diagnostic
Urine/Plasma Osmolality	>1.3	<1.1	Non-Diagnostic
Urine Creatinine/ Plasma Creatinine	>40	<20(ATN) >40(GN)	<20
Urine Urea/ Plasma Urea	>8	<3(ATN) >8(GN)	<3
Urine Sodium	<20 mEq/L	>40 mEq/L	Non-Diagnostic
Fractional Excretion of Sodium	<1%	>2%(ATN) <1%(GN)	>2%

ATN = Acute Tubular Necrosis; GN = Glomerulonephritis

Examination of Urinary Sediment: The <u>most important</u> finding is the presence of <u>casts</u>. Erythrocyte casts are strongly suggestive of acute glomerulonephritis; <u>granular casts indicate acute glomerulonephritis</u> or acute tubular necrosis.
Pre-Renal: There is hypoperfusion of the kidney and avid reabsorption of sodium and water by the nephrons. The most important diagnostic criterion of pre-renal uremia is response to treatment (Editorial, Brit. Med. J., pgs. 1333-1335, June 7, 1980). As body fluids deficits are made up, the <u>flow</u> of urine <u>increases</u> and the plasma concentration of <u>urea will fall</u>.
Renal Parenchymal: Damaged tubules fail to reabsorb solutes and water normally. See <u>Acute Glomerulonephritis Panel</u> for tests for glomerular diseases.
Ultrasound is useful to detect post-renal obstruction; if ultrasound is not available, intravenous pyelography or retrograde pyelography may be considered. However, these procedures carry risk; <u>intravenous pyelography-contrast induced acute tubular necrosis</u> if serum creatinine exceeds 4 mg/dl; <u>retrograde pyelography-general anaesthesia</u> and should <u>not</u> be done in the presence of <u>urinary tract sepsis</u>.

RENAL STONE ANALYSIS

SPECIMEN: Specimen should be washed free of tissue and blood and submitted in a clean, dry container.

METHOD: The stone is sawed or split in half and examined under a dissecting microscope. Analysis is done on identifiable layers. Qualitative chemical analysis may be done for calcium, magnesium, oxalate, phosphate, urates and cystine; anions may be analyzed using spectroscopic procedures and cations by atomic absorption. Other methods of analysis include crystallographic analysis using a polarizing microscope, X-ray diffraction and infrared spectroscopy.

INTERPRETATION: Each year, one out of 1,000 people develop kidney stones. The conditions that are associated with renal stone formation are listed in the next Table; (The incidence of the various types are given in parentheses):

| Conditions Associated with Renal Stone Formation ||
Type	Conditions

Calcium Containing Stones (70%)
    Pure Calcium Oxalate (33%)
    Pure Calcium Phosphate (10%)
    Mixtures of Calcium Oxalate
        and Calcium Phosphate (27%)
	In majority cause unknown; normal urinary calcium and oxalate excretion; possibly mild transient changes in urinary oxalate excretion. (See CALCIUM EXCRETION TESTS)
	Hypercalciuric States (>250mg/24 hours in female and >300mg/24 hours in the male)
	        Idiopathic; Normocalcemic (Most Common Hypercalciuric Condition)
	        Primary Hyperparathyroidism
	        Vitamin D Intoxication
	        Hyperthyroidism
	        Renal Tubular Acidosis (Type I or Distal Tubular Type)
	        Sarcoidosis

Magnesium Ammonium Phosphate (Triple Phosphate) (20%) (Struvite)	Recurrent Urinary Tract Infections with Urea-Splitting Organisms, particularly Proteus

Uric Acid (5%)	Hyperuricosuric States:
	    Gout
	    Secondary to Lymphoproliferative and Myeloproliferative Disorders
	Persistently Acid Urine:
	    Chronic Diarrhea; Ileostomy

Cystine (3%)	Renal Tubular Defects of Cystine, Lysine Arginine and Ornithine

Oxalate	Hyperoxaluric States (Rare)
	    Increased Ingestion
	    Pyrodoxine (Vitamin B-6) Deficiency
	    Primary Hyperoxaluria
	    Methoxyflurane Anesthesia
	    Secondary to Ileal Disease

It is important to differentiate the various stone compositions. There are four main types of stones; these are calcium, struvite, uric acid and cystine. Two-thirds of all renal stones are composed of either calcium oxalate or calcium oxalate mixed with calcium phosphate. A protein called matrix substance A has been found in all types of renal stones. Struvite stones are most likely caused by infection; the most common cause is urea-splitting bacteria, Proteus mirabilis. If the stone can be completely removed and infection stopped, the patient will have a very low rate of recurrence (Silverman, D.E. and Stamey, T.A., Medicine 62, 44, 1983). In a pediatric population, stones of infection, e.g., calcium phosphate and magnesium ammonium phosphate, are the most common type. (Walther, P.C., Lamm, D. and Kaplan, G.W., Pediatrics 65, 1068-1072, 1980).

For the work-up of patients with calcium renal stone formation, see CALCIUM EXCRETION TEST.

RENAL STONE PANEL

In the United States, renal stone disease accounts for one hospitalization per 1,000; there are numerous office and emergency-room visits per year for renal colic. Renal stone disease is one of the most common urologic problems in clinical practice. The types and incidence of renal calculi are given in the next Table:

Types of Renal Calculi	
Type	Incidence (%)
Calcium	70
Pure Calcium Oxalate	33
Pure Calcium Phosphate	10
Mixtures of Calcium Oxalate and Calcium Phosphate	27
Magnesium Ammonium Phosphate (Struvite; Triple Phosphate)	20
Uric Acid	5
Cystine	3
Miscellaneous: Xanthine; Silicates; etc.	Few Percent

Two-thirds of all renal stone are composed of either calcium oxalate or calcium oxalate mixed with calcium phosphate. There are four main types of stones; these are calcium, struvite, uric acid and cystine. A protein called matrix substance A has been found in all types of renal stones; (Williams, H.E., Nepholithiasis, N. Engl. J. Med. 290, 33-38, 1974, Williams, H., Hosp. Pract. 20-27, Dec. 1980; Editorial, Brit. Med. J. 282, pg. 5, Jan. 3, 1981; Coe, F.L. and Favus, M.J., "Idiopathic Hypercalciuria in Calcium Nephrolithiasis," in Disease-A-Month, XXVI, Sept. 1980).

If a patient develops a calcium containing stone of unknown cause and passes the stone, there is a 50 to 80 percent chance that the patient will never develop another stone.

Most stones pass out of the normal urinary tract, but if retained may act as a nidus for further deposition. When stones pass into the ureter, intense pain in the flank is produced with pain radiating toward the groin. In a pediatric population, stones of infection, e.g., calcium phosphate and magnesium ammonium phosphate, were the most common type, (Walther, P.C., Lamm, D. and Kaplan, G.W., Pediatrics 65, 1068-1072, 1980).

The types and causes of renal calculi are given in the next Table:

Types and Causes of Renal Calculi	
Type	Causes
Calcium Containing Stones	See Calcium, Total, Serum
	See Calcium Excretion Test
Magnesium Ammonium Phosphate	Recurrent Urinary Tract Infections with Urea-Splitting Organisms, particularly Proteus
Uric Acid	Hyperuricosuric States:
	Gout
	Secondary to Lymphoproliferative and Myeloproliferative Disorders
	Persistently Acid Urine: Chronic Diarrhea; Ileostomy
Cystine	Renal Tubular Defects of Cystine, Lysine Arginine and Ornithine
Oxalate	Hyperoxaluric States (Rare)
	Increased Ingestion
	Pyridoxine (Vitamin B-6) Deficiency
	Primary Hyperoxaluria
	Methoxyflurane Anaethesia
	Secondary to Ileal Disease

Calcium Stones: Most calcium stone formers have normal serum calcium levels, (see Calcium Excretion Test).

Hypercalcemic conditions associated with recurrent stone formation are as follows: hyperparathyroidism, vitamin D intoxication, hyperthyroidism, sarcoidosis and distal hereditary renal tubule acidosis. In renal tubular acidosis, distal tubular type, there is an apparent defect in calcium transport allowing excessive calcium to leak into the urine.

Magnesium Ammonium Phosphate: There may be increased ammonia production in patients with chronic urinary tract infections.

Uric Acid: Increased uric acid excretion occurs in <u>gout</u> and in <u>malignancies</u> associated with increased cell turnover such as occurs in lymphoproliferative and myeloproliferative disorders.

<u>Cystine</u>: Excessive excretion of cystine may occur secondary to faulty renal tubular reabsorption.

Oxalate: Increased oxalate excretion occurs in the conditions listed in the previous Table. Patients with jejunoileal bypass, ileal disease and ileal resection have an enhanced absorption of dietary oxalate via the colon. Increased fatty acids secondary to malabsorption results in increased complex formation with calcium ions. Decreased free calcium ions within the lumen of the gut results in oxalate combining into more soluble sodium oxalate; concommitently, there is the presence of unabsorbed bile salts leading to increased colonic permeability of oxalate and hyperoxaluria (Doffins, J.W. and Binder, H.J., N. Engl. J. Med. <u>296</u>, 298-301, 1977).

The <u>initial</u> laboratory studies in the workup of patients with renal calculi are listed in the next Table:

Initial Laboratory Studies in Patients with Renal Calculi
Urinalysis - Look for Crystals
Urine Culture if there is Evidence of Infection on Urinalysis
Blood Studies: Serum Calcium, Serum Phosphate and Serum Uric Acid
Plain Film of the Abdomen

Urinalysis: Urinalysis is particularly important. There are three ways in which urinalysis may be of assistance. Alkaline urine suggests the syndrome of <u>renal tubular acidosis</u> or urinary tract infection with <u>urea-splitting organisms</u>; evidence of <u>chronic infection</u> in the urinary tract suggests <u>triple phosphate stones</u>; <u>crystalluria</u> may help identify the cause of stone disease, e.g., cystine or uric acid crystals. However, calcium oxalate stones may be found in the urine of patients who have no history of stone formation. The appearance of crystals in the urine is illustrated in the next Figure:

Crystals in the Urine				
"Envelope" Oxalate	"Coffin Lids" Magnesium Ammonium Phosphate	"Thorn Apple" Ammonium Biurate	"Hexagonal Plates" Cystine	"Flat-Notched Plates" Cholesterol

Urine Culture: See URINE CULTURE
As already noted, magnesium ammonium phosphate stones are associated with <u>recurrent urinary tract infections</u> with urea-splitting organisms, especially <u>proteus</u>. A major complication of stone formation in the urinary tract is urinary tract infection.

Blood Studies: Serum <u>calcium</u> and <u>phosphorus</u> are done to detect <u>hypercalcemia</u>; serum <u>uric acid</u> is done to detect <u>hyperuricemia</u> as in gout.

Plain Film of the Abdomen: Diagnosis may be made by X-ray if the stones are radiopaque; <u>radiopaque</u> stones are composed of calcium and magnesium salts and comprise 80% to 95% of stones. The <u>most common radiolucent stone</u> is composed of uric acid.

RENIN ACTIVITY, PLASMA

SPECIMEN: Record information on the status of the patient on the requisition form as follows: Patient fasting, or regular diet, or salt restricted diet, and/or diuretics; supine or upright, patient ambulated for ___ hours specimen obtained following catheterization of _____ vein.

The following schedule may be used in evaluating a patient in an office or outpatient setting. Discontinue all medication for one week before the test. At 6 P.M., midnight and 6 A.M., the patient takes 40 mg of the diuretic furosemide (Lasix) orally. Patient remains in an upright position for 3 hours prior to test. Blood specimens for renin and aldosterone are obtained. Draw blood into a chilled syringe; place blood into a chilled purple(EDTA) top tube; keep specimen on ice after collection; transport immediately to the laboratory. In the laboratory, keep specimen on ice or refrigerated until plasma is separated; centrifuge specimen and collect plasma. Freeze plasma immediately.

REFERENCE RANGE: Normal plasma renin levels, in different conditions, are shown in the next Table; (Met Path):

Normal Plasma Renin Levels in ng/ml/hour			
Dietary Salt Intake (mEq of Na⁺)	Recumbent (Overnight or six Hours)	Upright (4 hours)	Upright plus Diuretic
Normal Salt Intake (100-180 mEq)	0.5 - 1.6	1.9 - 3.6	
Low Salt for 4-10 days	2.2 - 4.4	4.0 - 8.1	6.8 - 15.0

To convert conventional units in ng/ml/hr to international units in ng/liter/sec., multiply conventional units by 0.2778.

Salt restricted diets and/or diuretics increase supine and upright renin levels.

METHOD: Plasma renin activity(PRA) is determined by measuring the rate of angiotensin I generated by plasma renin without addition of exogenous renin or angiotensin. The reaction is as follows:

Angiotensinogen —Renin→ Angiotensin I.

Angiotensin I is assayed by radioimmunoassay(RIA).

This method for measuring PRA depends not only on the concentration of renin but also on the concentration of substrate in the specimen.

INTERPRETATION: Renin levels are determined in the differential diagnosis of hypertension. Renin is an enzyme which is synthesized in the juxta-glomerular cells lining the afferent arterioles of glomeruli of the kidney. Renin is released by change in pressure or change in stretch of the afferent arteriole. The action of renin is illustrated in the next Figure:

Action of Renin

Congestive Heart Failure
Cirrhosis of the Liver
Edema
Nephrosis
Hypoproteinemia
Malignant Hypertension
Renal Artery Stenosis
→ KIDNEY

Liver
↓
Angiotensinogen (M.Wt. 57,000) An Alpha-2 Globulin
—Renin→ Angiotensin I (Decapeptide) + Resid. Prot.

Renin Released From Kidney (Renin-35,000 to 40,000 Proteolytic Enzyme)
↓

N-Term.-Leu-Leu-C-Term. —Resin→ N-Terminal Asp-Arg-Val-Tyr-Leu-His-Pro-Phe-His-Leu + Resid.
AA-1--- 10 11   1   2   3   4   5   6   7   8   9   10 Prot.
                                            C-Terminal

LUNG

Angiotensin I (Decapeptide) —Converting Enzyme / Lung→ Angiotensin II + 2 Amino Acids (Octapeptide)

Angiotensin is a potent <u>vasoconstrictor</u> and stimulates the <u>release of aldosterone</u> from the adrenal cortex.

The relative levels in serum and urine of renin and aldosterone in patients with primary and secondary hyperaldosteronism are shown in the next Table:

| Aldosterone and Renin Levels in Primary and Secondary Aldosteronism |||
Aldosterone	Aldosterone Level	Renin Level
Primary (60% Adenoma)	↑(Start Here)	↓
Secondary	↑	↓(Start Here)

The combination of <u>elevated aldosterone and low plasma renin activity is</u> practially pathognomonic of primary aldosteronism. See ALDOSTERONE, SERUM.

Plasma renin may be <u>low, normal or elevated in essential hypertension.</u>

Plasma renin is <u>increased</u> in patients on <u>oral contraceptives.</u>

Reticulocyte
Count

RETICULOCYTE COUNT

<u>SPECIMEN:</u> Whole Blood: Lavender (EDTA) top tube.

<u>REFERENCE RANGE:</u> 0.5-1.5% in adults and children; 2-6% in full-term infants falling to adult level at 2 weeks.

<u>METHOD:</u> Use mixture of blood plus reticulocyte stain such as New Methylene Blue N. The reticulocyte count is usually expressed as a percentage of the erythrocyte count.

<u>INTERPRETATION:</u> The reticulocyte count is an index of erythropoiesis and is <u>increased in the following conditions:</u>

Increased Reticulocyte Count
Hemolytic Anemia, and Hemolysis
Acute Blood Loss
Response of Anemia to Specific Therapy;
e.g., Iron, Vitamin B_{12}, Folic Acid
Infiltrative Marrow Disorders

In <u>hemolytic</u> anemias, the reticulocyte <u>count</u> should be corrected for <u>low</u> <u>hematocrit</u> and increased <u>survival time</u> of the premature reticulocytes. Under normal conditions, 1% of reticulocytes are released daily into the peripheral blood; these cells remain in the circulation for <u>one day</u> before they lose their RNA reticulum and are converted into young erythrocytes. If red blood cell production in the bone marrow is increased (<u>hypoxia</u> or <u>erythropoietin</u> <u>stimulation</u>), premature delivery of large young reticulocytes occurs; these young reticulocytes survive for about <u>two</u> days, or twice as long as normal cells. In order to correct the reticulocyte count, for <u>both</u> lower <u>hematocrit</u> and <u>young reticulocytes</u>, the following formula (Finch) is used:

$$\text{RETICULOCYTE PRODUCTION INDEX} = \frac{\text{Reticulocytes (\%)} \times \dfrac{\text{Patient's Hematocrit}}{45}}{2 \text{ (Reticulocyte Maturation Time)}}$$

The reticulocyte count, corrected for the patient's <u>abnormal hematocrit</u> <u>only</u>, is as follows:

$$\text{CORRECTED RETICULOCYTE COUNT} = \text{Reticulocytes (\%)} \times \frac{\text{Patient's Hematocrit}}{45}$$

<u>Infiltrative Marrow Disorders:</u> In infiltrative marrow disorders, immature red blood cells may "escape" into the circulation from the marrow and from sites of production in the sinusoids of the liver and spleen.

The reticulocyte count is <u>decreased</u> in the following conditions:

Decreased Reticulocyte Count
Marrow Aplasia
Replacement of Erythroid Precursors by Leukemic Cells
Infections
Toxins
Drugs: (more than 100 Drugs)
Renal Disease
Disorders of Maturation:
Iron Deficiency Anemia
Megaloblastic Anemia
Sideroblastic Anemia
Anemia of Chronic Disease

RETINAL-BINDING PROTEIN(RBP)

Specimen: Red top tube, separate serum

Reference Range: Adult: 3-6 mg/dl; Children: 1.5-3.0 mg/dl; sharp increase in RBP at puberty

Method: RIA; Nephelometry

Interpretation: Retinol-binding protein(RBP) is synthesized and stored in the liver; it circulates in the plasma complexed to thyroxine-binding pre-albumin. Vitamin A is transported by the RBP from its storage site in the liver to peripheral organs. The plasma concentration of RBP is regulated by vitamin A status; in vitamin A deficiency, RBP molecules are not secreted from the liver. The causes of decreased level of serum RBP are given in the next Table; (Rask, L. et al., Scand. J. Clin. Lab. Invest. 40, Suppl. 154, 45-61, 1980):

Causes of Decreased Level of Serum RBP
Vitamin A Deficiency
Fat Malabsorption Syndromes
(1) Deficiency of Pancreatic Digestive Enzymes:
Chronic Pancreatitis
Cystic Fibrosis
Pancreatic Carcinoma
Pancreatic Resection
(2) Impairment of Intestinal Absorption:
Celiac Disease (Coeliac Disease, Non-Tropical Sprue, Gluten-Sensitive Enteropathy)
Rare Causes:
Tropical Sprue
Abetalipoproteinemia
Lymphangiectasis
Intestinal Lipodystrophy
Amyloidosis
Lymphoma
Surgical Loss of Functional Bowel
(3) Deficiency of Bile:
Extrahepatic Bile Duct Obstruction
Intrahepatic Disease
Cholecystocolonic Fistula
Dietary Deficiency of Vitamin A
Insufficient Synthesis of RBP
Chronic Liver Diseases
Cystic Fibrosis of the Pancreas
Protein-Energy Malnutrition

Vitamin A Deficiency: Vitamin A is a fat soluble vitamin and it decreases in the fat malabsorption syndromes; in vitamin A deficiency, RBP molecules are not secreted by the liver.

Insufficient Synthesis of RBP: Since RBP is synthesized in the liver, conditions involving the liver, such as chronic liver disease, cause decrease in this protein. Cystic fibrosis is associated with fat malabsorption and thuis vitamin A deficiency and decrease in RBP occurs; in addition, cystic fibrosis is associated with insufficient synthesis of the carrier protein.

In protein-energy malnutrition, the liver may contain significant amounts of vitamin A but due to a diminished protein synthesis, an insufficient amount of RBP molecules are synthesized and symptoms of vitamin A deficiency occur in peripheral tissues. The concentrations of RBP and pre-albumin in serum are significantly reduced in protein-energy malnutrition (Rask, L. et al., Scand. J. Clin. Lab. Invest. 40, Suppl. 154, 45-61, 1980).

REYE'S SYNDROME

The abnormal laboratory values in Reye's Syndrome are listed in the next Table:

Abnormal Laboratory Values in Reye's Syndrome
Elevated Transaminases(SGOT, SGPT)
Hyperammonemia
Hypophosphatemia

Hypoglycemia tends to be observed more often in patients less than 2 years of age.

The definitive test for Reye's syndrome is liver biopsy; however, liver biopsy is seldom necessary because history of viral infection and laboratory values are sufficient for diagnosis.

RHABDOMYOLSIS, PANEL

A screen for rhabdomyolysis is shown in the next Table; (Liu, E.T., et al., Arch. Intern. Med. 143, 154-157, 1983):

Screen for Rhabdomyolysis
Enzymes:
Creatine Phosphokinase(CPK)
Lactate Dehydrogenase(LDH)
Serum Glutamic Oxalacetic Transaminase(SGOT)
Hypocalcemia
Serum Myoglobin
Urinary Myoglobin

Urinary myoglobin may be absent when serum myoglobin is markedly elevated.

RHEUMATOID FACTOR (RF, RA)

SPECIMEN: Red top tube, separate serum. (No anticoagulant.)
REFERENCE RANGE: Screen: Negative; Titer: Negative
METHOD: Latex agglutination with latex coated with heat denatured IgG
INTERPRETATION: The rheumatoid factor is an immunoglobulin (usually an IgM) present in the serum of patients with rheumatoid arthritis (and some other conditions, especially connective tissue diseases) and which reacts or binds to the Fc fragment of IgG in the same patient's serum.

Although RF may be present in a wide variety of conditions, the titer of RF is usually significantly higher in rheumatoid arthritis as compared to other conditions. Conditions, other than rheumatoid arthritis, in which RF may be elevated, are listed in the next Table:

Conditions Associated with Elevated Rheumatoid Factor (RF)
"Collagen"-Vascular Diseases
Felty's Syndrome
Sjögren's Syndrome
Systemic Lupus Erythematosus
Scleroderma
Polyarteritis Nodosa
Infectious Conditions: Kala-azar; Leprosy; Tuberculosis; Syphilis;
Bacterial Endocarditis; Viral Hepatitis; Chronic Hepatic Diseases
Myocardial Infarction
Renal Disease
Malignancy
Thyroid Diseases
Normal Sera

Typically, the rheumatoid factor is absent when the clinical signs and symptoms of rheumatoid arthritis first appear. Patients in which rheumatoid factor is demonstrable early in the course of rheumatoid arthritis have a greater risk of developing articular destruction and having sustained disabling disease.

ROCKY MOUNTAIN SPOTTED FEVER(RMSF) ANTIBODIES

SPECIMEN: Red top tube, separate serum

REFERENCE RANGE: A fourfold increase in antibody titer between acute- and convalescent-phase serum specimens by complement fixation(CF), indirect fluorescent antibody(IFA), indirect hemagglutination, latex agglutination, or microagglutination; or a single convalescent titer 1:16 or higher(CF) or 1:64 or higher(IFA) in a clinically compatible case.

METHOD: Complement fixation(CF); indirect fluorescent antibody(IFA); indirect hemagglutination; latex agglutination; or microagglutination. The Weil-Felix reaction, agglutination with Proteus vulgaris antigen, OX-2, gives false positive reactions and is not reliable.

INTERPRETATION: Rocky Mountain Spotted Fever is caused by Rickettsia rickettsii (an intracellular parasite) and is transmitted by a tick bite, the parasite enters the blood stream, and then invades vascular endothelial cells, causing systemic vascular injury. The incidence of RMSF, by state, (cases per 100,000 population) in decreasing order, is as follows: Oklahoma (6.9), North Carolina (3.4), South Carolina (2.5), Arkansas (1.8), Georgia (1.2), Virginia (1.1) and Tennessee (1.1) (Morbidity and Monthly Weekly Report 33, 188-190, April 13, 1984). Note the high incidence in the eastern states which border the Appalachian mountains; it is rarely seen in the Rocky Mountains. The vast majority of cases occur between April 30 and September 30. About 50 percent of patients are under 20 years of age. Symptoms are as follows: fever (98%), headache (90%), rash on torso (90%) and rash on palms or soles of feet (70%).

Ticks can be readily removed if they are first covered with petroleum jelly e.g. Vaseline; this may serve to diminish their air supply (Pediatric Alert, 8, No. 10, 40, May 12, 1983).

Rickettsial antibodies are first detectable within 7 to 10 days after onset of symptoms. However, there is a fatality rate of about 10 percent and treatment with tetracycline or chloramphenicol must be started before the results of these tests for antibodies are obtained.

Rapid diagnosis is done by identification of Rickettsia in biopsies of skin lesions. A full-thickness section of skin is obtained, placed in 0.85 percent saline, and transported to the laboratory on ice. Frozen tissue sections are cut and fixed in acetone at -70°C until examined. Fluorescein-labeled antiserum to Rickettsia rickettsii is used for the direct identification of Rickettsia in the skin lesion. Antiserum is available from the Viral and Rickettsial Products Branch, Centers for Disease Control. (Lab Report for Physicians, 4, 73-74, Oct. 1982; Walker, D.H. Cain, B.G. and Olmstead, M.P., Am. J. Clin. Path. 69, 619-623 (1978); Westerman, E.L., Arch. Int. Med. 142, 1106-1107 (1982); MacCaughelty, T.C., Ernst, G. and Hottz, A.S., Crit. Care Med. 10, 124-126 (1982); Kaplowitz, L.G. et al., Arch. Intern Med. 143, 1149-1151, 1983).

LDH is invariably elevated in Rocky Mountain Spotted Fever; this is probably secondary to the diffuse vasculitis that occurs in this disease.

ROTAVIRUS ANTIGEN, FECES

SPECIMEN: Fecal specimen

REFERENCE RANGE: Negative

METHOD: ELISA, enzyme-linked immunosorbent assay, (Rotazyme, Abbott Laboratories, Diagnostic Division, North Chicago, Illinois)

INTERPRETATION: Rotavirus appears to be the most frequent cause of viral gastroenteritis accounting for 60 to 80 percent of cases in infants and young children, most often in the 3 month to 2 year age group. Infection occurs primarily in the winter.

The incubation period is about 3 days. Profuse watery diarrhea lasts for 2 to 11 days and may be associated with vomiting (1 to 5 days) and fever (2 to 4 days); dehydration may require hospitalization. The laboratory findings include the following: Increase in urea nitrogen reflecting dehydration; decrease in serum sodium reflecting gastrointestinal loss; acidosis due to bicarbonate loss in the stool and/or lactate production. When hospitalized, the patient should be isolated. Treat patient by replacing lost fluids and correcting electrolyte abnormalities. Milk intolerance may develop after resolution of the gastroenteritis due to a deficiency in disaccharidases in regenerating gastrointestinal epithelium.

In neonates, asymptomatic infants are occasionally positive for rotavirus. However, infection by rotavirus is infrequently a cause of serious diarrheal disease in normal newborns (Totterdell, B.M. et al., Arch. Dis. Child. 51, 924-928, 1976); this relative protection is explained by the presence of maternal antibody or other unknown factors. Breast feeding may decrease the severity of rotavirus infection.

Neonatal rotavirus infection does not confer immunity against reinfection but does protect against the development of clinical disease during reinfection (Bishop, R.F. et al., N. Engl. J. Med. 309, 72-76, 1983); serious infections have occurred in immunocompromised children and adults.

The incidence of serious infection in adults is low and the infection is mild.

Reference: Yolken, R.H.; Check Sample, Cont. Ed. Prog. ASCP, Advanced Clin. Chem. No. ACC 82-4, March 1983.

RPR (RAPID PLASMA REAGIN),TEST FOR SYPHILIS ANTIBODIES

SPECIMEN: Red top tube, separate serum; cord blood.
REFERENCE RANGE: Non-reactive
METHOD: The RPR uses an alcoholic extract of beef heart and cholesterol, lecithin and alcohol to give reproducible qualitative and quantitative agglutination reactions for the detection of antibody to treponemal infection.
INTERPRETATION: The RPR is a screening test for antibodies in the serum of patients with syphilis. The reactivity of nontreponemal and treponemal tests in untreated syphilis is given in the next Figure; (Henry, J.B. [ed.] Todd-Sanford-Davidsohn, Clinical Diagnosis by Laboratory Methods, 16 ed., 1979, pg. 1890, Vol. II, W. B. Saunders Co., Phila. Pa):

Reactivity of Nontreponemal and Treponemal Tests in Untreated Syphilis

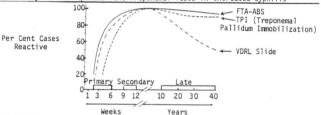

The percent positive for RPR in each stage of untreated syphilis is as follows: Primary Stage (Chancre stage): 30% after one week; 90% after three weeks; Secondary Stage: 100%; Tertiary Stage: 90%; Latent Stage: may be unreactive.

This test is used to follow the effects of therapy; with successful therapy, the value of RPR will tend to fall and will become negative in the majority of patients if treatment is given in the primary or secondary stages. However, treatment during late syphilis is infrequently associated with reversion of a reactive RPR to negative.

Some hospitals routinely perform RPR testing on all cord blood; this is done as a screen for congenital syphilis. A positive test would have to be correlated with the maternal test. If a positive RPR is found on cord blood, current recommendations are to follow the titer monthly; a rising titer indicates an active and ongoing infection.

False-positive reactions occur in a variety of acute and chronic conditions as listed in the next Table:

Conditions Causing False-Positive Using Non-Treponemal Tests for Antibodies Produced in Response to Syphilis
Lupus Erythematosus
Infectious Diseases: Malaria; Infectious Mononucleosis; Infectious Hepatitis; Leprosy; Brucellosis; Atypical Pneumonia; Typhus; Related Treponemal Infections; Pregnancy

False-positive tests for syphilis, in patients with systemic lupus erythematosus, are caused by anti-DNA antibodies that cross-react with cardiolipin (Koike, T. et al., Clin. Exp. Immunol. 56, 193-199, 1984).

RUBELLA (GERMAN MEASLES) ANTIBODIES, IgM AND IgG

SPECIMEN: Red top tube, separate serum. A single serum specimen is required for immune IgG status. For suspected infection, two serum specimens are required: the acute serum specimen should be obtained while the rash is present; convalescent serum is obtained 7 to 14 days later.

REFERENCE RANGE: The presence of IgM antibodies indicates current infection; a fourfold rise in IgG antibody titer is also indicative of an acute rubella infection; a stable or diminishing antibody titer is indicative of a past rubella infection.

METHOD: IgM: Enzyme immunoassay (EIA) or RIA (False-positive reactions may occur due to rheumatoid factor.) IgG: EIA; Hemagglutination inhibition (HI); passive hemagglutination; complement fixation.

INTERPRETATION: Rubella or German measles is a mild acute viral illness that most commonly occurs in school-age children; however, maternal rubella infections acquired during the first sixteen weeks of gestation are associated with fetal malformations and sequalae such as abortions, stillbirths, and congenital anomalies. Fifty percent of infants will be affected if disease occurs in the first four weeks of gestation, dropping to 10% at 20 weeks. The serum of a congenitally infected infant contains actively acquired IgM antibody, and passively acquired IgG antibody; in one year IgG antibody is present.

It is important to determine the immune status of women of childbearing age. An IgG antibody titer indicates immunity; a susceptible woman should be immunized and is advised not to conceive for two to three months following immunization.

Programs have evolved to help prevent personnel working in hospitals from being sources of rubella infections (nosocomial rubella transmission). The Public Health Services Advisory Committee on Immunization Practices (ACIP) recommended in 1978 "to protect susceptible female patients and female employees, persons working in hospitals and clinics who might contact rubella from infected patients or who, if infected, might transmit rubella to pregnant patients should be immune to rubella." The American Academy of Pediatrics' Committee on Infectious Disease concurred and made similar recommendations (Pediatrics 65, 1182-1184, 1980). In the study on employee rubella screening programs in Arizona Hospitals, data indicate slow and partial compliance with recommendations for rubella screening in hospitals (Sacks, J.J. et al., JAMA 249, 2675-2678, May 20, 1983).

SALICYLATE [ASPIRIN, SALICYLIC ACID, ACETYLSALICYLIC ACID (ASA)], BLOOD

SPECIMEN: Red top tube; serum may be stored in refrigerator at 4°C for at least one week. Lavender (EDTA) or green (heparin) top tubes with collection of plasma; salicylate is stable in plasma stored at 4°C for 2 months. There is no difference in salicylate values obtained from serum or plasma.

REFERENCE RANGE: Therapeutic level: 15-30mg/dl; Half-Life (Adults): 2-4.5 hours; Half-Life (Children): 2-3 hours; Time to Peak Plasma Level: 1-2 hours; Time to Steady State, Adults: 10.0-22.5 hours; Time to Steady State, Children: 10-15 hours. Toxicity may appear in the range of 20-30mg/dl. At >30mg/dl, toxicity will definitely occur. Aspirin is lethal at levels >70mg/dl. Salicylate blood levels do not correlate well with degree of intoxication in chronic salicylism.

METHOD: Colorimetrically by reaction with ferric ions; microfluorometric procedure; HPLC.

INTERPRETATION: (Atwood, S.J., The Pediatric Clinics of North America, 27, 871-879, 1980). In acute salicylate toxicity, the Done nomogram (Done, A.K., Pediatrics 26, 800-807, 1960) can be used to predict severity of intoxication from a single determination of serum salicylate, provided that the time of ingestion is known and sufficient time has elapsed between ingestion and sampling of blood (six hours); the Done nomogram is shown in the next Figure:

Serum Salicylate Concentration and Expected Severity of Intoxication at Varying Intervals of Time Following Ingestion

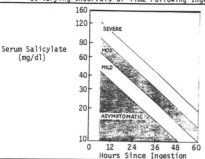

If the initial value is in a "safe" range, the determination of serum salicylate should be repeated in two to four hours; then, if the serum salicylate does not rise between tests, the child is considered not to be at risk for serious intoxication. Laboratory data suggestive of aspirin are a positive urine by clinitest (reducing substances) and a negative glucose oxidase and positive ferric chloride test on boiling urine.

Laboratory tests, other than serum salicylate levels, that should be used to monitor complications of salicylate toxicity are given in the next Table:

Laboratory Tests to Monitor Possible Complications of Salicylate Toxicity

Tests	Monitor Complications
Arterial Blood Gases (pH, PCO_2, HCO_3^-)	Respiratory Alkalosis; Metabolic Acidosis; Hypermetabolism
Serum Electrolytes (Na^+, K^+, Cl^-, CO_2 Content)	Anion Gap Increased; (Na^+)−(Cl^-+CO_2 Content): Due to Increased Lactic Acid, Ketones and Salicylic Acid. Inappropriate Fluid Retention: Detect the Effects of Bicarbonate Diuresis following Therapy with Bicarbonate for Acidosis
Serum and Urine Osmolality	Inappropriate Secretion of Antidiuretic Hormone
Blood Glucose	Blood Glucose May be Low, Normal or Elevated; Blood Glucose Should be Maintained at Elevated Levels to ensure Adequate Supply of Glucose to Brain
Prothrombin Time (PT)	If Prolonged, Vitamin K Therapy Indicated

Following acute ingestion, children quickly pass through the stage of pure respiratory alkalosis and present with some degree of metabolic acidosis. The toxic properties of salicylates result principally from direct stimulation of the respiratory center (respiratory alkalosis) and influence on metabolic pathways (metabolic acidosis).

Symptoms of hypocalcemic tetany, secondary to alkalosis, may develop (see Calcium, Ionized (Free).

Reference: Temple, A.R., Pediatr. Supp. 62, 873-876, Dec. 1978.

SALINE WET MOUNT FOR TRICHOMONAS VAGINALIS AND CANDIDA ALBICANS

<u>SPECIMENS AND METHOD</u>: Obtain a specimen of vaginal discharge and place on a clean slide; mix with <u>one or two drops</u> of <u>normal saline</u>. <u>Cover</u> the specimen with a clean <u>glass cover slip</u>. Examine the specimen under low power and high dry for the <u>active motile flagellates</u> of Trichomonas vaginalis.

<u>INTERPRETATION</u>: Vaginitis may be caused by Trichomonas vaginalis; the <u>most sensitive</u> method of detecting Trichomonas vaginalis is <u>culture</u> but it is expensive. Saline wet mount is less sensitive but inexpensive.

Candida albicans may also be diagnosed by saline wet mount <u>but potassium hydroxide</u> preparation is the <u>preferred method</u>.

Besides Trichomonas vaginalis and Candida albicans, another cause of vaginitis is <u>Gardnerella vaginalis</u>.

SARCOIDOSIS PANEL

A list of tests that may assist in the diagnosis of sarcoidosis is given in the next Table; (Sharma, Om P., Arch. Intern. Med. <u>143</u>, 1418-1419, 1983):

Tests of Sarcoidosis

Test	Sensitivity (%)	False Positives
Kveim-Siltzbach(K-S) Test	97	Regional Enteritis
		Infectious Mononucleosis
		Chronic Lymphatic Leukemia
		Nonspecific Lymphadenopathy
Serum Angiotensin-Converting Enzyme(ACE)	60-80	Leprosy
		Gaucher's Disease
		Primary Biliary Cirrhosis
		Silicosis
		Miliary Tuberculosis
		Lymphoma
		Extrinsic Allergic Alveolitis
		Talc Granulomatosis
		Asbestosis
		Fibrosing Alveolitis
		Chronic Lymphocytic Pneumonitis
		Coccidioidomycosis
		Berylliosis
Serum Lysozyme Activity	40-50	See LYSOZYME, SERUM
Transcobalamin II	50	Gaucher's Disease
		Lymphoproliferative Disorder
		Myeloma
		Waldenström's Macroglobuline-mia
		Liver Disease
Gallium Citrate GA-67 Lung Scan	High Sensitivity but Low Specificity	
Bronchoalveolar Lavage	Normal:>90% Alveolar Macrophages; <10% Lymphocytes; <1% Polymorphonuclear Leukocytes. Sarcoidosis: Increase in T⁻ Lymphocytes in Lavage Fluid	

In the Kveim-Siltzbach(K-S) test, a saline suspension of human sarcoidal spleen or lymph node is injected intracutaneously; in patients with sarcoidosis, a nodule develops in two to six weeks. A biopsy specimen of the nodule demonstrates characteristic non-caseating granulomatous reaction. <u>The basic problem in this test is the unavailability of potent biologically active antigen for injection</u>.

SCHILLING TEST, I, II, III

SPECIMEN: Twenty-four hour urine for assay of Co-57
REFERENCE RANGE: Normal >7% excretion in 24 hours
METHOD: Co-57; Half-life = 270 days; decay mode, EC, gamma
INTERPRETATION: The Schilling urinary excretion test is used to differentiate intrinsic factor deficiency from small bowel (ileum) malabsorption of vitamin B_{12}. In part I, vitamin B_{12} (Co-57) is given alone; in Part II, vitamin B_{12} (Co-57) plus intrinsic factor are given; in Part III, tetracycline is given for 10 days followed by vitamin B_{12} (Co-57) by mouth.
Part I: The patient is given vitamin B_{12} (Co-57) by mouth. This is followed 2 hours later by an intravenous injection of nonradioactive vitamin B_{12}; this saturates the liver storage capacity for the vitamin; thus, the excess over that taken up by the liver is excreted in the urine. Urine is collected for 24 hours and radioactivity is measured. Interpretation of results is illustrated in the next Figure:

Results of Part I, Schilling Test

Vitamin B_{12} (Co-57) by Mouth; Collect Urine for 24 hours; Measure
Percentage of Administered Radioactivity in 24 Hour Urine

```
   0      5  7    10        15        20
   |------+--+-----|---------|---------|
   <---Low--->|<------------Normal------------>
```

1. Stomach-Intrinsic |Dietary Deficiency
   Factor Deficiency |Lack of B_{12} Binding Protein
   a. Pernicious anemia |Folate Deficiency
   b. Gastrectomy, total |
      or partial |
   c. Atrophic gastritis |
2. Intestine-(Ileum) |
   a. Loop Syndrome |
   b. Idiopathic Steatorrhea
   c. Tropical Sprue

Part II: If low excretion of Vitamin B_{12} (Co-57) is obtained, then, it is necessary to differentiate between Vitamin B_{12} deficiency due to inadequate secretion of intrinsic factor by the stomach from an abnormality in absorption by the ileum. Vitamin B_{12} (Co-57) is given along with intrinsic factor. This test is done 5 to 7 days after the first test. Interpretation of results is illustrated in the next Figure:

Results of Part II, Schilling Test

Vitamin B_{12} (Co-57) plus Intrinsic Factor by Mouth; Collect Urine for 24 Hours and Measure Radioactivity

```
   0    5  7    10        15         20
   |----+--+-----|---------|----------|----->
   <---Low--->|<--------------Normal------------>
```

Intestine (Ileum)	Stomach-Intrinsic Factor Deficiency
a. Loop Syndrome	a. Pernicious Anemia
b. Idiopathic Steatorrhea	b. Gastrectomy, Total or Partial
c. Tropical Sprue	c. Atrophic Gastritis

Part III: The third stage is performed on those patients who have low excretion of cobalt-57 labeled vitamin B_{12} in the urine with the first and second stage Schilling tests. The patient receives tetracycline, 250mgs, 4 times a day for 10 days prior to the test. Tetracycline is administered to reduce or alter intestinal bacteria in the possible blind loop syndrome.

Markedly improved excretion of labeled vitamin B_{12} after tetracycline confirms the presence of malabsorption secondary to interference with the normal absorptive process by the bacterial flora which compete for the vitamin B_{12} within the body.

S. Bakerman

The Schilling Test, Parts I, II and III are summarized in the next Table:

	Nutritional Megaloblastic Anemia	Pernicious Anemia	Gastrectomy Total or Partial	Loop Syndrome	Idiopathic Steatorrhea	Tropical Sprue
Schilling						
Dose alone (Part I)	Normal	Subnormal	Subnormal	Subnormal	50% of patients subnormal	Subnormal
Dose with I.F. (Part II)	Normal	Improvement	Improvement	No Improvement	No Improvement	No Improve.
Dose after antibiotics (Part II)	Normal	-	-	Improvement if the ileum is intact	No Improvement	Improve. in Many

SCHLICHTER TEST (SERUM BACTERICIDAL TITER)

SPECIMEN: Three specimens are required: two serum specimens (or other body fluid) and a bacterial isolate.
Serum Specimen: Two serum specimens are obtained; one specimen is collected at the trough level just prior to the next antibiotic dose and next specimen is collected at the peak, that is, 30 minutes post I.V. infusion or 1.5 hours post I.M. or oral administration. Use red top tubes; separate serum and refrigerate or freeze.
Microbiological Specimen: Request the laboratory to save the bacteriological isolate.
REFERENCE RANGE: Bactericidal activity should be observed equal to or greater than 1:8 dilution.
METHOD: Serially dilute the patient's serum with pooled normal human serum to maintain the protein-binding relationships. The serum is serially diluted in broth and then inoculated with a standard suspension of organisms as in the minimal bactericidal concentration(MBC) test; then the specimen is incubated and subcultured.
INTERPRETATION: The Schlichter test is the simplest and most convenient approach of establishing adequate antimicrobial therapy in patients with bacterial endocarditis and in osteomyelitis; the effectiveness of this test for guiding the adequacy of therapy in other conditions is not well established.

SELECTOGEN (see COOMBS, INDIRECT)

S. Bakerman

SEMEN ANALYSIS

SPECIMEN: An absolutely fresh specimen is required; thus, collection must be made at the site of examination. The subject is requested to abstain from sexual activity for 3 days prior to examination. The semen is collected via masturbation in a labeled dry small container such as a disposable plastic drinking glass. Do not refrigerate or freeze.

REFERENCE RANGE: The normal parameters for an ejaculation are given in the next Table:

Normal Parameters for an Ejaculation	
Parameter	Normal
Volume	2-5 ml.
Consistency	Gelatinous, high viscosity initially
Color	Light gray and opaque
Odor	Musty
Liquefaction	Complete in 15 min.; over 30 min. abnormal
pH	7.2-8.0; below 7.2 is abnormal
Specific Gravity	About 1.033
Sperm Count	70-150 million/ml.
Sperm Motility	80% or more active; less than 80% active is abnormal
Sperm Morphology	80-90% normal forms
Normal Heads	80% or more
Cytology	A few crystals and epithelial cells

The American Society for the Study of Sterility has proposed a volume of 2-5 ml., a minimum count of 60 million/ml., and at least 60% active forms as adequate for fertility. Others in this field feel that these figures are too high and that a minimum sperm count of 30-40 million/ml. with 40% or more active forms is adequate. Morphology is classified as poor when less than 60% are normal.

METHOD: Appearance, volume, color, odor, microscopic examination, sperm count, morphology with stain.

INTERPRETATION: The testicle contains the two functional units, the seminiferous tubules and the Leydig cells; the seminiferous tubules are responsible for spermatogenesis and fertility while the Leydig cells are responsible for the production of testosterone and potency. The only objective finding that correlates with impotence is a serum testosterone level below the normal range for an adult male (Spark, R.F., Ann. Int. Med. 98, 103-104, 1983). The integrity of the adult male reproductive system is tested by measuring parameters of ejaculation. If the "normal" values are achieved or exceeded, then the entire hormonal system is normal.

The causes of male impotence are given in the next Table; (Seag, M.F. et al., JAMA 249, 1736, 1983):

Causes of Male Impotence	
Causes	Percent
Medications	25
Diuretics	
Antihypertensives	
Vasodilators	
Diabetes Mellitus	9
Nondiabetic Endocrine Abnormalities	20
Primary Hypogonadism	10
Hypothyroidism	5
Hyperthyroidism	1
Hyperprolactinemia	4
Alcoholism	7
Psychogenic	14
Stroke or Other Disease of CNS	7

Sperm counts are always performed to assess vasectomy sterilization procedures. The sperm count should drop significantly within several days, and by 2 weeks postoperatively, only a few dead sperm are found.

SEPTIC SHOCK TEST PANEL

Septic shock test panel is shown in the next Table:

Septic Shock Test Panel
Blood Cultures: A total of 3-4 blood culture bottle set (2 bottles/set, 5 ml blood/bottle) collected in a 24 hr. period, 3 different venipunctures at different sites, preferably at 1 hr. intervals; accuracy is greater than 90%.
Blood Gases: pH, PCO_2, PO_2
Electrolytes: Na^+, K^+, Cl^-, CO_2 Content
Lactate
Glucose
Disseminated Intravascular Coagulopathy(DIC); if clinical signs and symptoms of DIC develop, determine PT, fibrinogen level, and platelet count. See Disseminated Intravascular Coagulopathy(DIC) Panel

SERUM GLUTAMIC OXALACETIC TRANSAMINASE (SGOT)

Synonym: Aspartate Amino-transferase, (AST)
SPECIMEN: Red top tube, separate serum; green (heparin) top tube may be used, separate plasma; stable for 3 to 4 days at room temp., 7 to 10 days at 4°C and 6 months frozen. Interference: hemolysis will give spuriously elevated results; turbidity and icterus may interfere.
REFERENCE RANGE: Adult, 7-45U/liter; Newborn, 7-62U/liter decreasing to adult values at 3 months.
METHOD:

$$\text{Alpha-Ketoglutarate + Aspartate} \xrightarrow[\text{Pridoxal Phosphate}]{\text{GOT}} \text{Glutamate + Oxalacetate}$$

$$\text{Oxalacetate + NADH + H}^+ \xrightarrow[\text{Dehydrogenase}]{\text{Malate}} \text{NAD}^+ + \text{Malate}$$

INTERPRETATION: SGOT has a widespread tissue distribution. It is elevated in a wide variety of conditions as shown in the next Table:

Elevation of Serum Glutamic Oxalacetic Transaminase (SGOT)
Liver Diseases: Hepatitis; Giant Cell Hepatitis in Infants; Cholestasis, Intrahepatic and Extrahepatic; Alcoholism
Heart: Acute Myocardial Infarction; Acute Myocarditis (any cause)
Skeletal Muscle: Skeletal Muscle Diseases; Trauma
Red Blood Cells: Hemolytic Anemia (Severe); Megaloblastic Anemia
Other: Malignancy; Infectious Mononucleosis; Congestive; Heart Failure; Acute Renal Infarction; Acute Pulmonary Infarction; Acute Pancreatitis (Occ.); Tissue Necrosis; Third Degree Burns; Convulsions; Eclampsia; Heparin Therapy

SGOT is primarily used to detect and monitor liver parenchymal disease.
Very high values, over 500, may be found in liver disease, large necrotic tumors, congestive heart failure and shock.
SGOT values are greatly increased in acute liver damage, e.g., viral hepatitis, and toxic damage. Moderate elevation occurs in cirrhosis, metastatic cancer, obstructive jaundice and infectious mononucleosis. There is slight to moderate elevation of serum levels in congestive hepatomegaly.
The SGOT level is elevated in almost 98 percent of patients with acute myocardial infarction.
The serum levels are elevated in muscle disease including certain forms of muscular dystrophy, crush injury and gangrene of muscle and in dermatomyositis.
Drugs which may cause serum elevations include oral contraceptives, aspirin, isoniazid, codeine, cortisone and heparin.
SGOT is elevated in 27% of patients receiving heparin; the mean maximal increase is 3.1 times baseline value. The SGOT returned to normal in 80% of patients after heparin therapy is discontinued and in 20% during therapy (Dukes, G.E. et al., Ann. Int. Med. 100, 646-650, 1984).
SGOT is decreased in patients with pyridoxine deficiency (B-6, beriberi) and in patients with chronic dialysis.

SERUM GLUTAMIC PYRUVATE TRANSAMINASE (SGPT)

Synonym: Alanine Amino-transferase (ALT)

<u>SPECIMEN</u>: Red top tube, green (heparin) top tube may be used; serum stable for 3 days at room temp., 7 to 10 days at 4°C, and 6 months frozen. Interference: excess hemolysis.

<u>REFERENCE RANGE</u>: <u>Adult</u>, 7-35U/liter; <u>Newborn</u>, 6 to 62U/L decreasing to adult range in several months.

<u>METHOD</u>:

Alpha-Ketoglutarate + Alanine $\xrightarrow{GPT}$ Glutamate + Pyruvate

Pyruvate + NADH(340nm) + H$^+$ $\xrightarrow{LDH}$ Lactate + NAD$^+$

<u>INTERPRETATION</u>: SGPT is elevated in the diseases listed in the next Table:

Elevation of Serum Glutamate Pyruvate Transaminase (SGPT)
Liver Disease
Congestive Heart Failure
Infectious Mononucleosis
Acute Myocardial Infarction
Acute Renal Infarction
Skeletal Muscle Disease
Acute Pancreatitis
Heparin Therapy

SGPT is elevated in 60% of patients receiving heparin; the mean maximal increase is 3.6 times baseline value. The SGPT returned to normal in 80% of patients after heparin therapy is discontinued and in 20% during therapy.

SEXUAL ASSAULT DATA SHEET

A sexual assault data sheet, appropriate for emergency room setting, as an aid in the guidance in the collection of evidence, the obtaining of microbiologic cultures, prophylaxis against veneral disease and the prevention of pregnancy, is as follows:(Long, W.A. et al., Pediatrics 72:738-740,Nov.1983).

Sexual Assault Data Sheet

I. History
  A. Presentation in emergency room
    1. Date seen _____
    2. Time seen _____ A.M. _____ P.M.
    3. Mode of entry: police _____ friend _____ family _____
       self-referral _____ other _____
  B. Date of assault _____ A.M. _____ P.M.
  C. Time of assault _____ A.M. _____ P.M.
  D. Circumstances of assault (including postassault activity, changes of clothing, bathing, douching. Record evidence of torn clothing, bruises, blood or semen stains): _____
    _____
  E. Menarche _____
  F. Last menstrual period _____
  G. Method of birth control _____
  H. Current medications; yes _____ no _____
     specify _____

II. Physical Examination
  A. General appearance (include the emotional state, behavior of patient. Document areas of obvious trauma by photograph or diagram): _____
    _____
  B. T ___ P ___ BP ___ Wt ___ Pubertal stage (Tanner) ___
  C. Evidence of trauma: _____
    _____
  D. Description of clothing: torn _____
                             blood-stained _____
                             semen-stained _____
                             normal _____
  E. Description of perineum: normal _____
                             laceration _____
                             ecchymosis _____
                             bleeding _____
  F. Pelvic examination: Vagina _____
                         cervix _____
                         uterus _____
                         adnexa _____
                         rectum _____

III. Laboratory Evaluation	Done	Not Done	Results
A. Wet preparation of vaginal fluid for motile sperm and T. vaginalis	___	___	___
B. Vaginal washing for			
a. Acid phosphatase	___	___	___
b. ABH agglutinogen	___	___	___
C. Culture of vagina for GC	___	___	___
D. Culture of anus for GC	___	___	___
E. Culture of oropharynx for GC	___	___	___
F. Culture of urethra for GC	___	___	___
G. Serologic test for syphilis	___	___	___

Laboratory Evaluation (Continued)	Done	Not Done	Results
H. Pregancy test (pubertal females)			
I. Wood's lamp for semen	___	___	___
J. Hair combing of pubis	___	___	___
K. Fingernail scrapings	___	___	___
L. Serum sample frozen and saved for future testing	___	___	___

IV. Therapy   Dose Given

    A. Antibiotic prophylaxis (in accordance with current CDC recommendations):
       Tetracycline, 50 mg/kg/day for older children and 500 mg 4 times a day for adolescents, for ten days. Should not be given to patients who are pregnant or those allergic to tetracycline.
               OR   ___

       Doxycycline, 100 mg twice a day for 10 days (for adolescents).
               OR   ___

       Ampicillin, 3.5 g, or amoxicillin, 3.0 g as a single dose, taken with 1.0 g of probenecid, for adolescents.
       Younger children can be given 50 mg/kg of probenecid. Contraindicated in patients allergic to penicillin.

    B. Tetanus toxoid as indicated according to Public Health recommendations.   ___

    C. Pregnancy prevention for pubertal females: Ovral, 4 tablets in 2 divided doses 12 hours apart.   ___

V. Reported to police: date_____ time _____

VI. Disposition and follow-up:_____

Reference: Sarles, R.M., Pediatr. Rev. 4, 93 (1982).

SKELETAL MUSCLE PANEL
SPECIMEN: Red top tube, separate serum
    (a) Creatine Phosphokinase (CPK) and Isoenzymes
    (b) Serum Aldolase
    (c) Myoglobin, Urine (Random Specimen)

Creatine Phosphokinase (CPK) is found in high concentration in skeletal muscle and the heart; CPK and aldolase are increased in the serum in patients with muscle necrosis from any cause, such as trauma, and in the myopathies. CPK is not elevated in patients with neurogenic disorders.

Aldolase is not specific for skeletal muscle. It was assayed more frequently in the past prior to the time that assay of CPK became readily available. It is usually not necessary to determine both CPK and aldolase; CPK alone is usually adequate to detect and follow patients with muscle disease. Assay of urine myoglobin is especially useful when there is massive necrosis of skeletal muscle, and the patient is at risk for the renal tubular necrosis and acute renal failure.

The conditions involving skeletal muscle that are associated with an increase in CPK are shown in the next Table:

Skeletal Muscle Conditions and Elevated Serum CPK
Muscle Necrosis
Polymyositis
Mixed Connective Tissue Disease
Dystrophies such as Duchenne Type
Metabolic Myopathies: Hypothyroidism, Alcoholism, Malignant Hyperpyrexia

About 15 to 20 percent of patients with polymyositis have cardiac involvement; thus CPK-MB may be elevated. Almost 3/4 of patients with mixed connective tissue disease (MCTD) have muscle pain, muscle tenderness and weakness with elevated CPK and aldolase and with abnormal electromyograms consistent with inflammatory myositis. In muscular dystrophy, (Duchenne's), CPK values are markedly elevated at an early age and gradually decrease with progression of the disease; both CPK-MM and CPK-MB are increased. CPK is the most reliable means of identifying female muscular dystrophy carriers. In hypothyroidism, the mean value of CPK is seven times the post-treatment value.

The laboratory abnormalities in malignant hyperpyrexia are as follows: acidosis, hypoxemia, hyperkalemia, myoglobinemia and myoglobinuria and elevated CPK (Stehling, L. and Brown, D., Diagnostic Medicine, pgs. 59-64, May/June 1983).

SMOOTH MUSCLE ANTIBODY (SEE ANTI-SMOOTH MUSCLE ANTIBODY)

SODIUM, EXCRETION FRACTION OF FILTERED SODIUM)

SODIUM, SERUM

SPECIMEN: Red top tube, separate serum; or green (heparin) top tube, separate plasma.

REFERENCE RANGE: Adult: 135-148mmol/liter; Premature (cord): 116-140mmol/liter; premature: 128-148mmol/liter; newborn (cord): 126-166mmol/liter; newborn: 134-144mmol/liter; infant: 139-146mmol/liter; child: 138-145mmol/liter. The following are life-threatening values and, after confirmation, preferably by repeat determination on a different specimen, should be telephoned to the responsible nursing staff, or if unavailable, to the responsible physician so that corrective therapy may be immediately undertaken: serum sodium: <120mmol/liter; serum sodium: >155mmol/liter.

METHOD: Ion-specific electrode or flame emission photometry.

INTERPRETATION:

Hyponatremia: (Flear, C.T.G. et al., "Hyponatremia: Mechanisms and Management", The Lancet, pgs. 26-31, July 4, 1981): The causes of decreased serum sodium (hyponatremia) are shown in the next Table:

Causes of Hyponatremia (Simon, R.P. and Freedman, D.D., Geriatrics, pgs, 71-83, June 1980)

(1) Without renal sodium loss (urine sodium <5mmol/24hrs; normal, 40-200mmol/24hrs)
 (a) Dilutional Hyponatremia
  Congestive Heart Failure
  Cirrhosis with Ascites
  Edema associated with Hypoalbuminemia
   Nephrotic Syndrome
   Malnutrition
   Chronic Debilitating Disease
   Malabsorption
  Excessive Hypotonic Intravenous Fluid Administration
  Excessive Water Intake (Psychogenic Polydipsia)
 (b) Nonrenal Sodium Loss
  Vomiting (normal, Na^+ 20-120mmol/liter; mean, 40mmol/liter)
  Diarrhea
  Sweating (normal, $Na^+ \leq$ 60mmol/liter)
(2) With Renal Sodium Loss (Urine Sodium > 5mmol/24hrs.)
 (a) Enforced Diuresis
  Diuretic Drug Therapy
  Diabetes Mellitus with Glycosuria
  Hypercalciuria
  Radiographic Contrast Media
 (b) Renal Tubular Damage
  Renal Tubular Acidosis
  Aminoglycoside Antibiotics (e.g. Gentamicin)
 (c) Renal Disease
  Chronic Renal Disease, especially Tubulo-Interstitial
  Pyelonephritis-Acute or Chronic
  Obstructive Uropathy
  Polycystic Kidney Disease
 (d) Hormonal Abnormalities
  Adrenal Insufficiency
  Inappropriate Antidiuretic Hormone (ADH) Secretion
(3) Artifactual: Hyperlipidemia; Elevated Blood Glucose; Hyperproteinemia

If the patient is excreting less than 5mmol/liter of sodium over 24 hours in the absence of renal failure and sodium loss is not ascribed to excessive vomiting, diarrhea or sweating, the hyponatremia has resulted from an excess of extracellular water.

When urinary sodium content is greater than 5mmol/24 hours, a renal disorder is causative or contributory, and total body sodium content is depleted.

Clinical symptoms of hyponatremia depend both on the level of the serum sodium and how rapidly the decrease occurs. An acute change in Na^+ from normal range to 125mmol/liter may be associated with symptoms; a gradual change in Na^+ from normal range to 110mmol/liter may be asymptomatic.

A decrease in serum Na^+, relatively impermeable solute or an increase in extracellular water will cause water to move into cells, yielding cell swelling and intracellular hypotonicity. The symptoms of hyponatremia are primarily neurologic.

381

S. Bakerman

The plasma sodium concentration indicates the balance between salt and water and, by itself, gives no certain information about overall salt deficiency or excess.

Hyponatremia occurs as a result of water retention or sodium loss or both. Beer has a sodium concentration of 1.5 to 10.0mmol/liter and dilutional hyponatremia due to beer-drinking is probably common. Bodily depletion of sodium may be due to gastrointestinal loss, e.g., diarrhea and vomiting, diuretic therapy, sweating, hydroadrenalism or a variety of renal disorders. Urinary sodium concentration may be a useful indicator of salt depletion (usually less than 20mmol/24 hours or less than 10mmol/liter on a "spot" sample.) Drug-induced hyponatremia has two usual causes - diuretic induced and an ADH-like action of some drugs such as chlorpropamide and carbamazepine.

Hypernatremia: The causes of hypernatremia are given in the next Table:

Causes of Hypernatremia
Loss of Water
Insufficient Water Intake
Coma
Hypothalamic Lesion with Loss of Thirst Sensation
Excessive Sweating
Diarrhea
Renal Polyuria
Potassium Depletion
Hypercalciuria
Interstitial Nephritis
Fanconi's Syndrome
Extrarenal Polyuria
Diabetes Insipidus (Decreased ADH)
Nephrogenic Diabetes Insipidus
Diabetes Mellitus
High Protein Tube Feedings
Osmotic Diuretics
Increased Salt Intake (i.e., I.V. Saline)
Excessive Mineralocorticoids
Hyperaldosteronism
Cushing's Syndrome
Adrenogenital Syndrome
11-Beta Hydroxylase Deficiency
17-Hydroxylase Deficiency

The causes of hypernatremia have been discussed in a CPC in the N. Engl. J. Med. (Fang, L.S. and Young, R.H., N. Engl. J. Med. 308, pgs. 148-149, 1983).

Severe hypernatremia is uncommon; the conditions that are underlined in the Table are more commonly observed clinically.

Hypernatremia usually indicates water depletion sometimes with excessive sodium intake.

A common cause of hypernatremia in pediatric cases is enteric disease; in enteric disease, hypernatremic dehydration occurs. Dehydration occurs secondary to diarrhea, vomiting, anorexia and failure of water intake.

Hypernatremia elevates serum osmolality and results in intracellular dehydration as water shifts into the extracellular space; cells shrink. The effects are mediated, pathologically, via marked brain cell shrinkage, yielding mechanical trauma intracranially. Vascular damage is extensive; venous and capillary congestion is prominent and subarachnoid and intracerebral hemorrhages occur with cortical venous thrombosis and areas of venous infarction. Neurological symptoms occur in more than 50 percent of patients. Therapy is directed at the cause; replace water I.V. 5% dextrose or 0.45% saline or colloid when there is concomitant severe dehydration.

SODIUM, URINE

SPECIMEN: A random specimen may be used but a timed specimen, i.e., 8, 12 or 24 hours urine collection is preferred. Collect 24 hour urine as follows: Refrigerate specimen as it is collected. Instruct the patient to void at 8:00 A.M. and discard the specimen. Then collect all urine including the final specimen voided at the end of the 24 hour collection, i.e., 8:00 A.M. the next morning. Transport to lab and refrigerate specimen at 4°C.

REFERENCE RANGE: Dependent on intake. Infant: 0.3-3.5 mmol/day; Child: 40-180 mmol/day; Adult: 40-210 mmol/day.

METHOD: Ion-selective electrodes, flame photometry.

INTERPRETATION: Urinary sodium varies with intake, state of hydration, influence of drugs such as diuretics, abnormalities of renal perfusion, glomerular filtration or tubular function.

The major diagnostic value of urinary sodium determination is evaluation of patients with conditions listed in the next Table:

Major Diagnostic Value of Urinary Sodium Determination
Acute Oliguria
Hyponatremia
Volume Depletion

The level of urinary sodium in different conditions is given in the next Table:

Interpretation of Urinary Sodium	
Urinary Sodium	Interpretation
0-10 mmol/liter	Extra-Renal Sodium Loss (Gastrointestinal or Sweat Loss)
	Prerenal Azotemia
	Severe Volume Depletion
	Edematous States (Congestive Heart Failure, Liver Disease, Nephrotic Syndrome)
>10 mmol/liter	Acute Tubular Necrosis
	Inappropriate ADH Syndrome
	Adrenal Insufficiency (Addison's Disease)
	"Renal Salt Wasting"

Determination of urinary sodium can be extremely helpful if interpreted with knowledge of the clinical situation; this can be appreciated by the level of urinary sodium as shown in the next Table; (Harrington, J.T. and Cohen, J.J., N. Engl. J. Med. 293, 1241-1243, 1975):

Interpretation of Urinary Sodium Levels*		
Diagnostic Problem	Urinary Value	Primary Diagnostic Possibilities
Acute Oliguria	Na^+, 0-10 mmol/liter	Prerenal Azotemia
	Na^+, >30 mmol/liter	Acute Tubular Necrosis
Hyponatremia	Na^+, 0-10 mmol/liter	Severe Volume Depletion; Edematous States (Congestive Heart Failure, Liver Disease, Nephrotic Syndrome)
	Na^+, > Dietary Intake	Inappropriate Antidiuretic Hormone(ADH) Secretion; Salt-Losing Nephritis; Adrenal Insufficiency
Volume Depletion	Na^+, 0-10 mmol/liter	Extra-Renal Sodium Loss
	Na^+, > 10 mmol/liter	"Renal Salt Wasting" or Adrenal Insufficiency

*For purposes of this table, it is assumed that the patient is not receiving diuretics.

Acute Oliguria: Prerenal: There is hypoperfusion of the kidney and avid reabsorption of sodium and water by the nephrons. The most important diagnostic criterion of prerenal uremia is response to treatment (Editorial, Brit. Med. J., pgs. 1333-1335, June 7, 1980).

S. Bakerman

Renal Parenchymal: Damaged tubules fail to reabsorb solutes and water normally.
    The level of urinary sodium is used to differentiate prerenal and renal causes of oliguria as illustrated in the next Figure: (Espinel, C.H., JAMA 236, 579, 1976):

Urinary Sodium in Patients with Acute Tubular Necrosis and Patients
with Prerenal Azotemia

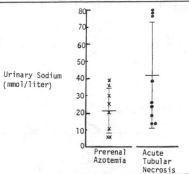

    Note that when a cut-off of urinary sodium of 30 mmol/liter is used, some patients with prerenal are above that level, and some patients with acute tubular necrosis are below that level. The excretion fraction of filtered sodium is a more sensitive and specific test to differentiate prerenal azotemia from acute tubular necrosis.
Hyponatremia: The concentration of sodium in the urine tends to be proportional to the serum sodium concentration. However, there are three different circumstances whereby a low serum sodium is associated with sodium greater than 30 mmol/liter; these are IADH syndrome, salt-losing nephritis and Addison's disease.

SOMATOMEDIN-C

SPECIMEN: Lavender(EDTA) top tube; separate plasma into plastic tube and freeze

REFERENCE RANGE: (Nichols Institute): The changes in somatomedin-C concentration in the serum in normal adults, acromegalic adults and hypopituitary children, are shown in the next Figure:

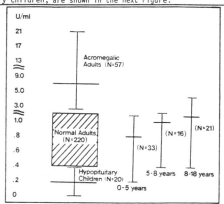

SOMATOMEDIN-C RIA VALUES IN NORMAL AND HYPOPITUITARY INDIVIDUALS									
Group	Number	Mean SM-C	SD	Females		Males		#<0.25	#.25–.39	#>0.4
		U/ml		N	Mean SM-C	N	Mean SM-C			
Normal 0-5 years	33	0.79	0.44	19	0.88	14	0.67	3	2	28
Normal 5-8 years	16	1.05	0.54	8	1.36	8	0.74	1	0	15
Normal 8-18 years	21	1.78	1.02	11	2.22	10	1.28	0	2	19
Normal Adults 18-64 years	220	1.17	0.63	115	1.21	105	1.12	0	3	217
Hypopituitary Children 2-18 years	20	0.17	0.11	5	0.23	15	0.15	16	3	1

METHOD: RIA

INTERPRETATION: The assay of somatomedin-C levels are useful clinically for the reasons listed in the next Table:

Clinical Uses of Somatomedin-C Assay

Normal Serum Somatomedin-C Levels: Virtually Rule Out Deficiencies of Growth Hormone Thus Eliminating Need for Growth Hormone Stimulation Tests.

Low Serum Somatomedin-C Levels: Requires G.H. Stimulation Tests for Work-up of Subjects with Possible G.H. Deficiency.

High Serum Somatomedin-C Levels: Compatible with Diagnosis of Acromegaly (Clemmons, D.R. et al., N. Engl. J. Med. 301, 1138, 1979).

Growth hormone acts on skeletal tissues indirectly by stimulating production of the serum peptides, somatomedins.

(1) Deficiency of Growth Hormone(GH): Growth hormone deficiency is usually idiopathic. The deficiency may be isolated or it may be accompanied by a deficiency of other pituitary hormones. The prevalence of growth hormone deficiency is about 1 child in 5,000.

(a) Somatomedins: Growth hormone acts on skeletal tissues indirectly by stimulating production of the serum peptides, somatomedins; a major source of somatomedins is the liver. Some of the other effects of growth hormone may be mediated through the action of cortisol.

Somatomedins were originally designated as the sulfation factor or thymidine factor. The somatomedins have a molecular weight between 6,000 to 9,000 daltons; somatomedin A is a neutral peptide; somatomedin C is a basic peptide. The somatomedins have insulin-like effects and the plasma concentrations are, to a large extent, under the influence of growth hormone.

SPERM COUNT (see SEMEN ANALYSIS)

STANDARD TEST FOR SYPHILIS (STS)-CSF

SPECIMEN: Cerebral spinal fluid; minimum volume 0.5ml; cause for rejection is excessive hemolysis or turbidity.
REFERENCE RANGE: Nonreactive
METHOD: The STS uses an alcoholic extract of beef heart and cholesterol, lecithin and alcohol to give reproducible qualitative and quantitative agglutination reactions for the detection of antibody to treponemal infection There is a slide test and a tube test.
INTERPRETATION: The STS is a screening test for antibodies in the CSF of patients with syphilis. Increased leukocytes and/or protein is suggestive of CNS involvement. This test on CSF gives a high percentage of false negatives; in neurosyphilis, this test is reactive in all cases of general paresis, but in only 70% of tabetics.

The FTA-ABS test is not used for CSF testing because it lacks sensitivity.
See SYPHILIS TEST FOR ANTIBODIES.

STAT TOXICOLOGY PANEL

Stat toxicology tests are listed in the next Table (Saxena, K., Res. and Staff Physician, pgs. 47-57, March, 1983):

Stat Toxicology Tests
Salicylate
Alcohol
Barbiturate Screens
Acetaminophen
Iron
Volatile Screen: Methanol, Ethylene Glycol, Isopropanol
Methemoglobin Determination in Nitrate or Nitrite Ingestion
Carbon Monoxide

Tricyclic Antidepressants: A QRS duration of 100msec or more suggests the possibility of a serious overdose of tricyclic antidepressants.

STONE ANALYSIS (see RENAL STONE ANALYSIS)

STOOL CULTURE AND SENSITIVITY

SPECIMEN: Rectal swab; fresh random stool in stool container or Culturette;
specimen must be less than 4 hours old when delivered to the laboratory. Do not
refrigerate. Blood cultures for Salmonella, Shigella and Campylobacter should
be obtained in those patients with moderate to severe illness and those with
high fever.

REFERENCE RANGE: Negative

METHOD: In patients with suspected Clostridium difficile colitis, stools for
diagnostic toxin assays are done.

INTERPRETATION: The usual causes of bacterial diarrhea are given in the next
Table:

Usual Causes of Bacterial Diarrhea
Staphlococcus Aureus
Salmonella
Shigella
Clostridium Difficile
Campylobacter Jejuni
Yersinia Enterocolitica

Salmonella (nontyphoidal) is usually food borne; the most frequent sources
are poultry and eggs. The incubation period ranges from 6 to 24 hours.
Salmonella typhi may also cause constipation; causes are often associated with
recent foreign travel.

Shigella is readily spread among members of a family or group. The
incubation period varies from 6 to 120 hours.

Yersinia may cause the following: enterocolitis, acute mesenteric adenitis,
arthritis, erythema nodosa or septicemia.

Enterocolitis is the most common manifestation with fever, diarrhea and
abdominal pain lasting one to three weeks. The typical patient is less than
five years of age. These infections are usually self-limited.

There are two Campylobacter pathogenic to man; one is Campylobacter jejuni
and the other is Campylobacter fetus; the latter is associated with systemic
infections. The former, Campylobacter jejuni, is associated with
gastroenteritis. The clinical picture is that of acute gastroenteritis with
fever and severe abdominal pain. Diarrhea usually follows within twenty-four
hours after onset of symptoms with stools that are watery, mucoid or bloody.
The course is variable, and mild disease, which may resemble viral
gastroenteritis, can be treated with supportive therapy only. In more severe
cases, or in relapsing cases, erythromycin (or tetracycline) is the drug of
choice. (Marrow, H., Personal communication.)

STOOL ELECTROLYTES

SPECIMEN: 24 hours stool specimen

REFERENCE RANGE: Chloride:6-17 mmol/24 hours; (6 to 17 mmol/liter); Sodium: 7.8±2.0 mmol/24 hour; Potassium: 18.2±2.5 mmol/24 hour.

METHOD: Collect 24 hour stool specimen; obtain 30 gram sample, add 100 ml of water and stir for 30 minutes. It is necessary to prevent clogging of aspiration devices of electrolyte machines. Therefore, obtain 40 ml of the suspension, centrifuge, collect supernatant and set aside; then, add 40 ml of distilled water to the residue, mix, centrifuge, collect supernatant and set aside. Repeat extractions for a total of five times. Collect fluids and bring to a volumne of 250 ml in a volumetric flask; (Caprilli, R. et al., Scand. J. Gastroenterol. 13, 331-335, 1978).

INTERPRETATION: Chloride is increased in the feces in the following conditions: congenital chloride diarrhea(CCD); acquired chloride diarrhea (Kaplan, B.S. and Vitullo, B., J. Pediatr. 99, 211-214, 1981); secondary chloride-losing diarrhea (Aaronson, I., Arch. Dis. Child. 46, 479, 1971); idiopathic proctocolitis(IPC) (Caprilli, R. et al., Scand. J. Gastroent. 13, 331-335, 1978); cholera.

CCD and acquired chloride diarrhea are characterized by large concentrations of fecal chloride, metabolic alkalosis, hyponatremia, hypokalemia, hypochloremia, and almost no chloride in the urine; CCD is a more severe illness. Patients with cholera have metabolic acidosis rather than alkalosis.

The mechanism of development of hypochloremic metabolic alkalosis is illustrated in the next Figure:

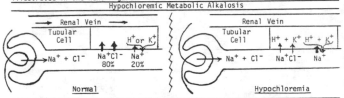

Hypochloremic Metabolic Alkalosis

Normally, 80% of the $[Na^+]$ in the renal tubular lumen is reabsorbed while accompanied by $[Cl^-]$ and 20% is exchanged for $[K^+]$ and $[H^+]$.

When there is hypochloremia, the amount of $[Na^+]$ that is resorbed accompanied by $[Cl^-]$ is reduced and more $[Na^+]$ is exchanged for $[K^+]$ and $[H^+]$. When $[Na^+]$ is exchanged for $[K^+]$ and $[H^+]$, the loss of $[H^+]$ represents a loss of acid; thus, the patient becomes alkalotic, hence hypochloremic alkalosis; with concomitant loss of $[K^+]$ with $[H^+]$, hypokalemia may occur.

STOOL FAT (see FECAL FAT)

STOOL LEUKOCYTES

SPECIMEN: Mucus or liquid stool

REFERENCE RANGE: Negative

METHOD: A small fleck of mucus or liquid stool is carefully and thoroughly mixed with an equal amount of methylene blue stain solution (reticulocyte stain) or Wright stain and examined for the presence of leukocytes.

INTERPRETATION: This test helps to separate treatable bacterial diseases from diarrhea due to viruses, bacterial toxins or parasites; interpretation is given in the next Table:

Fecal Leukocytes in Diarrhea	
Leukocytes Present	Leukocytes Absent
Shigella	Bacterial Toxins
Salmonella	Staphylococcus
E. Coli, Invasive Strains	E. Coli
Yersinia	Clostridium Perfringens
Ulcerative Colitis	Vibrio Cholerae
	Giardia Lamblia
	Entamoeba Histolytica
	Dientamoeba Fragilis
	Viruses

Satterwhite, T.K. and Dupont, H.L., JAMA 236, 2662-2664, 1976.

Screening of feces, using the Hemoccult test, has been proposed because of the ambiguities that may arise in the test for fecal leukocytes; the Hemoccult test is usually positive when fecal leukocytes are present (Vögtlin, J. The Lancet, pg. 1204, Nov. 19, 1983).

STREPTOCOCCAL ANTIBODIES (see STREPTOZYME, ANTI-STREPTOCOCCAL-O, ASO TESTS and ANTIDESOXYRIBONUCLEASE B)

STREPTOZYME

SPECIMEN: Red top tube, remove serum

REFERENCE RANGE: <100 STZ units

METHOD: (Wampole Laboratories) Streptozyme is a commercial reagent consisting of formalin-treated sheep erythrocytes that are coated with group A streptococcal antigens, DNase, streptokinase, streptolysin O, hyaluronidase and NADase. Serum diluted 1:100 is mixed on a glass slide with a drop of Streptozyme and agglutination is observed macroscopically.

INTERPRETATION: This is a test for antibody in a patient's serum to multiple different streptococcal antigens, the extracellular product of Group A streptococcus. A significant rise of antibody titer results in a positive test. It is positive in the conditions listed in the next Table:

Positive Streptozyme Test
Post-Streptococcal Glomerulonephritis
Acute Rheumatic Fever
Streptococcal Infections of Pharynx and Skin
Bacterial Endocarditis

The antibodies measured by streptozyme increase more rapidly and appear in the blood earlier than ASO following infection with group A streptococci.

Antibodies increase 1-3 weeks after onset and decrease 8-10 weeks after uncomplicated infection. The sensitivity of the test in patients with streptococcal infection is 95 percent.

SWEAT CHLORIDE (TEST FOR CYSTIC FIBROSIS)

SPECIMEN: Volume of sweat desired, 200mg; minimum volume, 100mg. The specimen can be stored in the refrigerator at 4°C for 1 week if the container is tightly sealed against evaporation; otherwise, the container may be stored in a freezer. Sweat testing is not considered accurate until the third or fourth week of life. Infants younger than six weeks may not sweat sufficiently; the leg may be used for sweat collection in these infants.

REFERENCE RANGE: Sweat chloride and sweat sodium $\leq$ 60mEq/liter. Sweat chloride > 70mEq/liter or sodium > 70mEq/liter is consistent with the diagnosis of cystic fibrosis (98% of cases). Children who have values of sweat chloride or sweat sodium between 50 and 70mEq/liter should have repeated sweat tests. During the first two or three postnatal days chloride values may be as high as 80mmol/liter and returning to normal levels on the third or fourth day. Normal adults (>21 years of age) have a range of values for chloride of 10-70mEq/liter.

A gap of more than 30mmol/liter between the sodium and chloride values indicates an error in calculation, analysis or contamination.

Sweat chloride and sweat sodium are elevated in up to 98% of persons with cystic fibrosis.

METHOD: A kit for assay may be obtained from Schwachman, H., Cystic Fibrosis Clinic, The Children's Hospital Medical Center, 300 Longwood Avenue, Boston, Mass. 02115 or Farrall Instrument Co., P.O. Box 1037 Grand Island, Neb. 68802, Tel. #308-384-1530. (Schwachman, H. and Mohmoodian, A., "Quality of Sweat Test Performance in the Diagnosis of Cystic Fibrosis", Clin. Chem. 25, 158-161, 1979). The method used is the standard quantitative pilocarpine iontophoresis test. The sweat test is done as follows: (1) Local stimulation of sweat with pilocarpine iontophoresis. Pilocarpine is a cholinomimetric; one of its parasympathomimetric effects is its potent ability to produce sweat (2) Collection of the sweat on Curity gauze pads (3) Analysis of chloride by titration and sodium and potassium by flame photometry.

INTERPRETATION: Cystic fibrosis is the most common serious genetic disease in the white population in the United States; it is transmitted as an autosomal recessive trait, with a carrier rate among Caucasians of 1 in 20 and an incidence of about 1 in 1800 live births. The carrier rate in the black population is one in 60 to one in 100; therefore, one in 3600 to one in 10,000 black couples is at risk; their risk is one in four; in 1967, in the black population, there were 17 new cases of cystic fibrosis in the United States; there were 571,000 births of black infants.

The classical clinical triad consists of malabsorption due to pancreatic insufficiency, chronic suppurative lung disease and failure to thrive. Chronic respiratory disease is the major cause of morbidity and mortality. The sweat test is the definitive test for cystic fibrosis and should be done on subjects with the following conditions (Schwachman, H., Instructions accompanying Iontophoresis Kit):

1. Infants who pass their initial meconium late, i.e., after approximately 30 hours of age
2. Intestinal obstruction in the newborn
3. Failure to thrive in infancy or childhood
4. All siblings of patients with cystic fibrosis, including the newborn
5. All patients suspected of having cystic fibrosis or celiac disease
6. Infants and children with steatorrhea or chronic diarrhea
7. Infants with rapid respirations and retraction with chronic cough
8. Infants with diagnosis of asthma
9. Infants with hypoproteinemia, especially when on soy bean formula
10. Infants who show hyperaeration on chest x-ray or have atelectasis
11. Infants with hyperprothrombinemia, including major hematologic disease
12. Infants and children with rectal prolapse
13. All patients suspected of having disaccharidase intolerance or diagnosed as having celiac disease
14. Infants and children who taste salty
15. Patients with recurrent pneumonia, chronic atelectasis, chronic pulmonary disease, bronchiectasis or chronic cough
16. Children and adults with nasal polyposis even though they may be allergic

17. Children and young patients (non-alcoholic) with cirrhosis of the liver and portal hypertension
18. Infants of a CF parent; the obligate heterozygote
19. Finally, for any parents who request a sweat test on their child

Abnormally high values are found in cystic fibrosis and may be found in glucose-6-phosphate dehydrogenase deficiency, hypothyroidism, glycogen storage disease, untreated adrenal insufficiency and malnutrition.

Newborns with cystic fibrosis have elevated serum levels of the pancreatic enzyme trypsin; the feasibility of mass screening of newborns using assay of serum trypsin is being investigated.

Ref.: David, T.J. and Phillips, B.M., The Lancet, pgs. 1204-1206, Nov. 27, 1982; Editorial, The Lancet, pgs. 1196-1197, Nov. 27, 1982; Gunby, P., Arch. Intern. Med. 914, 631-634, April, 1983.

SYNOVIAL FLUID ANALYSIS
SPECIMEN: Specimens: Divide specimen into 4 aliquots as follows:
(1) Red top tube for gross appearance and mucin clotting
(2) Heparin tube for microscopic examination
(3) Sterile tube for culture
(4) Heparin tube for chemical analysis
REFERENCE RANGE: The constituents in normal synovial fluid are shown in the next Table:

Constituents of Normal Synovial Fluid	
Constituent	Synovial Fluid
Protein	1-3g/dl
Albumin	55-70%
Hyaluronate	0.3-0.4g/dl
Glucose	70-110mg/dl
Uric Acid	
Males	2-8mg/dl
Females	2-6mg/dl
Lactate	10-20mg/dl (1-2mmol/liter)

Normally about 0.1 to 2.0 ml of synovial fluid is present in the knee joint. Synovial fluid protein is less than that of plasma. Hyaluronate is a polymer of repeating disaccharide units and is not found in the plasma; it gives synovial fluid its high viscosity. Glucose and uric acid concentrations are like that of plasma; lactate concentration is like that of venous blood. Synovial fluid does not normally clot.

METHOD: Mucin Clot (Ropes) Test: The mucin clot test reflects polymerization of hyaluronate. In the mucin clot test, a few drops of synovial fluid are added to 20ml of 5% acetic acid. Normally, a mucin clot forms within 1 minute; a poor clot indicates inflammation.

INTERPRETATION: Diseases and synovial fluid analysis are given in the next Table:

Diseases and Synovial Fluid Analysis				
Disease	Appearance	Mucin Clot	Glucose	Cells/cumm
Normal	Straw, Clear	Good	70-110 mg/dl	10-600 25% polys
Osteoarthritis	Yellow, Clear	Good to Fair	Decreased (-10)	1000 50% Lymphs
Traumatic Arthritis	Cloudy, Bloody	Good to Fair	Decreased (-10)	20,000 Many RBCs
Pseudogout	Slightly Yellow Cloudy	Fair to Poor	Decreased (-20)	6000 75% polys
Gout	Yellow-White, Milky	Fair to Poor	Decreased (-20)	10,000 75% polys
Rheumatic Fever	Yellow, Slightly Cloudy	Good to Fair	Decreased (-10)	10,000 50% polys
Bacterial Arthritis	Grey-Red, Turbid	Poor	Decreased (-60)	80,000 90% polys
Tuberculous Arthritis	Yellow, Cloudy	Poor	Decreased (-60)	25,000 50% polys
Rheumatoid Arthritis	Yellow-Green Cloudy	Fair to Poor	Decreased (-30)	8,000-40,000 70% polys
Lupus Erythematosus		Good to Fair	Decreased (-10to20)	3,000 10% polys

S. Bakerman

Crystals: <u>Pseudogout</u>: Using polarized light, the <u>calcium pyrophosphate
dihydrate crystals</u> appear as rods, rectangles or rhomboids and weakly negative
birefringent.

<u>Gout</u>: Crystals are seen in about 90% of patients during attacks of acute gouty
arthritis and in about 75% between attacks. Using <u>polarized</u> light urate
crystals appear as birefringent rods or needles, <u>strongly positive birefringent</u>.
Note whether crystals are intracellular or extracellular. Intracellular
crystals suggest urate as cause of acute arthritis. Incubate with uricase to
confirm impression (McCarty, D.J. and Hollander, J.L., Ann. Intern. Med. <u>54</u>,
1961).

SYPHILIS, ANTIBODIES

SPECIMEN: Red top tube, separate serum
REFERENCE RANGE: Non-reactive or negative
METHOD: See RPR, STS and FTA-ABS (Indirect Fluorescence Test)
INTERPRETATION: There are <u>two</u> general <u>procedures</u> for the detection of
antibodies produced in response to treponemal pallidum infection; these are
listed in the next Table:

Tests for Detection of Antibodies Produced
in Response to Treponema Pallidum

(1) Use of Non-Treponemal Derived Substances to Precipitate Antibody:
    (a) VDRL (Venereal Disease Research Laboratory)
    (b) RPR (Rapid Plasma Reagin)
    (c) ART (Automated Reagin Test)
    (d) STS (Standard Test for Syphilis)
(2) Use of Antigen Derived from T. Pallidum to Precipitate Antibody:
    (a) FTA-ABS (Fluorescent Treponemal Antibody-Absorbed Test)
    (b) MHA-TP (Microhemagglutination-Treponema Pallidum)

The RPR cannot be used with cerebrospinal fluid (CSF); the STS can be used
to test antibody in CSF.

(1) The non-treponemal tests are used as <u>screening tests</u> and for <u>following
therapy</u>; <u>quantitation</u> is done to <u>follow the response to therapy</u> and to <u>detect</u>
reinfection.

These tests use extracts of <u>beef heart</u> (cardiolipin) to give reproducible
qualitative and quantitative reactions to antibody produced following infection
with treponemal pallidum. It is not known whether the antibodies are produced
in response to antigens in T. pallidum or to antigens resulting from the
interaction of T. pallidum with tissue. <u>False-positive reactions</u> occur in a
variety of acute and chronic conditions as listed in the next Table:

Conditions Causing False-Positive Using Non-Treponemal Tests
for Antibodies Produced in Response to Syphilis

        Lupus Erythematosus
        Malaria
        Infectious Mononucleosis
        Infectious Hepatitis
        Post-Vaccination States
        Leprosy
        Brucellosis
        Atypical Pneumonia
        Miliary Tuberculosis
        Typhus
        Related Treponemal Infections
        Yaws
        Pinta
        Pregnancy

False-positive tests for syphilis, in patients with <u>systemic lupus
erythematosus</u> are caused by <u>anti-DNA antibodies</u> that <u>cross-react with
cardiolipin</u> (Koike, T. et al., Clin. Exp. Immunol. <u>56</u>, 193-199, 1984).

(2) Use of Antigen Derived from T. Pallidum: These tests are most commonly used to determine whether the results of a non-treponemal test are due to syphilis or due to a condition causing false-positive. These tests may also be used to detect syphilis in patients with negative nontreponemal test results but with clinical evidence of late syphilis.

The MHA-TP test is less sensitive than the FTA-ABS test in primary untreated syphilis; the sensitivity of the MHA-TP test is 65% to 70%, and the FTA-ABS has a sensitivity of 80% greater. However, the MHA-TP test has the following characteristics: simplicity, lower overall cost, and reliability. If the non-treponemal test is positive, then the MHA-TP test should be done.

The treponemal tests do not indicate the patient's response to treatment; they are of doubtful value in the diagnosis of active neurosyphilis.

T. pallidum has not been cultivated in-vitro; it is grown intratesticularly in rabbits; thus, antigens are contaminated with rabbit tissue.

False positive FTA-ABS test results have been obtained in patients with increased or abnormal globulins, with pinta, yaws and bejell and with lupus erythematosus and during pregnancy.

The Centers for Disease Control are able to perform a specific IgM antibody test by the fluorescent treponema antigen-antibody test. Rheumatoid factor is removed from the serum with a column. A positive IgM antibody is indicative of recent and ongoing infection.

The reactivity of nontreponemal and treponemal tests in untreated syphilis is given in the next Figure; (Henry, J.B. [ed.] Todd-Sanford-Davidsohn, Clinical Diagnosis by Laboratory Methods, 16 ed., 1979, Vol. II pg. 1890, W. B. Saunders Co., Phila. Pa):

Reactivity of Nontreponemal and Treponemal Tests in Untreated Syphilis

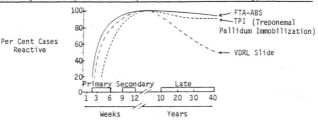

Per Cent Cases Reactive — FTA-ABS — TPI (Treponemal Pallidum Immobilization) — VDRL Slide. Primary Secondary Late. Weeks 1 3 6 9 12 — Years 10 20 30 40.

Incubation Period: Two to 10 weeks; during this period, there is no clinical or serological evidence of infection. T. pallidum is multiplying at the site of inoculation and is invading the lymphatics and blood stream.

Primary Stage: Appearance of lesion (chancre) at the site of inoculation; chancres usually heal spontaneously within a few weeks.

Secondary Stage: Generalized skin and mucous membrane lesions; multiple organs may be involved including the central nervous system. These manifestations disappear usually after several weeks. About 10 percent of patients with recognized secondary syphilis have clinical evidence of liver disease (Koff, R.S. in Case Records of Massachusetts General Hospital 309, 35-44, 1983).

Late Syphilis: 5 to 20 years after infection; the most frequent complications involve lesions in the central nervous and cardiovascular systems and gummatous lesions of skin, bone and viscera.

Congenital Syphilis: The spirochete, in a mother who has active syphilis, crosses the placenta and causes a spirochetemia in the fetus, usually around 6-7 months gestation. Congenital syphilis occurs in less than 0.05% of all pregnancies.

T AND B LYMPHOCYTE ENUMERATION

SPECIMEN: Green top (heparin) top tube and deliver to laboratory immediately.
REFERENCE RANGE: B-lymphocytes: 10 to 18 percent; T-lymphocytes: 80 to 90 percent; Null cells(L-cells, Non-T, Non-B): 0.5 to 1 percent.
METHOD: Kits are available using fluorescent or peroxidase and monoclonal antibodies (See 1984 "Gold Book," published by Laboratory Management 22, No. 2, Feb. 1984). Rosette formation and immunofluorescence assays. Lymphocytes are isolated using a Ficoll-Hypaque gradient; Assay for B-lymphocytes: Surface immunoglobulins are detected by fluorescein-labeled antihuman immunoglobulin, either polyvalent or monospecific; Assay for T-lymphocytes: Suspend cells in diluant; dilute an aliquot of sheep red cells; incubate sheep red cells and lymphocytes; obtain a drop of suspension and observe for rosette formation. T-lymphocytes form spontaneous nonimmunoglobulin-mediated rosettes with sheep red blood cells.
INTERPRETATION: B-lymphocytes are derived from bone marrow and T-lymphocytes are derived from the thymus. B- and T-cell lymphoid proliferations are given in the next Table:

B- and T- Cell Lymphoid Proliferations	
Lymphoid Proliferation	Cell Type
Chronic Lymphocytic Leukemia(CLL).........	B-Cell, usually-monoclonal and IgM class; T-Cell rarely.
Acute Lymphoblastic Transformation.....Usually same as CLL. of CLL	
Acute Lymphoblastic Leukemia(ALL).........	Usually no markers; that is, null cells; 25% T-Cell; rarely B-Cell(2%).
"Hairy Cells" of Leukemic................	Controversial-Monocytes in One Study
Reticuloendotheliosis	Study and B in Another; Also T.
Infectious Mononucleosis..................	T-Cell(circulating Cell); B-Cell (infected Cell) may be present in Lymphoid Tissue.

Cells from about 20-30 percent of patients with null cell ALL have characteristics of pre-B-cells in that they contain intracytoplasmic IgM but lack surface Ig.
Primary Immunodeficiency Disorders: B- and T-cell immunodeficiency disorders are given in the next Table:

B- and T-Cell Immunodeficiency Disorders		
Immunodeficiency Disorder	Blood Lymphocytes	Major Immune Defect
Bruton-Type X-Linked Infantile Agammaglobulinemia	Absent B-Cells	B-Cell Defect
Severe Combined Immuno- Deficiency Disorders(SCID)	Decreased B-Cells Decreased T-Cells	B- and T-Cell Defects
Thymic Dysplasia (DiGeorge Syndrome)	Decreased T-Cells Increased B-Cells	T-Cell Defect
Ataxia Telangiectasia	Decreased T-Cells Increased B-Cells	B- and T-Cell Defects
Wiskott-Aldrich Syndrome	Decreased T-Cells Dec. to Inc. B-Cells	B- and T-Cell Defects

General Reference: Gupta, S. and Good, R.A., Seminars in Hematology 17, No. 1, 1-29, Jan. 1980; Foucar, K. and Goeken, J.A., Laboratory Medicine 13, No. 7, 403-414, July, 1982.
T-cell receptors apparently consist of two chains, held together by disulfide bonds; each chain has a molecular weight of about 40,000 (Science, 221, 444-446, 1983).

S. Bakerman

TEGRETOL (see CARBAMAZEPINE)

TEICHOIC ACID ANTIBODY TITER
SPECIMEN: Red top tube, separate serum and refrigerate
REFERENCE RANGE: Titer less than 1:8
METHOD: Counterimmunoelectrophoreis (CIE)
INTERPRETATION: Antibody to teichoic acid is increased in infection with
Staphylococcus aureus. Interpretation of teichoic acid antibody results is as
follows; (Smith-Kline Clinical Laboratories):
1. Antibodies are often negative during the first two weeks following
   onset of Staphylococcal infection; therefore, if Staphylococcal
   infection or endocarditis is a significant concern, and if antibodies
   are negative early, they should be repeated.
2. The incidence of antibodies relates to the seriousness of the
   infection.

	% Positive	% with Titer >1:2
Staphylococcal Endocarditis	87	72
Staphylococcal Bacteremia with Metastatic Infection	50	35
Uncomplicated Staphylococcal Bacteremia	8	0
Non-bacteremic Staphylococcal Infection	58	10
Normals, Estimated	5	0

3. Occasional positive results occur in patients with other types of
infections, collagen or neoplastic diseases, and these are usually in
undiluted serum only.

TERMINAL DEOXYNUCLEOTIDYL TRANSFERASE(TdT)
SPECIMEN: Bone marrow; peripheral blood; cerebrospinal fluid.
REFERENCE RANGE: Negative
METHOD: Bone marrow and peripheral blood are treated with 0.1M NH_4Cl to
lyse erythrocytes; the mononuclear fraction is spread on a glass slide by
cytocentrifuge; fix smear and maintain at room temperature in a dessicator for
48 to. 72 hours. Layer rabbit antibody to TdT over the cytocentrifuge smear and
incubate for 30 minutes; wash slide with phosphate buffered saline, and add
fluorescein tagged antirabbit IgG for 30 minutes. Wash slides and mount.
INTERPRETATION: The presence of terminal deoxynucleotidyl transferase(TdT),
an intranuclear enzyme, in blast cells may have diagnostic, therapeutic and
prognostic significance. It is found in a high percent of the cells of patients
with acute lymphocytic leukemia(ALL); it is observed in very few cells during
remission of ALL. Patients with chronic myelocytic leukemia may have
"lymphoblastic" crisis with presence of TdT in cells.
    The presence of TdT identifies blast cells as probably lymphoid in nature;
TdT positive cells are sensitive to steroids and vincristine while TdT negative
cells are resistant to steroids.
    (Morse, E.E. et al., Ann. Clin. Lab. Sci. 13, No. 2, 128-132, 1983).

TESTOSTERONE, FREE

<u>SPECIMEN</u>: Red top tube, separate serum promptly, store at -20°C; or green (heparin) top tube, separate plasma and store at -20°C until assay.

<u>REFERENCE RANGE</u>: <u>Adult Males</u>: 9-30 ng/dl; <u>Females</u>: The average free testosterone concentration in girls rises on pubertal maturation; by midpuberty, it reaches adult level (Moll, G.W. and Rosenfield., R.L., J. Pediatr. <u>102</u>, 461-464, 1983). Adult values vary with methodology; Mayo Clinic Laboratories, 0.3-1.9 ng/dl.

<u>METHOD</u>: Radioimmunoassay with extraction and equilibrium dialysis.

<u>INTERPRETATION</u>: Free testosterone is the non-protein-bound fraction of circulating testosterone and is probably the physiologically active form; only 1 to 3% of plasma testosterone is in the free state. The bound testosterone is bound to a <u>specific</u> binding protein, testosterone-estradiol binding globulin; there is <u>non-specific</u> binding to albumin.

In the diagnostic workings of <u>hirsutism</u>, due to androgen-secreting <u>ovarian</u> or <u>adrenal neoplasms</u>, <u>total testosterone</u> and <u>dehydroepiandrosterone sulfate</u> are the <u>most useful markers</u> in that these neoplasms hardly ever occur without peripheral elevation of either steroid. Measurement of free testosterone is not necessary in these conditions.

Usually, there is a <u>positive linear correlation</u> between the level of total plasma testosterone and free testosterone. However, <u>hyperthyroidism</u> and <u>hyperestrogenic</u> states, such as pregnancy and oral contraceptives, are characterized by <u>increased testosterone-estradiol binding globulin</u>, <u>increased total testosterone and normal free testosterone</u>.

There are <u>relatively few conditions</u> in which assay of free testosterone is required or recommended. <u>The measurement of plasma free testosterone before and after administration of dexamethasone appears to be the most sensitive single method for detecting polycystic ovary syndrome in adolescence</u> (Moll, G.W. and Rosenfield, R.L., J. Pediatr. <u>102</u>, 461-464, 1983).

The clinical significance of the use of free testosterone for the diagnosis of hirsutism is controversial. Although plasma free testosterone concentration was found to be elevated in every hirsute woman studied by Paulson, J. D. et al., (Am. J. Obstet. Gynecol. <u>128</u>, 851, 1977), other studies indicate that free testosterone is elevated in only half of hirsute women (Schwartz, U. et al., Fertil. Steril. <u>40</u>, 66-72, 1983; Wu, C.H., Obstet. Gynecol. <u>60</u>, 188-194, 1982).

The drug, <u>danazol</u>, an <u>ethinyl testosterone derivative</u>, is frequently used in the clinical <u>treatment</u> of <u>endometriosis</u>. Free testosterone is markedly elevated in patients during danazol therapy, and the effects of danazol could be explained by increased levels of free testosterone (Nilsson, B. et al., Fert. Steril. <u>39</u>, 505-509, 1983).

TESTOSTERONE, TOTAL, SERUM

SPECIMEN: Hormone levels undergo rapid and large oscillations; obtain three equally-spaced samples taken at about 10 min. intervals and pool the three specimens. Red top tube, separate serum; store frozen; or green (heparin) top tube, separate plasma and freeze.

REFERENCE RANGE: Adult male: 350-800ng/dl; in healthy men, serum testosterone remains in the normal range even with advancing age (Sparrow, D. et al., J. Clin. Endocrinol. Metab. 51, 508-113, 1980). Adult female: 10-60ng/dl; Prepubertal children: <10ng/dl; Pregnancy: 75-300ng/dl.

METHOD: RIA

INTERPRETATION: This test is useful in the evaluation of hypogonadism in the male and hirsutism and virilization in females.

Males: The integrity of the adult male reproductive system may be tested by measuring parameters of ejaculation (see SEMEN ANALYSIS). Serum testosterone is decreased in (a) testicular failure (hypergonadotropic hypogonadism) and (b) pituitary failure (hypogonadotropic hypogonadism): (see LUTEINIZING HORMONE).

There are two conditions in which male sexual dysfunction is associated with elevated serum total testosterone level; these are hyperthyroidism and syndromes of androgen resistance. Hyperthyroidism causes an increase in testosterone binding globulin and more testosterone binds to this globulin; the free testosterone is normal. When hyperthyroidism is treated successfully, testosterone binding globulin returns to normal and potency is restored (Spark, R.F. et al., JAMA 243, 750-755, 1980).

Syndromes of androgen resistance are also associated with increased serum testosterone; in this syndrome, there may be an abnormality of the androgen receptor of the 5-alpha-reductase enzyme which is responsible for the conversion of testosterone to the active metabolite, dihydrotestosterone.

Females: Hirsutism in the female can be of adrenal or ovarian origin and may be caused by excess androgens. Testosterone is a good indicator of ovarian function while dehydroepiandrosterone-sulfate (DHEA-S) is a good indicator of adrenal function. Plasma testosterone over 200ng/dl is usually indicative of an ovarian abnormality (Maroulis, G.B., Fertil. Steril. 36, 273, 1981).

The clinical causes of hirsutism are given in the next Table:

Causes of Hirsutism
Common
Polycystic Ovary Syndrome
Idiopathic
Rare
Adrenal Origin: Cushing's Syndrome; Congenital Adrenal Hyperplasia; Androgen-secreting Tumors
Ovarian Origin: Hilus Cell Tumor; Androblastoma; Teratoma
Hypothyroidism
Acromegaly
Drugs
Phenytoin; Diazoxide; Minoxidil; Menopausal mixtures containing Androgens, Corticosteroids (rarely)

An algorithm based upon the dexamethasone-suppression test may be useful for the diagnostic approach to hirsutism (Hatch, R. et al., Am. J. Obstet. Gynecol. 140, 815-830, Aug.1, 1981). Stein-Leventhal Syndrome (Polycystic Ovaries): The Stein-Leventhal Syndrome is the most common hormonal cause of hirsutism. Evidence of menstrual irregularity may point towards polycystic ovarian disease. Borderline elevations of the androgens, testosterone and 17-ketosteroids are present; raised LH levels and increased LH/FSH ratios are consistent with this disease.

Galactorrhea should be excluded since there is an association of hirsutism with hyperprolactinemia.

THEOPHYLLINE (AMINOPHYLLINE)

SPECIMEN: Red top tube, separate serum and refrigerate.
REFERENCE RANGE: Therapeutic: 10 to 20mcg/ml (10 to 20mg/liter). Toxicity:
>20mcg/ml (20mg/liter). The serum half-life for different age groups is given
in the next Table:

Change of Plasma Half-Life with Age	
Age Group	Half-Life (Hours)
Adult (Healthy Non-Smoker)	8.7 (4 to 16)
Adult (Healthy Smoker)	4.4 (3.0 to 9.5)
Children (1 to 9)	3.7 (2 to 10)
Premature Infants	30
Term Infants (under 3 Months)	24

Factors that Alter Aminophylline Clearance (in Terms of Half-Life)

Decreased Clearance:
  Patients with cardiac decompensation or hepatic cirrhosis:
    Half-life: 20 to 30 hours
  Neonates and Infants
    Half-Life: Prematures - about 30 hours
              Normal Newborns - about 24 hours
              Infants beyond Neonatal Period - gradually decreases
                from Newborn value to Childhood levels by year one.
  Concomitant Drug Therapy: Erythromycin, Cimetidine
Increased Clearance:
  Cigarette or marijuana smokers; Half-life: 4.4 hours
  Children one to nine years old
    Half-Life: 3.7 hours mean (2 to 10 hours range); half-life
      increases with age beyond nine years to adult values at age 16.

The plasma half-life is altered by the following conditions: Tobacco
smoking: half-life shortened; chronic liver disease, congestive heart failure
and severe obstructive pulmonary disease: half-life prolonged.

Drugs that alter the metabolism of theophylline are given in the next
Table:

Drugs and Metabolism of Theophylline	
Decreased Clearance (Prolonged Half-Life)	Increased Clearance (Decreased Half-Life)
Cimetidine (Tagamet)	Phenytoin
Propranol (Inderal)	Barbituates
Allopurinol (Zyloprim, and others)	
Erythromycin	

Serum concentrations of theophylline should be monitored in patients taking
these drugs concurrently with theophylline, especially when these drugs are
introduced or discontinued (The Medical Letter 26, 1-3, Jan. 6, 1984).

To convert traditional units in mg/liter to international units in
micromol/liter, multiply traditional units by 5.550.

METHOD: RIA; EMIT; HPLC; SLFIA(Ames); Nephelometry(ICS); Fluorescence
Polarization(Abbott).

INTERPRETATION: Theophylline is used as a potent bronchodilator for relief of
acute asthmatic symptoms, as a "prophylactic" agent for controlling the symptoms
and signs of chronic asthma and to control premature infant apnea.

Aminophylline is theophylline ethylene diamine; aminophylline and other
theophylline salts can be given orally or parenterally. Theophylline can be
given only orally.

Dosage of Aminophylline for Acute Therapy: For acute therapy, an intravenous
loading dose of about 6mg/kg aminophylline, if ideal body weight, will usually
yield a serum concentration in the range 10 to 15mcg/ml in patients who have
previously taken no theophylline. The intravenous loading dose should be
infused over a period of at least 30 minutes to minimize the probability of
serious arrhythmias.

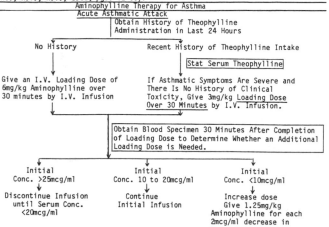

A protocol for aminophylline therapy for acute asthmatic attack is given in the next Figure; (Hendeles, L. and Weinberger, M. in Individualizing Drug Therapy, Vol. I, Taylor, W.J. and Finn, A.L., eds., Gross, Townsend and Frank, Inc., N.Y., N.Y., 1981, pgs. 32-65):

Aminophylline Therapy for Asthma

Acute Asthmatic Attack

Obtain History of Theophylline Administration in Last 24 Hours

No History

Give an I.V. Loading Dose of 6mg/kg Aminophylline over 30 minutes by I.V. Infusion

Recent History of Theophylline Intake

Stat Serum Theophylline

If Asthmatic Symptoms Are Severe and There Is No History of Clinical Toxicity, Give 3mg/kg Loading Dose Over 30 Minutes by I.V. Infusion.

Obtain Blood Specimen 30 Minutes After Completion of Loading Dose to Determine Whether an Additional Loading Dose is Needed.

Initial Conc. >25mcg/ml

Discontinue Infusion until Serum Conc. <20mcg/ml

Initial Conc. 10 to 20mcg/ml

Continue Initial Infusion

Initial Conc. <10mcg/ml

Increase dose Give 1.25mg/kg Aminophylline for each 2mcg/ml decrease in serum concentration

Blood Specimen: (Loading Dose in Hospital) A blood sample for aminophylline assay should be drawn prior to loading dose if the patient has a history of theophylline therapy; then obtain a blood specimen 30 minutes after completion of loading dose.

Maintenance Dose: The aminophylline infusion rate necessary to attain a serum level of 10mcg/ml for different conditions is as follows (Powell, R., School of Pharmacy, Univ. N.C.):

Aminophylline Infusion Rate

Adults:

Characteristic	0.5mg/kg/hr x Factor	Aminophylline Infusion Rate
Nonsmoker	0.5mg/kg/hr x 1 =	0.5 (mg/kg/hr)
Smoker (within last 3 mos > 1 pack/day) Plus ages 10-16	0.5mg/kg/hr x 1.6 =	0.8 (mg/kg/hr)
CHF (edema, x-ray)	0.5mg/kg/hr x 0.4 =	0.2 (mg/kg/hr)
Pneumonia, Viral Respiratory Infection	0.5mg/kg/hr x 0.4 =	0.2 (mg/kg/hr)
Cirrhosis	0.5mg/kg/hr x 0.4 =	0.2 (mg/kg/hr)
Severe hepatic obstruction (PEER ≤ 100L/min or PCO_2 > 45mmHg)	0.5mg/kg/hr x 0.8 =	0.4 (mg/kg/hr)

Example. The aminophylline infusion rate in a 60kg male who smokes 2 packs of cigarettes per day and currently has signs and symptoms of CHF: Aminophylline infusion rate = 0.5mg/kg/hr · 60kg · 1.6 · 0.4 = 19.2mg/hr

Premature Infants and Neonates: Aminophylline Infusion Rate = 0.19mg/kg/hr

Infants 4 to 52 Weeks: Aminophylline Infusion Rate = 0.37 + (0.012 x Age in Weeks)

Children (1 to 9 Years of Age): Aminophylline Infusion Rate = 1.0mg/kg/hr

Blood Specimen (Inpatient Maintenance Dose): Obtain a blood specimen 4 to 8 hours after start of maintenance I.V. infusion and 18 to 30 hours after starting the infusion and at 24 hour intervals thereafter for possible adjustment of the infusion rate until the infusion is discontinued. As a function of results, the dose may be altered appropriately.

Outpatient Maintenance Dose: Maintenance dose to achieve a peak level of 10 to 20mg/L is given in the next Table (Powell, R.):

Maintenance Dose of Theophylline or Aminophylline (mg/kg/day) Ideal Body Weight

	Theophylline (mg/kg/day)	Aminophylline (mg/kg/day)
Adult - Nonsmoker	10	12
Smoker	15	19
Cirrhotic	4	5
Children (1-8)	12-24	15-30
Children (9-12)	10-20	12.5-25
Children (12-16)	8-18	10-22.5

Blood Specimens (Outpatient Maintenance Dose): Blood specimens should be obtained at about 3 day intervals and at the time of peak concentration or about 2 to 4 hours after a scheduled dose depending on the product formulation. Concentrations of 5 to 10 mcg/ml may be effective for some patients, particularly children.

Toxicity: Toxicity usually is seen when serum levels rise above 20mg/liter. Early signs of toxicity are nausea, vomiting and headache. Toxic reactions are listed in the next Table:

Toxic Reactions of Theophylline
Gastrointestinal
Cardiovascular
Central Nervous System: Agitation, Irritability, Insomnia, Tremor, Seizures
Occasionally: Hyperthermia, Diuresis, Ketosis, Hyperglycemia

An acute oral overdose should be treated by induced emesis, followed by administration of activated charcoal and a purgative.

THERAPEUTIC PHERESIS

Therapeutic pheresis (hemapheresis) is a procedure in which whole blood is withdrawn from the patient, a portion of the blood separated and retained (e.g., leukocytes, platelets, plasma) and the remainder retransfused into the patient. Therapeutic leukapheresis is used to reduce the level of circulating leukocytes. It is used to remove leukocytes in newly diagnosed patients with acute leukemia, and in some patients with chronic myelogeneous or lymphocytic leukemia.

Therapeutic Plateletpheresis: Therapeutic plateletpheresis is used to reduce the level of circulating platelets. It is used to remove platelets in patients with thrombocytosis (usually associated with myeloproliferative syndrome) who may be at risk from thrombosis. The platelet count can be reduced by approximately 50% in three to four hours.

Therapeutic Plasmapheresis: Plasmapheresis is used to remove a wide variety of substances. A list of conditions, for which there are established indications for therapeutic plasma exchange, is given in the next Table; (Shumak, K.H. and Rock, G.A., N. Engl. J. Med. 310, 762-771, March 22, 1984):

Established Indications for Therapeutic Plasma Exchange		
Disorder	Clinical Indications	Pathogenic Substance Removed
Hyperviscosity, Syndrome	Relief of Acute Syndrome; Prophylaxis	Monoclonal Immunoglobulin
Cold Antibody-Type Autoimmune Hemolytic Anemia	Life-Threatening Hemolysis	Red-Cell Autoantibody
Post-Transfusion Purpura	At Diagnosis	Platelet Antibody (Anti-Pl^{A1})
Myasthenia Gravis	Preparation for Thymectomy Relief of Weakness	Autoantibody to Acetylcholine Receptors
Goodpasture's Syndrome	At diagnosis in Conjunction with Immunosuppression	Antibody to Glomerular Basement Membrane
Factor VIII Antibody Unresponsive to Factor VIII Therapy	Serious Hemorrhage or Preparation for Surgery	Antibody to Factor VIII
Refsum's Disease	Adjunct to Dietary Therapy	Phytanic Acid
Poisoning	Toxicity	Dependent on Poison

Conditions that are commonly treated with plasma exchange but not meeting one or more assessment criteria are given in the next Table; (Shumak, K.H. and Rock, G.A., N. Engl. J. Med. 310, 762-771, Mar. 22, 1984):

Conditions Commonly Treated with Plasma Exchange but Not Meeting One or More Assessment Criteria
Maternal Alloimmunization to Rh(D) Antigen
Thrombotic Thrombocytopenic Purpura
Idiopathic Thrombocytopenic Purpura
Cryoglobulinemia
Preparation for ABO-Incompatible Bone-Marrow-Transplantation
Familial Hypercholesterolemia
Acute and Chronic Guillain-Barre Syndrome
Multiple Sclerosis
Rheumatoid Arthritis
Systemic Lupus Erythematosus
Rapidly Progressive Glomerulonephritis
Renal-Allograft Rejection
Biliary Cirrhosis

The serious risks associated with plasma exchange include anaphylactoid reactions, pulmonary microvascular occlusion syndrome and infection (Rubenstein, M. et al., Am. J. Med. 75, 171-174, 1983). Plasma exchange is very expensive; the cost per procedure is $400 to $1,600 in the United States.

THERAPEUTIC DRUG MONITORING

Multiple Drug Doses: Drug accumulation, following multiple oral doses, is illustrated in the next Figure; (The Bio-Science Handbook of Clinical and Industrial Toxicology, 1st Ed., March 1979):

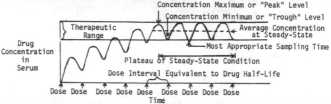

Drug Concentration - Time Relationship for Multiple Oral Dose

After administration of a single dose, drug levels increase and then decline. If a second dose is administered before complete elimination of the first dose, the drug levels achieved by the second dose will be higher. A third dose produces higher levels and an increased elimination.

When the rate of oral administration remains constant, the rate of elimination will increase until it is in equilibrium with the rate of administration; at this time, the steady state is established. The time required to reach the steady state is approximately five half-lives. This phenomenon occurs in reverse when drug dosing is discontinued. A blood level measured after four doses, administered at half-life intervals, will reflect approximately 94% of the ultimate steady-state drug level (Bio-Science Handbook). Approximately five half-lives are required for the nearly complete removal of the drug from the body.

Note that in the previous Figure, drug administration causes drug concentration to fluctuate between a maximum (peak) and a minimum (trough) level. Therefore, timing of specimen in relationship to drug administration is important. Also, note that the interval between doses of the drug is equal to the drug half-life.

If the dosing interval, at steady state, is longer than the half-life of the drug, then the relationship that is obtained is shown in the next Figure:

Dosing Interval, at Steady State, Longer than Half-Life

There are large fluctuations between peak and trough levels.
If the dosing interval, at steady state, is shorter than the half-life of the drug, then the relationship that is obtained is shown in the next Figure:

Dosing Interval, at Steady State, Shorter than Half-Life

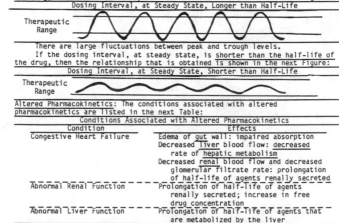

Altered Pharmacokinetics: The conditions associated with altered pharmacokinetics are listed in the next Table:

Conditions Associated with Altered Pharmacokinetics	
Condition	Effects
Congestive Heart Failure	Edema of gut wall: impaired absorption Decreased liver blood flow: decreased rate of hepatic metabolism Decreased renal blood flow and decreased glomerular filtrate rate: prolongation of half-life of agents renally secreted
Abnormal Renal Function	Prolongation of half-life of agents renally secreted; increase in free drug concentration
Abnormal Liver Function	Prolongation of half-life of agents that are metabolized by the liver

The routine screening procedures recommended for <u>elderly patients</u> in whom use of psychotropic drugs is being considered are <u>liver function tests</u> and <u>renal function tests</u>. Since older persons have a decreased muscle mass, they have a decreased rate of creatinine production; a reduced rate of creatinine clearance may coexist with serum creatinine in the normal range (Koch-Weber, J. N. Engl. J. Med. <u>308</u>, 134-138, 1983).

<u>Determination of Half-Life of a Drug:</u> Collect at least 3 timed specimens at "appropriate" timed intervals (e.g., 30-60 minutes apart or other appropriate times) and plot log concentration versus time.

<u>Sampling Time:</u> Basic principles of sampling time are as follows:

<u>Toxicity:</u> A single stat specimen is required; determine peak levels.

<u>Determination of Half-Life:</u> Obtain <u>two samples</u> drawn at precise times after the drug is administered.

<u>Intravenous and Intramuscular Medication:</u> Sample one-half to one hour after administration, e.g., Gentamicin.

<u>THERAPEUTIC RANGE; PANIC VALUE:</u>

THERAPEUTIC RANGE; PANIC VALUE		
Drugs	Therapeutic Range	Panic Value
Antiepileptic Drugs		
Carbamazepine	6-10mcg/ml	>20mcg/ml
Ethosuximide	40-100mcg/ml	>200mcg/ml
Phenobarbital	15-30mcg/ml	>60mcg/ml
Phenytoin	10-20mcg/ml	>40mcg/ml
Primidone	8-12mcg/ml	>24mcg/ml
Valproic Acid	50-100mcg/ml	>200mcg/ml
Digoxin	0.5-2.0ng/ml	>2.0ng/ml
Digitoxin	10-32ng/ml	>35ng/ml
Disopyramide	2-5mcg/ml	>7mcg/ml
Lidocaine	1.6-5.0mcg/ml	>9mcg/ml
Lithium	0.5-1.5mEq/Liter	>2.0mEq/liter
Procainamide	4.0-8.0mcg/ml	>12mcg/ml
Quinidine	2.3-5.0mcg/ml	>6mcg/ml
Theophylline	10-20mcg/ml	>25mcg/ml
Tricyclic Antidepressants		
Imipramine	150-300ng/ml	400ng/ml
Amitriptyline	120-250ng/ml	400ng/ml
Desipramine	150-300ng/ml	400ng/ml
Nortriptyline	50-150ng/ml	200ng/ml
Protriptyline	50-150ng/ml	200ng/ml
Doxepin	150-300ng/ml	400ng/ml

<u>AMINOGLYCOSIDES:</u> The aminoglycosides are listed in the next Table:

Aminoglycosides
Streptomycin
Kanamycin
Gentamicin
Tobramycin
Amikacin
Netilmicin
Sisomicin

The aminoglycosides are often the first choice in the treatment of <u>serious</u> infections due to <u>gram-negative bacilli</u>; the aminoglycosides for current clinical use for <u>parenteral therapy</u> are streptomycin, kanamycin, gentamicin, tobramycin and amikacin. Neomycin is no longer used for parenteral use because of nephrotoxicity. (Suffix "mycin" derived from Streptomyces species; "micin" derived from Micromonospora.)

Toxicity and untoward effects are listed in the next Table; (Edson, R.S. and Keys, T.F., Mayo Clin. Proc. <u>58</u>, 99-102, 1983):

Toxicity of Aminoglycosides
<u>Major:</u>
Renal Toxicity
Vestibular Toxicity
Auditory Toxicity
Neuromuscular Blockade
<u>Minor:</u> Skin Rash, Drug-Induced Fever

Therapeutic
Drug Monitoring

ORAL ANTICONVULSANT DRUGS

A comparison of the properties of anticonvulsants is given in the next Table (Davidson, D.L.W., Brit. Med. J. 286, 2043-2045, 1983; Chadwick, D., Brit. J. Hosp. Med. pgs. 421-424, Dec. 1983):

Oral Anticonvulsant Drugs			
Anticonvulsant	Maintenance Dose Range	Ther. Range mcmol/liter(mcg/ml)	Half-Life(Hours)
Phenytoin (Dilantin)	(A)150-600 mg (start at 100 mg twice daily)	40-80 (10-20)	(A)10-60 (Longer Higher Dose)
	(C)5-8 mg/kg		(C)5-12
Carbamazepine Chronic (Tegretol) Treatment	(A)300-600 mg (start at 100 mg twice daily, incr. over 4 wks)	25-50 (6-12)	25-50 Initially 10-30
	(C)9-20 mg/kg		(C) 5-30
Clonazepam (Rivotril)	(A)1.5-8 mg (start at 0.5 mg twice daily dose,incr. over 2-4 wks)	Not generally available	20-60
	(C)0.5-6 mg		
Ethosuximide (Zarontin)	(A)500-2000 mg (start at 250 mg twice daily)	300-750 (40-100)	(A)30-70
	(C)250-1000 mg		(C)30
Phenobarbitone	(A)60-240 mg (start at 30 mg daily)	80-180 (15-40)	(A)70-120
	(C)30-180 mg		(C)40-70
Primidone (Mysoline)	(A) 250-1500 mg (start at 125 mg daily) (C)125-1000 mg (low dose initially)		(A) 4-11 (Phenobarbitone Metabolite as above)
Valproate	(A)400-2600 mg (start at 200 mg twice daily)	350-700 (50-100)	(A)6-15
	(C)20-30 mg/kg		(C)4-14

A = Adults; C = Children

Choice of anticonvulsant drugs is given in the next Table; (Davidson, D.L.W., Brit. Med. J. 286, 2-43-2045, 1983; Scott, A.K., Brit. Med. J., 288, 986-987, March 31, 1984):

Choice of Anticonvulsant Drugs		
Type of seizure	First Choice	Alternatives
Partial(focal) Including Temporal Lobe and Jacksonian	Carbamazepine, Phenytoin	Clonazepam, Valproate, Primidone, Phenobarbitone
Tonic-clonic (Grand Mal)	Valproate, Phenytoin, Carbamazepine	Clonazepam, Primidone, Phenobarbitone
Absence Only (Petit Mal)	Ethosuximide	Valproate, Clonazepam
Absence and other Seizures	Valproate	Clonazepam, Ethosuximide plus Phenytoin or Carbamazepine, Nitrazepam
Myoclonus	Valproate, Clonazepam	Nitrazepam
Status Epilepticus	Intravenous Diazepam or Clonazepam	Chlormethiazole, Phenytoin, or Thiopentone Intravenously, Paraldehyde Intramuscularly

The indications for anticonvulsant serum level monitoring are given in the next Table; (Chadwick, D., Brit. J. Hosp. Med., pgs. 421-424, Dec. 1983):

Indications for Anticonvulsant Serum Level Monitoring
Use of Phenytoin
Polypharmacy
In Patients in whom Assessment of Toxicity is Difficult
In Patients with Renal or Hepatic Disease
Assessing Compliance
During Controlled Studies of Anticonvulsants

CARDIOACTIVE DRUGS

A comparison of the properties of cardioactive drugs is given in the next Table:

Cardioactive Drugs			
Drug	Use	Ther. Range	Half-Life
Digoxin	Congestive Heart Failure; Atrial Fibrillation; Supraventricular Tachy.	0.5-2 ng/ml	1.6 days (If Renal & Hepatic Function is Normal.)
Lidocaine	Prevention and Treat. of Ventricular Arrhythmias	2-5 mcg/ml	74-140 minutes
Procainamide plus N-Acetyl Procainamide (NAPA)	Ventricular Arrhythmias that are Resistant to Lidocaine; Prophylaxis against Ventricular Arrhythmias following Acute Myocardial Infarction.	10-30 mcg/ml	Procainamide: 3 Hours NAPA: 7 Hours
Quinidine	Prophylaxis and Treatment of Atrial and Ventricular Arrhythmias; Slows Heart Rate.	2.0-5.0 mcg/ml	6 Hours

THERAPEUTIC PLASMA EXCHANGE (SEE THERAPEUTIC PHERESIS)

THROAT CULTURE FOR C. DIPHTHERIA

SPECIMEN: Obtain specimen before antimicrobial chemotherapy is started. Depress the tongue to expose the pharynx. Use a sterile cotton or dacron swab. Commercially available sterile "Culturette" may be used. Rub vigorously over the posterior pharynx, tonsils, and tonsillar fossae; avoid the tongue, lips, and buccal mucosa.

Also obtain a specimen for staining with alkaline methylene blue.

REFERENCE RANGE: Negative.

METHOD: Culture on Loeffler's medium

INTERPRETATION: Infection by Corynebacterium diphtheriae is rare but should be suspected in patients who have pharyngitis but have not been immunized. Patients may not have a diphtheritic membrane.

S. Bakerman

THROAT CULTURE FOR GROUP A BETA-HEMOLYTIC STREPTOCOCCI ONLY

<u>SPECIMEN:</u> Obtain specimen before antimicrobial chemotherapy is started. Depress the tongue to expose the pharynx. Use a sterile cotton or dacron swab. Commercially available sterile "Culturette" may be used. Rub swab vigorously over the posterior pharynx, tonsils and tonsillar fossae; avoid the tongue, uvula, lips and buccal mucosa.

The specimen may be treated in one of two ways. Transport Culturette to the laboratory as soon as possible; if transport is delayed, refrigerate specimen; or the swab may be inoculated onto <u>sheep blood agar plate.</u> Sheep blood agar is the <u>best</u> medium for detecting Group A beta hemolytic streptococci.
<u>REFERENCE RANGE:</u> No beta hemolytic streptococci isolated.
<u>METHOD:</u> The method of inoculation of <u>sheep blood agar</u> is as follows: Initially, run the <u>swab</u> over approximately <u>one-sixth</u> of the agar. Then using a sterile loop, <u>streak</u> the primary inoculum onto about <u>one-half</u> of the plate in 10 to 20 to-and fro-strokes. The plate is then turned <u>90°</u> and the <u>plate</u> is <u>again</u> streaked with the same loop. <u>Incubate</u> at 35° to 37° overnight preferably under anaerobic conditions. A GasPak jar is available from BBL Microbiology Systems (P.O. Box 243, Cockeysville, Md. 21030).

When group A beta hemolytic streptococci were incubated <u>anaerobically,</u> <u>98 perent</u> of 149 strains of group A were detected while <u>only 63 percent</u> were detected when the <u>plates</u> were stabbed and incubated <u>aerobically</u> (Dykstra, M.A. et al., J. Clin. Microbiol. <u>9,</u> 236, 1979).

<u>Group A beta hemolytic streptococci</u> may be differentiated from <u>nongroup A</u> beta hemolytic streptococci as follows: Inoculate the organism onto a 5 percent sheep blood agar plate and place a <u>bacitracin disk</u> of 0.04 units of bacitracin at the center; incubate the plate for 18 to 24 hours at 35° to 37°C. Group A beta hemolytic streptococci are susceptible to 0.04 units of bacitracin while other groups of beta hemolytic streptococci are usually resistant.
<u>INTERPRETATION:</u> The most common cause of bacterial pharyngitis is <u>group A beta hemolytic streptococci</u> (GABHS) and most physicians differentiate <u>group A beta hemolytic streptococci</u> from nonstreptococcal pharyngitis.

Patients with pharyngitis are at increased risk for development of <u>acute</u> rheumatic fever and <u>acute glomerulonephritis.</u>

The frequency of isolation by age of GABHS, as a <u>cause of pharyngitis,</u> is given in the next Table; (Levy, M.L., et al., Medical Clinics of North America, <u>67,</u> 153-171, 1983):

| Frequency by Age of Isolation of Group A Beta Hemolytic Streptococci in Patients with Pharyngitis ||
Age	Frequency (%)
< 3 years	3 to 15
Children	35 to 50
15 to 35	10 to 20
Over 35	5

Beta-hemolytic streptococci are usually susceptible to penicillin or penicillin derivatives; therefore, susceptibility studies are usually not routinely performed. If a patient is allergic to penicillin, erythromycin might be given.

THROAT CULTURE FOR NEISSERIA GONORRHOEAE ONLY (see GONORRHEA CULTURE)

THROMBIN CLOTTING TIME

<u>SPECIMEN:</u> Blue (citrate) top tube; fill tube; separate plasma and refrigerate if necessary. Determination must be done within 2 hours from the time of collection. Do <u>not</u> perform test on patients who are on <u>heparin</u> therapy.
<u>REFERENCE RANGE:</u> Clot formation in 12-18 seconds
<u>METHOD:</u> Add low concentration of thrombin to citrated plasma and measure the time necessary for clot formation.
<u>INTERPRETATION:</u> The thrombin clotting time is a measure of the rate of conversion of <u>fibrinogen to fibrin</u> and <u>fibrin polymerization.</u> The causes of prolonged thrombin time are listed in the next Table:

Prolonged Thrombin Time
Heparin
Severe Hypofibrinogenemia (<80mg/dl)
Inhibitors of Fibrin Polymerization
Fibrin Degradation Products (FDP)
Paraproteins
Dysfibrinogenemia

This is used as a rapid semi-quantitative test for intravascular clotting and fibrinolysis.

The presence of heparin in plasma as a cause of prolongation of the thrombin time can be evaluated by using the Protamine Sulfate Corrected Thrombin Time.

Thrombin Clotting Time,
Protamine Sulfate

THROMBIN CLOTTING TIME, PROTAMINE SULFATE CORRECTED

SPECIMEN: Blue (citrate) top tube; fill tube; separate plasma and refrigerate if necessary. Determination must be done within 2 hours from the time of collection.

REFERENCE RANGE: Correction of prolonged thrombin time due to heparin.

METHOD: Protamine sulfate is added to test plasma and thrombin clotting time is determined.

INTERPRETATION: Protamine sulfate will fully correct a prolonged thrombin time due to heparin up to 2.5U/ml but will have little or no effect on prolonged thrombin time due to hypofibrinogenemia or inhibitors of fibrin polymerization such as fibrin degradation products (FDP), paraproteins or dysfibrinogenemia.

Thrombolytic Therapy
Monitoring, Panel

THROMBOLYTIC THERAPY MONITORING, PANEL

The tests that are done to monitor thrombolytic therapy (streptokinase or urokinase) are given in the next Table; (Sharma, G.V.R.K. et al., N. Engl. J. Med. 306, 1268-1276, 1982):

Tests to Monitor Thrombolytic Activity	
Tests	Comment
Euglobulin Lysis Time	Most sensitive test but is not available in most laboratories
Thrombin Time	Next most sensitive test
Partial Thromboplastin Time(PTT)	Less Sensitive
Prothrombin Time(PT)	Less sensitive
Fibrin Degradation Products(FDP)	Less sensitive

The timing of the tests is given in the next Table; (Sharma, G.V.R.K. et al., already cited):

Timing of Testing and Objective
Before Therapy
Detect and correct coagulation defects (by means of thrombin, partial thromboplastin, and prothrombin times)
Determine base line or control for fibrinolysis (any of the above tests; euglobulin lysis time or fibrin(ogen) degradation products if patient has been receiving heparin)
Drug Therapy (3-4 hr after start)
Use same test(s) used for establishing base line or control
After Therapy
Use partial thromboplastin time if heparin therapy is to begin
Objective
Ensure establishment of fibrinolytic state:
Determine whether values have marked prolongation over control value
Administer streptokinase:
If values not changed from control at 3-4 hr, give another loading dose (250,000 units) and continue infusion (standard, 100,000 units per hr).
Recheck in another 3-4 hr. If still no change, discontinue streptokinase and switch to urokinase, or begin heparin therapy.

S. Bakerman

THYROID PANEL
    (a) Total Thyroxine(T-4)
    (b) T-3 Resin Uptake(T-3RU)
    (c) Free Thyroxine Index(FTI) or Free Thyroxine
    (d) Thyroid Stimulating Hormone(TSH)
    (e) Triiodothyronine(T-3)
    (f) TSH following Thyroid Releasing Hormone(TRH)
    (g) Thyroid Anti-Thyroglobulin Antibody
    (h) Thyroid Anti-Microsomal Antibody
    (i) Thyroglobulin
    (j) Thyroxine Binding Proteins(TBP)
    (k) Radioactive Iodine (or Technetium) Uptake(RAIU)

HYPERTHYROIDISM (Red top tubes, separate serum): In hyperthyroidism, total
T-4, T-3 resin uptake (T-3RU), free thyroxine index (FTI) or free thyroxine are
elevated. Serum total T-3 should be done if free thyroxine is in the borderline
range. In T-3 thyrotoxicosis there is increased T-3 but normal T-4 and free
T-4. The best single test for differentiation of hyperthyroidism from
euthyroidism is measurement of TSH following injection of TRH; inject TRH
(Protierline Thypinone), 7micrograms/kilogram, and obtain serum specimens for
TSH assay at times 0, 15, 30, 45 and 60 minutes; in hyperthyroidism, there is no
rise in plasma TSH after TRH, and in euthyroidism, there is a rapid rise in TSH
of 2 to 20 microunits/ml from the 0 time TSH value.
    The three most important causes of thyrotoxicosis with increased or
occasionally normal RAIU are given in the next Table:

Thyrotoxicosis with Increased or Occasionally Normal RAIU
Toxic Diffuse Goiter (Grave's Disease)
Toxic Multinodular Goiter (Plummer Syndrome)
Toxic Adenoma

Causes of hyperthyroidism with reduced RAIU are listed in the next Table:

Hyperthyroidism with Decreased RAIU
Thyroiditis
Grave's Disease with Acute Iodide Loading
Thyrotoxicosis Factitia (Surreptitious Intake of Thyroid Hormone)
Antithyroid Drugs (Propylthiouracie or Methimazole)

Thyroid antibodies (antibodies against thyroglobulin and microsomal
antigen) may be present in patients with hyperthyroidism.

HYPOTHYROIDISM (Red Top Tubes, Separate Serum): In primary hypothyroidism,
total T-4, T-3 resin uptake (T-3RU) and free thyroxine index (FTI) or free
thyroxine are depressed. Measurement of TSH is the most sensitive test for
detecting primary hypothyroidism; TSH is markedly elevated in primary
hypothyroidism. It is important to screen patients who are on lithium therapy
with thyroid function tests, total T-4, free thyroxine index (FTI) and TSH,
because of the high incidence, 4% to 13%, of hypothyroidism associated with
lithium therapy. In thyroid screening in the newborn in Utah, the rate of
hypothyroidism was one in 3,800 live births; the rate of hypopituitarism was one
in 27,600 live births; the incidence of congenital absence of thyroid binding
globulin was one in 41,000 live births (Buehler, B.A., Ann. Clin. Lab. Science
13, 5-9, 1983).
    Thyroid antibodies (antibodies against thyroglobulin and microsomal
antigen) may be present in patients with hypothyroidism, especially when
hypothyroidism is associated with thyroiditis, especially Hashimoto's
thyroiditis. Significant elevation of serum cholesterol occurs in
hypothyroidism.
    In secondary (pituitary) and tertiary (hypothalamic) hypothyroidism, TSH is
low. To differentiate secondary from tertiary, inject TRH (7micrograms/kilogram)
and measure TSH at time 0, 15, 30, 45 and 60 minutes; if TSH increases then
hypothyroidism is due to hypothalamic failure; if TSH remains low, then the
hypothyroidism is due to pituitary failure.
THYROIDITIS: In Hashimoto's thyroiditis, most patients are metabolically
normal during the early stages of the disease; hypothyroidism is found in 20% to
50% and up to 15% early in the disease. High titers of the thyroid
autoantibodies (antibodies against thyroglobulin and microsomal antigen) are
usually present; antimicrosomal antibodies are present more often than
antithyroglobulin antibodies (Baker, B.A., Am. J. Med. 74, 941-944, 1983).

408

In Riedel's struma, the patient is <u>hypothyroid</u>; <u>autoantibodies</u> are present as seen in Hashimoto's disease.

In <u>granulomatous thyroiditis</u>, at least <u>50% of the patients</u> are hyperthyroid, and thyroid autoantibodies are present. <u>However, unlike classic thyrotoxicosis, RAIU is very low</u>. The thyrotoxic state usually only lasts for a short time.

In <u>thyroiditis</u>, the erythrocyte sedimentation rate (ESR) is elevated; cholesterol may be elevated if the patient is hypothyroid.

<u>THYROID NODULE</u>: A comparison of plans for diagnosis of thyroid nodules is given in the next Table; (Solomon, D.H. and Keeler, E.B., Ann. Intern. Med. <u>96</u>, 227-231, 1982; "Thyroid Nodule? Is Surgery Indicated," Patient Care, Pgs. 119-153, June 15, 1983):

			Comparison of Plans for Diagnosis of Thyroid Nodules		
Plan	Characteristics	Missed Malignancies Per 1,000 Nodule	Operations for Benign Lesions Per 1,000 Nodules	Average cost Per Nodule	Average Cost Per Correct Diagnosis
A	Aspiration → Scan	18	66	$1,685	$224
B	Scan → Ultrasonography	9	517	$3,945	$517
C	Scan → Aspiration	18	66	$1,769	$305

Plan A is most economical but had 18 false negative results and only 66 operations were performed on these patients.

<u>Scan</u>: <u>Seventeen percent</u> of <u>cold</u> nodules are <u>malignant</u>; <u>9 percent</u> of <u>hypofunctioning</u> and <u>normofunctioning</u> (warm) nodules. Autonomously functioning <u>hot nodules are almost never carcinomas</u>; however, <u>hot nodules</u> constitute only approximately 5 percent of solitary nodules.

<u>Confirming the Hot Nodule</u>: Use a thyroid hormone suppression test. Perform an I-123 scan. Give 75mcg/day liothyronine sodium (Cytomel, 25mcg/8 hours) for 7 to 10 days. Liothyronine suppresses TSH. Perform another I-123 scan. The autonomous nodule will not be suppressed by the lack of TSH stimulation but will appear as a hyperfunctioning spot on the scan. Do <u>not</u> perform this test in the elderly, patients with heart disease, or patients who are thyrotoxic.

<u>THYROID CANCER</u>: Many of the malignant tumors, especially follicular carcinomas, take up <u>radioactive iodide</u>. Calcitonin is often secreted by patients with medullary carcinomas; in the <u>familial</u> form of <u>medullary carcinoma</u>, tests provocative of calcitonin secretion, that is, infusion of calcium or pentagastrin or both, may be done.

THYROID FUNCTION TESTS-SUMMARY:

Thyroid Function Tests in Various Conditions						
Total T-4	T-3RU	Free T-4 or FTI	T-3	TSH	TSH Response to TRH	Interpretation
↓	↓	↓	↓	↑	Hyperresponse	Primary Hypothyroidism
N or ↓	↓	↓	N or ↓	↑	Hyperresponse	Primary Hypothyroidism with increased TBG
N or slt ↓	N or slt ↓	N or slt ↓	N or slt ↓	slt ↑	Hyperresponse	Borderline Primary Hypothyroidism
↓	↓	↓	N or ↓	↓	---	Secondary Hypothyroidism
↓	↓	↓	N	N	N	Optimal Therapy of Primary Hypothyroidism with T-3
N	N	N	N	N	N	Optimal Therapy of Primary Hypothyroidism with T-4
↑	↑	↑	↑	N	Blunted Response	Hyperthyroidism
↑	↑	↑	↑	N	Blunted Response	Hyperthyroidism; Decreased RAIU plus Autoantibodies = Thyroiditis; Surreptitiously Induced Thyroiditis
N	N	N	↑	N	Blunted Response	T-3 Thyrotoxicosis
N or ↑	N or ↑	↑	N or ↑	N	Blunted Response	Hyperthyroidism with TBG Deficiency

TSH Response to TRH: See Thyroid Stimulating Hormone

Thyroid test results in the euthyroid pregnant woman are given in the next Figure:

Laboratory Findings in the Euthyroid Pregnant Woman	
Test	Finding
Total Serum T-4	Increased - due to increase in TBG (Estrogens)
T-3 Resin Uptake	Decreased - due to increase in TBG (Estrogens)
Total Serum T-3	Increased - due to increase in TBG (Estrogens)
Free T-3	Normal

Thyroid Function Tests in Hospitalized Patients: Thyroid function test values in patients with euthyroid sick syndrome, primary hypothyroidism, and primary hypothyroidism and concomitant illness are given in the next Table; (Morley, J.E. et al., JAMA 249, 2377-2379, May 6, 1983):

Thyroid Function Test Values in Patients With Euthyroid Sick Syndrome, Primary Hypothyroidism, and Primary Hypothyroidism and Concomitant Illness			
Thyroid Function Test	Euthyroid Sick Syndrome	Primary Hypothyroidism	Primary Hypothyroidism and Concomitant Illness
Thyroxine (T4)	Normal or ↓	↑	↓
Triiodothyronine (T3) uptake (T3RU)	↑	↑	---
Triiodothyronine (T3)	↓	Normal or ↓	↓
Thyrotropin (TSH)	Normal	↑	↑ (or Normal)*
Free T4 (by dialysis)	↑ Normal ↓	↑	↓
Reversed T3	↑	↓	↑ Normal ↓
TSH response to thyrotropin-releasing hormone	Normal or ↓	↑	↑

*Particularly when dopamine is being administered.

The most common abnormality in euthyroid sick syndrome is a decrease both in the total and the free T-3; this is due to a diminished conversion of T-4 to T-3 by the peripheral tissues. With severe nonthyroidal illness, T-4 also decreases, due both to a decrease in plasma protein binding and to an increased rate of peripheral metabolism; the lowering of T-4 is an especially grave prognostic sign (Slag, M.F. et al., JAMA 246, 2702-2706, 1981). The biological mechanisms of thyroid hormone action and an explanation of the paradoxical discrepancy between low plasma hormone concentrations and the apparent absence of clinical hypothyroidism with nonthyroidal illness have been reviewed (Oppenheimer, J.H., Resident and Staff Physician, pgs. 15s-29s, Jan. 1983).

In a recent study, the serum T-4 index was increased in 11.7 percent of unselected consecutive admissions to a medical service (Gooch, B.R. et al., Arch. Intern. Med. 142, 1801-1805, 1982).

THYROXINE, TOTAL (T-4, TETRAIODOTHYRONINE)

SPECIMEN: Adult: Red top tube, separate serum. Newborn: In the usual procedure, blood specimens obtained by heel prick are taken on the third day after birth. Special filter paper is provided by the laboratory as illustrated in the next Figure:

Filter Paper for Blood Specimens for Thyroid Testing in the Newborn

All three circles are soaked with blood; be sure blood soaks through. Patients that have values of total T-4 less than 7mcg/dl are retested by TSH. Reference: Sadler, W.A. and Lynskey, C.P., Clin. Chem. 25, 933-938 (1979).

REFERENCE RANGE: Adults: 5-13mcg/dl; fetus: 4-16mcg/dl; amniotic fluid: 0.54(non-detected-1.75mcg/dl); cord: 6-17mcg/dl; 24 hours: 16-26mcg/dl; 48 hours: 12-2mcg/dl; 3-5 days: 9-20mcg/dl; 2 weeks-months: 7-15mcg/dl

To convert conventional units in mcg/dl to international units in nmol/liter, multiply conventional units by 12.87.

METHOD: RIA

INTERPRETATION: Thyroxine (T-4) is synthesized in the thyroid gland; T-4 is released from thyroglobulin, the storage form of T-4, following proteolytic hydrolysis of the thyroglobulin molecule. The proteolytic enzymes are stimulated by the thyroid stimulating hormone, TSH. When thyroxine is released into the blood, it combines with the serum proteins, thyroxine binding proteins. Only one molecule out of 10,000 molecules of T-4 is free T-4; the free-T-4 is the physiologically active T-4.

The causes of elevation of total T-4 are given in the next Table:

Causes of Elevation of Total T-4
Hyperthyroidism
Increased Thyroxine Binding Proteins:
Pregnancy
Drugs:
Estrogens, e.g., oral contraceptives or
estrogens used in osteoporosis or menopause
Clofibrate
Perphenazine
Genetic increase of thyroxine binding proteins
Acute hepatitis

The causes of decreased total T-4 are given in the next Table:

Causes of Decreased Total T-4
Hypothyroidism
Competition with T-4 for Binding Sites on Thyroxine Binding Proteins:
Dilantin; Phenytoin; Salicylates in high doses; Phenylbutazone
Decreased Thyroxine Binding Proteins
Androgens; Renal Failure; Nephrosis; Malnutrition; Active Acromegaly;
Corticosteroids (large doses); Major illness; Genetic decrease of TBP

Total T-4 may be increased or decreased in euthyroid sick people.

T-4 can be considered to be a prohormone for T-3 production. About 80 percent of the circulating T-3 is formed following monodeiodination of T-4 in peripheral tissue, especially the liver and kidney, by 5'deiodinase.

About 33 percent of the T-4 that is secreted daily undergoes peripheral monodeiodination to T-3; another 40 percent undergoes monodeiodination in the inner ring to produce reverse T-3.

Endogeneous antibodies may combine with T-4 causing spurious values for T-4 (Trimarchi, F. et al., J. Clin. Endocrinol. Metab. 54, 1045-1050, 1982).

T-3 RESIN UPTAKE TEST (T-3RU)
SPECIMEN: Red top tube, separate serum
REFERENCE RANGE: 23-35%. To convert conventional units in percent to
international units in l, multiply conventional units by 0.01.
METHOD: RIA
INTERPRETATION: The T-3 resin uptake test gives a quantitative measure of the
unsaturated binding sites on the thyroxine binding proteins. The measurement of
total T-4 plus T-3RU gives the free thyroxine index.
    The conditions associated with elevation of T-3RU are given in the next
Table:

Conditions Associated with Elevation of T-3RU
Hyperthyroidism
Competition with T-4 for Binding Sites on Thyroxine Binding Proteins:
Dilantin
Salicylates in high doses
Phenylbutazone
Decreased Thyroxine Binding Proteins:
Androgens
Renal Failure
Nephrosis
Malnutrition
Active Acromegaly
Corticosteroids (large doses)
Major Illness
Genetic Decrease of TBP

    The conditions associated with decreased T-3RU are given in the next
Table:

Conditions Associated with Decreased T-3RU
Hypothyroidism
Increased Thyroxine Binding Proteins
Pregnancy
Estrogen Effect, i.e., Oral contraceptives;
or estrogens in osteoporosis or menopause
Clofibrate
Perphenazine
Genetic Increase of TBP
Acute Hepatitis

THYROXINE INDEX (CALCULATED) (FTI, T-7, T-12)
SPECIMEN: Red top tube; minimum volume of serum, 2ml.
REFERENCE RANGE: 0.8-2.4ng/dl. To convert conventional units in ng/dl to
international units in pmol/liter, multiply conventional units by 12.87.
METHOD: RIA
INTERPRETATION: The serum level of the free thyroxine index reflects the
serum level of the free thyroxine. The free thyroxine index is obtained by
multiplying the value for the T-4 by the value for the T-3 resin uptake.
    The free thyroxine index is elevated in hyperthyroidism and decreased in
hypothyroidism.
    Phenytoin is a drug that causes decreased levels of free thyroxine index
and free thyroxine.
    There are two genetic conditions in which there is an elevated FTI and
normal measured levels of free T-4 in euthyroid subjects. In one type, there is
an abnormal T-4 binding protein (Ruiz, M. et al., N. Engl. J. Med. 306, 635-639,
1982); in the other type, there is increased binding of T-4 to immunoreactive
TBPA (Moses, A.C. et al., N. Engl. J. Med. 306, 966-969, 1982).

THYROXINE, FREE (FREE T-4)

<u>SPECIMEN</u>: Red top tube; separate serum.
<u>REFERENCE RANGE</u>: 1.3 to 3.8ng/dl in children and adults; in infants under one month of age, normal values are higher. To convert <u>conventional</u> units in ng/dl to <u>international</u> units in pmol/liter, <u>multiply conventional units by 12.87</u>.
<u>METHOD</u>: RIA
<u>INTERPRETATION</u>: Free thyroxine is <u>elevated</u> in <u>hyperthyroidism</u> and <u>reduced</u> in <u>hypothyroidism</u>.

Thyroid values altered in thyroid states and euthyroid sick people are given in the next Table; (Blum, M., Hosp. Pract. pgs. 105-116, Oct. 1981):

Thyroid Values in Altered Thyroid States and Euthyroid Sick People							
	T-4	Free T-4 Dialysis	Free T-4 Index	T-3	Reverse T-3	TSH	TRH Test
Normal	N	N	N	N	N	N	N (Doubling of baseline level of TSH)
Euthyroid: Sick T-3	N	N	N or ↓	↓	↑	N	N
Euthyroid: Sick T-4	↓	N or ↓	↓	↓	↑	N	N
Euthyroid: Sick T-4	↑	↑	N or ↑	N	↑	N	N
Hyperthyroid: Sick	↑ or N	↑	N or ↑	↑ or N	↑	↓	No response
Hypothyroid, Primary	↓	↓	↓	↓	↓	↑	Hyper-responsiveness
Hypothyroid, Secondary	↓	↓	↓	↓	↓	↓	No response
Hypothyroid, Tertiary	↓	↓	↓	↓	↓	↓	Variable, Blunted Response

THYROID STIMULATING HORMONE (TSH)

<u>SPECIMEN</u>: <u>Adults</u>: Red top tube, separate serum; store serum in refrigerator.
<u>Newborn</u>: In the usual procedure, blood specimens obtained by heel prick are taken on the <u>third day</u> after birth. Special filter paper is provided by the laboratory as illustrated in the next figure:

Filter Paper for Blood Specimens for Thyroid Testing in the Newborn

All three circles are soaked with blood; be sure blood soaks through.
Patients that have values of total T-4 less than 7mcg/dl are retested by TSH.
Reference: Sadler, W.A. and Lynskey, C.P., Clin. Chem. <u>25</u>, 933-938 (1979).
<u>REFERENCE RANGE</u>: Units of TSH in mU/liter.
   <u>Adults</u>: ND - 10; <u>Fetus</u>: ND - 20; <u>Amniotic Fluid</u>: 1.5 (ND-3.7);
   <u>Cord</u>: 9.5 (1-20); <u>24 Hours</u>: 20; <u>48 Hours</u>: 18; <u>3-5 Days</u>: 6;
   <u>2 Wks - Months</u>: 4 (ND-10); <u>Adults</u>: ND - 10
ND = non detectable
<u>METHOD</u>: RIA
<u>INTERPRETATION</u>: Measurement of the TSH is the <u>most sensitive</u> test for detecting primary hypothyroidism; TSH is markedly <u>elevated</u> in <u>primary hypothyroidism</u>. Assay of TSH is used to <u>monitor</u> the effectiveness of thyroxine therapy in hypothyroidism.
   In secondary <u>hypothyroidism</u> and in <u>hyperthyroidism</u>, TSH levels are <u>low</u>. Levels in normal individuals are also low and it is not usually possible to differentiate low values in normals from low values in secondary hypothyroidism and hyperthyroidism.

S. Bakerman

Thyroid Stimulating
Hormone (TSH)

Measurement of <u>TSH</u> following <u>thyroid releasing hormone</u> (TRH; Protierline
Thypinone supplied by Abbott Laboratories) is used in the evaluation of several
types of thyroid disease as listed in the next Table; (Ontjes, D.A., North
Carolina Medical Journal 44, 211-214, 1983):

<table><tr><td>Applications of TRH as a Stimulation Test</td></tr></table>

Diagnosis of Hyperthyroidism When Borderline Values for Thyroid
    Tests Are Obtained
Differentiation of Secondary versus Tertiary Hyperthyroidism:
    Pituitary versus Hypothalamus

The test for adults is done as follows: The patient should be in a supine
position. A serum specimen is obtained for serum TSH assay. Then TRH
(Protierline Thypinone), supplied by Abbott Laboratories, Chicago, Illinois in
1ml single-dose ampules containing 500mcg of drug and costing approximately $30,
is rapidly injected intravenously over 30-60 seconds in a dosage of 7
<u>micrograms/kilogram</u>. Venous specimens are obtained in 15, 30, 40 and 60
minutes following injection and are assayed for TSH.
<u>Diagnosis of Hyperthyroidism; Borderline Values for Thyroid Function Tests</u>: The
responses of TSH to injection of TRH are shown in the next Figure:

Response of TSH to TRH in Hyperthyroidism
(Diagnostic Pathways, Damon Corporation, 1890)

No response is found in the conditions listed in the next Table:

Conditions - No Response of Serum TSH to I.V. TRH

    Thyrotoxicosis
    Euthyroid Grave's Disease
    Euthyroid Nodular Goiter
    Secondary Hypothyroidism

In euthyroid adults, TRH given intravenously in adults will elicit a <u>rapid</u>
<u>rise</u> in TSH of 2 to 20 microunits/ml (usually 10 to 15 microunits/ml) from the
basal TSH level. This rise begins within 1 to 2 minutes, with the <u>peak</u> TSH rise
occurring <u>25-30 minutes after injection</u>. The TSH level begins to return to the
baseline soon after this peak. Patients with chronic renal failure who seem to
be euthyroid (often with low levels of thyroid hormones) will show a blunted TSH
response to TRH; this is the <u>only non-thyroidal disease</u> associated with an
abnormally low TSH response (Watts, N.B., Advanced Clinical Chemistry No.
ACC81-3, 10, Number 2, 1981).

The TRH stimulation test rules out the diagnosis of mild or subclinical
<u>hyperthyroidism</u> if TRH infusion causes a rise in plasma TSH; <u>in hyperthyroidism,</u>
<u>there is no rise in plasma TSH after TRH infusion.</u>
<u>Differentiation of Secondary versus Tertiary Hypothyroidism; (Pituitary versus</u>
<u>Hypothalamus)</u>: The TRH is useful in differentiating causes of secondary
hypothyroidism, that is hypothyroidism due to a pituitary lesion versus
hypothyroidism due to a hypothalamic lesion. In these patients, TSH is low. A
control level of serum TSH is obtained; then TRH is injected intravenously and
TSH is measured at 15 minute intervals as illustrated in the next Figure:

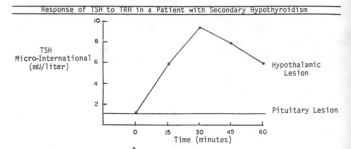

Response of TSH to TRH in a Patient with Secondary Hypothyroidism

TSH
Micro-International
(mU/liter)

Hypothalamic
Lesion

Pituitary Lesion

Time (minutes)

+
Injection of TRH

As shown in the previous illustration, serum TSH increased <u>ten times</u> its
<u>control level in 30 minutes</u>. The results indicate that the patient's
hypothyroidism is due to a <u>lesion in the hypothalamus</u>. If the serum TSH does
not increase, the lesion is in the pituitary.

THYROID STIMULATING HORMONE (TSH) - FOLLOWING TRH
   (see TSH)

THYROID STIMULATING HORMONE (TSH) RECEPTOR
   ANTIBODIES

<u>SPECIMEN:</u> Red top tube, separate serum; store serum frozen and forward to
reference laboratory in plastic tube on dry ice.
<u>REFERENCE RANGE:</u> Depends on method.
<u>METHOD:</u> Radioreceptor assay which measures <u>inhibition</u> of radiolabelled
thyrotropin(TSH) binding to thyroid plasma membranes.
<u>INTERPRETATION:</u> There is considerable evidence that the <u>hyperthyroidism</u> of
<u>Grave's disease</u> is due to <u>antibodies to the TSH receptor</u> (Rees Smith, B.,
Thyrotropin Receptor Antibodies. Receptors and Recognition. Series B, Vol. <u>13</u>,
Chapman and Hall, London, 1981).
    Thyroid stimulating hormone(TSH) receptor antibodies are elevated in about
80 percent of patients with Grave's disease (both treated and untreated) and
about 10 percent of patients with Hashimoto's disease (Ginsberg, J. et al.,
Clin. Endocrinol. <u>19</u>, 305-311, 1983; Shewring, G. and Rees Smith, B., ibid <u>17</u>,
173-179, 1982; de Bruin, T.W.A. et al., Acta. Endocrinologica <u>100</u>, 245-251,
1982).

TRIIODOTHYRONINE (T-3)

<u>SPECIMEN:</u> Red top tube, separate serum; or green (heparin) top tube, separate
plasma.
<u>REFERENCE RANGE:</u> Units of T-3 in ng/dl
    Adults: 60-200; <u>Fetus:</u> 15-126; Amniotic Fluid: 30 (25-35);
    <u>Cord:</u> 50 (10-90); <u>24 Hours:</u> 300 (220-400); <u>48 Hours:</u> 180 (130-230);
    <u>3-5 Days:</u> 130 (50-210); <u>2 Wks - Months:</u> 160 (160-240)
<u>METHOD:</u> RIA. To convert <u>conventional</u> units in ng/dl to <u>international</u> units
in nmol/liter, multiply conventional units by 0.01536.
<u>INTERPRETATION:</u> T-3 is <u>increased</u> in hyperthyroidism; it is <u>decreased</u> in
hypothyroidism but T-3 has a <u>low sensitivity</u> in the diagnosis of
hypothyroidism.

The reasons for the assay of T-3 are given in the next Table:

Assay of T-3 (Nuclear Medical Laboratories)
Diagnosis of Hyperthyroidism
Detection of T-3 Thyrotoxicosis
Detection of Recurrent Hyperthyroidism following previously adequate treatment with Blocking Drugs, Radioactive Iodine or Surgery
Assessment of Acute Changes in Thyroid Hormone Secretion (Within 24 hours of Antithyroid Drug Therapy, the T-3 levels should decrease from baseline level)

(a) T-3 has a high discriminant value in the diagnosis of hyperthyroidism (Evered et al., J. Clin. Path. 29, 1054, 1976).

(b) Early in the course of hyperthyroidism, the serum T-3 may be elevated before serum T-4.

(c) Patients may have clinical signs and symptoms of hyperthyroidism but have normal serum T-4, normal free T-4 and normal TBP; these patients may have T-3 thyrotoxicosis.

Since stress, starvation and several pharmacologic agents inhibit T-4 conversion to T-3, then serum T-3 levels may be normal in sick hyperthyroid patients and may be reduced in sick euthyroid patients.

Endogenous antibodies may combine with T-3, causing spuriously elevated or depressed values for serum T-3, depending on the RIA method. (Neeley, W.E. and Alexander, N.M., Clin. Chem. Check Sample 23, No. 1, 1983; DeBaets, M. et al., Clin. Chim. Acta 118, 293-301, 1982).

THYROXINE BINDING GLOBULIN (TBG)
SPECIMEN: Red top tube, separate serum.
REFERENCE RANGE: 16-24mcg/dl (binding capacity for T-4). To convert conventional units in mcg/dl to international units in nmol/liter, multiply conventional units by 12.87.
METHOD: Radioactive T-4 is added to the serum to saturate the TBG. Unbound T-4 is removed with charcoal. The bound radioactive thyroxine is counted and reflects the TBG.
INTERPRETATION: The causes of reduced concentrations of serum TBG are given in the next Table:

Causes of Reduced Thyroxine-Binding Globulin(TGB)
Congenital TBG Deficiency
Acquired:
Nephrotic Syndrome
Marked Hypoproteinemia
Hepatic Disease
Severe Acidosis
Androgenic or Anabolic Steroid Therapy
Aspirin in Large Doses

The incidence of congenital TBG deficiency is 1 in 13,000 (Dussault, J.H. et al., J. Pediatr. 92, 274-277, 1978); these patients have normal free T-4.

The causes of increased concentration of serum TBG are given in the next Table:

Causes of Increased Thyroxine-Binding Globulin(TBG)
Congenital TBG Increase
Acquired:
Pregnancy
Oral Contraceptive
Liver Disease
Perphenazine Therapy
Hypothyroidism

Congenital TBG increase has an X-linked mode of inheritance; these patients have normal free T-4. Estimated incidence is 1 in 40,000 (Viscardi, R.M. et al., N. Engl. J. Med. 309, 897-899, 1983).

During pregnancy, TBG levels are doubled and remain so until term; several weeks postpartum, TBG levels return to prepregnancy levels.

THYROID, ANTIMICROSOMAL ANTIBODY (MICROSOMAL ANTIBODY)

SPECIMEN: Red top tube, separate serum and store serum at -20°C. The specimen should be rejected if there is excessive hemolysis, or if an anticoagulant is used.

REFERENCE RANGE: Titer less than 1:10; varies with laboratory.

METHOD: Immunofluorescent technique on monkey substrate

INTERPRETATION: The test for serum microsomal antibody is useful in detection and confirmation of autoimmune thyroiditis, such as Hashimoto's thyroiditis and subacute thyroiditis. This test should be used in conjunction with the test for thyroid anti-thyroglobulin antibody since one antibody may be present in the absence of the other. The approximate percentages of patients showing the presence of the two major thyroid antibodies in various diseases are given in the next Table; (Taylor, D.G. and Nakamura, R.M. Clinical Immunology, No. Check List - 1, Am. Soc. Clin. Path, 1977):

Clinical Diagnosis and Percentage of Patients with Thyroid Autoantibodies			
	Thyroglobulin Antibody		Microsomal
Diagnosis	Titers 1:10	High Titers >1:1000	Antibody
Hashimoto's Thyroiditis	30% - 100%	25%	80%
Subacute Thyroiditis	35% - 70%	0	40%
Grave's Disease	35% - 85%	10%	80%
Primary Myxedema	60% - 80%	"common"	80%
Nontoxic Nodular Goiter	20% - 50%	0	40%
Adenoma	10% - 30%	?	?
Carcinoma	15% - 60%	8%	2% - 20%
Juvenile Thyroiditis	30% - 60%	?	90%

High titers of both antibodies (>1:1000 thyroglobulin antibody; >1:100 microsomal antibody) are, with rare exceptions, seen only in Hashimoto's thyroiditis, primary myxedema, and thyrotoxicosis. The antimicrosomal antibody was positive in 94 percent of patients with cytologically proven Hashimoto's thyroiditis; however, only 24 percent had positive antithyroglobulin antibody results (Baker, B.A. et al., Am. J. Med. 74, 941-944, 1983).

A negative result in both tests usually but not necessarily excludes the diagnosis of struma lymphomatosa. Thyroglobulin antibody levels may be low due to the formation of thyroglobulin-antibody complexes immune complexes in-vivo. Patients with Riedel's struma usually show high titers of both antibodies.

Autoantibody tests lack sensitivity and specificity. Occasionally tests for thyroid autoantibodies are negative in patients who show clinical evidence of thyroid autoimmunity as detected by thyroid biopsy. Lack of specificity is observed in that antibodies may be detected in the serum of patients with no clinically apparent disease of the thyroid (Peter, J.B., Diagnostic Medicine, pgs. 19-25, July-Aug. 1981).

The erythrocyte sedimentation rate (ESR) is prolonged in thyroiditis.

Other autoimmune disorders, such as Sjögren's Syndrome, lupus erythematosus, rheumatoid arthritis, may be positive for thyroid antibodies.

Diagnosis of thyroid autoimmune disease should be made in conjunction with thyroid function tests and if necessary, biopsy.

THYROID ANTI-THYROGLOBULIN ANTIBODY (THYROGLOBULIN ANTIBODY)

SPECIMEN: Red top tube, separate serum and store serum at -20°C. The specimen should be rejected if there is excessive hemolysis, or if an anticoagulant is used.

REFERENCE RANGE: Negative; positives are reported as titer.

METHOD: Tanned cell hemagglutination

INTERPRETATION: The test for serum thyroid anti-thyroglobulin is useful in detection and confirmation of autoimmune thyroiditis, such as Hashimoto's thyroiditis and subacute thyroiditis. This test should be used in conjunction with the test for anti-microsomal antibody since anti-microsomal antibody may be present in the absence of anti-thyroglobulin antibody. See Thyroid, Antimicrosomal Antibody (Microsomal Antibody)

THYROGLOBULIN

SPECIMEN: Red top tube, separate serum
REFERENCE RANGE: Up to 50ng/ml
METHOD: RIA
INTERPRETATION:
Serum Thyroglobulin in Thyroid Cancer: Thyroglobulin is the storage form of the
thyroid hormones, T-4 and T-3. Serum thyroglobulin is elevated in patients with
differentiated thyroid carcinoma, e.g., papillary carcinomas, follicular
carcinomas and mixed papillary follicular carcinoma. Serum thyroglobulin is
normal or undetectable in subjects with medullary carcinoma of the thyroid.
    Serum thyroglobulin is also elevated in benign conditions, e.g., non-toxic
goiter, Grave's disease, subacute thyroiditis, endemic goiter; therefore, assay
of serum thyroglobulin is not useful in the diagnosis of thyroid cancer.
    Serum thyroglobulin measurement is useful in the post-treatment followup of
patients with differentiated thyroid cancer in that it is an indicator of tumor
recurrence and metastatic disease (Van Herle, A.J., et al., N. Engl. J. Med.
301, 307-314, 1979; Van Herle, A.J. and Uller, R.P., "Elevated Serum
Thyroglobulin, A Marker of Metastases in Differentiated Thyroid Carcinomas,"
J. Clin. Invest. 56, 272, 1975; Ashcraft, M.W. and Van Herle, A.J., Am. J. Med.
71, 806-814, 1981; Black, E.G. et al., The Lancet, pgs. 443-445, Aug. 29, 1981).
    Thyroglobulin cannot be assayed in the presence of anti-thyroglobulin
antibodies; thyroglobulin antibodies are present in a high percent of patients
with carcinoma of the thyroid (Bates, Harold, Personal Communication).

THYROCALCITONIN (see CALCITONIN)

TOBRAMYCIN (NEBCIN)

SPECIMEN: Red top tube, separate serum and freeze. Obtain serum specimens as
follows:
(1) 24-48 hours after starting therapy if loading dose is not given.
(2) 5 to 30 minutes before I.V. tobramycin (trough)
(3) 30 minutes following completion of a 30 minute I.V. infusion of tobramycin
    (peak)
REFERENCE RANGE: Recommended Dose: 3-5 mg/kg/day, I.M. or I.V. Therapeutic
Range: 4-10mcg/ml; Toxic Level Peak: >12mcg/ml; Trough: <2mcg/ml; Toxic Level,
Trough: >2 mcg/ml.
METHOD: RIA; EMIT
INTERPRETATION: Tobramycin is an aminoglycoside antibiotic (amikacin,
tobramycin and kanamycin) which is used frequently in hospitals to treat
patients who have serious gram-negative bacterial infections, especially
septicemia and staphylococcal infections, untreatable with penicillins.
    The first serum level of tobramycin is obtained when tobramycin has reached
steady-state serum concentrations; steady-state is reached in 24 to 48 hours
after starting therapy if the patient has not received a loading dose (steady
state = 5-7 drug half-lives; half-life of tobramycin is 2 to 3 hours). The
following times should be recorded on the laboratory requisition form and on the
patient's chart:
    Trough Specimen Drawn_____(Time) (5 to 30 minutes before Tobramycin)
    Tobramycin Started_____(Time)
    Tobramycin Completed_____(Time) (30 minutes I.V. infusion)
    Peak Specimen Drawn_____(Time) (30 minutes after I.V. Tobramycin)
    The three main toxic side effects of tobramycin are ototoxicity,
nephrotoxicity and neuromuscular blockage (Smith, C.R. et al., N. Engl. J. Med.
302, 1106, May 15, 1980); it is important to control the dose given by
monitoring peak and trough levels of the drug, particularly in patients with any
degree of renal failure. As renal function declines, drug half-life increases
to up to 50 hours.
    To minimize risk of toxicity, it has been recommended that peak levels not
exceed 10mcg/ml and that trough levels should fall between 1 and 2mcg/ml.
    Tobramycin is eliminated exclusively by renal excretion; excessive serum
concentrations may occur and lead to further renal impairment. The renal damage
is to the renal proximal tubules and is usually reversible if discovered early.
    Ototoxicity is usually due to vestibular damage and is often not
reversible.

TORCH (ANTIBODIES TO TOXOPLASMA, RUBELLA, CYTOMEGALOVIRUS, HERPES SIMPLEX)

SPECIMEN: Red top tube, separate serum; IgG: <u>acute and convalescent sera are required</u>; IgM: single specimen.

For <u>congenital infections</u>, serial sera from <u>both</u> the <u>mother</u> and <u>infant</u> are required.

REFERENCE RANGE: <u>Rising</u> specific IgG antibody <u>titers</u> on serial determinations indicates infection; the <u>presence</u> of specific <u>IgM</u> antibody indicates active disease.

METHOD: See specific tests

INTERPRETATION: Torch tests are tests for detecting <u>antibodies</u> to <u>toxoplasma</u>, <u>rubella</u>, <u>cytomegalovirus</u> and <u>herpes simplex</u>. Infections of the mother in early pregnancy are associated with <u>fetal malformations</u> and sequalae such as abortions, stillbirths and congenital anomalies: The natural history of herpes simplex virus in pregnant women has been studied (Harger, J.H. et al., Am. J. Obstet. Gynecol. <u>145</u>, 784-791, Apr. 1, 1983). The presence of specific IgG antibody prior to pregnancy indicates immunity; <u>rising IgG antibody titers and/or the presence of IgM antibody indicates active disease)</u>

Antibodies in "normal" infants decrease during the first several months of life because they are passively acquired by placental transfer. <u>Active infection in the neonate</u> is indicated by antibody levels that are unchanged or increase in serial sera over several months. <u>The absence of antibody in the mother rules out congenital infection.</u>

The presence of <u>IgM antibodies</u> in the infant indicates active disease.

TOXOPLASMOSIS ANTIBODY

SPECIMEN: Red top tube, separate serum; <u>IgG</u>: acute and convalescent sera are required. IgM: single specimen.

REFERENCE RANGE: A four-fold rise in <u>IgG</u> antibody titers on serial determinations indicates active disease. Different laboratories report different diagnostic titers: <u>Mayo Medical Laboratories</u>: Titer <1:16, no previous infection except for ocular infection; titer between 1:16 to 1:256 is prevalent in the general population; a titer between 1:256 to 1:1024 indicates recent infection; a titer >1024 indicates <u>active infection</u>. <u>North Carolina State Laboratory</u>: Indirect hemagglutination: diagnostic titer, 1/256; indirect fluorescent antibody diagnostic titer, 1/256; indirect fluorescent antibody-IgM, diagnostic titer, 1/64.

False Positives: Antinuclear antibodies may cause a <u>false positive reaction</u> in the conventional and IgM indirect fluorescent antibody test, and <u>rheumatoid factor</u> may cause a false positive IgM indirect fluorescent antibody test; <u>dye tests</u> and <u>complement fixation tests</u> do <u>not</u> give false positive results (Burkhart, T., personal communication; Remington, J.S. and Krick, J.A., N. Engl. J. Med. <u>298</u>, 550, 1978).

METHOD: Indirect hemagglutination(IgG) and indirect immunofluorescence (IgG and IgM)

INTERPRETATION: <u>Toxoplasma gondii</u> is a protozoan obligate intracellular parasite. The sources of toxoplasma are <u>uncooked meat</u> and domestic animals. <u>Toxoplasmosis in the pregnant female</u> may cause spontaneous <u>abortion</u> or <u>disease in the infant</u> with symptoms developing after birth (<u>congenital toxoplasmosis</u>).

The usual changes in IgM and IgG antibodies to T. gondii following infection are as follows: <u>IgM</u> antibodies appear within <u>one week</u>, peak in <u>three to four weeks</u> and disappear in <u>three to four months</u>. <u>IgG</u> antibodies appear in about three weeks, <u>peak in two months</u> with the <u>immunofluorescent</u> test, in <u>four to six months</u> by the <u>hemagglutination</u> test.

Pregnant Female: Transplacental transmission of toxoplasma to the fetus occurs if the mother first acquires acute toxoplasmosis during pregnancy. <u>It does not repeat in subsequent pregnancies.</u> Fetal parasitemia and generalized disease develop in about 50% of such pregnancies. In 10% of cases, the mother shows evidence of acute gestation illness. When the maternal infection occurs in the <u>third trimester</u>, the fetal infection rate is high, but the disease in the newborn is mild or subclinical. When the maternal infection occurs in the <u>second trimester</u>, fetal disease is devastating.

A <u>positive hemagglutination test</u>(IgG) in <u>early pregnancy</u> indicates that the patient was exposed to toxoplasmosis in the <u>past</u> and is now <u>immune</u>. If the hemagglutination test is negative, then the <u>immunofluorescent test</u> should be

done; a positive IgG immunofluorescent test indicates immune status; a positive
IgM immunofluorescent test indicates active disease; if both the
hemagglutination and immunofluorescent tests are negative, repeat the
immunofluorescent test every three months. A four-fold rise in IgG titer, an
IgG titer >1024 or the appearance of IgM antibodies indicate active infection.
Newborn: Detection of active infection in the newborn is more difficult in that
the immunofluorescent test for IgM antibody on cord blood is not sensitive; most
neonates who have been infected in-utero are born without detectable IgM
antibody, and IgG antibody may be present in the neonate by passive transfer
from immune mother. The laboratory diagnosis in the neonate of acquired
toxoplasmosis in-utero is done by demonstrating a rising IgG titer over a
several month period.
Ref.: Lab. Report for Physicians 3, 9-11, Feb. 1981; Beattie, C.P., Lancet 1,
873 (1980); Stagno, S., Am. J. Dis. Child. 134, 635-637 (1980); Stray-Pedersen,
B., Am. J. Dis. Child. 134, 638-642, (1980).

TOURNIQUET TEST
SPECIMEN: This test is done at the bedside.
REFERENCE RANGE AND METHOD: Inspect arm for petechiae. Inflate a blood
pressure cuff on the arm to a pressure midway between systolic and diastolic
pressures. After 5 minutes, the cuff is deflated and the arm distal to the cuff
inspected for the appearance of new petechiae over the next 5 minutes. The
grading is as follows: 4+ = confluent petechiae; 3+ = non-confluent petechiae;
2+ = large number of petechiae; 1+ = minimal change.
INTERPRETATION: A positive tourniquet test is seen in the conditions listed
in the next Table:

Positive Tourniquet Test
Primary Vascular Abnormalities:
Scurvy
Collagen Vascular Disease
Decreased Number of Platelets
Qualitative Platelet Disorder

A positive tourniquet test and a normal bleeding time is suggestive of a
primary vascular abnormality, that is, scurvy or collagen vascular disease.

TOXI-SCREEN (see DRUG SCREEN)

TRANSFERRIN
SPECIMEN: Red top tube, separate serum; avoid hemolysis
REFERENCE RANGE: 200-400mg/dl. To convert conventional units in mg/ml to
international units in g/liter, multiply conventional units by 0.01.
METHOD: Gel diffusion or light scattering
INTERPRETATION: Transferrin is elevated in the serum in the conditions listed
in the next Table:

Causes of Elevated Serum Transferrin
Iron Deficiency Anemia
Oral Contraceptives
Late Pregnancy
Viral Hepatitis

Transferrin is decreased in the serum in the conditions listed in the next
Table:

Causes of Decreased Serum Transferrin
Chronic Infections
Malignancy
Iron Poisoning
Hemolytic Anemia (sometimes Normal)
Hemochromatosis (sometimes Normal)
Nephrosis
Kwashiorkor
Thalassemia

TRICYCLIC ANTIDEPRESSANTS (INCLUDES AMITRIPTYLINE, NORTRIPTYLINE, IMIPRAMINE (TOFRANIL), DESPIPRAMINE, PROTRIPTYLINE, DOXEPIN AND DESMETHYLDOXEPIN)

SPECIMEN: The prerequisites for therapeutic monitoring are dosage must be stable for at least ten days and dosage cannot be changed or missed. Specimens for monitoring therapeutic response should be drawn just prior to the first dose of the day or in the morning if single bedtime doses are taken. These sampling times are at least six hours, but preferably 10-15 hours after the last dose. Specimens for toxic overdose are obtained anytime. Blood: Green (heparin) top tube, remove plasma from red cells within two hours. There is a substance in some vacutainer tubes that displaces tricyclic antidepressants from plasma protein. The tricyclic antidepressants may then enter the red blood cells yielding artifactually low plasma values. Urine: 50ml random urine.

REFERENCE RANGE:

Therapeutic plasma ranges for tricyclic antidepressants are given in the next Table:

Therapeutic Plasma Ranges for Tricyclic Antidepressants			
Drug	Active Metabolite	Therapeutic Ranges (ng/ml)	Half-Life (hours)
Amitriptyline (Elavil Endep and Triavil)	Nortriptyline	120-250	17-40
Nortriptyline (Aventyl)		50-150	18-93
Imipramine (Tofranil and Presamine)	Desipramine	150-250	9-24
Desipramine (Norpramin and Pertofrane)		150-250	14-76
Doxepin (Sinequan and Adapin)	Desmethyldoxepin	150-300	8-25

Although there is individual variation in response to these drugs, the frequency of adverse effects increases sharply at levels above the therapeutic range.

Therapeutically effective concentrations are approximately 200ng/ml except for nortriptyline (100ng/ml). A level of tricyclics above 600ng/ml may produce serious side effects and extremely toxic concentrations exceed 1000ng/ml. Toxic levels are lower for nortriptyline and protriptyline. These levels include the parent compounds and active metabolite.

METHOD: Gas liquid chromatography (GLC); high-pressure liquid chromatography (HPLC); radioimmunoassay (RIA). The method used must measure the active metabolites for amitriptyline (nortriptyline) and the active metabolite for imipramine (desipramine).

INTERPRETATION: Tricyclic antidepressant drugs are used primarily for the treatment of acute episodes of endogeneous depression and for long term treatment to prevent recurrence and relapse.

The tricyclic antidepressants most frequently prescribed are amitriptyline, imipramine and doxepin; these same drugs are most frequently involved in suicidal attempts (Priest, R.G. et al., J. Int. Med. Res. (3), 8, suppl. (3), 8-13, 1980). At levels greater than 1000ng/ml, these drugs are toxic to the nervous system causing grand mal seizures, convulsions and respiratory arrest, and the heart causing ventricular arrhythmias, e.g., ventricular tachycardia and fibrillation. Orthostatic hypotension may develop. Total tricyclic concentrations reflect the severity of overdose; however, the best correlation is with an ECG finding of QRS prolongation. In the absence of severe impairment of myocardial performance, depressed patients with preexisting heart disease can be effectively treated with tricyclic antidepressants without an adverse effect on ventricular rhythm or hemodynamic function (Veith, R.C. et al., N. Engl. J. Med. 306, 954-959, 1982).

Tricyclic antidepressants may also cause vasospasm in some patients, and the vasospasm may be dose related (Appelbaum, P. and Kapoor, W., Am. J. Psychiatry 140, 913-915, July 1983).

The most common adverse effects of tricyclic antidepressants are anticholinergic, e.g., dry mouth, decreased gastrointestinal motility, mydriasis, urinary hesitancy or retention and tachycardia.

Treatment of mild toxicity is supportive; moderate and severe overdoses require respiratory support, anticonvulsants, physostigmine and beta-blockers;

and cardioversion and pacing may be necessary (Spiker, D.G. et al., Clin.
Pharmacol. Ther. 18, 539-546, 1976; Rose, J.B., Clinical Toxicology 11, 391-402,
1977). Sodium bicarbonate has been used to treat arrhythmias (Manoguerra, A.S.,
Crit. Care Quart. 4, 43-52, Mar. 1982).

Withdrawal symptoms (nausea, dizziness, headache, increased perspiration
and salivation) have been reported after stopping treatment with tricyclics
(Dilsaver, S.C. et al., Am. J. Psychiatry 140, 249, 1983); dosage should be
tapered when tricyclics are discontinued.

Dosage Guidelines: One dose daily before bedtime because of the long half-
life, and the sedative and anticholinergic side effects. One-quarter of the
average dose is given and the dose is gradually increased to full average dose
over a period of seven to ten days. When a stable dose is maintained for ten
days, about 60 percent of the patients will have expected therapeutic response.

Lack of therapeutic response may be due to doses that are too low or too
high, or to unresponsive type of depression or non-compliance. A serum assay
should be done, and the dose adjusted appropriately; a therapeutic response will
be obtained in an additional 25-30 percent of the patients.

Patients with low serum albumin have a higher percentage of free to bound
drug ratio (protein binding, 90 percent) and thus therapeutic or toxic reactions
occur at a lower serum level. The half-life of the tricyclics is accelerated by
barbituates and smoking; the half-life is decreased by corticosteroids,
phenothiazines, haloperidol and disulfiram.

Metabolism: The tricyclics are metabolized in the liver; monodemethylation to
active metabolites and hydroxylation and conjugation to inactive metabolites
occur (Baer, D.M. Interpretation of Drug Concentrations, Am. Soc. Clin. Path.,
1981; Gram, L.F. et al., Therapeutic Drug Monitoring, 4, 17-25, 1982;
Braithwaite, R.A., ibid., 27-31, 1982; Smith, R.K. and O'Mara, K., J. Family
Pract. 15, 247-253, 1982).

TRIGLYCERIDES

SPECIMEN: Red top tube, separate serum. Serum triglycerides are done on serum specimens from a patient who has fasted for 12-14 hours (overnight). The patient may drink all the water that he wishes.

REFERENCE RANGE:

Age	Triglycerides (mg/dl)
0-19	10-140
20-29	10-140
30-39	10-150
40-49	10-160
50-59	10-190

Age Changes

To convert conventional units in mg/dl to international units in mmol/liter, multiply conventional units by 0.01129.

METHOD:

Breakdown of triglycerides to glycerol plus fatty acids:

$$\text{Triglycerides} \xrightarrow{\text{Lipase, Protease}} \text{Glycerol + Fatty Acids}$$

or

$$\text{Triglycerides} \xrightarrow{\text{Ethanol(KOH)}} \text{Glycerol + Fatty Acids}$$

Glycerol is assayed using two consecutive kinase reactions followed by a dehydrogenase reaction:

(1) $\text{Glycerol + ATP} \xrightarrow{\text{Glycerokinase(GP)}} \text{Glycerol-1-Phosphate + ADP}$

$\text{ADP + Phosphoenolpyruvate (PEP)} \xrightarrow{\text{Pyruvate Kinase}} \text{ATP + Pyruvate}$

$\text{Pyruvate + NADH + H}^+ \xrightarrow{\text{Lactate Dehydrogenase(LDH)}} \text{Lactate + NAD}^+$

(2) In another method, the first reaction is the same as that given above; that is:

$\text{Glycerol + ATP} \xrightarrow{\text{Glycerokinase(GP)}} \text{Glycerol-1-Phosphate + ADP}$

$\text{Glycerol-1-Phosphate + NAD}^+ \xrightarrow{\text{Glycerol-1-Phosphate}} \text{Phosphate + Dihydroxyacetone + NADH + H}^+$

(3) $\text{Glycerol + NAD}^+ \xrightarrow[\text{Dehydrogenase (GDH)}]{\text{Glycerol}} \text{Dihydroxyacetone + NADH + H}^+$

INTERPRETATION: The causes of elevation of serum triglycerides are given in the next Table:

Causes of Elevation of Serum Triglycerides
Non-fasting Specimen
Primary Hyperlipoproteinemia
(Types I, IIb, III, IV, V)
Secondary Hyperlipoproteinemias:
Diabetes Mellitus
Acute Alcoholism
Oral Contraceptives
Nephrotic Syndrome
Chronic Renal Failure
Steroids
Acute Pancreatitis
Gout
Hyperlipidemia in Gram Negative Infections
Glycogen Storage Disease

Triglycerides are insoluble in blood, and thus they do not circulate free in the serum but are transported as chylomicrons and pre-beta liporoteins (very low density lipoproteins (VLDL)). 80% of chylomicrons are triglycerides; 55% of pre-beta lipoproteins are triglycerides. It takes 10-12 hours to clear the blood of chylomicrons after a meal; peak lipidemia is reached in 3 to 5 hours and persists for another 6 to 8 hours. If cholesterol and chylomicrons are normal and triglycerides are elevated, the most likely primary hyperlipidemia is Type IV.

Secondary hyperlipoproteinemias with elevation of serum triglycerides are very common e.g., <u>diabetes mellitus</u>, <u>acute alcoholism</u>, <u>nephrotic syndrome</u> and use of <u>oral contraceptives.</u>

In most women taking oral contraceptive steroids there is a small but definite increase in serum lipids. However, in about 1/3 of the women, taking oral contraceptive steroids, the triglyceride level exceeds the upper limit of normal. The increase in the triglyceride is related to the estrogen but not the progesterone content of the contraceptive drug. Patients on oral contraceptives who are most "susceptible" to hyperlipemia are the following: obese patients, patients showing evidence of carbohydrate intolerance, patients having pre-existing hyperlipemia, and patients having a family history of hypertriglyceridemia. Caution should be exercised in the use of contraceptive agents and other estrogenic compounds in patients with pre-existing hyperlipemia, e.g., women with Type I, IV and Type V hyperlipoproteinemia, because elevated plasma triglyceride may trigger an attack of pancreatitis.

TRYPSIN, IMMUNOREACTIVE, SERUM

<u>SPECIMEN:</u> Red top tube, separate serum; store sera at -20°C until assay.

Assay may be done on dried blood-spots collected for neonatal screening of other inborn errors. <u>Dried blood specimens stored at room temperature lose half their immunoreactivity over a period of three months</u> (Heeley, A.F. et al., Arch. Dis. Childhood 57, 18-21, 1982).

<u>REFERENCE RANGE:</u> Adult: 20 to 80 ng/ml; reference range dependent on kit used.

<u>METHOD:</u> Trypsin radioimmunoassay kit, such as the Trypsik kit, Damon Diagnostics, Needham Heights, Mass. In the disc method, the disc is incubated with antibody and I-125 labelled trypsin. Blood disc method, (see King, D.N. et al., Lancet 2, 1217-1219, 1979 and Heeley, A.F. et al., Arch. Dis. Childhood 57, 18-21, 1982; Crossley, J.R. et al., Clin. Chem. Acta. 113, 111-121, 1981).

<u>INTERPRETATION:</u> Immunoreactive trypsin(IRT) is trypsinogen, a precursor of trypsin; it is synthesized in the pancreas. IRT may be useful in the diagnosis of cystic fibrosis and chronic pancreatitis with steatorrhea.

<u>Cystic Fibrosis:</u> <u>The concentration of immunoreactive trypsin(IRT) is increased in the blood of infants with cystic fibrosis(CF) during the first few months of life</u>; it is thought that, in these patients, the pancreatic duct is progressively blocked with initial "back-leakage" of acinar contents into the plasma. The level of IRT subsequently decreases to values below normal. The elevation of IRT is characteristic of newborn CF, whether or not they have residual exocrine pancreatic function, e.g., measurable stool trypsin activity (Crossley, J.R. et al., Clin. Chem. Acta 113, 111-121, 1981).

<u>The incidence of false negative test results is 10% to 25%; the incidence of false positive test results is 0.1% to 0.3%.</u> The IRT test may be particularly useful during the first month of life when the results of the sweat test are not reliable. (For Discussion see Ad Hoc Committee Task Force on Neonatal Screening. Cystic Fibrosis Foundation: Neonatal Screening for Cystic Fibrosis: Position Paper. Pediatrics 72, 741-745, 1983; Holtzman, N.A., Pediatrics 73, 98-99, 1984; Farrell, P.M., Pediatrics 73, 115-117, 1984).

The blood spot IRT assay performed in the neonatal period is an excellent screening test for detection of cystic fibrosis (Wilchen, B. et al., J. Pediatrics 102, 383-387, 1983). The cost of the screening is about $0.70 per infant screened; the cost of the reagents is about $0.15 per infant.

<u>Chronic Pancreatitis with Steatorrhea:</u> IRT is <u>decreased</u> in patients with <u>chronic pancreatitis with steatorrhea</u>; it is normal in patients with <u>chronic pancreatitis without steatorrhea</u> and in patients with steatorrhea but with normal pancreatic function (Jacobson, D.G. et al., N. Engl. J. Med. 310, 1307-1309, May 17, 1984). It is also decreased in patients with cystic fibrosis who are older than 1 year of age and in patients with alcohol-induced chronic pancreatitis (Andriulli, A. et al., Dig. Dis. Sci. 26, 532-537, 1980).

TUMOR MARKERS PANEL (SEE CANCER MARKERS PANEL)

UREA NITROGEN, BLOOD (BUN; UREA NITROGEN, BLOOD)

SPECIMEN: Red top tube, separate serum; or green (heparin) top tube, separate plasma

REFERENCE RANGE: 5-20mg/dl. To convert conventional units in mg/dl to international units in mmol/liter, multiply conventional units by 0.3570.

METHOD:

$$Urea + H_2O \xrightarrow{Urease} 2NH_3 + CO_2$$

$$NH_3 + Alpha\text{-}Ketoglutarate + NADH + H^+ \xrightarrow[\text{Dehydrogenase}]{Alpha\text{-}Ketoglutarate} L\text{-}Glutamate + NAD^+$$

INTERPRETATION: Urea is increased in the serum in conditions listed in the next Table:

Causes of Elevation of Serum Urea
Renal Disease
Prenal Azotemia
G.I. Hemorrhage
G.I. Obstruction
Shock
Tissue Necrosis
Third Degree Burns
Fever, Protracted
Dehydration
Diarrhea
Diabetic Coma
Congestive Heart Failure
Addison's Disease
Steroid Therapy
High Protein Diet
Post-Renal
Tetracycline (Results in Net Catabolism)

The BUN is increased in both renal and prerenal azotemia. Serum creatinine and creatinine clearance is needed to differentiate pre-renal and renal azotemia.

The three general causes of an increased serum urea concentration are decreased glomerular filtration rate, an increased load of urea for excretion from the diet or tissue metabolism, and an increased tubular reabsorption of urea.

Causes of raised urea concentrations in a hospital population are given in the next Table (Morgan, D.B., Carver, M.E. and Payne, R.B., Brit. Med. J. 2, 929-932, 1977):

Causes of Elevated Serum Urea in a Hospital Population	
Cause	Incidence (Percent)
Congestive Heart Failure	36
Dehydration	12
Post-operation	6
Hypotension	3
Acute Renal Failure	2
Chronic Renal Failure	3
Increased Urea Load	2
Obstructive Renal Disease	1
Combined Causes	9
Unclassified	26

A plasma urea concentration of greater than 180mg/dl is usually found only in renal disease.

Urea is decreased in the serum in conditions listed in the next Table:

Causes of Decrease in Serum Urea
Inappropriate ADH
Liver Disease
Overhydration
Anabolic Hormones
Malnutrition
Normal Pregnancy

URIC ACID, SERUM

SPECIMEN: Red top tube, separate serum; stable for 3 days at room temperature.

REFERENCE RANGE: Adult Females: 2.5-6.2mg/dl; Adult Males: 3.5-8.0mg/dl; One Month to 12 years: 2.0-7.0mg/dl. To convert conventional units in mg/dl to international units in mcmol/liter, multiply conventional units by 59.48.

METHOD:

(1) Measure Decrease in Absorption at 293nm.

$$\text{Uric Acid} + O_2 + H_2O \xrightarrow{\text{Uricase}} \text{Allantoin} + H_2O_2 + CO_2$$
(293nm)

(2) Measure Hydrogen Peroxide

$$\text{Uric Acid} + O_2 + H_2O \xrightarrow{\text{Uricase}} \text{Allantoin} + H_2O_2 + CO_2$$

$$H_2O_2 + \text{Reduced Chromogen} \xrightarrow{\text{Peroxidase}} \text{Oxidized Chromogen} + H_2O$$

INTERPRETATION: Uric acid is increased in the serum in the conditions listed in the next Table:

Causes of Increased Serum Uric Acid
Gout
Hematologic Conditions
Leukemia
Lymphoma
Hemolytic Anemia
Megaloblastic Anemia
Infectious Mononucleosis
Polycythemia Vera
Chronic Renal Disease(Renal Failure)
Drug-Induced:
Thiazides
Salicylates
Pyrazinamide(PZA)
Cytotoxics
Tissue Necrosis
Malnutrition of All Types
Therapeutic Radiation
Alcohol
Lead Poisoning
Glycogen Storage,Type I
Lactic Acidosis
Toxemia of Pregnancy
Psoriasis(Active)
Lesch-Nyhan Syndrome

Increased concentration of uric acid is found in gout, in conditions involving increased cellular destruction, such as leukemia and lymphoma. In treating leukemia, uric acid must be monitored because massive destruction of cancer cells by cytotoxic agents causes large amounts of uric acid. In conditions whereby increased lactic acid is produced, the lactic acid and uric acid compete for renal excretory sites and serum uric acid is increased.

Using 7.0 mg/dl as cutoff for men, serum uric acid has a sensitivity of 90% and a specificity of 95% for the diagnosis of gout.

Uric acid is decreased in the serum in conditions listed in the next Table:

Causes of Decreased Serum Uric Acid
Severe Alcoholism
with Liver Disease
Chronic Debilitating Disease
Renal Tubular Defects
Salicylates(Large)
Allopurinol
Probenecid
X-Ray Contrast Media
Glyceryl Guaiacolate
Wilson's Disease
Hemochromatosis
Xanthine Oxidase Deficiency

URINALYSIS ROUTINE

SPECIMEN: 10ml fresh urine

REFERENCE RANGE: Color, Straw; Turbidity, clear; Sp.Gr. 1.001-1.035. Dipsticks, pH 4.5-7.5; Protein, Negative; Sugar, Negative; Acetone, Negative; Bile, Negative; Hemoglobin, Negative; Positive dipstick tests are confirmed as follows: Protein, Sulfosalicylic acid; Coomassie blue dye binding method; Sugar, Clinitest; Acetone, Acetest; Bilirubin, Ictotest. Microscopic: WBC, 1-2 hpf; RBC, 0-1 hpf; Cast, 0-1 hyaline, occasional granular, Lpf; hpf = high powered field; lpf = low powered field. Bacteria, rare.

METHOD: Appearance, Sp.Gr. by Refractometer; Dipstick; Microscopic.

INTERPRETATION: Routine urinalysis requires a random specimen; an early morning specimen is preferred. With specimens for routine analysis, avoid preservatives and refrigeration. Analyze specimen within two hours after collection. If a urine specimen is left standing at room temperature for 5 hours, it will become alkalinized; it is not suitable for culture; the erythrocytes, if present, will decompose and urinary casts will disintegrate.

If immediate examination of urine is impractical, refrigeration is the preferred form of preservation; refrigeration does not affect the appearance of casts; it does not affect the urine osmolality and does not destroy erythrocytes.

The normal ranges for 24 hour urine volume are as follows:

Age	Volume(ml/day)
Full Term Newborn	15-60
Two Months	250-450
Six-Eight Months	400-500
1-2 Years	500-600
2-4 Years	600-750
5-7 Years	650-1000
8-15 Years	700-1500
Adult	1000-1600

Routine urinalysis usually consists of gross observation of the specimen, use of dip sticks and microscopic analysis.

APPEARANCE: Cloudiness due to phosphates (alkaline urine) and urates (acid urine) is normal. Cloudiness and color in urine may be associated with the conditions given in the next Table:

Cloudiness in Urine
Blood (pink, red or brown)
Myoglobin
Leukocytes
Mucous
Urobilin (Hemolytic Anemia, Parenchymal Liver Disease)
Conjugated Bilirubin (Parenchymal Liver Disease, Biliary Tract Obstruction)

SPECIFIC GRAVITY: Urea (20 percent); chloride (25 percent), sulfate and phosphate contribute most to normal urine. The measurement of specific gravity is done using a hydrometer (urinometer); other measurements that reflect specific gravity are measurement of refractive index using a refractometer and measurement of osmolality using an osmometer and the measurement of specific gravity using dip-stick test strips. The specific gravity may range from 1.003 to 1.030 depending on fluid intake. A specific gravity of 1.023 or more indicates normal urine concentrating ability.

MULTIPLE REAGENT STRIP: The multiple reagent strip is dipped into the urine and read; the color indicates the pH range or the concentration of the substance. pH: Urine is normally acid; the distal tubular cells exchange $[H^+]$ for sodium of the glomerular filtrate and the urine becomes acid because of the exchange. The usual pH of urine is pH = 6. The pH of the urine indicates acid-base status e.g., acid urine in acidosis and alkaline urine in alkalosis. However, the urine is alkaline in renal tubular acidosis; in this condition, tubular ability to form ammonia and exchange $[H^+]$ for $[Na^+]$ is defective.

PROTEIN: Protein is found in the urine in the conditions listed in the next Table:

Protein in Urine
Nephrotic Syndrome
Nephritic Syndrome
Toxic Nephropathies
Renal Tubular Diseases
Nephrosclerosis
Polycystic Kidney Disease
Severe Venous Congestion of Kidney
Pyelonephritis
Pre-eclampsia of Pregnancy
Postural Proteinuria
Other Conditions: Hemorrhage, Salt Depletion and Febrile Illness

The upper limit of normal for urine protein is about 150mg/24hr. The protein lost in greatest amount in renal disease is <u>albumin</u>; however, when glomerular damage is present, larger proteins may be found in the urine.

With glomerular damage, proteins with molecular weight greater than 60,000 appear in the urine. With tubular damage, low molecular weight proteins, which are normally absorbed in the proximal tubule, appear in the urine.

Bence-Jones protein is present in the urine in about 20 percent of proven cases of multiple myeloma in the absence of a serum paraprotein (Hobbs, J.R., Essays in Med. Biochem. 1, 105, 1975).

The test material on reagent strips is <u>not</u> as sensitive to <u>globulins</u> nor to Bence-Jones protein as to albumin.

The nephrotic syndrome is by far the most common <u>cause of decrease in immunoglobulin levels</u>. IgG and IgA concentrations are decreased and IgM concentration is usually normal. The synthesis of alpha-2 macroglobulin is switched on; this may be due to albumin loss and decreased oncotic pressure in the sinusoids; these proteins accumulate in the blood due to their high molecular weight, and thus they are not lost in the urine. The increased beta-lipoprotein (low density lipoprotein) accounts for the increase in serum cholesterol in the nephrotic syndrome.

GLUCOSE: Normally, no detectable glucose is present in the urine; the conditions in which glucose is found in the urine are given in the next Table:

Excess Glucose in Urine
Diabetes Mellitus
Other Endocrine Disorders
(Pituitary and Adrenal Diseases):
Cushing's Syndrome or Hyperadrenocorticism
Acromegaly
Pheochromocytoma
Hyperthyroidism
Pancreatic Disorders: Hemochromatosis; Pancreatitis;
Carcinoma of the Pancreas
Central Nervous System Disorders
Disturbances of Metabolism
Burns, Infection, Fractures, Myocardial Infarction,
Uremia, Liver Disease, Glycogen Storage Diseases, Obesity
Drugs: Thiazides, Corticosteroids, ACTH, Birth Control Pills
Renal Tubular Dysfunction relates to inability to absorb glucose

The test material on the reagent strip(glucose oxidase) is specific for glucose; however ascorbic acid(vitamin C) inhibits the test.

Glucose passes through the glomerulus into the glomerular filtrate and is reabsorbed into the blood through the proximal convoluted renal tubules. Glucose is elevated in the urine when glucose is elevated in the serum or when there is a lower renal threshold for glucose.

The renal threshold for glucose is a serum concentration of 180mg/dl; however, there is wide individual variation in the renal threshold for glucose. Patients with significant glycosuria have increased specific gravity due to increased glucose in the urine.

KETONES: The test material on reagent strips measures acetone and aceto-acetic acid but does not detect beta-hydroxybutyric acid; the strips detect 5 to 10mg/dl of ketones in urine. Ketonuria is found in conditions listed in the next Table:

Ketonuria
Uncontrolled Diabetes Mellitus
Non-Diabetic Ketonuria: In children -
Acute Fibrile Illnesses
Toxic States with Vomiting or Diarrhea
Alcoholics
Some Weight Reducing Diets

BLOOD: The test reagent strip material detects hemoglobin and myoglobin; it detects 0.05 to 0.3mg hemoglobin/dl of urine; 0.3/mg hemoglobin is equivalent to 10 lysed red blood cells per microliters. The test does not detect intact red blood cells. However, when red blood cells enter the urine, hemolysis usually occurs, and free hemoglobin is released in the urine. Conditions in which hemoglobin appear in the urine are given in the next Table:

Hemoglobin in the Urine
Hemolytic Anemias
Renal Disease
Glomerulonephritis
Lupus Nephritis
Calculi
Tumor
Acute Infection
Tuberculosis
Infarction
Renal Vein Thrombosis
Trauma
Hydronephrosis
Polycystic Kidney
Acute Tubular Necrosis
Malignant Nephrosclerosis
Lower Urinary Tract Disease
Acute and Chronic Infection
Calculus
Tumor
Stricture

The most common causes of hematuria are stones (20%) malignant neoplasm (15%), urethrotrigonitis, bacterial infection (10%), prostatic hypertrophy (10%) and glomerulonephritis (Abuelo, J.G., Arch. Intern Med. 143 #5, 967-970, 1983).

BILIRUBIN: The bilirubin that appears in the urine is bilirubin diglucuronide. The test material in the reagent strip is sensitive to 0.2mg/dl to 0.4mg/dl. The conditions in which positive tests are obtained are liver disease, obstructive biliary tract disease and the congenital hyperbilirubinemias, e.g., Dubin-Johnson and Rotor types.

LEUKOCYTES: The number of leukocytes is indicated in urine by a color reaction on the reagent strip. It exploits the esterase activity of leukocytes: Leukocytes liberate esterase; esterase acts on the substrate on the dipstick. The substrate is an indoxyl carbonic acid ester from which indoxyl is liberated; indoxyl is unstable and oxidizes to the blue compound, indigo when exposed to atmospheric oxygen. The presence of leukocytes indicates pyelonephritis or inflammation involving other structures in the urinary tract.

MICROSCOPIC EXAMINATION OF URINE: The following values obtained on examination of sediment of a centrifuged urine specimen are considered normal: white blood cells, 3 to 5/hpf; hyaline casts, rare to 1/low pf; red blood cells 0 to 2/hpf, (hpf = high power field). The number of casts present are counted per low power microscopic field (10x objective lens).

URINE CULTURE

SPECIMEN "Clean Catch" Method: The patient should be supplied with a sterile urine collection cup and instructed as follows: First morning specimen is preferred. 1. Wash hands thoroughly. 2. Wash penis or vulva using downward strokes. 3. Start to urinate directly into toilet or bedpan - stop - position container and take urine sample. 4. Screw cap on securely without touching the inside rim. 5. Keep specimen refrigerated. The urine should be transported to the laboratory and refrigerated until processed. Urine specimens must be cultured within two hours (preferably within one hour) of collection. If a specimen cannot be cultured immediately, it may be held in a refrigerator (not frozen) for up to 48 hours.

Catheterization of the urinary bladder is not benign; a single, short-time catheterization causes bacteriuria in 1% to 5% of patients. This risk may be higher in patients who have urinary tract abnormalities (Klein, R.S., Mayo Clin. Proc. 54, 412, 1979). Indwelling bladder catheters account for more than 500,000 nosocomial infections per year in United States hospitals. Apparently, sealed catheter systems can reduce infection and mortality among hospitalized patients (Platt, R. et al., Lancet 1, 893, Apr. 23, 1983). Bacteria gain entry into the catheterized bladder by two routes: migration from the collection bag or the catheter drainage tube junction within the catheter lumen, or they may ascend into the periurethral mucous sheath outside the catheter. Migration of bacteria extraluminally in the periurethral space is the major pathway for entry into the bladder and that meatal colonization by bacteria is a major risk factor (Garibaldi, R.A. et al., N. Engl. J. Med. 303, 316-318, 1980).

Collection of urine in children less than three years of age is a particularly difficult problem. In one way, urine is collected by use of a bag with adhesive placed over the genitalia; however, this leads to a high level of contamination.

Suprapubic aspiration is another method for the collection of urine and may be useful for obtaining uncontaminated urine specimens in children. This procedure minimizes the risk of introducing bacteria into an uninfected bladder and any growth indicates infection.

REFERENCE RANGE: Formerly, a bacteria count of 10^5 or more organisms per ml of urine indicated the presence of infection. Recently, that number has been revised downward (Stamm, W.E. et al., N. Engl. J. Med. 307, 463-468, 1982; Stamm, W.E. et al., N. Engl. J. Med. 304, 956, 1981; Stamm, W.E. et al., 303, 409, 1980).

On suprapubic aspiration, 150 or more bacteria per ml is a significant count in bladder urine.

METHOD: For bacteriologic studies, an unspun, first morning, midstream, clean catch specimen preceded by cleaning of the external genitalia, is used for culture for identification and antibiotic sensitivity.

Most laboratories estimate the bacterial count in urine by streaking the surface of the plates with wire loops calibrated to deliver 0.01 or 0.001ml. Other techniques include the filter paper and dip slide methods (Cohen, S.N. and Kass, E.H., N. Engl. J. Med. 277, 176, 1967).

A presumptive diagnosis of bacteriuria may be made by microscopic examination of the urine (Farrar, W.E., Medical Clinics of North America, 67, 187-201, 1983). Unspun Specimen: One or more bacteria per oil immersion field of the unstained centrifuged sediment. Spun specimens yield a higher positive bacterial detection rate than unspun specimens on microscopic examination. (Wallach, J., JAMA, 248, 1509, Sept. 24, 1982). A 12ml aliquot of urine is centrifuged at 1500rpm for three minutes at 400rcf (relative centrifugal force); the supernatant is decanted and the sediment is resuspended in 1ml of urine. The sediment is observed on a slide. Examine the sediment under high power and look for rods and cocci; rods are relatively easy to identify. Twenty or more bacteria per high power field may indicate urinary tract infection.

Gram Stain: Allow a drop of urine to air-dry on a microscopic slide; heat fix and stain with Gram stain or methylene blue stain (Wright's stain); Gram stain is done as follows:

Cover slide for 1 minute with 1% crystal violet; wash with water; cover with Gram's iodine for 30 seconds; decolorize with 95% ethyl alcohol or acetone until no more dye runs off slide; counterstain with dilute safranin for 30 seconds; wash and dry.

INTERPRETATION: The etiologic agent most commonly identified in patients with pyelonephritis is Escherichia coli. There are numerous other urinary pathogens, e.g., Proteus vulgaris, Pseudomonas aeruginosa, Klebsiella pneumoniae, Salmonella, the enterococci and the hemolytic staphylococci. About 80 percent of urinary infections in children are caused by Escherichia coli. Proteus species account for 50-80 percent of infections in boys. Staphylococcus albus causes about 40 percent of symptomatic urinary infections in children of both sexes 11 to 16 years. With anatomical lesions of the urinary tract, there is a higher incidence of infections caused by Klebsiella-Enterobacter, Proteus and Pseudomonas.

Renal cortex abscesses almost invariably follow either an apparent or occult staphylococcal bacteremia, although pyuria is usually absent; affected patients generally have no underlying renal disease.

Abscesses of the renal medulla often are associated with underlying kidney disease such as renal or ureteral calculi or pyelonephritis. Pyuria and gram-negative organisms on smears are typical of medullary abscesses.

Perinephric abscesses are seen most commonly in association with ureteral calculi, hydronephrosis, non-calculous renal infections and diabetes. The most common organisms are the gram-negatives.

UROBILINOGEN, URINE

SPECIMEN: Fresh random urine, 20ml; any clean container, no preservatives; do not expose specimen to light
REFERENCE RANGE: 0.5 to 1.0 Ehrlich Units
METHOD: Specimens are screened with Urobilistix. If greater than 1.0 Ehrlich units, Watson-Schwartz semi-quantitative test is run.
INTERPRETATION: This test is used as a liver function test and for the differential diagnosis of obstructive jaundice and hemolytic disease of the newborn; urobilinogen is increased in the urine in hemolytic anemias but is decreased in obstructive liver disease, especially in patients with complete obstruction. The test material on the reagent strips is sensitive to 0.1 Ehrlich unit per dl of urine. The test material (p-dimethylamino-benzaldehyde) is not specific for urobilinogen; it reacts with substances in the Watson-Schwartz reaction, e.g., porphobilinogen and the drugs sulfosoxazole and p-amino-salicylic acid. Increased urobilinogen is found in the conditions listed in the next Table:

Increased Urobilinogen in Urine
Hemolytic Anemias
Liver Disease

UROPORPHYRINOGEN I SYNTHETASE

SPECIMEN: 1ml blood in EDTA or Heparin
REFERENCE RANGE: Females: 8.1-16.8nmol/sec/liter; males: 7.9-14.7nmol/sec/liter; indeterminate: 6.0-8.0nmol/sec/liter; definitive acute intermittent porphyria: less than 6.0nmol/sec/liter.
METHOD: Measurement of rate of synthesis of uroporphyrin from porphobilinogen.
INTERPRETATION: Decreased uroporphyrinogen I synthetase activity may be found in erythrocytes of carriers of the genetic defect of acute intermittent porphyria in the absence of clinical or chemical manifestations of the disease.

VALIUM (see DIAZEPAM)

VALPROIC ACID (VPA, DEPAKENE)

SPECIMEN: Red top tube, separate serum

REFERENCE RANGE: Therapeutic: 40-100mcg/ml; Time to Obtain First Specimen
(Steady State): 2-4 days after starting therapy; Half-Life: 8-15 hours; Trough:
just before next dose

METHOD: EMIT; GLC; Fluorescence Polarization (Abbott)

INTERPRETATION: Valproic acid is used in the treatment of grand mal and petit
mal epilepsy in children. Dosage guidelines are shown in the next Table:

Dosage and Blood Specimen Guidelines	
Dosage	Blood Specimens
Adults: 15 to 45mg/kg/day	Two to four days following initiation
Children: 15 to 100mg/kg/day	of therapy; obtain blood specimens
Dosage 3-4 times/day	two hours post-drug (peak) and just
	before next dose

Valproic acid serum levels are obtained for purposes as follows; (Cloyd,
J.C. and Leppik, I.E., Individualizing Drug Therapy 2, 88-108, 1981):
  (a) Two to four days following any change in valproic acid dosage or
      dosing interval.
  (b) On addition or withdrawal of other antiepileptic drugs.
  (c) Decrease in seizure control.
  (d) Physical or laboratory signs of toxicity.
  (e) Noncompliance suspected.

The most common side effect of valproic acid treatment is gastrointestinal upset; this may be avoided by taking valproate with food; it may be
decreased by using enteric-coated tablets. The most common adverse reaction is
hepatotoxicity; this is rare but liver function tests should be performed every
two months for six months after starting valproate. Side effects include hair
loss, weight gain and tremor.

VANILLYLMANDELIC ACID (VMA), URINE

SPECIMEN: Add 25ml of 6N HCl to container prior to collection. 24 hour
urine: Instruct the patient to void at 8:00 A.M. and discard the specimen.
Then, collect all urine including the 8:00 A.M. specimen at the end of the 24
hour collection period. Refrigerate jug as each specimen is collected.
Following collection, add 6N HCl to pH 1 to 2 (Do not use boric acid).

REFERENCE RANGE: Adults: less than 8mg/24 hour.

Age	VMA (mg/gm Creatinine)
1-12 months	1.40-15.0
1- 2 years	1.25- 8.0
2- 5 years	1.50- 7.5
5-10 years	0.50- 6.0
10-15 years	0.25- 3.25

To convert conventional units in mg/gm creatinine to international units in
mcmol/gm creatinine, multiply conventional units by 5.046.

METHOD: Solvent extraction followed by adsorption onto silica gel column,
washing, elution from column, diazotization and further solvent extraction.

INTERPRETATION: Urinary VMA levels are elevated in patients with pheochromocytoma and neuroblastoma.

VDRL (see RPR-SERUM AND STS-CSF)

VIRAL CULTURE AND SEROLOGY (ANTIBODIES)

<u>SPECIMENS:</u> Specimens for viral culture and for antibodies should be collected as indicated in the next Figure:

Specimens for Viral Culture and Serology (Antibodies)

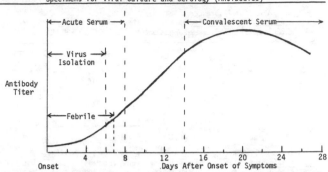

<u>Specimens for Viral Antibodies:</u> Collect both <u>acute</u> and <u>convalescent</u> blood specimens; the <u>convalescent</u> specimen is obtained 7 to 30 days after the acute serum is collected. Use red top tubes; separate sera from clot and refrigerate sera.

<u>INTERPRETATION:</u> A four-fold or greater rise in antibody titers between the acute and convalescent phase sera or the presence of antibody of the IgM class indicates active infection.

The common pathogenic viruses, the culture specimens and serology are given in the next Table; (Lennette, D.A., et al., in Manual of Clinical Microbiology, 3rd ed., Am. Soc. for Microbiology, Wash., D.C., 1980, pgs. 760-778):

Pathogenic Viruses, Culture and Serology

Viruses	Culture Specimens	Serology
Neonatal Infection Profile TORCH (Congenital Infection)		
Toxoplasmosis (Protozoa)		
Rubella (Measles)	Not Cultured	Serum
Cytomegalovirus	Urine, Throat Swab	Serum
Herpes Simplex	Throat Swab, Visicle Fluid	Serum
Cutaneous and Mucous Membrane Disease		
Vesicular		
Herpes Simplex	Vesicle Fluid and Throat Swab	Serum
Varicella Zoster		
(Chicken Pox in Children,		
Shingles in Adults)		
Maculopapular		
Adenovirus	Throat Swab, Rectal Swab	Serum
Rubella Virus (Measles)	Not Cultured	Serum
Rubeola Virus (German Measles)	Not Cultured	Serum
Parotitis		
Mumps	Throat Swab and Urine	Serum
Respiratory Tract Infections		
Postnatal Pneumonitis		
Adenovirus Group	Throat Swab	Serum
Respiratory Syncytial	Throat Swab	No serology
Rubeola (Measles)	Throat Swab	Serum
Varicella Zoster	Throat Swab	Serum
Upper Respiratory Tract Infections		
Rhinovirus	Throat Swab	No Serology
(Mycoplasma Pneumoniae)	Throat Swab	Serum
Parainfluenza	Throat Swab	Serum
Adenovirus	Throat Swab	Serum
Enterovirus	Throat Swab, Rectal Swab	
Lower Respiratory Tract Infections		
Influenza A	Throat Swab	Serum
Influenza B	Throat Swab	Serum
Adenovirus	Throat Swab	Serum
Parainfluenza 1,2,3	Throat Swab	Serum
Respiratory Syncytial Virus	Throat Swab	Serum
(Mycoplasma)		Serum
Psittacosis (Chlamydiae)		Serum
Pleurodynia		
Coxsackievirus	Throat Swab, Rectal Swab	Serum
Cardiac Profile (Myocarditis/Pericarditis)		
Coxsackie (B1 through B6)	Throat Swab, Rectal Swab	Serum
Central Nervous System Infections		
Mumps	Throat Swab and Urine	CSF and Serum
Enterovirus	Throat Swab, Rectal Swab	No Serology
Herpes Simplex Virus	Spinal Fluid, Throat Swab	CSF and Serum
Arbovirus	Not Cultured	Serum
California Virus		
Western Equine Encephalitis		
St. Louis Equine Encephalitis		
Immunosuppressed or Immunodeficient		
Cytomegalovirus	Urine, Throat Swab	Serum
Epstein-Barr Virus		Serum
Herpes Simplex Virus	Throat Swab	Serum
Varicella Zoster Virus	Vesicle Fluid	Serum
Enteritis		
Rotovirus	Test for Antigen (ELISA)	Serum
Norwalk Agent	Agent not Recoverable	---

435

VISCOSITY, SERUM

SPECIMEN: Red top tube, separate serum

REFERENCE RANGE: <1.8 centipoises; symptoms occur when viscosity >4 centipoises.

METHOD: Viscosity is measured using a cone rotated in the serum; the torque necessary to overcome the viscous resistance of the serum is measured in centipoises.

INTERPRETATION: Viscosity varies as a function of hematocrit level, red cell size and deformability, plasma proteins, fibrinogen concentrations and presence of abnormal plasma proteins. Hyperviscosity syndromes are not infrequently observed in Waldenström's macroglobulinemia and multiple myeloma when the serum concentration of immunoglobulins is high.

Hyperviscosity of blood may be due not only to elevated serum immunoglobulins as in macroglobulinemia and myelomatosis but also to increased number of cells (polycythemia or leukemia) or to increased resistance of cells to deformation (sicklemia or spherocytosis).

Diseases associated with hyperviscosity are given in the next Table:

Diseases Associated With and Causes of Hyperviscosity	
Diseases	Causes of Hyperviscosity
Monoclonal Gammopathies especially in Waldenström	Increased Protein
Polycythermia; Leukemia	Increased Number of Cells
Sickle Cell Anemia; Other Hemoglobinopathies; Pyruvate Kinase Deficiency; Burr-Cell Formation; Hereditary Spherocytosis	Increased Resistance of Cells to Deformation

Patients with hyperviscosity may present with mucous membrane hemorrhages. The increased viscosity may result, in-vivo, in a decreased flow rate of blood and ischemic changes in the tissues with secondary necrosis and hemorrhage.

The most common cause of hyperviscosity in adults is Waldenström's macroglobulinemia.

Neonatal Polycythemia and Viscosity: Neonatal polycythemia has been associated with an increased risk of neurologic and motor abnormalities (Black, V.D. et al., Pediatrics 69, 426-431, 1982); there is decreased cerebral blood flow and cardiovascular, respiratory sequalae, necrotizing enterocolitis and even acute renal failure(Herson, V.C. et al., J. Pediatrics 100, 137-139, 1982).

Hyperviscosity occurs in up to five percent of all neonates; four percent had polycythemia while one percent had hyperviscosity without polycythemia (Wirth, F.H. et al, Pediatrics 63, 833, 1979).

At birth, the hematocrit is relatively high; viscosity of the blood increases as the hematocrit rises. The relationship between viscosity and hematocrit is shown in the next Figure; (Barum, R.S., J. Pedatrics 69, 975, 1966):

Viscosity versus Hematocrit

Relative Viscosity

Hematocrit

S. Bakerman   Viscosity

   Viscosity increases linearly with peripheral venous hematocrits up to 65
percent; then the curve becomes exponential. A recommendation for screening
hematocrits in the neonates is given in the next Figure; (Ramamurthy, R.J. and
Brans, Y.W., Pediatrics 68, 168-174, 1981):

Hematocrit and Exchange Transfusions

        Capillary Hematocrit
             ↓ > 69 Percent
        Peripheral Venous Hematocrit
             ↓ > 64 Percent
        Insert Umbilical Vein Catheter
             | > 62 Percent
             | or
             ↓ Viscosity at 11.5 sec. Shear Rate > 14.0 Shear Rate
        Consider Partial Exchange Transfusion

   Kamamurthy, R.J. and Brans, Y.W. (Pediatrics 68, 168-174, 1981) recommend
screening of all neonates by measuring capillary hematocrit followed by
peripheral venous hematocrit levels in cases in which the capillary hematocrit
level is greater than 69 percent. If the peripheral venous hematocrit is
greather than 64 percent, an umbilical vein catheter should be inserted and the
umbilical vein hematocrit level and, preferably, viscosity should be determined.
If the umbilical vein hematocrit level is greater than 62 percent and/or
viscosity at 11.5/sec. shear rate is greater than 14.0 cps, then partial
exchange transfusion should be considered. The usual volume for exchange is
determined by the equation:

$$\frac{\text{Current Venous Hematocrit - Desired Venous Hematocrit}}{\text{Current Venous Hematocrit}} \times \text{Weight(kg)} \frac{(80 \text{ ml})}{\text{kg}}$$

   (Black, V.D. and Lubchenco, L.O., Pediatric Clinics of North America 29,
1137-1148, Oct. 1982.)
   Cord blood hematocrit greater than 56 is associated with a markedly
increased risk of neonatal polycythemia (Shohat, M. et al., Pediatrics 73, 7-10,
1984).

VITAMIN A(RETINOL)

SPECIMEN: Overnight fast. Red top tube, separate serum. Forward specimen frozen in plastic vial on dry ice or wet ice to reference laboratory.

REFERENCE RANGE: Vitamin A: 125-400 IU/dl.

METHOD: High performance liquid chromatography(HPLC)

INTERPRETATION: Vitamin A retinols direct normal differentiation of epithelia and affects the plasma membrane. Vitamin A is a fat soluble vitamin; it is decreased in conditions listed in the next Table:

Causes of Decreased Vitamin A
Fat Malabsorption Syndromes
(1) Deficiency of Pancreatic Digestive Enzymes:
Chronic Pancreatitis
Cystic Fibrosis
Pancreatic Carcinoma
Pancreatic Resection
(2) Impairment of Intestinal Absorption:
Celiac Disease (Coeliac Disease, Non-Tropical Sprue,
Gluten-Sensitive Enteropathy)
Rare Causes:
Tropical Sprue
Abetalipoproteinemia
Lymphangiectasis
Intestinal Lipodystrophy
Amyloidosis
Lymphoma
Surgical Loss of Functional Bowel
(3) Deficiency of Bile:
Extrahepatic Bile Duct Obstruction
Intrahepatic Disease
Cholecystocolonic Fistula
Dietary Deficiency

The incidence of vitamin A deficiency in Crohn's disease is 11 percent (Driscoll, R.H. and Rosenberg, I.H., Med. Clin. of North America 62, 185-201, 1978).

Vitamin A deficiency may lead to blindness and epithelial metaplasia with replacement of mature differentiated cells of various epithelial tissue by more primitive squamous and keratinizing cells. There is increased respiratory infections, plugging of salivary and pancreatic ducts and increased incidence of bladder stones.

Increased vitamin A is associated with excessive ingestion, or impaired disposal caused by diabetes mellitus, chronic nephritis or myxedema.

The effects of excess vitamin A are as follows: acute overdose: elevation of intracranial pressure and skin dequamation; chronic excess: increased intracranial pressure leading to hydrocephalus, skin desquamation and other dermatological findings; bone pain.

Vitamin A has been found to reduce the growth of chemically-induced tumors, especially those of epithelial origin (Sporn, M.B., Hospital Practice, pgs. 83-98, Oct. 1983).

Retinoids is a generic term that includes all natural and synthetic analogues of vitamin A(retinol).

VITAMIN B-6 (PYRIDOXAL PHOSPHATE)

SPECIMEN: Lavender(EDTA) top tube; separate plasma and freeze. Protect from light by wrapping in foil.

REFERENCE RANGE: 3.6-18.0 ng/ml.

METHOD: Enzymatic assay using the radiolabeled substrate tyrosine-1-^{14}C and the enzyme, tyrosine apodecarboxylase; $C^{14}O_2$ is liberated.

INTERPRETATION: Vitamin B-6 comprises a group of water-soluble vitamins which includes pyridoxine, pyridoxal and pyridoxamine. The vitamin B-6 group functions as coenzymes; participating in the metabolism of amino acids, and the breakdown of glycogen to glucose-1-phosphate.

Deficiency: Conditions associated with deficiency of vitamin B-6 deficiency are given in the next Table:

Conditions Associated wtih Vitamin B-6 Deficiency
Chronic Alcoholism
Malnutrition
Malabsorption
Pregnancy
Gestational Diabetes

Excess: There is a health fad of taking large doses of vitamin B-6; vitamin B-6 has been used to promote muscle growth and relieve premenstrual swelling. However, nerve damage may occur with difficulty in walking, and loss of feeling in arms and legs. Recovery takes one to two years and may not be complete.

VITAMIN B$_{12}$

SPECIMEN: Fasting; red top tube, separate and freeze serum in a plastic tube. This test should not be done in patients who have recently received radioisotopes therapeutically or diagnostically (Schilling test); severe hemolysis is cause for rejection of specimen.

REFERENCE RANGE: 200-800pg/ml; indeterminate, 100-200pg/ml; early vitamin B$_{12}$ deficiency, 100-150pg/ml; deficiency, <100pg/ml.

METHOD: RIA. To convert conventional units in pg/ml to international units in pmol/liter, multiply conventional units by 0.7378.

INTERPRETATION: Causes of decreased vitaminB$_{12}$ are given in the next Table:

Causes of Decreased Serum Vitamin B$_{12}$
Inadequate Diet: Strict Vegetarianism
Inadequate Absorption:
Deficient or Defective Intrinsic Factor:
Pernicious Anemia
Total Gastrectomy: Gastritis
Small Bowel Disease:
Ileal Resection or Bypass; Blind Loop Syndrome with Abnormal Gut Flora
Malabsorption; Tropical Sprue; Crohn's Disease; Fish Tapeworm
(Diphyllobothrium Latum)
Pancreatic Insufficiency
Interference with Vitamin B$_{12}$ Absorption:
Drugs: Neomycin, Metformin, Colchicine, Ethanol, p-Aminosalicylic Acid
Fish Tapeworm (Diphyllobothrium Latum)
Anticonvulsants
Dietary Folic Acid Deficiency
Multiple Myeloma
Rare Congenital Disorders: Orotic Aciduria; Transcobalamin II Deficiency;
Defective Intrinsic Factor Production

Nutritional Deficiency: Subjects who eat no food of animal origin are liable to develop B$_{12}$ deficiency; the largest group of vegetarians are religious Hindus.

Ancillary evidence of vitamin B$_{12}$ deficiency include macrocytosis, atrophic gastritis and a positive Schilling test.

The causes of increased serum vitamin B$_{12}$ are given in the next Table:

Causes of Increased Serum Vitamin B$_{12}$
Acute Hepatitis
Myeloproliferative Diseases
Acute and Chronic Granulocytic Leukemia
Myelomonocytic Leukemia
Polycythemia Vera
Oral Contraceptives

Fairbanks, V.F. and Elveback, L.R., "Tests for Pernicious Anemia, Serum Vitamin B$_{12}$ Assay," Mayo Clin. Proc. 58, 135-137, 1983.

VITAMIN C (ASCORBIC ACID)

SPECIMEN: Two gray (oxalate) top tubes, separate plasma and freeze plasma.

REFERENCE RANGE: 0.2 - 2.0 mg/dl. To convert conventional units in mg/dl to international units in micromol/liter, multiply conventional units by 56.78.

METHOD: Ascorbic acid is oxidized to dehydroascorbic acid which is coupled with 2,4-dinitrophenylhydrazine to form 2,4-dinitrophenylhydrazone which is rearranged in H_2SO_4 to a compound that is measured at 515nm (Carr, R.S. et al., Anal. Chem. 55, 1229-1232, 1983).

INTERPRETATION: Vitamin C is a water soluble vitamin which is necessary for the preservation of capillary integrity; it may function to maintain normal venous endothelium. Vitamin C is involved in the hydroxylation of proline to hydroxyproline; hydroxyproline occurs in collagen. Thus, vitamin C is essential for collagen synthesis; in addition, it is also associated with the metabolism of the mucopolysaccharide ground substance of connective tissue.

Severe ascorbic acid deficiency (scurvy) is characterized by connective tissue changes with delayed wound healing, poor scar strength, edema, hemorrhage, and bone weakness, sometimes leading to fractures. In addition, there may be anemia relating to the role of ascorbic acid in absorption of dietary iron and folic acid metabolism.

Long-stay (years) psychiatric patients are at increased risk for development of vitamin C deficiency (Thomas, S.J. et al., J. Plant Foods, 4, 191-197, 1982).

Ascorbic acid is a strong reducing agent; this property results in interferences with laboratory determinations as given in the next Table; (Woolliscroft, J.A., Disease-A-Month, 29, No. 5, pg. 21, Feb. 1983):

Interference of Megadoses of Vitamin C with Laboratory Determinations	
Test	Result
Multistix Test for Blood in Urine	False Negative
Hemoccult Test for Blood in Stool	False Negative or Delay
Urine Glucose: Test-Tape, Clinistix or Labstix	False Negative
Blood Tests:	
Glucose	Elevated
Uric Acid	Elevated
Cholesterol	Elevated

It is recommended that patients refrain from ascorbic acid supplementation for 48-72 hours prior to testing for occult blood. Ascorbic acid interferes with those tests that are based on oxidation-reduction reactions, that is, those reactions that produce hydrogen peroxide as listed in the previous Table.

In the tests for blood in urine or stool, hemoglobin functions as a peroxidase as follows:

$$H_2O_2 + \text{Reduced chromogen} \xrightarrow{\text{Hemoglobin}} \text{Oxidized Chromogen} + H_2O$$
$$\text{(Color)}$$

In the tests for glucose in urine or blood using the glucose oxidase method, hydrogen peroxide is generated; hydrogen peroxide then reacts with reduced chromogen, in the presence of peroxidase, to form oxidized chromogen. In the test for uric acid and cholesterol, hydrogen peroxide is formed.

VITAMIN D (25-HYDROXYVITAMIN D)

SPECIMEN: Green (heparin) top tube; separate plasma; place plasma in plastic vial and send frozen on dry ice.

REFERENCE RANGE: Deficiency: < 15ng/ml; <u>Winter</u>: 14 to 42ng/ml; <u>Summer</u>: 15-80ng/ml. Values obtained during the summer are usually higher than those obtained in the winter. To convert <u>conventional</u> units in ng/dl to <u>international</u> units in nmol/liter, <u>multiply conventional units by 2.599.</u>

METHOD: HPLC: Competitive Protein Binding.

INTERPRETATION: 25-Hydroxyvitamin D determination is useful in the work-up of patients with suspected rickets or osteomalacia; its plasma level is low in vitamin D deficiency and elevated in vitamin D excess. 25-Hydroxyvitamin D is synthesized in the liver; the steps in the synthesis of the biologically active form of vitamin D(1,25 DiOH-cholecalciferol) are shown in the next Figure:

Synthesis of Biologically Active Vitamin D
7-Dehydro-cholesterol $\xrightarrow[\text{U.V.Light}]{\text{Skin}}$ Chole-calciferol (CC) $\xrightarrow[\text{+25-OH}]{\text{Liver}}$ 25-OH-CC $\xrightarrow[\text{+1-OH PTH}]{\text{Kidney}}$ 1,25 DiOH-CC

Vitamin D_3 (cholecalciferol) is obtained in the diet or is synthesized in the skin from 7-dehydrocholesterol; vitamin D_3 is converted to 25-OH vitamin D in the liver and finally converted to $1,25-(OH)_2$ vitamin D (calcitriol) in the kidney. $1,25-(OH)_2$ vitamin D has 100 times the activity of 25(OH) vitamin D. The causes of 25-hydroxyvitamin D deficiency are listed in the next Table:

Causes of 25-Hydroxyvitamin D Deficiency	
Cause	Mechanism
Lack of Sunlight	Failure to Convert 7-Dehydrocholesterol to D_3 (Cholecalciferol)
Poor Diet	Lack of D_2, (Ergocalciferol) in Diet
Malabsorption	Failure to Absorb Fat Soluble Vitamins and Loss of 25-(OH)D_3 because of its Enterohepatic Circulation
Chronic Renal Failure	Reduced 1-Hydroxylation in Kidney, Elevated Serum Phosphate, Secondary Hyperparathyroidism, Acidosis
Anticonvulsant Therapy	Induction of Hepatic Microsomal Enzymes and Subsequent Inactivation of 25-(OH)D_3
Vitamin D Dependency Rickets	Type I: Congenital (Autosomal Dominant) Lack or Decrease of Renal 1-Hydroxylase Activity
Others (Liver Disease Hyperthyroidism, Diabetes)	Reduced Vitamin D Metabolites

25-Hydroxyvitamin-D is increased in 25-OH-D deficient patients after even brief exposure to ultraviolet radiation (Adams, J.S., N. Engl. J. Med. <u>306</u>, 722-725, 1982).

VITAMIN E (TOCOPHEROL)

SPECIMEN: Overnight fast. Red top tube, separate serum. Forward specimen frozen in plastic vial on dry ice or wet ice to reference laboratory.
REFERENCE RANGE: 0.550-1.750 mg/dl. To convert conventional units in mg/dl to international units in mcmol/liter, multiply conventional units by 23.22.
METHOD: High performance liquid chromatography(HPLC).
INTERPRETATION: Vitamin E (tocopherol) is a fat-soluble vitamin as are vitamins A, D, and K. Vitamin E is associated with polyunsaturated fatty acids and is present in most conventional diets. There are relatively few conditions that respond to vitamin E therapy (Roberts, H.J., JAMA 246, 129-131, 1981). Nutritional inadequacy or frank deficiency of vitamin E is found only in patients with various genetic or acquired diseases, with the exception of premature children, in whom the deficiency may be iatrogenous. In adult conditions, in which vitamin E has a claimed efficacy, an unknown pharmacologic effect is probably involved (Bieri, J.G., et al., N. Engl. J. Med. 308, 1063-1071, 1983).

The possible uses of vitamin E are listed in the next Table; (Oski, F.A., N. Engl. J. Med. 303, 454-455, 1980; Bieri, J.G. et al., ibid, 308, 1063,1983):

Possible Uses of Vitamin E	
Category	Examples
Correct a Deficiency State e.g., Malabsorption Disorders	Neuropathic and Myopathic Abnormalities Hereditary Abetalipoproteinemia: Retinitis Pigmentosa, Myopathy and Cerebellar Dysfunction. Modest Shortening of Red-Cell Life Span in Cystic Fibrosis Hyperaggregability of Platelets in Patients with Biliary Atresia
Anti-Oxidant	Infants Exposed to Prolonged Oxygen Administration in the Treatment of Respiratory-Distress Syndrome: Retrolental Fibrodysplasia; Bronchopulmonary Dysplasia Counter Cardiotoxic Effects of the Chemotherapeutic Agent Doxorubicin
Defense Against Free Radicals (Hematolog.Disorder)	Hereditary Hemolytic Anemias due to Deficiency of Glutathione Synthetase
Premature Infants	Hemolytic Anemia of Low-Birth Weight Infants

In neonatal hyperbilirubinemia, intramuscular administration of alpha-tocopherol to infants weighing 1000 to 1500 g resulted in a significant decrease in the duration of phototherapy (Gross, S.J. Pediatrics 64, 321-323, 1979). Supplementation may be beneficial if the reduction in duration of phototherapy is observed in future studies of infants (Bieri, J.G. et al.).

High dose vitamin E does not decrease the rate of chronic hemolysis in glucose-6-phosphate dehydrogenase deficiency (Johnson, G.J. et al., N. Engl. J. Med. 308, 1014-1017, 1983).

Vitamin E is decreased in conditions causing fat malabsorption (see FAT MALABSORPTION PANEL). The most common cause of vitamin E deficiency is cystic fibrosis. Neurologic abnormality may be associated with vitamin E deficiency and malabsorption. In children, neurologic abnormalities can be detected by clinical examination after only 18 to 24 months of malabsorption; in adults, neurologic abnormalities can be detected after at least 10 to 20 years of fat malabsorption reflecting vitamin E stores (Skol, R.J. et al., Gastroenterology 85, 1172-1182, 1983; Sokol, R.J., Ann. Intern. Med. 100, 769, May, 1984).

High dose vitamin E is associated with a wide variety of disorders (Roberts, H.J., JAMA 246, 129-131, 1981); the more serious ones are as follows: thrombophlebitis; pulmonary embolism; hypertension; severe fatigue; gynecomastia in men and women; breast tumors. Cohen, M.H. (N. Engl. J. Med. 289, 980, 1983) described the onset of several epidsodes of fatigue following vitamin E therapy with relief following withdrawal of the vitamin; he lists, in order, causes of fatigue as follows: depression, menopausal syndrome, hypervitaminosis E, anemia and hypothyroidism. Other clinical disorders and laboratory abnormalities induced by high dose vitamin E are given in the article by Roberts.

Rich sources of vitamin E are the vegetable oils, soybean, corn, cottonseed and safflower oils. The average daily diet contains 8 to 11 mg of alpha-tocopherol; the recommended dietary allowance of vitamin E is 8 mg of alpha-tocopherol for women and 10 mg for men.

VITREOUS HUMOR, POSTMORTEM

The changes in vitrous humor, postmortem, are shown in the next Table:

Examination of Vitreous Humor	
Substance	Comment
BUN	Constant for 30 hours
Creatinine	Changes slightly
Sodium	Constant for 30 hours
Calcium	Changes slightly
Magnesium	Changes slightly
Chloride	Constant for 30 hours
Glucose	Falls

The constituents in the vitreous humor are relatively stable as compared to blood values; therefore, analysis of vitreous humor may be done to reflect certain body constituents during life.

VOLATILES (SCREENING TESTS)

SPECIMEN: Red top tube, separate serum and store at 4°C.
REFERENCE RANGE: Serum osmolality: 285-295mosm/liter; pH 7.4; Anion gap: 15-20mmol/liter; Acetone: negative
METHOD: Osmolality; pH; electrolytes; acetone.
INTERPRETATION: There are simple screening tests that may be done to reflect the presence of methanol, ethylene glycol or isopropanol; these are illustrated in the next Figure:

Screening Tests for Volatiles

The contribution of 100mg/dl of the volatiles to the serum osmolality is given in the next Table:

Volatiles and Osmolality	
Volatile (100mg/dl)	Osmolality Increased (mosm/liter)
Ethanol	22
Isopropanol	17
Ethylene Glycol	16
Methanol	31

The volatiles increase serum osmolality between 16 and 31 mosm/liter thus increasing the measured serum osmolality from 285-295mosm/liter to more than 300mosm/liter.

WATSON-SCHWARTZ

SPECIMEN: Fresh random urine; keep specimen protected from light during transit to lab by wrapping in foil; freeze until ready for assay.

REFERENCE RANGE: Negative

METHOD: Reaction with p-dimethylaminobenzaldehyde; porphobilinogen, urobilinogen and substituted indoles all react with p-dimethylamino-benzaldehyde to give a red color. If a red color is obtained, the compounds are identified on the basis of solubility; porphobilinogen is soluble in aqueous solvents while urobilinogen is soluble in organic solvents, e.g., chloroform and butanol.

INTERPRETATION: Porphobilinogen is increased in the conditions listed in the next Table:

Increase in Porphobilinogen
Acute Intermittent Porphyria (AIP)
Variegate Porphyria (VP)
Hereditary Coproporphyria (HIP)

Porphobilinogen is not increased in porphyria cutanea tarda (PCI) or lead poisoning.

Urobilinogen is increased in the conditions listed in the next Table:

Increase in Urobilinogen
Hemolytic Anemias
Liver Disease

WHITE BLOOD COUNT (WBC) AND DIFFERENTIAL

SPECIMEN: Lavender(EDTA) top tube or microtube containing EDTA. The specimen is stable for up to 24 hours in the refrigerator.

Prepare blood film as follows: Place small drop of blood from needle on slide. Allow the blood to spread at junction of spreader slide, and then push the spreader slide, at a 45° angle, smoothly and quickly. Allow the film to air dry.

METHOD: Automatic Methods: (1) Selective cytochemical stains in a liquid milieu; sensing device measures light scatter and absorption (Technicon D/90); (2) Stained blood smears; automated microscope with computerized morphologic and tintorial criteria for cell identification (Hematrak); (3) Unstained cells are classified by phase microscopy on the basis of size and refractive index in a liquid milieu; classified on basis of size and density (Coulter Electronics).

Manual Methods: Total leukocyte count and absolute eosinophil count; Unopette technic (Becton, Dickinson & Co., Rutherford, N.J.) may be used.

Manual differential counts: The smear is scanned under low-power magnification for an area of reasonable cell distribution, and then examined under oil immersion magnification; the percent distribution of the various types of leukocytes, in a sample of 100-200 cells, is obtained.

REFERENCE RANGE: White cells (thousands/cu mm), white cell groups (percentage) with age, are listed in the next Table:

White Cells (Thousands) and White Cell Groups (Percentage) with Age; Mean Values								
White Cell	Age							
	Birth	2 Days	14 Days	3 Mos.	1 Yr.	4 Yrs.	8-21 Yrs.	Adult
White Cells/cu mm (Thousands)	15 (9-30)	21	11 (5-20)	9.5	9.0 (6-18)	8.0 (5-15)	8.0 (4.5-13.5)	7.5 (4.5-11)
White Cell Groups (Percentages)								
Polymorphnuclear (Neutrophils)	45	55	36	35	40	50	60	60
Lymphocytes	30	20	53	55	53	40	30	32
Monocytes	12	15	8	7	5	8	8	4
Eosinophils	2	4	2	2	1	1	1	3
Basophils	1	1	1	1	1	1	1	1
Immature White Cells	10	5	-	-	-	-	-	-

POLYMORPHONUCLEAR NEUTROPHILES:

REFERENCE RANGE: (Altman, P.L. and Dittmer, D.S. eds., Blood and Other Body Fluids, Bethesda, Maryland 1961, Fed. Am. Soc. Exp. Biol.):

	Polymorphonuclear Neutrophils		
	Thousands/cu mm		
Age	Total	Segmented	Band
Birth	11.0 (6-26)	9.4	1.6
12 Hours	15.5 (6-28)	13.2	2.3
24 Hours	11.5 (5-21)	9.8	1.7
1 Week	5.5 (1.5-10.0)	4.7	0.8
4 Weeks	3.8 (1.0-9.0)	3.3	0.5
1 Year	3.5 (1.5-8.5)	3.2	0.3
10 Years	4.4 (1.8-8.0)	4.2	0.2
20 Years	4.4 (1.8-7.7)	4.2	0.2

INTERPRETATION: Conditions associated with an increase in neutrophilic leukocytes are given in the next Table:

Conditions Associated with Increased Neutrophilic Leukocytes

Physiologic: Newborn; Pregnancy (especially near term); delivery; emotional disturbances; nausea and vomiting; strenuous physical exercise; and exposure to ultraviolet light; cold; severe stress; and heat

Pathologic:

    Acute Infections: Bacterial; Some Viral; Mycotic; Spirochetal; Rickettsial and Parasitic Infections

    Acute Inflammatory Disorders: Acute Rheumatoid Arthritis; Rheumatic Fever; Vasculitis; Myositis; Hypersensitivity Reactions

    Metabolic Disturbances: Uremia; Diabetic Acidosis; Eclampsia; Thyroid Storm

    Hematologic Disorders: After Hemorrhage; Hemolytic Anemias; Leukemias; and Myeloproliferative Disorders

    Tissue Nacrosis: Burns; Myocardial Infarction; Gangrene; Carcinoma; and Sarcoma

    Drugs and Toxins: Heparin; Digitalis; Epinephrine; Lithium; Histamines

    Stress: Allergies

Conditions associated with a decrease in neutrophilic leukocytes are given in the next Table:

Conditions Associated with Decreased Neutrophilic Leukocytes

Infections:

    Bacterial: Typhoid Fever; Paratyphoid; Brucellosis; and Septicemia (Mainly Gram Negative)

    Viral: Hepatitis; Infectious Mononucleosis; Measles; Rubella; Influenza; Chicken Pox; Colorado Tick Fever

    Other: Protozoa (Especially Malaria); Overwhelming Infections of any kind.

    Myeloid Hypoplasia: Aplastic Anemia; Vitamin B_{12} and Folic Acid Deficiency; Agranulocytosis; Space-Occupying Bone Marrow Lesions (in Leukemia, Myelofibrosis, and Metastatic Carcinoma)

    Chemical and Physical Agents: Bone Marrow Depressants (Radiation, Cytotoxic Drugs, Benzene); Drug Reactions, e.g., Chloramphenicol Phenothiazines

    Other Conditions: Collagen-Vascular Diseses, especially Lupus Erythematosus; Rheumatoid Arthritis; Infectious Mononucleosis; Hypersplenism, e.g., Liver Disease, Storage Diseases

LYMPHOCYTES:
REFERENCE RANGE: The mean values and reference ranges in thousands, per cu mm, with age, are as follws: Birth: 5.5 (2.0-11.0); 1 Week: 5.0 (2.0-17); 4 Weeks: 6.0 (2.5-16.5); 6 Months: 7.3 (4.0-13.5); 1 Year: 7.0 (4.0-10.5); 2 Years: 6.3 (3.0-9.5); 4 Years: 4.5 (2.0-8.8); 6 Years: 3.5 (1.5-7.0); 10 Years: 3.1 (1.5-6.5); 16 Years: 2.8 (1.2-5.2); 21 Years: 2.5 (1.0-4.8).
INTERPRETATION: Conditions associated with an increase in lymphocytes are given in the next Table:

Conditions Associated with Lymphocytosis
Infectious Diseases:
Viral: Hepatitis, Infectious Mononucleosis, Cytomegalovirus Infections, Herpes Zoster and H. Simplex, Chicken Pox, Viral Pneumonia, and Measles
Bacterial: Whooping Cough, Mumps, Brucellosis, Typhoid and Paratyphoid, Tuberculosis (Occasionally), Secondary and Congenital Syphilis, Toxoplasmosis, Infectious Lymphocytosis
Chronic Inflammatory Conditions:
Ulcerative Colitis
Immune Diseases: Serum Sickness, Idiopathic Thrombocytopenic Purpura
Metabolic:
Hypoadrenalism
Hyperthyroidism (Occasionally)
Blood Diseases:
Lymphocytic Leukemia(Acute and Chronic), Aplastic Anemia, Agranulocytosis, Heavy-Chain Disease, Multiple Myeloma, Felty's Syndrome, Banti's Syndrome

Conditions associated with a decrease in lymphocytes are given in the next Table:

Conditions Associated with Decreased Lymphocytes
Acute Infections and Illnesses: Associated with Increased Plasma Corticosteroids; ACTH; Epinephrine
Increased Corticosteroids: Cushing's Syndrome; Corticosteroid Therapy
Immunodeficiency Syndromes: Congenital Defects of Cell Mediated Immunity; Immunosuppressive Medication
Defects of Lymphatic Circulation: Intestinal Lymphangietasia; Thoracic Duct Drainage; Disorders of Intestinal Mucosa
Severe Debilitating Diseases: Miliary Tuberculosis; Hodgkin's Disease; Lupus Erythematosis; Terminal Carcinoma; Renal Failure

MONOCYTES:
REFERENCE RANGE: The mean values and reference ranges in the thousands per cu mm, with age, are as follows: Birth: 1.0 (0.40-3.1); 12 Hours: 1.2 (0.4-3.6); 24 Hours to 1 Week: 1.10 (0.2-3.1); 2 Weeks: 1.0 (0.2-2.4); 4 Weeks: 0.7 (0.15-2.0); 2 Months: 0.65 (0.1-1.8); 4 Months: 0.6 (0.1-1.5); 12 Months: 0.5 (0.05-1.1); 10 Years: 0.35 (0-0.8); Adult: 0.3 (0-0.8).
INTERPRETATION: Conditions associated with an increase in monocytes are given in the next Table:

Conditions Associated with Increased Monocytes
Infections:
Bacterial Infections: Tuberculosis; Subacute Bacterial Endocarditis; Syphilis, and Brucellosis
Viral Infections: Hepatitis, Mumps
Parasitic Diseases: Malaria; Kala-Azar
Other: Rickettsial Infections; Mycotic; Protozoal Infections
Hematologic Disorders:
Preleukemic States:
Leukemia: Chronic Myelomonocytic; Acute Monocytic; Chronic Myelogeneous
Lymphomas: Hodgkin's Disease and Non-Hodgkin Lymphomas
Histiocotyic Medullary Reticulosis
Myeloproliferative Disorders: Myelosclerosis; Agnogenic Myeloid Aplasia; Polycythemia Vera
Hemolytic Anemias
Collagen-Vascular Disease: Periarteritis Nodosa; Lupus Erythematosus; Rheumatoid Arthritis
Other: Ulcerative Colitis; Regional Ileitis; Cirrhosis; Malignancies; Hand-Schüller-Christian Disease

EOSINOPHILS:

REFERENCE RANGE: The mean values and reference ranges in thousands per cu mm, with age, are as follows: Birth: 0.4 (0.02-0.85); 24 Hours: 0.45 (0.05-1.00); 4 Weeks: 0.30 (0.07-0.90); 6 Months: 0.30 (0.07-0.75); 1 Year: 0.30 (0.05-0.70); 10 Years: 0.20 (0-0.60); 21 Years: 0.20 (0-0.45)

INTERPRETATION: The causes of increased blood eosinophils are listed in the next Table; (Editorial, The Lancet 1, 1417-1418, June 25, 1983):

Causes of Increased Blood Eosinophils
Parasitic Diseases: Especially with tissue invasion. Trichinosis; Visceral Larva Migrans (Toxocara Cunis or T. Cati), and Strongyloides
Allergic Diseases: Bronchial Asthma and Seasonal Rhinitis (Hay Fever)
Skin Disorders: Atopic Dermatitis; Eczema; Acute Urticarial Reactions; Pemphigus
Pulmonary Eosinophilias: Loeffler's Syndrome; Pulmonary Infiltration with Eosinophilia(PIE Syndrome); Tropical Pulmonary Eosinophilia caused by Microfilariae.
Infectious Diseases in Immunodeficient Children
Certain Vasculitic and Granulomatous Diseases
Connective Tissue Diseases such as Polyarteritis Nodosa, Lupus Erythematosus and Eosinophilic Fasciitis
Many Types of Drug Eruptions
Malignant Tumors: Bronchogenic Carcinoma, Hodgkin's Disease and T-Cell Leukemias
Hypereosinophilic Syndrome(HES)(The Lancet 1, 1417-1418, June 25, 1983)

The characteristics of HES are discussed in an editorial.

BASOPHILS:

REFERENCE RANGE: The mean values and reference ranges, in thousands per cu mm, with age, are as follows: Birth: 0.10 (0-0.65); 12 Hours: 0.10 (0-0.50); 24 Hours: 0.10 (0-0.30); 1 Week to 8 Years: 0.05 (0-0.20); 9 Years to Adult: 0.04 (0-0.20).

INTERPRETATION: The conditions causing basophilia are listed in the next Table:

Causes of Basophilia
Chronic Hypersensitivity Reactions to Food, Drugs or Inhalants
Myeloproliferative Disorders: Polycythemia Vera, Chronic Granulocytic Leukemia, and Basophilic Leukemia
Mast Cell Disease (Urticaria Pigmentosa)
Chronic Hemolytic Anemias
Ulcerative Colitis
Myxedema

WHOLE BLOOD CLOT LYSIS (see CLOT LYSIS TIME)
Whole Blood
Clot Lysis

WHOLE BLOOD CLOTTING TIME (SEE CLOTTING TIME, LEE-WHITE)
Whole Blood
Clotting Time

XYLOSE

SPECIMEN: Adult patient is prepared as follows: Nothing by mouth except water after midnight; the patient empties his/her bladder between 8:00 A.M. and 9:00 A.M. and then takes 25g of D-xylose (or significantly lower dose, 5g) dissolved in 250ml of tap water. Then, the patient takes an additional 750ml of water; the time is recorded after the patient finishes drinking the water. All urine is collected without preservation for five hours; keep urine refrigerated during collection. Mix the specimens and measure and record volume. Freeze until time of assay.

Children (10kg weight to 9 years old): The child takes nothing by mouth for six hours. Between 8:00 A.M. and 10:00 A.M., draw 1ml of blood into a red top tube. The patient is given 5.0gm of D-xylose dissolved in 100- 200ml of water. The time is recorded. After exactly one hour, obtain 1ml of blood in a red top tube. Separate serum from both tubes and freeze in plastic vials.

Infants (< 10kg body weight): The child takes nothing by mouth for 4 hours. Between 8:00 A.M. and 10:00 A.M., draw 1ml of blood into a red top tube. The patient is given 5% solution of D-xylose (10ml/kg). The time is recorded. After exactly one hour, obtain 1ml of blood in a red top tube. Separate serum from both tubes and freeze in plastic vials.

REFERENCE RANGE: Urine: 5g of D-xylose in 5 hour urine collection. Blood: Children (10kg weight to 9 years old): >20mg/dl at one hour. Infants (<10kg body weight): >15mg/dl at one hour. To convert conventional units in mg/dl to international units in mmol/liter, multiply conventional units by 0.06661.

METHOD: Reaction of D-xylose with phloroglucinol. (Clin. Chem. 25, 1440-1443, 1979).

INTERPRETATION: This test is done to differentiate malabsorption caused by intestinal problems versus malabsorption caused by pancreatic insufficiency. Pancreatic enzymes are not required for absorption of xylose. D-xylose is absorbed chiefly from the upper small intestine, especially the jejunum. The mechanism for absorption is by diffusion facilitated by a carrier system; D-xylose is excreted by the kidney.

The causes of decreased absorption of D-xylose are given in the next Table:

Causes of Decreased Absorption of D-xylose
Celiac Disease (Coeliac Disease, Nontropical Sprue, Gluten-Sensitive Enteropathy)
Other Causes:
Tropical Sprue
Abetalipoproteinemia
Lymphangiectasis
Intestinal Lipodystrophy
Amyloidosis
Lymphoma
Scleroderma and other "Collagen" diseases
Whipple's Disease
Surgical Removal of Jejunum

Changes associated with malabsorption are in the next Table:

Changes Associated with Malabsorption	
Change	Cause
Osteomalacia	Vitamin D Deficiency
Prolonged Prothrombin Time	Vitamin K Deficiency
Megaloblastic Anemia	Folate Deficiency

The results of this test are modified if renal function is abnormal.

ZARONTIN (see ETHOSUXIMIDE)

ZINC PROTOPORPHYRIN, BLOOD

SPECIMEN: Lavender (EDTA) top tube; ZPP is stable for 4 days at room temperature.

REFERENCE RANGE: MetPath

Blood Lead (mcg/dl)	Zinc Protoporphyrin (mcg/dl)	Comments
<29	<79	Normal
30-49	81-147	Monitor carefully
50-79	148-255	Lead poisoning likely
>79	>255	Lead poisoning definite

METHOD: Hematofluorometer; extraction followed by HPLC

INTERPRETATION: Zinc protoporphyrin is increased in the conditions listed in the next Table:

Increase in Zinc Protoporphyrin
Lead Poisoning
Iron Deficiency anemia
Anemia of Chronic Disease
Erythropoietic Protoporphyria

Lead Poisoning: The measurement of zinc protoporphyrin in red blood cells is often used as a screening test for lead poisoning. Zinc protoporphyrin is formed from free protoporphyrin as illustrated in the next Figure:

Pathway for Synthesis of Zinc Protoporphyrin

"Activated   Succinyl
Glycine"   +   CoA

$\downarrow$ ALA Synthetase

2 Delta-Aminolevulinic Acid

$\downarrow$ ALA Dehydrase

Porphobilinogen

$\downarrow$ Uroporphyrinogen Synthase
$\downarrow$ Uroporphyrinogen Isomerase

Uroporphyrinogen III

$\downarrow$ Uroporphyrinogen Decarboxylase

Coproporphyrinogen III

$\downarrow$ Coproporphyrinogen Oxidase

Protoporphyrinogen III

$\downarrow$

Protoporphyrin III
$+Fe^{+2}$ $\rightarrow$ Ferrochelatase $\xrightarrow{+Zn^{++}}$ Zinc Protoporphyrin

$\downarrow$
Heme

Lead inhibits the enzyme ferrochelatase, thus leading to the accumulation of "free" protoporphyrin(FPP); FPP is in fact not "free" but in the erythrocyte, chelates zinc, and forms zinc protoporphyrin. The level of red blood cell zinc protoporphyrin(ZPP) is directly related to the level fo lead in the blood.

Iron Deficiency Anemia: In iron deficiency anemia, a lack of iron supply to the developing red cell impairs heme synthesis and results in an accumulation of protoporphyrin in erythrocytes.

The measurement of free erythrocyte protoporphyrin(FEP) is useful in patients with latent iron deficiency anemia before changes in peripheral blood, in patients who have received iron therapy, and in the differential diagnosis of iron deficiency anemia from that of beta-thalassemia. Normal FEP is 35 ± 50mcg/dl RBC; levels greater than 100mcg/dl indicate overt iron deficiency anemia (Cook, J.D., Seminars in Hematology 19, 6-18, 1982).

The FEP-hemoglobin ratio (mcg protoporphyrin/g hemoglobin) is a sensitive index of iron deficiency erythropoiesis. The mean protoporphyrin-hemoglobin ratio is 16 with an upper limit of 32 (Labbe, R.F. et al., Clin. Chem. 25, 87-92, 1979).

The FEP provides information similar to that provided by the transferrin saturation. FEP increases after several weeks of iron deficiency erythropoiesis; it returns to a normal level weeks after iron therapy. Transferrin saturation is affected erratically following iron therapy and is affected by acute viral illnesses. The FEP:hemoglobin ratio is not affected by these conditions (Thomas, W.J. et al., Blood 49, 455-462, 1977).

S. Bakerman

ZINC, SERUM

SPECIMEN: Mayo Medical Laboratories: Equipment for venipuncture: One disposable plastic syringe, two appropriately cleansed Sarstedt® syringes and one Monoject® needle.
1. Draw 3ml of blood through the Monoject® needle into a regular disposable plastic syringe. The purpose of this is to rinse the needle.
2. Discard the blood and the syringe.
3. Utilizing the Monoject® needle already in place, slowly draw the required volume of blood into the first Sarstedt® syringe. Cap the syringe.
4. After adequate clotting, centrifuge the specimen in the Sarstedt® syringe.
5. After centrifugation, pour (do not transfer with pipette) the serum into the second Sarstedt® syringe (5ml adequate for multiple requests).
6. Cap the second syringe and ship to Mayo Medical Laboratories at room temperature.

Reference: Moody, J.R. and Lindstrom, R.M.: Selection and cleaning of plastic containers for storage of trace element samples. Anal. Chem. 49, 2264 (1977).
REFERENCE RANGE: 75-125mcg/dl. To convert conventional units in mcg/dl to international units in mcmol/liter, multiply conventional units by 0.1530.
METHOD: Atomic absorption spectrometer; anodic stripping voltammetry
INTERPRETATION: Zinc is the second most common trace element in the body; there are over 90 zinc metalloenzymes. Causes of decreased serum levels of zinc are given in the next Table:

Causes of Decreased Serum Levels of Zinc
Dwarfism
Acrodermatitis Enteropathica
Pathological Conditions:
Acute Tissue Injury
Chronic Liver Disease
Sickle Cell Disease
Some Cancers, especially Carcinoma of the Bronchus
Patients who have received total Parenteral Nutrition
for Several Weeks

Subnormal serum zinc levels are found in patients with cancer of the lung and colon but usually not in association with other tumors.

There are a number of variables in the determination of serum zinc levels; these variables include: (1) contamination of samples; (2) serum albumin concentration; and (3) prolonged occlusion of blood vessels which may occur during phlebotomy producing localized increases in serum zinc levels. A new approach involves the determination of serum zinc levels following ingestion of a challenge dose of zinc (Zinc Tolerance Test) (Sullivan, J.F., Jetton, M.M. and Burch, R.E., "A Zinc Tolerance Test", J. Lab. Clin. Med. 93, 485-492, 1979; Capel, I.D., Spencer, E.P., Daivies, A.E. and Levitt, H.N., "The Assessment of Zinc Status by the Zinc Tolerance Test in Various Groups of Patients", Clin. Biochem. 15, 257-260, 1982). In this test, the subject ingests 200mg of zinc sulfate (equivalent to 50mg of the metal) in a small quantity of aqueous solution after an overnight fast. Blood is drawn before and 2, 4 and 6 hours after ingestion.

Zinc deficiency is often associated with delayed wound healing.

INTERPRETIVE
LABORATORY DATA, INC.

Post Office Box 7066
Greenville, NC 27835-7066
919-756-6113

MAIL ORDER PRICES

$19.50 per book
10% discount with order of 5 or more
20% discount with order of 25 or more
Add $1.50 postage & handling (all orders)

NOTES

NOTES

NOTES

NOTES

NOTES

NOTES

NOTES

NOTES